Histories of Scientific Observation

Histories of Scientific Observation

EDITED BY LORRAINE DASTON
AND ELIZABETH LUNBECK

The University of Chicago Press
Chicago and London

Lorraine Daston is director at the Max Planck Institute for the History of Science, Berlin, and visiting professor in the Committee on Social Thought at the University of Chicago.
Elizabeth Lunbeck is the Nelson Tyrone, Jr. Professor of History and professor of psychiatry at Vanderbilt University.

The University of Chicago Press, Chicago 60637
The University of Chicago Press, Ltd., London

Printed in the United States of America

20 19 18 5

ISBN-13: 978-0-226-13677-6 (cloth)
ISBN-13: 978-0-226-13678-3 (paper)
ISBN-13: 978-0-226-13679-0 (e-book)
ISBN-10: 0-226-13677-9 (cloth)
ISBN-10: 0-226-13678-7 (paper)

Library of Congress Cataloging-in-Publication Data

Histories of scientific observation / edited by Lorraine Daston and Elizabeth Lunbeck.
p. cm.
Includes index.
ISBN-13: 978-0-226-13677-6 (cloth : alk. paper)
ISBN-13: 978-0-226-13678-3 (pbk. : alk. paper)
ISBN-10: 0-226-13677-9 (cloth : alk. paper)
ISBN-10: 0-226-13678-7 (pbk. : alk. paper) 1. Observation (Scientific method)—History. 2. Science—Methodology—History. I. Daston, Lorraine, 1951– II. Lunbeck, Elizabeth.
Q174.8.H57 2011
507.2′3—dc22 2010023093

♾ The paper used in this publication meets the minimum requirements of the American National Standard for Information Sciences—Permanence of Paper for Printed Library Materials, ANSI Z39.48-1992.

Contents

PART 3 Observing in New Ways: Techniques

Ingenious inquirers invent strategies, instruments, and regimens to make invisible, inaccessible, or intractable phenomena observable.

PART 4 Observing New Things: Objects

Observation defines its objects by how it studies them. Choosing how to observe in some cases dictates what is observed.

PART 5 Observing Together: Communities

Scientific observation is almost always collective, carried out by inquirers dispersed in space and time. Communities are required to organize, motivate, and coordinate individual observers.

Introduction

Observation Observed

LORRAINE DASTON AND ELIZABETH LUNBECK

Observation is the most pervasive and fundamental practice of all the modern sciences, both natural and human. It is also among the most refined and variegated. Observation educates the senses, calibrates judgment, picks out objects of scientific inquiry, and forges "thought collectives."[1] Its instruments include not only the naked senses, but also tools such as the telescope and microscope, the questionnaire, the photographic plate, the glassed-in beehive, the Geiger counter, and a myriad of other ingenious inventions designed to make the invisible visible, the evanescent permanent, the abstract concrete. Where is society? How blue is the sky? Which way do X-rays scatter? Over the course of centuries, scientific observers have devised ways to answer these and many other riddles—and thereby redefined what is under investigation by the way in which it is investigated. Observation discovers the world anew.

Yet scientific observation lacks its own history: why? Countless studies in the history and philosophy of science treat one or another aspect of observation: observation through telescope and microscope, observation in the field or in the laboratory, observation versus experiment, theory-laden observation. But observation itself is rarely the focus of attention and almost never as an object of historical inquiry in its own right. Observation seems at once too ubiquitous, too basic, and altogether too obvious to merit a history. One might well wonder whether a history of observation wouldn't simply be the history of science in its vast entirety—or the still more vast history of experience.

This book challenges these assumptions by showing what a history of scientific observation might look like, at least in its broad outlines, from the fifth to the late twentieth century. It is a long, surprising, and epistemologically significant history, full of innovations that have enlarged the possibilities of perception, judgment, and reason. It is also a history of how experience has been shaped and sharpened to scientific ends: how the senses have been schooled and extended; how practices for recording, correlating, and displaying data have been developed and refined; and how the private experiences of individuals have been made collective and turned into evidence.

Although this book is the first to attempt such a history of scientific observation, conceptualized as an epistemic category in its own right and illustrated with examples from many centuries and disciplines, it joins a larger and older project to write the history of experience, both quotidian and scientific. Our way has been paved by two decades of remarkable research in cultural history, gender history, and the history of scientific practices—research that was in turn stimulated by historical sociology and cultural anthropology.[2] Following the lead of cultural critics such as Raymond Williams and Michel Foucault, a generation of cultural historians has addressed topics previously believed to be beyond the purview of history, such as the body and sexuality. They showed how not only ideas and social meanings but also the lived experience of these and other purportedly universal aspects of human existence had been dramatically and diversely molded by historical context.[3] Intersecting with these trends in cultural history and inspired by historian Joan W. Scott's call for a history that would do justice to "radically different social experiences,"[4] the history of gender pioneered a history of experience understood as collectively structured by social and political circumstances (as opposed to both timeless perception and individual subjectivity). More recently, historians of art and music have joined cultural historians in researching the ways in which the conditions of daily life, instruments and architectures, and webs of meaning have patterned the ways people in past times and places have seen, heard, and even smelled: a history of the senses.[5]

These histories of experience have been more concerned with the creation of meaning than of knowledge. Nonetheless, their resolutely historicist, practice-oriented approaches have pointed the way for historians of science more interested in experience as a way of knowing than as a way of feeling. Taking their cue from many of the same sources in social history and anthropology that propelled the cultural turn in general history, historians of scientific practices have created a rich literature that documents how abstract categories like "experiment" and "classification" were anchored in concrete practices: for example, how the first scientific laboratories drew upon the skills and furnishings of the workshops of early modern artisans, or how late nineteenth-century astronomers employed women whose eyes had been trained to discern the tints of embroidery skeins to classify stellar spectra. Thanks to the efforts of historians of culture, gender, science, and the senses, experience may still be overwhelmingly immediate, ceaseless, and inescapable, but it is no longer timeless.

If all experience is to some extent refined, framed by context and circumstance, scientific experience is still more deliberate and cultivated, and each of its distinctive forms has a history.[6] Like experiment, observation is a highly contrived and disciplined form of experience that requires training of the body and mind, material props, techniques of description and visualization, networks of communication and transmission, canons of evidence, and specialized forms of reasoning. And like scientific experiment, scientific observation emerged, flourished, and diversified under specific historical conditions, as the essays in this volume show. Even though there was never a time before experience, there was a time before the scientific experiment—and the scientific observation: these were forms of "learned experience"[7] that had to be crystallized out of vernacular practices and conceptualized as evidence and proof.

Why then has the history of scientific observation not received the attention paid to other forms of scientific experience such as the experiment—or for that matter, to the history of the senses more generally? The answer to this question lies with the ways in which scientists and philosophers have defined scientific observation and experiment in contradistinction to one another during the nineteenth and twentieth centuries. When observation and experiment first emerged as forms of scientific experience in the late sixteenth and seventeenth centuries, two words that had rarely been coupled in the Middle Ages, the Latin *observatio* and *experimentum* (and their vernacular cognates), became closely intertwined. Throughout the eighteenth and early nineteenth centuries, observation and experiment were understood to work hand in hand: observation suggested conjectures that could be tested by experiment, which in turn gave rise to new observations, in an endless cycle of curiosity. Observation and experiment were conceived as partners; insofar as one was granted pride of place over the other, observation was often favored as more fundamental and more fertile in novelty. Observation discovered and discerned; experiment tested and proved.[8]

But starting in the 1820s, prominent scientific writers began to oppose observation to experiment and to vaunt the prestige of the latter over the former. In this new scheme of things, experiment was active and observation passive: whereas experiment demanded ideas and ingenuity on the part of a creative researcher, observation was reconceived as the mere registration of data, which could, some claimed, be safely left to untrained assistants. The reasons for this shift in philosophical precept (in contrast with actual scientific practice) were complex, but prominent among them was the fear that overly engaged scientists might contaminate observation with their own preferred theories. The British astronomer John Herschel, writing in 1830, was alarmed by the "mixture of theory in the statement of observed fact. . . . There is no greater fault (direct falsification of fact excepted) which can be committed by an observer." The "perfect observer" should command "every branch of knowledge" relevant to the subject, but Herschel in the next breath called for an army of amateur observers to enlist as foot soldiers in the cause of science, "to observe regularly and methodically some particular class of facts" that were ready at hand and to fill out standardized forms.[9] His friend and colleague, the mathematician Charles Babbage, hoped that inventions like the vernier would render special talents obsolete in scientific observation: "[G]enius marks its tract, not by the observation of quantities inappreciable to any but the acutest senses, but by placing Nature in such circumstances, that she is forced to record her minutest variations on so magnified a scale, that an observer, possessing ordinary faculties, shall find them legibly written."[10] This program to deskill scientific observation was driven by anxieties about how more sophisticated researchers might be tempted (in Babbage's terminology) to "forge," "hoax," "trim," or "cook" the data.[11]Although skilled, sophisticated observation was praised, numerous mid-nineteenth-century scientists worried that skill and sophistication might open the door to subjectivity or even fraud.

Yet some kind of guiding idea was a prerequisite for all productive scientific inquiry. The British philosopher and scientist William Whewell, writing in reply to his colleagues Herschel and Babbage, acknowledged that "the ideas which enable us to combine facts into general propositions, do commonly operate in our minds while we are still engaged in the office of observing." But Whewell too insisted that generalization was and ought to be "in the order of nature, posterior to, and distinct from, the process of observing facts."[12] The solution, at least in principle if not in practice, was to sharply segregate idea-driven experiment, in which the scientist, according to the French physiologist Claude Bernard, posed audacious questions to nature, from open-minded or even blank-minded observation, in which the scientist (or better, the untrained and therefore unprejudiced assistant) patiently recorded nature's answers verbatim: "The observer no longer reasons; he registers."[13]

The nineteenth-century methodological opposition of active, reasoning experiment to passive, registering observation left its stamp on twentieth-century philosophy of science, which continued to be much exercised by the problem of the relationship between theory and observation.[14] Members of the Vienna Circle, including Rudolf Carnap and many of his followers in the post–World War Two generation of philosophers of science in the logical empiricist tradition, formulated "protocol sentences" that would render observations in a language as close as possible to the raw data of perception—and therefore, it was hoped, as far as possible from the tendentious claims of the theories observations were meant to test.[15] Other philosophers, most influentially N. R. Hanson and Thomas Kuhn, countered that scientific observation was inevitably and necessarily "theory-laden," because the naive and trained eye see differently: "The infant and the layman can see: they are not blind. But they cannot see what the physicist sees; they are blind to what he sees."[16] In the debates sparked by Kuhn's *Structure of Scientific Revolutions* (1962; 2nd ed. 1970), the stakes were high, and the denial of neutral observation was branded as creeping relativism: in the words of the philosopher Jerry Fodor, "The thing is: if you don't think that theory neutral observation can settle scientific disputes, you're likely to think that they are settled by appeals to coherence, or convention or—worse yet—by mere consensus."[17] Yet even those who, like Hanson and Kuhn, conceived of scientific observation as something more than bare retinal impressions still described it as a kind of enriched perception, drawing on the work of Gestalt and cognitive psychologists,[18] rather than as the fusion of perception, judgment, memory, and reasoning that early modern philosophers of observation had theorized.

Against this background of philosophical assumptions that equated scientific observation with passive perceiving and registering, and socially demoted observers to the level of amateurs and assistants, the history of scientific observation seemed an unpromising project. But once these assumptions have themselves been historicized, the specificity, complexity, and variety of observation as a form of scientific experience, so evident to its past and present practitioners, becomes a beckoning topic of historical inquiry. Moreover, the career of observation sheds light on the histories of other forms of scientific experience—especially and essentially that of experiment, which has been paired with observation since the seventeenth century, whether as partner, complement, or opposite. Finally, the history of the practices devised by scientific observers enriches the cultural history of the senses: the education of the eye of the astronomer or the nose of the chemist or the tongue of the botanist are among the most ambitious and best-documented attempts to calibrate perception and judgment.

This book is the first attempt to give scientific observation its own history—or indeed, multiple histories—from the Middle Ages through the late twentieth century, with episodes drawn from many sciences, including meteorology and medicine, natural history and economics, astronomy and psychology. No single-volume treatment of such a large and central and yet heretofore unexamined topic can aspire to be comprehensive. Our aim is to open up a new area of research, not to exhaust it. But in order to prove that scientific observation *has* a history—and to give some indication of its richness—the scope of the book must be long with respect to chronology and broad with respect to disciplines. Despite this range, this book is regrettably restricted to the Western tradition; we fully expect other traditions to yield equally copious historical harvests in this domain.

The book's structure corresponds to that of the long history of observation: a survey of how scientific observation coalesced as practice and concept in the medieval and early modern period, followed by case studies that spotlight how scientific observation spread and diversified in the natural and human sciences thereafter. Part 1, "Framing the History of Scientific Observation, 500–1800," provides an overview of the *longue durée* of observation as practice, word, and epistemic category in Europe from medieval through early modern times. In order to appreciate just how, when, and why observation became a recognized and cultivated form of scientific experience by the late seventeenth century, one must also know something about what came before and after: the three essays correspond to origins, emergence, and consolidation. This part charts how scattered and unnamed practices associated more with divination than with science emerged as a respected and indeed essential form of scientific inquiry and then spread like wildfire to almost all disciplines. It provides the framework for the focused studies in parts 2–5 by explaining how the myriad ways of observing described therein came to count as scientific.

The individual chapters in parts 2–5 hold up to the light the many facets of scientific observation: its sites (the field, observatory, and laboratory—but also the household and the consulting room); its instruments (from the dissecting scalpel to the strobe—but also the notebook and the table of data); its images (the botanical illustration, the photograph—but also the pencil sketch); its personae (the virtuoso, the adventurer, the correspondent). It is characteristic of modern scientific observation to invent new ways of probing, recording, and fixing its objects of inquiry, but these technologies never supplant the observer, whose senses, judgment, and acuity are always essential to the integrity of the observation.

Amidst all of this variety, however, several themes cut across particular contexts of time, place, and discipline. The essays are grouped to draw attention to these themes: evidence, techniques, objects, communities. This sequence retraces the progression from the use of observation to generate conjectures and fortify belief, to the refinement of techniques in order to enlarge the range of observation and sharpen its focus, to the ways in which these inventive and subtle techniques bring new objects into the fold of the observable, and finally to how dispersed observers are orchestrated to serve collective empiricism. Within the parts, the chapters explore episodes in the history of observation that combine a detailed account of just how observations were made and used in a particular context with reflections on analytical themes that are a first attempt to give shape as well as substance to the history of scientific observation since the early modern period.

Although the essays in this volume are individually authored, the volume as a whole was conceived from the outset as a collective undertaking. No one scholar, no matter how erudite and industrious, could hope to do justice to the full range and richness of the history of observation in the human and natural sciences, much less to the ways in which that history has unfolded over centuries. The volume is collective not only in involving seventeen historians with differing nationalities and specialties; it is also collective in the way it was produced and organized. It results from a sustained collaboration rather than a single conference. Under the auspices of the History of Scientific Observation project at the Max Planck Institute for the History of Science in Berlin, the group met four times over a period of three years (June 2006–November 2008) for intensive discussion of drafts, which were then revised in light of the group's recommendations. (In addition, the authors of the essays in part 1 held several extra coordinating meetings.) The fruits of this collective effort to weld the volume into a whole are to be found in the part introductions and also in the individual essays as cross-references in text and notes, many of which cut across the parts.

Throughout its long history, observation has always been a form of knowledge that straddled the boundary between art and science, high and low sciences, elite and popular practices. As a practice, observation is an engine of discovery and a bulwark of evidence. It cultivates the senses of the connoisseur and straitens the judgment of the savant. It is pursued in solitude but also in the company of thousands. As a product, observations have been accumulated anonymously over millennia but also authored singly by individuals eager to secure priority and fame. They have been preserved in proverbs, in chronicles, in diaries, in archives, in learned journals, and in computer data banks. As a way of life, observation has been pursued by shepherds in fields, astronomers in towers, monks in monasteries, ladies and gentlemen in country seats, doctors at the bedside, physicists in the laboratory, and psychoanalysts behind the couch. The very word "observation" is suggestively ambiguous: at once a process, a product, an all-consuming pursuit. This book observes observation in its own spirit: open to possibilities for new knowledge in the most unexpected places.

Notes

1. The term comes from what is still the most useful (and almost only) philosophical meditation on observation, Ludwik Fleck, *Enstehung und Entwicklung einer wissenschaftlichen Tatsache. Einführung in die Lehre vom Denkstil und Denkkollektiv* (Basel: B. Schwabe, 1935), translated as *Genesis and Development of a Scientific Fact*, trans. Fred Bradley and Thaddeus J. Trenn, ed. Thaddeus J. Trenn and Robert K. Merton (Chicago: University of Chicago Press, 1979).

2. Norbert Elias, *Über den Prozess der Zivilisation. Soziogenetische und psychogenetische Untersuchungen*, 2nd ed., 2 vols. (Bern and Munich: Francke, 1969); Clifford Geertz, *Local Knowledge: Further Essays in Interpretive Anthropology* (New York: Basic Books, 1983); Emily Martin, *The Woman in the Body: A Cultural Analysis of Reproduction* (Boston: Beacon Press, 1987).

3. Seminal works include Raymond Williams, *Keywords: A Vocabulary of Culture and Society*, rev. ed. (New York: Oxford University Press, 1985); Michel Foucault, *Histoire de la sexualité*, 3 vols. (Paris: Gallimard, 1976–86); and, especially for the history of the body, Barbara Duden, *Geschichte unter der Haut. Ein Eisenacher Artzt und seine Patientinnen um 1730* (Stuttgart: Klett-Cotta, 1983); Caroline Walker Bynum, *Holy Fast and Holy Feast: The Religious Significance of Food to Medieval Women* (Berkeley: University of California Press, 1987); and Michel Feher, Ramona Naddaff, and Nadia Tazi, eds., *Fragments for a History of the Human Body* (New York: Zone, 1989).

4. Joan W. Scott, "Gender: A Useful Category of Historical Analysis," *American Historical Review* 91 (1986): 1053–75, on 1055.

5. Examples include (for cultural history) Alain Corbin, *Le miasme et la jonquille: l'odorat et l'imaginaire social au XVIIIe–XIXe siècles* (Paris: Aubier Montaigne, 1982); idem, *Les cloches de la terre: paysage sonore et culture sensible au XIXe siècle* (Paris: A. Michel, 1994); Constance Classen, *Worlds of Sense: Exploring the Senses in History and across Cultures* (New York: Routledge, 1993); and Robert Jütte, *Geschichte der Sinne. Von der Antike bis zum Cyberspace* (Munich: Beck, 2000). Mark M. Smith, *Sensing the Past: Seeing, Hearing, Smelling, Tasting, and Touching in History* (Berkeley and Los Angeles: University of California Press, 2007), provides an overview and bibliography of recent literature in this burgeoning field.

6. There now exists a substantial literature on the history of early modern scientific experience. Book-length studies published since 2000 include (for earlier bibliography, see the essays in part 1 of this volume): Barbara J. Shapiro, *A Culture of Fact: England, 1550–1720* (Ithaca: Cornell University Press, 2000); "Fatti: storie dell'evidenza empirica," ed. Simona Cerutti and Gianna Pomata, special issue of *Quaderni storici* 108, no. 3 (2001); Pamela H. Smith, *The Body of the Artisan: Art and Experience in the Scientific Revolution* (Chicago: University of Chicago Press, 2004); Gianna Pomata and Nancy G. Siraisi, eds., *Historia: Empiricism and Erudition in Early Modern Europe* (Cambridge, Mass.: MIT Press, 2005); Jean-Claude Passeron and Jacques Revel, eds., *Penser par cas* (Paris: Éditions de l'École des Hautes Études en Sciences Sociales, 2005); Johannes Süßmann, Susanne Scholz, and Gisela Engel, eds., *Fallstudien: Theorie—Geschichte—Methode*, Frankfurter Kulturwissenschaftliche Beiträge, Bd. 1 (Berlin: Trafo, 2007); Christoph Hoffmann, *Unter Beobachtung. Naturforschung in der Zeit der Sinnesapparate* (Göttingen: Wallstein, 2006); Jutta Schickore, *The Microscope and the Eye: A History of Reflections, 1740–1870* (Chicago: University of Chicago Press, 2007); and David Aubin, Charlotte Bigg, and H. Otto

Sibum, eds., *The Heavens on Earth: Observatory Techniques in the Nineteenth Century* (Durham, NC: Duke University Press, 2010).

7. "Erfahrung," in Johann Georg Wald, *Philosophisches Lexikon* [4th ed., 1775] (Hildesheim: Georg Olms, 1968), cols. 1082–84.

8. See the essays by Katharine Park, Gianna Pomata, and Lorraine Daston in part 1 of this volume for a more detailed history of the changing relationships of observation and experiment.

9. John F. W. Herschel, *A Preliminary Discourse on the Study of Natural Philosophy* [1830] (Chicago: University of Chicago Press, 1987), 131–34. Herschel's vision of centralized networks of standardized volunteer observers was to some extent realized in nineteenth-century meteorology: Fabien Locher, "Le nombre et le temps: la météorologie en France (1830–1880)" (Ph.D. diss., École des Hautes Études en Sciences Sociales, 2004); Katharine Anderson, *Predicting the Weather: Victorians and the Science of Meteorology* (Chicago: University of Chicago Press, 2005).

10. Charles Babbage, *Reflections on the Decline of Science in England and on Some of Its Causes* (London: B. Fellowes, 1830), 169.

11. Ibid., 176–83.

12. William Whewell, *The Philosophy of the Inductive Sciences*, 2 vols. (London: John W. Parker, 1840), 2: 503.

13. Claude Bernard, *Introduction à l'étude de la médicine expérimentale* [1865], ed. François Dagognet (Paris: Garnier-Flammarion, 1966), 51–56, 70–71, on 55.

14. Ian Hacking, *Representing and Intervening: Introductory Topics in the Philosophy of Natural Science* (Cambridge: Cambridge University Press, 1983), 149–85, provides a succinct, critical overview.

15. Rudolf Carnap, *Der logische Aufbau der Welt* [1928], 4th ed. (Frankfurt am Main: Ullstein Materialen, 1979), §§66–69, pp. 90–95. The literature on *Elementarerlebnisse*, "protocol sentences," "observation sentences," and other variants is vast, but see Dudley Shapere, "The Concept of Observation in Science and Philosophy," *Philosophy of Science* 49 (1982): 485–525; W. V. Quine, "In Praise of Observation Sentences," *Philosophy of Science* 90 (1993): 107–16; and W. F. Brewer and B. L. Lambert, "The Theory-Ladenness of Observation," *Philosophy of Science* 68 (2001): 176–86, for a sense of the long and ongoing debate.

16. Norwood Russell Hanson, *Patterns of Discovery: An Inquiry into the Conceptual Foundations of Science* [1958] (Cambridge: Cambridge University Press, 1972), 17. See also Thomas S. Kuhn, *The Structure of Scientific Revolutions* [1962], 2nd ed. (Chicago: University of Chicago Press, 1970), 111–29.

17. J. Fodor, "Observation Reconsidered," *Philosophy of Science* 51 (1984): 23–43, on 42.

18. In both cases, the reception of Gestalt and cognitive psychology was heavily influenced by a reading of Ludwig Wittgenstein's *Philosophical Investigations* (1953): Hanson, *Patterns of Discovery*, 180–81 n1 and passim; Kuhn, *Structure of Scientific Revolutions*, 112–13.

PART ONE

Framing the History of Scientific Observation, 500–1800

Epistemic histories can seldom be told by following just words *or* practices *or* ideas. It is characteristic of this sort of history that practices, concepts, and terms that now cohere in a single dense word like "observation" have in other periods also existed, but existed apart, or in constellations and contexts that take the modern mind by surprise: monastic timekeeping, humanist concordances, and personal diaries are as much part of the history of scientific observation as astronomical instruments. The three framing essays in part 1 chart the slow and convoluted history of scientific observation as practice, word, and concept over more than a millennium and thereby set the stage for the case studies in the parts that follow, in which observation as an epistemic category could be taken for granted. The aim of part 1 is to survey *how* observation came to be taken for granted, acknowledged, and cultivated as an essential part of the natural and human sciences.

The three essays are tightly coordinated, and although each may be read singly, together they tell a continuous story, albeit one punctuated by surprising twists and turns. In the Middle Ages (covered in Katharine Park's essay), observation can be discerned as a set of *practices*, often involving diligent, attentive watching and waiting, but rarely as a word or as an idea that attracted learned reflection as a distinct form of knowledge. From the late fifteenth through the early seventeenth century (covered in Gianna Pomata's essay), collections of observations (often signaled by the Latin word *observationes* in their titles) became, especially in medicine and astronomy, a flourishing *epistemic genre*, that is, a standardized textual format with recognizable conventions of style and content. During the seventeenth and eighteenth centuries (covered in Lorraine Daston's essay), observation as word and practice spread to many disciplines in the human and natural sciences and also became a concept: an *epistemic category* acknowledged and analyzed by philosophers as well as practitioners, an essential method for gaining knowledge. Practice, word, and idea had finally emerged and converged.

The focus of these three framing essays is the Latin West. Although the authors are acutely aware that developments in Mesopotamian and Greco-Roman antiquity and in medieval Islamic cultures form the essential background for their story, particularly in the contexts of astronomy, astrology, and medicine, length constraints dictate that coverage even within the Latin European tradition and its vernacular successors be selective.

Park's essay describes how in the Latin Middle Ages the word *observatio* was used mostly in the context of what the ancient Roman philosopher Cicero had called "natural divination": how farmers, mariners, and other outdoor laborers watched the skies and the weather in search of the correlations still preserved in proverbs like "Red in the morning/Sailors take warning." Observation was not an activity associated with natural philosophy, or indeed with any branch of learning, with the important exception of astronomy and astrology. Many practices that from a modern (or even early modern) viewpoint seem to be clear examples of the observation of nature were instead designated by the terms *experimentum* (a trial or test) or *experientia* (cumulative experience)—two words often used almost as synonyms, because they both referred to results that could not be deduced from first principles, but rarely coupled with *observatio* (observation or observance). Until the late fifteenth century, when a handful of astronomers initiated sustained observational regimens, observations were for the most part scattered, occasional, and unpublished, noted in the margins of documents meant to serve other purposes, rather than an end in themselves.

Pomata's essay charts the emergence of the observation as a named form of scientific experience in the sixteenth and early seventeenth centuries and as a new genre of publication, especially for physicians. The collections of notable cases, drawn both from their own practices and from the medical literature stretching back to antiquity, published by medical men in the late sixteenth and seventeenth centuries, transformed *observationes*, a word that was increasingly featured in learned book titles, into a category of "learned experience," with its own standards and conventions. Drawing on both critiques of the empirical basis of astrology and a revival of Hellenistic philosophical and medical defenses of experience over theory, the new epistemic genre sharply distinguished observation from conjecture—in stark contrast to an ancient and medieval tradition that had associated observation with the conjectural arts such as divination. By the turn of the seventeenth century, the words "observation" and "experiment" were increasingly conjoined, both designating recourse to experience as opposed to rationalist systems.

Daston's essay takes up the story in the early seventeenth century and traces how the primarily medical genre of *observationes* spread to other sciences, entered into a partnership with another new form of scientific experience, the experiment, ramified into an ever more sophisticated array of practices, and, by the late eighteenth century, was celebrated as the core of scientific logic. Not only physicians and astronomers but also naturalists, chemists, and physicists cultivated the observational arts of note taking, channeling attention, making tables, and synthesizing data from multiple sources into both word and image. Collectives of observers were organized by means of questionnaires, epistolary networks, and state and commercial initiatives. Observation became a topic of philosophical reflection as an epistemic category, and savants in many disciplines aspired to become "geniuses of observation."

The intertwined careers of "observation" and "observance" is a theme that runs through all three essays: whether practiced by medieval monks timing the offices of prayer by the stars, or sixteenth-century physicians monitoring the minutest details of the course of a disease, or eighteenth-century naturalists squinting at insects for days on end, observation was always a way of life as well as a way of knowing.

1

Observation in the Margins, 500–1500

KATHARINE PARK

Historians of medieval science have long labored to temper its reputation as a bookish enterprise focused on the transmission and interpretation of canonical works. Formal education in natural philosophy and medicine, pursued in the schools of religious orders and private masters as well as in universities, was indeed dominated by reading and textual commentary: learned writers, the focus of this essay, were generally satisfied with an account of the natural world grounded in the writings of a relatively small number of earlier authors (Aristotle in natural philosophy, Galen in medicine, Ptolemy in astronomy, and so forth). For the most part, they saw themselves as refining or filling in gaps and blank spaces in this account.[1] Yet those blank spaces were not negligible. Some involved phenomena that had been treated inadequately or ignored by ancient writers—the workings of the human uterus, the causes of the rainbow, the behavior and properties of plant and animal species indigenous to northern Europe—while others involved reconciling or even choosing between conflicting opinions of authoritative writers on topics as various as the role of women in generation and the motions of the celestial pole. In these cases, as in many others, knowledge could not be derived from first principles but had to be determined through direct, sensory engagement with animals, plants, and minerals, or with physiological, optical, meteorological, and astronomical phenomena.

In one sense, observation itself was marginal to the Aristotelian notion of science (*episteme* in Greek, *scientia* in Latin translation) that dominated the learned study of nature in the later Middle Ages: certain knowledge consisting of general causal explanations based on deduction from first principles. In the view of late medieval natural philosophers, the phenomena that demanded empirical study were those that could not be explained by general causes, requiring instead careful attention to particular effects, by reason of their complexity (e.g., the motions of the heavenly bodies); their contingency (e.g., the weather, the course of disease in individual patients); their dependence on the singularity of specific forms rather than the manifest qualities inherent in matter (e.g., the occult powers of healing springs or the lodestone); and their mind-boggling multiplicity (e.g., the wildly varying appearances and behaviors of different species of animals, plants, and minerals).[2] There were powerful practical motives to study phenomena of this sort. Physicians needed to know the therapeutic properties of plants, animals and minerals; astrologers required accurate planetary positions to make reliable analyses and predictions; engineers hoped to harness mysterious but powerful natural forces such as magnetism; and the weather was a matter of concern to everyone from ships' captains to municipal officials to those who lived off the land.

Although historians of science have characterized medieval scholars' work in these areas as based on observation,[3] the blanket use of this term blurs what was to contemporaries an important distinction between experience (*experimentum* or *experientia*) and observation (*observatio*).[4] These two activities involved two quite separate kinds of phenomena and corresponding forms of natural inquiry. This essay aims to untangle the concepts and activities identified with experience in medieval Europe from those identified with observation, focusing on the latter, which has to date received little attention as either practice or epistemic category. I begin by identifying the elements that characterized the observation of natural phenomena as conceived by ancient and medieval Latin writers. I then describe the changing landscape of observational practice in medieval Latin Europe between the sixth and the fifteenth century, focusing on two main contexts, the early medieval monastery and the late medieval city, and emphasizing the dramatic transformation wrought by the gradual assimilation of knowledge from the Islamic world beginning in the late eleventh century. Finally, I describe the emergence in Western Europe of new forms of observational practice, first in weather science and then in the field of astronomy more narrowly defined.

Experience, Observation, and the Arts of Conjecture

At the heart of the medieval concept of experience lay the notion of test or trial.[5] The *experimentum*, a well-established genre of scientific writing in medieval Europe, was typically a set of directions—usually a medical, magical, or artisanal formula—purportedly derived from and tested by experience, including both purposeful experience and trial and error. More broadly, the term corresponded to knowledge of singular, specific, or contingent phenomena that could not be grasped by deductive reasoning, as well as the process by which such knowledge was obtained.[6] In most cases the invocation of testing was formulaic and referred to trials supposedly performed by others and compiled from earlier texts or oral report. But some later medieval records of *experimenta* or *experientiae* included firsthand descriptions of investigations undertaken by the author. These might involve forms of artificial contrivance that modern historians would call "experimental," as in the case of the manipulations of lodestones performed by the late thirteenth-century French engineer Peter of Maricourt or the use by the early fourteenth-century Dominican Dietrich of Freiberg of prisms and spherical vessels filled with water to study the refractive phenomena that produced the rainbow.[7] Other accounts of *experientia*, in contrast, described knowledge derived from the use of the unassisted senses, such as the remarks on the morphology and behavior of raptors in *On the Art of Hunting with Birds*, composed in the mid-thirteenth century by the Holy Roman Emperor Frederick II, a fanatical falconer.

Frederick explicitly invoked experience in his prologue, where he assessed the truth of certain statements in Aristotle's *History of Animals*:

> I have followed Aristotle where it was appropriate. For on many topics, and especially concerning the natures of certain birds, as I have discovered by experience [*experientia*], he seems to depart from the truth. For this reason I do not follow the prince of philosophers in all things, for he rarely or never practiced falconry, whereas I have always enjoyed and practiced it. Regarding many topics that he discussed in his book *On Animals*, he says that certain people said such and such, but possibly neither he himself nor those whom he cites saw the thing in question, for certain conviction [*fides*] does not proceed from hearsay.[8]

Typical of the errors Frederick II attributed to Aristotle were several propositions regarding the migration of birds: for example, that the calls of migrating cranes, heron, geese, and ducks are expressions of physical effort rather than forms of communication and that flocks follow a single leader throughout their migration.

In appealing to experience—his own and that of other experts—to impugn Aristotle's own teachings, Frederick was himself making use of an important Aristotelian concept, whose epistemic function Aristotle had discussed most famously in the *Posterior Analytics* (100a4–9) and *Metaphysics* (980a27–981a16); there he described experience (*empeiría* in the Greek, *experimentum* in medieval Latin translation) as a form of knowledge intermediate between the raw sensation of phenomena and the higher reaches of theoretical and practical knowledge; the latter involved familiarity with causes, where men of experience dealt primarily with practical applications.[9] In contrast, no specific concept of observation emerges in Aristotelian epistemology; in *History of Animals*, for example, Aristotle used the Greek term *tērēsis* and its cognates (*observatio* and its cognates in medieval Latin translation) to describe the behavior of animals, most notably the spider as it lay in wait for prey.[10] As this passage suggests, the root meaning of observation was watching and attentive waiting, rather than test or trial. And indeed, throughout the medieval period, observation in the context of natural inquiry was primarily identified with a tight bundle of distinctly non-Aristotelian sciences: the science of the stars, which included celestial timekeeping, positional astronomy, judicial astrology, and weather prediction, and which will as a result be the focus of this essay.[11]

The roots of the medieval Latin understanding of observation as part of natural inquiry lay in a small number of Roman texts. The earliest of these was Cicero's *On Divination*, written in the first century BCE, where divination was defined as "an art on the part of those who, having learned old things by observation [*observatione*], seek new things by conjecture [*conjectura*]."[12] Cicero's definition emphasized the predictive aims of observation, which characterized fields in which causal explanations were lacking, so that causal reasoning could not be relied on to divine the future course of things. Instead, conjecture operated by extrapolation: faced with a striking phenomenon and wishing to know what might follow on it, the interpreter of natural phenomena searched for similar events in oral or written memory that might offer guidance as to what to expect.[13] For this reason, the conjectural arts were based on the long-term tracking of correlations between unusual natural phenomena—flights of birds, anomalies in the entrails of animals, unusual meteorological occurrences, remarkable constellations of stars and planets—and other kinds of events of immediate relevance to human beings, such as epidemics, battles, or the death of political leaders. Only after these past correlations ("old things") had been diligently noted and recorded over untold periods of time could one predict what would happen ("new things") on the basis of particular natural signs. Key to this cumulative process were repetition and a long temporal baseline. In the words of Quintus, Cicero's brother and interlocutor in book 1 of the dialogue, "in every field of inquiry great length of time employed in continued observation begets an incredible body of knowledge [*incredibilem scientiam*], . . . since repeated observation makes it clear what follows from what and what is a sign of which thing."[14]

In book 2, Cicero himself (as interlocutor) endorsed this characterization of the role played by observation in conjectural knowledge, while making the case for the "natural" divinatory sciences—navigation, farming, and medicine—where the antecedent signs involved long-term cyclical phenomena related to the heavens and the weather, or the consistent effects of particular therapeutic interventions, rather than capricious events such as the color of the liver of a particular pig. Just as the captain of a ship learned by observation to predict a coming storm from cloud patterns or to set a course by the stars, so the farmer developed rules of thumb that told him when to perform particular activities such as sowing and harvesting, and the doctor learned to project the future course of illness or the future effects of particular therapeutic interventions. While not infallible, these predictions provided adequate guidance for everyday practice.[15]

This general association of observation with the cycles of the heavens and the seasons, as well as with the interpretation of natural signs, was echoed in Pliny's *Natural History*, composed in the first century CE. Like Cicero, Pliny identified observation with medicine, navigation, and farming. For example, observation tells the farmer when to plant particular crops: "For [the sowing of navew and turnip] is properly done between the holidays of two deities, Neptune and Vulcan, and as a result of careful observation it is said that these seeds give a wonderfully fine crop if they are sown on a day that is as many days after the beginning of the period specified as the moon was old when the first snow fell in the preceding winter"—a fine example of a "subtle" correlation that would take unfathomable lengths of time to discern.[16] Pliny used the term *observationes* to refer to both the original process of tracking correlations and the practical rules derived from them, such as the "observation" that eating an odd number of boiled and grilled snails is particularly good for stomach problems.[17] Like Cicero, too, he presented observation as a collective and largely anonymous process, associated with the early, originary phases of natural knowledge, long ago and far away. For example, navigation by the stars was invented by the Phoenicians and astronomy by the Babylonians, who recorded 720 years' worth of celestial observations on "baked bricks."[18] (Because of the phenomenal lengths of time required, observational records had to be kept in durable media—cuneiform tablets or stone inscriptions, like those used by the Egyptians.) At the same time, Pliny emphasized that literacy was not necessarily a prerequisite for the development of conjectural knowledge; generations' worth of observation might be orally transmitted. For example, although the observations on which the art of farming is based were originally made by illiterate peasants, "observation was not less ingenious among them than theory is now."[19]

Comparing Pliny's uses of *observatio* with his uses of *experimentum* and *experientia* makes clear how differently the two functioned as sources of sense-based knowledge. Observations were the product of cumulative, long-term practices of noting and recording, while experience usually took the form of onetime interventions or of things noted casually in the course of daily life. Observation was associated above all with the distant past, while experience was often recent or ongoing. Observation was concerned with predictions and conjectures, while experience was concerned with effects. Experience, as a process of testing, often required physical manipulation (e.g., procedures to determine if a particular type of precious stone was genuine) or the creation of some kind of contrivance (e.g., the well dug at Syene to prove that the sun was directly overhead at midday on the summer solstice there).[20] Observation, in contrast, generally involved no interaction with the object on the part of the observer beyond patient and careful attention and the recording of its results as traces in memory or on the written page. Finally, experience usually related to testing and thus confirming or disproving claims or precepts, unlike observation, which provided the raw material from which general claims and precepts were made.

Medieval Christian scholars inherited these distinctions from Cicero and Pliny, as well as from Augustine, who in his *Letter to Januarius* (ca. 400) drew on Cicero's endorsement of "natural" divination. Augustine contrasted the legitimate celestial observations of farmers and navigators with the activities of judicial astrologers, whose attempts to predict the future based on the motions of the heavens smacked of paganism and a denial of the direct dependence of all events on God's will:

> Who could fail to understand the great difference between heavenly observations [*observationes*] focused on the weather, like those observed by farmer and sailors, or those observed by navigators and by travelers through the pathless and lonely deserts of the South, in order to determine where they are and how they should direct their course, . . . and the pointless practices of men who observe the stars not to know the weather, nor to determine ways through unfamiliar regions, nor to keep time, nor to understand their use in spiritual similitudes, but to pry into [future] events as if they were already fated?[21]

Rather than sparking a burst of learned interest in cosmography or celestial navigation, however, the principal effect of this text and others like it was to discourage the practice of judicial astrology in early medieval Europe and to eliminate one of the principal motivations for the detailed study of the planetary positions. This was part of a fundamental reorientation in the meanings and uses of observation toward religious ends.

Observation as Observance, 500–1100

In addition to referring to the empirical study of phenomena, "observation" in classical Greek and Latin had a second, normative sense, as Gianna Pomata emphasizes: the observance of precepts based on that study, such as the rules regarding the sowing of turnips and eating of snails cited by Pliny, or the rules that allowed Augustine's navigators and travelers to find their way.[22] In early medieval Latin writing, this normative sense of *observatio* swamped the epistemic sense and acquired, in addition, a strongly religious cast, except for occasional usage in a few technical fields with a strongly prescriptive focus, such as medicine or grammar;[23] elsewhere it was used to describe both Christian and "superstitious" (i.e., Jewish and pagan) forms of religious practice, particularly those associated with divination.[24] At the same time, references to observation as a source for natural knowledge dropped away, with one significant exception: throughout the early and central Middle Ages one finds a steady trickle of allusions to observation of the heavens related either to critiques of judicial astrology, following Augustine, or to timekeeping practices, where the word blended the meanings of calculatory rule, attention to the motions of heavenly bodies, and orderly religious observance.[25] Indeed, for the period before the early twelfth century, virtually all of the textual evidence concerning *observatio* as an epistemic practice relates to timekeeping in the context of Christian ritual.

Throughout the early medieval period, Christian timekeeping was of two main sorts, horological and calendrical. The former involved timing the prayers that were the principal focus of monastic observance, while the latter involved establishing the dates of Easter and its dependent festivals, which were based on the lunar cycle and therefore varied from year to year.[26] The system of monastic "hours" (prayer services) that became canonical in the Latin West was codified by Benedict of Norcia in his sixth-century *Rule*, where he prescribed eight daily offices: Matins, Prime, Terce, Sext, Nones, Vespers, Compline, and Vigils or Nocturns. Because the times of these services were anchored by dawn—Matins took place at sunrise, by which time the monks had to have finished Vigils, dispersed, and had time to reassemble—their timing and spacing changed over the course of the year.[27] While it was relatively easy to regulate the daytime offices using the sun's observed motion, the timing of Vigils and Matins forced monastic communities to rely principally on observation of the stars.[28] A generation after Benedict, the Frankish bishop Gregory of Tours wrote the earliest treatise codifying this practice. *On the Course of the Stars* explains how to tell time during the night hours by observing the times at which particular stars rise above the horizon.

Gregory gave simple instructions for each month: in January, for example, if you begin when a given star rises, you can sing a certain number of psalms before dawn.[29]

Because this method of telling time, by attending to the rising of particular stars, works best in a minimally built environment with an unobstructed view of the horizon, it had later to be modified to accommodate the visual restrictions of the medieval monastic cloister.[30] An eleventh-century text from an unidentified French abbey, for instance, took as frame of reference the architecture of the cloister viewed from a point inside the cloister garden, using the windows of the dormitory and refectory as a temporal scale. In the words of one typical entry, "On the Feast of the Circumcision, you should light the lamps when the bright star in the knee of [the constellation Artophiax] shows above the roof in the space between the first and second windows of the dormitory."[31]

Telling time by careful scrutiny of the heavens was a crucial part of the organization of monastic life, melding the two meanings of the term *observatio*: observation and observance. The regular, cyclical movements of the heavens expressed the order and harmony of Creation and were in turn reflected in the daily cycle of services and the annual cycle of Christian festivals. Time telling was a form of moral and spiritual discipline, as vividly expressed in the section of Peter Damian's *On the Perfection of Monks*, composed after 1067, that describes the duties of the monastery's timekeeper:

> Let the one responsible for marking the hours know that no one in the monastery must be less forgetful than him, for if he fails to keep the hour of any sacred office, by anticipating it or delaying it, he will disturb the order of every subsequent office. Therefore, let him not lose himself in stories, nor engage in long conversation with another, nor, finally, ask what those outside the monastery are doing, but—always intent on the responsibility entrusted to him, always attentive, always solicitous—let him observe [*meditetur*] the motion of the heavenly sphere, which never rests; the path of the stars; and the constant course of passing time. Let him learn the habit of singing psalms, if he wants to have a daily principle for telling time, so that whenever he cannot see the light of the sun or the variety of the stars on account of the density of clouds, he may measure them by the length of psalms, as if he were himself a clock.[32]

The use of the term *meditare* as a synonym for *observare* in monastic texts from this period underscores the spiritual dimensions of the discipline of astronomical observation: both words, together with *contemplari*, with which they were semantically associated, referred to the process whereby reflection on the physical objects of Creation led to knowledge of spiritual truths.[33] Observation in this sense functioned as a seamless part of monastic observance—attentive, orderly, and all-consuming, focused on the transcendental meanings of terrestrial experience—while the pious observer, immersed in the divine order, became an instrument for measuring the orderly passage of time.

The second form of early medieval Christian timekeeping, which involved computing the dates of Easter and other moveable feasts, posed greater technical problems than timing the daily round of prayer. Whereas the latter was simply and securely fixed by the diurnal circling of the sun and stars, the Christian calendar, anchored by Easter (vexingly defined as the first Sunday after the fourteenth day of the lunar month that falls on or after the spring equinox) required coordinating cycles of incommensurable length: the week, the Julian year, and the lunar month. This was the subject of the medieval discipline known as *computus*.[34] Observation had little part to play in this form of timekeeping; rather, *computus* focused on calculating the date of Easter arithmetically using tables for the relevant parameters. From its inception, this system was bedeviled by disagreements concerning the dates of the equinoxes, and these problems were exacerbated over time by the compounding of small discrepancies between the values in the tables and the actual motions of the sun and moon. By the eleventh century, there was more than two days' difference between the phases of the "ecclesiastical moon" (calculated from official tables) and the observable moon. This discrepancy generated commentary and confusion on the part of contemporaries, particularly on the occasion of spectacular events like solar and lunar eclipses, which were supposed to occur at the new and full moons, respectively. Thus an anonymous chronicler in the monastery of Saint André in Cambrai identified the widely observed solar eclipse of 2 August 1133 as a supernatural omen, since it happened two days before the official, that is, tabular, new moon.[35]

The growing concern with inconsistencies between observations and table-based predictions explains in part the eagerness with which Latin Christian scholars embraced the astral science then flourishing in the Muslim world. This knowledge was part of a much broader flow of learning from East to West that began in the late tenth and accelerated through the twelfth century, borne on a tide of texts, instruments, and travelers from Muslim lands, and it provided the tools for a dramatic change in observational practices in Europe. The Arabic sciences of the stars included an important body of knowledge devoted to religious timekeeping, as in the Latin West, as well as the whole range of secular astral sciences—astrological, physical, and mathematical—elaborated by Hellenistic authors, most notably Ptolemy (second century CE). Observation was important in both branches, since in Islam, unlike Christianity, the religious calendar was determined in practice by the observable, as opposed to the tabular, moon.[36] The secular astral disciplines supported an even more impressive tradition of observational practice in Islamic lands. This was reflected in the elaboration of Greek instruments (such as the armilla, the quadrant, the armillary sphere) and the development of new ones (such as the torquetum and various new kinds of astrolabes), as well as the creation of the first observatories under the sponsorship of rulers and other wealthy patrons. These were able to support systematic long-term programs of celestial observation that aimed at not only improving or revising the parameters of Ptolemaic theory but ultimately, as at the observatory of Maragha, founded in northern Persia in the thirteenth century, revising the basics of Ptolemaic theory itself.[37]

Latin European scholars sought out the fruits of this thriving Eastern tradition: both astronomical texts and tables and observational and calculatory instruments. Of the latter, the most exciting was the astrolabe, which greatly increased the precision with which celestial events and bodies could be located in time and space.[38] The astrolabe was quickly absorbed into the practice of Christian timekeeping, as is clear from a late tenth-century preface to a now lost treatise on the astrolabe, in which the late tenth-century author (most likely the archdeacon Lupitus of Barcelona) emphasized the utility of the instrument, and of astronomy in general, for regulating the times of prayer and the liturgical calendar, and for drawing the mind toward God through meditation on the order of Creation, enabling pious users "to attain the realm of invisible things by contemplation of the visible sphere."[39]

Over the course of the late tenth and eleventh centuries, however, this relationship to the stars as primarily objects of religious contemplation changed dramatically. The astrolabe continued to be used for sacred timekeeping, as is clear from a miniature in an early thirteenth-century psalter made most likely for Blanche of Castille (fig. 1.1, plate 1); this frontispiece, which shows an astronomer observing the heavens with an astrolabe, introduces the liturgical calendar at the beginning of the psalter.[40] But the instrument could also be employed in new, unsettling ways. In 1092, for example, Walcher, prior of the English monastery of Great Malvern, used an astrolabe to determine the elevation of the moon during a lunar eclipse, thus accurately fixing its time and therefore the time of the astronomical (as opposed to ecclesiastical) full moon; this allowed him to draw up an alternative set of lunar tables, more accurate than the ecclesiastical ones, which gave the exact time

FIGURE 1.1. Astronomers at work. Psalter of Blanche of Castille. Paris, Bibliothèque de l'Arsénal, MS 1186, fol. 1r (Paris, ca. 1226–32). The central figure is making an observation using an astrolabe. The figure on the left appears to be recording the result of the observation, while the one on the right consults a text, possibly a very early ephemerides or set of astronomical tables. By permission of Bibliothèque nationale de France.

of each new moon between 1036 and 1111.[41] Walcher's motives had nothing to do with the liturgical calendar—he described his tables in terms of their medical utility—and he mocked the fears of English monks, for whom an eclipse was a "terrifying prodigy" and a supernatural portent of disasters to come.[42] In his use of the astrolabe, his cultivation of precision, and his self-presentation as a well-traveled man of science committed to natural rather than portentous interpretations of celestial phenomena, Walcher was one of a number of late eleventh- and twelfth-century Latin writers for whom the newly imported Arabic knowledge of the heavens represented a compelling alternative to the early medieval Christian understanding of nature, with its hidebound relationship to textual authority, its lack of interest in the explanation of natural phenomena, and its Augustinian inclination to invoke supernatural rather than natural causes to explain the appearance of the physical world.

Walcher called his treatise on the lunar eclipse of 1092 *On an Experience of the Writer* despite what we would consider its observational cast. His use of the word *experientia* reflects the Plinian distinction between a onetime trial and the long-term tracking of cyclical phenomena. But it also corresponds to a new emphasis on the importance of testing textual—in this case tabular—authority against personal experience, which was invoked by Walcher and like-minded contemporaries as one of the most important lessons to be learned from Arabic science. In the words of Walcher's "master," the Aragonese astronomer Peter Alfonsi, paraphrasing in turn the *Great Introduction to the Science of the Judgment of the Stars* of Abu Ma'shar (ca. 850), "the art [of astronomy] cannot be understood except by direct experience [*experimentum*]."[43] This embrace of the language of experience by Abu Ma'shar and, indirectly, by Latin astronomers, was ultimately rooted in the natural philosophical works of Aristotle, newly translated into Latin in the twelfth century, for whom, as previously noted, experience rather than observation was a fundamental source of human knowledge about the natural world. In large part as a result of the increased circulation of Aristotelian texts in the twelfth and thirteenth centuries, the language of *experientia/experimentum* began to swamp *observatio*, itself not a particularly robust epistemic category, even in the sciences of the stars. (In contrast, contemporary descriptive writing on natural history, medicine, anatomy, mechanics, and the like referred only to experience, not to observation, except in its prescriptive sense.) This change corresponded to a change in the goals of the study of nature, which increasingly emphasized pursuit of the literal truth of the created order over its allegorical meaning, and the manipulation of natural forces over submission to the cycles that governed the physical world.

Observation as Inquiry, 1100–1500

In privileging the observable facts of the heavens, Peter Alfonsi and Walcher of Malvern differed from the earlier masters of *computus*, who asked of their tables only that they generate dates for major festivals that could be agreed on by all Latin Christians. This shift in orientation is reflected in the verbs with which the two groups described the work of observing the heavens. Whereas monastic writers used *observare*, *meditare*, and *contemplari* as rough equivalents, emphasizing the study of celestial phenomena as a search for spiritual meaning, Alfonsi, Walcher, and their twelfth- and thirteenth-century successors associated *observare* with verbs such as *inquirere*, *investigare*, *considerare*, and, increasingly, *experiri*, which reflected their goal to search out what the twelfth-century scholar Raymond of Marseilles called "the truth of the firmament itself."[44] Having a true picture of the heavens was important for many reasons, most of which had nothing to do with the liturgical calendar (though it was embarrassing for Christians to be mocked by Jews for getting the date of Easter astronomically wrong, as happened in Paris in 1291).[45] Alfonsi and Walcher emphasized the medical uses of this information, which gave guidance about when to let blood, cauterize, and lance abscesses, but other applications were equally important, including timekeeping and judicial astrology. One of the most important lessons Christian scholars took from newly translated Arabic and Greek works on the science of the stars was that celestial phenomena were not merely signs of future events, but their causes (subject always to God's will). The changing positions of the heavenly bodies influenced the fates of faiths and kingdoms as well as individual human lives.[46] As in the Islamic world, the men who engaged in this kind of work were for the most part formally trained professionals who supplemented their teaching and, sometimes, medical practice with a range of consultational services, astrological, calendrical, and cosmographical; their clients included popes, abbesses, queens, and kings, together, one presumes, with less illustrious patrons whose names they did not bother to record.

These practical motives were reinforced by theoretical ones, for a truthful understanding of the constitution of the heavens was also of prime importance to Aristotelian natural philosophy, which dominated the university curriculum from the thirteenth century on. Aristotle had described the universe as a great system of eight nested spheres, one for each of the planets (which included the sun and the moon) and one for the fixed stars. This system had been criticized and revised by Arabic scholars, who supplemented their study of Ptolemy with their own celestial observations. European masters could not avoid these astronomical and cosmological critiques of Aristotle's model.

Albertus Magnus, the most prolific and influential natural philosophical writer of the thirteenth century, appealed to observation—a concept conspicuously absent from *On the Heavens*, Aristotle's treatment of cosmology—in his commentary on that treatise. Although, like Cicero and Pliny, Albertus continued to associate observation above all with the foundational work of the Babylonians and Egyptians carried out in the distant past,[47] he argued that it had a continuing part to play in cosmology: the observations made by more recent Arabic masters, notably al-Battani, had allowed them to supplement and even correct the work of the ancients by demonstrating that either the cosmos was composed of ten rather than the traditional eight spheres; or, more likely, the eighth sphere, which carried the fixed stars, had multiple motions, including, in addition to its annual rotation, the much slower motions of precession and trepidation.[48] Albertus hypothesized that trepidation in particular had escaped the notice of both Aristotle and Ptolemy, whose observational baseline, long as it was, did not permit them to identify and chart it: "[F]alsehood is found out by true observations [*observationes*], . . . but since that motion requires a great deal of time on the part of the experiencer [*experienti*], it is perhaps for this reason that it was not perceived by the ancients."[49]

Albertus's association here of observation and experience reflects a melding of Ptolemaic astronomy, with its emphasis on the former, and Aristotelian natural philosophy, with its attention to the latter. At the same time, medieval astronomy remained a largely expository and calculatory field, organized around the elucidation of existing astronomical texts, tables, and canons (rules for the use of those tables) and the construction of calendars, almanacs, ephemerides, and horoscopes, which gave planetary positions for particular days and times, past, present, or future, based on those tables. When European astronomers turned their eyes and instruments to the skies, they did so for three main reasons: to intervene in cosmological and theoretical debates regarding problems like those relating to the eighth sphere;[50] to compare predicted planetary positions and the predicted times of celestial occurrences such as eclipses with physical reality; and to make practical judgments, such as astrological predictions or determinations of latitude and longitude at the request of founders and administrators of churches and abbeys. The liveliest center for work of this sort in the late thirteenth and early fourteenth centuries was Paris, where several generations of astronomers left a spotty body of records that give some sense of what empirical inquiry meant to a late medieval community of urban scholars, teachers, and consultants on astral matters.[51] The earliest serious observer of the heavens seems to have been William of Saint-Cloud, who was active at the French court—he compiled a calendar for Queen Marie, wife of Philip III—but about whom little else is known beyond his surviving works. These include an *Almanach of the Planets*, compiled circa 1292, whose preface contains calculations based on a number of William's own observations of solar and lunar eclipses; of conjunctions involving Saturn, Jupiter, and Mars; and of the altitude of the midday sun at several equinoxes and solstices.[52]

Considerably more is known about the observations of John of Murs a generation later, who, in addition to teaching at the University of Paris consulted on matters calendrical, astrological, and cosmographical for the heads of several religious houses as well as for Pope Clement VI. John's observations were less intellectually ambitious than William's. In 1319, for example, he made an observation of the time of the vernal equinox, which confirmed both William's own observation of the same event in 1290 and the reliability of the Alfonsine Tables.[53] John's *Exposition of the Intention of King Alfonso in His Tables* (ca. 1320) was an exercise in textual interpretation rather than of independent discovery and aimed not to correct but to "explain parameters and models . . . already found in the tables he had in hand."[54] In other words, while John, like most late medieval Latin astronomers, may have been in principle committed to the project pursued by his Arabic predecessors of refining and revising their knowledge of the heavenly motions by extending the temporal baseline of observation, in practice he worked in an "experimental" mode, comparing the results of his observations of eclipses, conjunctions, solstices, and equinoxes with those of calculations based on the Alfonsine Tables. Although he occasionally recorded differences between the two—most notably in the case of the solar eclipses of 1333 and 1337, both of which occurred early relative to the time predicted in the tables—he was much more hesitant than William to propose corrections of any of the relevant parameters, proposing instead a variety of explanations for the discrepancies, including arithmetical errors, errors in the commonly accepted values between the meridians of Toledo and Paris, and the possibility that solar eclipses were not, as he had assumed, simultaneously visible anywhere that they could be seen at all. In general, then, his first instincts were to "save the texts" when the phenomena did not cooperate.[55]

The vast majority of John's other known observations appear as scattered marginal notes in a 226-folio collection of astronomical texts and tables in his own hand.[56] The following record gives a sense of the format John generally used:

> In the name of God amen. In the year of His Incarnation 1326 in the pious abbey of the nuns of Fontevrault, in the archdiocese of Tours, by permission of the reverend lady Eleanor, abbess in Christ Jesus. I, J[ohn] of M[urs] and A. k.

> [*amicus karissimus?*] made an experience [*experientiam*] of the latitude of the region and the altitude of the equinoctial pole. . . . in the aforesaid year of our Lord, at noon on Wednesday, the 12th of March. . . . Carried out in obedience and reverence to the aforesaid lady.[57]

The verbal formulas in this note show that John did not view his observations as part of a long series of homogeneous—let alone anonymous—pieces of information, which gained meaning only in the aggregate, but as important and singular facts, certified not by checking or repetition, but by the presence of named witnesses of high standing and good character, furnished with the proper equipment, and precisely situated in place and time.[58] The quasi-notarial formulas in John's astronomical "experiences" confirm their epistemic function: they aimed either to test the reliability of the Alfonsine Tables or to assert John's own reliability as an expert consultant on matters such as the latitude of the abbey of Fontevrault. At the same time, the fact that they were confined to the margins of one of John's own reference works underscores the fundamentally individualistic nature of the observational enterprise in Latin Europe. The absence of evidence for long runs of celestial observations in thirteenth- and fourteenth-century Europe reflects the lack of institutions to support those observations on the model of the observatories of the Islamic world or of groups of masters and students engaged in a collaborative enterprise. Observatories like the one at Maragha, founded and funded by Ghengis Khan's grandson Hulegu, could house long-term observational projects such as the one undertaken by the Andalusian scholar al-Maghribi between 1262 and 1274.[59] In the Latin West, in contrast, observations of the heavenly bodies remained scattershot and individualistic, recorded in private documents and pursued for private ends. Even when observers convened on the occasion of dramatic occurrences such as the solar eclipse of 1337, as described by John of Murs—"for this experience [*experientia*] ten of us were present, many with good astrolabes"—there is no record of any systematic collaboration or comparison of results.[60]

By the same token, the observations engaged in by William of Saint-Cloud, John of Murs, and their contemporaries differed fundamentally from those of monastic timekeepers, for whom daily observations were integral to a life of ritual and spiritual observance. In contrast, the observations recorded by William and John (as far as we know) were relatively few: nine and fifteen, respectively, in the course of long and productive careers. Those observations were also punctual and intermittent, focused on special moments in the complicated dance of the heavenly bodies: the solstices, the equinoxes, the two or three days around the conjunction of Saturn and Jupiter's or Mars's entry into retrograde motion. This "experimental" character of late medieval astronomy—its focus on using isolated observations to evaluate existing tables and the calendars of planetary positions based on those tables—is reflected in the extreme paucity of serial records of raw astronomical data, the characteristic form, established by the Babylonians, of the observational report.[61]

The only coherent body of medieval records of serial observations pertained not to the motions and positions of celestial bodies but to the weather, which was the subject of a well-established branch of astral science that aimed to predict atmospheric conditions based on those motions and positions.[62] These observations took the form of brief descriptions of the day's weather inserted in the margins of almanacs and ephemerides, which supplied information regarding the absolute and relative positions of the planets for every day of the year. The ephemerides, laid out in tabular form with rows corresponding to the days of each month and columns corresponding to the positions of the planets in the zodiac, provided an ideal graphic framework for recording daily observations, which could be inserted in the margin next to the planetary information for the relevant date.[63] The earliest known document with annotations of this sort dates to the mid-thirteenth century,[64] and the practice of recording such observations seems to have gained steadily in importance over the course of the fourteenth and fifteenth centuries. Although the notes were often telegraphic—"clear day," "cloudy dry," and so on[65]—some observers went into more detail, including a learned fourteenth-century English cleric, William Merle, in a rudimentary treatise, *Observations [Considerationes] of the Weather at Oxford for Seven Years* (1337–44), which appears to be a fair copy of the same kind of marginal notes.[66] The entries for the month of March 1343 are typical of the whole:

> March: first, ice with a little snow and then a little rain. 2 small cloud. 3 and 4 a little ice with a little frost. . . . 27 stormy with a very great north wind and with hail, rain and snow several times over the course of the day; in the south there was a very large earthquake, which caused stones to fall from stone fireplaces, shaken by a very great commotion, in several places in Lyndsay [in Lincolnshire, where Merle was rector], and it lasted for the space of time in which it is possible clearly to say the Ave Maria. 28 stormy with hail, ice, rain and north and west wind. 30 more ice than on the 29th, with snow and wind. 31 more snow than on the 30th, with north and west wind, followed by notable snow.[67]

Records of this sort were almost certainly intended to enable their users to correlate observed weather conditions with the positions of the stars and planets in order to check and refine the meteorological predictions in almanacs and calendars; for example, Cuno, an otherwise unknown author of a mid-fourteenth-century treatise on weather science called *Judgments of the Impressions in the Air*, refers with great precision to weather events that he himself had witnessed (*experimentata*), although no weather diary by him is known to survive.[68] Another possible (and compatible) motive for these observations is suggested by Stuart Jenks's study of the owners of late medieval astrometeorological manuscripts, which revealed a striking interest in the subject among monks, friars, and provincial clerics, who otherwise had little in common with the professional practitioners that made up the bulk of readers and writers of astronomical and astrological texts.[69] The earliest compilers of daily weather observations before the years around 1500, when the practice spread to a wide range of learned professionals, seem mostly to have been regular or secular clerics: Merle, rector of the small parish of Driby, Lincolnshire; quite possibly Cuno, who describes himself as staying in the monastery of Saint Burgard in Würzburg,[70] and the anonymous author of an elaborate weather diary covering the years 1399–1405, once in the collection of the Dominican convent in Basel; and an early fifteenth-century canon of Bamberg Cathedral.[71] The clerical vocations of these men suggest that this form of observation, while part of a learned and highly technical discipline rooted in the science of the heavens, was nonetheless shaped by a daily reflection on the regular cycles of Creation that was continuous with the observant practices of Christian monks.

Practices and Genres

Toward the end of the fifteenth century, a few learned astronomers began to adopt practices that had more in common with those of contemporary students of the weather than those of "experiential" observers such as John of Murs; their observations became more frequent, more systematic, and less tied to rare events such as conjunctions and equinoxes, eclipses, and comets. In other words, by dint of repetition, astronomical *experientiae* mutated into *observationes* in a new sense—one that recalled the Plinian idea of tracking cyclical phenomena but retained the late medieval emphasis on the authority and identity of the observer. We can trace this shift in the work of three students of the heavens whose careers spanned the second half of the fifteenth century: Georg Peurbach, a teaching master at the University of Vienna and imperial astrologer to Frederick III; his student Johannes Regiomontanus, whose peripatetic career as a consultant and teaching master began in Vienna and ended in Rome; and Regiomontanus's disciple Bernhard Walther, a merchant and dedicated observer in Nuremberg. Records of the observations of all three men were printed in 1544, long after their deaths, by the Nuremberg astronomer and cosmographer Johannes Schöner, as described by Gianna Pomata in chapter 2.[72]

These records show a clear change between the observations carried out jointly by Peurbach and Regiomontanus, beginning in 1457, and those of Regiomontanus and Walther, whose collaboration began in 1472. Although, following the Latin translation of Ptolemy, all three use the language of *consideratio* and *observatio* rather than *experientia*, the former resemble John of Murs's notes more than a century earlier. Their form is narrative and descriptive, and they give the time, the place, and the names of those present, as in the case of Peurbach's and Regiomontanus's first recorded observation, of a total eclipse of the moon:

> In the year of our Lord 1457, after sunset on the third day of September, in Melk, Austria, near Vienna, Master Georg Peurbach and Johann of Königsberg [Regiomontanus] observed [*observaverunt*] a total eclipse of the moon when the sun and moon were in opposition. . . . This observation [*consideratio*] was in Melk, a fortified town in Austria, which is 11 German miles west of Vienna.

This note was followed by a set of calculations, which showed that the true time of opposition was eight minutes earlier than predicted by the Alfonsine Tables.[73] Like John of Murs, Peurbach and Regiomontanus focused on significant moments in the motions of the heavens, and their observations appear to have been both rare—one in 1457, two in 1460, at least in the records published by Schöner—and peripatetic, as they followed their patron Cardinal Bessarion from Vienna to Rome.[74] After Peurbach died in April 1461, Regiomontanus's records grew more frequent, but, with the exception of the solar observations carried out with Walther, they were still intermittent: the notes in Schöner include only 35 additional observations over the fourteen years before his own death in 1475.[75]

The solar observations begun by Regiomontanus in 1472, after he settled in Nuremberg and started to work with Bernhard Walther, are quite different in character; repeated records of the distance of the midday sun from the zenith, they were apparently intended to permit the accurate calculation of the daily position of the sun in the ecliptic over the course of the year in order to supply the basis for a new set of tables superseding those of Alfonso, and perhaps even to reform the reigning astronomical models.[76] Rather than merely cumulative, they became repetitive, and their novelty is evident from the format in Schöner's edition: telegraphic notes entered on a grid like those that structured the ephemerides and the weather observations in their margins.[77] Because the same thing was being measured by the same people in the same way in the same place using the same instrument (a wooden Ptolemaic ruler), descriptive detail was superfluous; except for the date and the day's measurement, the only additional notes pertained to special conditions, notably two particularly windy days that affected the stability and alignment of the instrument and, in one case, the special "diligence" of the observer. After Regiomontanus died, Walther acquired his notebooks and instruments, as Regiomontanus seems earlier to have acquired those of his own teacher Peurbach, and continued his master's labors for thirty years, with only two significant interruptions.[78] While this cumulative body of work, the product of a chain of three masters and disciples, was a far cry from the large-scale early modern observational projects analyzed by Gianna Pomata and Lorraine Daston (chaps. 2 and 3), it was much more robust in continuity and scope than anything found in the manuscripts of earlier Latin European astronomers.

Walther was a dedicated observer in the classical sense: his only surviving autograph is a weather diary in the margins of the 1474 edition of Regiomontanus's ephemerides.[79] His astronomical observations did not have the clocklike consistency of the more assiduous trackers of the weather, in part no doubt because his job as an agent of one of the major German commercial firms of the period took him on frequent trips outside Nuremberg. Nonetheless, he produced a body of data—almost 750 solar measurements—that dwarfed that of John of Murs, Peurbach, and Regiomontanus, and was to be surpassed in size and accuracy only by the work of Tycho Brahe in the late sixteenth century (as Daston describes in chap. 3). Furthermore, his long experience came to shape his sensibility in ways that anticipate the work of early modern observers. Over time he became increasingly sensitive to possible sources of error, especially his own care or "diligence," weather, and problems in aligning the vertical axis of his ruler to the zenith.[80] Through Walther's records we see the emergence of the habitus of an observer: consistent, disciplined, highly attuned to sources of error, and constantly seeking to improve the quality of his results. For three decades, until days before his death in 1504 at the age of seventy-four, he made regular observations at midday (for his solar observations) and in the evening or early morning (for his observations of the planets), though he seems to have tried, for understandable reasons, to avoid working in the middle of the night. Nonetheless, he regularly hauled his heavy instruments outside, set them up, and meticulously aligned them—an especially difficult task on windy days. In 1502, to make things easier, he had a small observing platform constructed outside one of the windows of his new house, but this generated problems of its own when the neighbors sued him for damage by falling objects.[81] We also catch glimpses of another novel element that characterized later observational practice, the pleasure and exaltation produced when the tedium of the observational routine was interrupted by the occasional spectacular sighting, such as the perfect conjunction of Mars and Saturn on September 5, 1477: "Oh, with what emotion did I see their coming together," wrote Walther, "since one was predicted to eclipse the other, but the event is very rare."[82]

All in all, the time and effort Walther committed to this project was unprecedented in European astronomy and had much more in common with the activities of Tycho (who had many more assistants) than with those of Peurbach or Regiomontanus.[83] Nonetheless, his aims remained in part traditional. For one thing, his solar observations may have been intended to help regulate the Nuremberg city clock.[84] Like Peurbach and Regiomontanus (and William of Saint-Cloud and John of Murs before them), he was also interested in using observation to test the values predicted by contemporary tables—in his case, Regiomontanus's ephemerides—and it seems quite possible that he hoped to reach the goal that had eluded Regiomontanus, of actually correcting the parameters of the Alfonsine Tables.[85] In any event, although Walther died without completing Regiomontanus's project, he ended up generating a body of new data that ultimately served as a resource to help ground the work of later theorists, including Nicholas Copernicus.[86]

In the later fifteenth century, this increasing engagement with the sensory details of natural phenomena was not confined to "observational" science—the science of the stars—but also characterized the "experiential" sciences, including medicine, alchemy, and natural history, as Gianna Pomata emphasizes in chapter 2. These focused not on long-term cyclical phenomena but on the properties of particular natural substances and species, especially those that were unique, variable, or contingent and therefore not amenable to deductive explanation. The careful attention paid by fifteenth-century artisans such as Leonardo da Vinci to the flow of water, the shapes of clouds, and the anatomy of the human body is well known, and would begin slowly to permeate learned Latin culture over the course of the sixteenth century. In the universities of the later Middle Ages, however, attention of this sort was rare: medicine, like natural philosophy, retained a nonempirical bent, privileging knowledge derived through abstract reasoning from first principles and based in books.[87] Information derived from the senses could not be rejected altogether, however, especially in the domain of *practica*, which, as its name suggests, dealt with the diagnosis of particular illnesses and their treatment using remedies derived from plants, minerals, and animals. Although the healing properties of many of these had been catalogued by ancient Greek and medieval Arabic writers, their work was by no means exhaustive: there were others (particularly northern ones) missing from their pharmacopeia, and even well-known substances might reveal previously uncataloged powers either on their own or when combined with each other in previously untried ways. For this reason, the academic medical discipline of *practica*, together with alchemy, with which it was sometimes associated, was one of the few arenas in which we see the beginnings of empirical research on terrestrial phenomena before the sixteenth century: *practica* not only served to prepare aspiring physicians for medical practice, but also drew on the often unpredictable information that this practice might reveal.

Such knowledge was "experiential," not "observational": it arose not from the systematic tracking of natural phenomena but from information derived through trial and error in particular cases—"fortuitous practical discoveries," in the words of Michael R. McVaugh.[88] This information was preserved in the form of what were known as *experimenta*, medical and alchemical recipes and short descriptions of successful procedures, such as this example from a compilation by Arnald of Villanova, who taught medicine at the University of Montpellier:

> Another *experimentum* for stomach cramps. If you suffer from stomach cramps that sometimes arise from an excess of burned or ascending choler [one of the four humors], take water of fennel and drink it in the morning on an empty stomach, and immediately you will be relieved by the power of that water. . . . It is very good for suppressing bile and dissolving the material by which such cramps are generated. And I tested [*expertus fui*] this on a certain cardinal, who suffered from this illness and was freed from it by this *experimentum.*[89]

Despite its (by Scholastic standards) shaky epistemic basis, knowledge of this sort was accepted as useful and important. It was cumulative, like observational knowledge, and even if no general rules or scientific theories could ever be based on it, it served as an invaluable pool of information for later practitioners.

Experimenta, singly or gathered together in collections like Arnald's, constituted a major genre of medieval scientific writing, surviving by the thousands in European manuscripts, both Latin and, later, vernacular, from the twelfth well into the seventeenth century. They testify not only to the understanding of practical medicine, like alchemy, as a field shaped in an ongoing way by experience, but also to the solid role of experience in Aristotelian epistemology, where it functioned as an fundamental type of knowledge. The status of observation was entirely different, as both a practice and an epistemic category: never theorized, or even acknowledged, as a form of human knowledge by Aristotle, it flourished only in a restricted domain, the science of the stars: astronomy, astrology, and astrometeorology. The relatively minor role played by observation in natural inquiry before the sixteenth century is reflected in the lack of any genre of *observationes* that would correspond to *experimenta* in the observational realm. With the exception of the astronomical *considerationes* of William Merle, the word is absent from titles and explicits. Furthermore, Merle's "treatise" is obviously only a fair copy of the typical form used to record observations: marginal annotations in texts falling into more established genres such as astronomical tables and, especially, ephemerides, where the day-by-day organization of these tables of planetary positions provided an ideal format for recording the day's weather. Similarly, we do not know for certain that the *Observationes* of Peurbach, Regiomontanus, and even Bernard Walther were so titled by their authors, rather than by their mid-sixteenth-century editor Schöner, who published the contents of what were clearly private notebooks rather than material, like medieval *experimenta,* intended for public circulation. It is in this sense that observation was "marginal" to medieval scientific inquiry—too informal to have its own genre, let alone to be a regular part of training and practice for astronomers and astrologers.

Notes

Many thanks to Catherine Davies and Josephine Fenger for invaluable research assistance, and to Daryn Lehoux, Craig Martin, Darrel Rutkin, Elaheh Kheirandish, and especially Richard Kremer, who generously helped me identify and work through materials for this paper and who are of course blameless for any errors in it.

1. Chiara Crisciani, "History, Novelty, and Progress in Scholastic Medicine," *Osiris* 6 (1990): 116–39, esp. 121–28. Much research remains to be done on the important contributions of vernacular and nonliterate researchers in such matters, which I cannot cover in this brief essay.

2. See Katharine Park, "Natural Particulars: Medical Epistemology, Practice, and the Literature of Healing Springs," in *Natural Particulars: Nature and the Disciplines in Renaissance Europe,* ed. Anthony Grafton and Nancy G. Siraisi (Cambridge, Mass.: MIT Press, 1999), 347–67; Lorraine Daston and Katharine Park, *Wonders and the Order of Nature, 1150–1750* (New York: Zone Books, 1998), chap. 3.

3. Cf. Pierre Michaud-Quantin, with M. Lemoine, *Études sur le vocabulaire philosophique du Moyen Âge* (Rome: Ateneo, 1970), 220: "lorsque l'on traduit *experientia* ou *experimentum* . . . il s'agit en fait pour les médiévaux d'observation pure et simple" (regarding the translation of *experientia* or *experimentum* . . . medieval writers were referring to observation pure and simple). The same is true of the otherwise very useful article by Danielle Jacquart, "L'observation dans les sciences de la nature au Moyen Âge: limites et possibilités," *Micrologus* 4 (1996): 55–75; and Jean A. Givens's impressive study, *Observation and Image Making in Gothic Art* (Cambridge: Cambridge University Press, 2005).

4. At least until the later Middle Ages, *experimentum* and *experientia* were used largely interchangeably; see Jacqueline Hamesse, "*Experientia/experimentum* dans les lexiques médiévaux et dans les textes philosophiques antérieurs au 14e-siècle," in *Experientia: X Colloquio internazionale, Roma, 4–6 gennaio 2001*, ed. Marco Veneziani (Florence: Olschki, 2002), 77–90, on 80–81 and passim.

5. Giacinto Spinosa, "Empeiría/*Experientia*: Modelli di 'prova' tra antichità, medioevo et età cartesiana," in *Experientia: X Colloquio internazionale, Roma, 4–6 gennaio 2001*, ed. Marco Veneziani (Florence: Olschki, 2002), 169–98.

6. The fundamental study is Jole Agrimi and Chiara Crisciani, "Per una ricerca su *experimentum-experimenta*: riflessione epistemologica e tradizione medica (secoli XIII–XV)," in *Presenza del lessico greco et latino nelle lingue contemporanee*, ed. Pietro Janni and Innocenzo Mazzini (Macerata: Facoltà di Lettere e Filosofia, Università degli Studi di Macerata, 1990), 9–49. See also Michael R. McVaugh, "Two Montpellier Recipe Collections," *Manuscripta* 20 (1976): 175–80, on 178.

7. Peter of Maricourt, *Epistula de magnete*, ed. Loris Sturlese, in *Opera* (Pisa: Scuola Normale Superiore, 1995); Dietrich of Freiberg, *De iride et de radialibus impressionibus*, 2.16.8–2.17.2, in *Opera omnia*, 4 vols., ed. Maria Rita Pagnoni-Sturlese, Rudolf Rehn, Loris Sturlese, and William A. Wallace (Hamburg: Felix Meiner, 1977–85), 4: 173–74.

8. Frederick II, *De arte venandi cum avibus*, ed. C. A. Willemsen (Leipzig: Insula, 1942), 1 (my translation).

9. See Agrimi and Crisciani, "Per una ricerca," 19–24.

10. See, e.g., Aristotle, *Historia Animalium*, 623a14 and 27.

11. On this cluster of ancient sciences, see Daryn Lehoux, "Observation and Prediction in Ancient Astrology," *Studies in History and Philosophy of Science* 35 (2004): 227–46; idem, *Astronomy, Weather, and Calendars in the Ancient World: Parapegmata and Related Texts in Classical and Near Eastern Societies* (Cambridge: Cambridge University Press, 2007), esp. chap. 3; Francesca Rochberg, *The Heavenly Writing: Divination, Horoscopy, and Astronomy in Mesopotamian Culture* (Cambridge: Cambridge University Press, 2004); and Liba Taub, *Ancient Meteorology* (London: Routledge, 2003), chap. 2. I have used the term "astral science" or "science of the stars" whenever possible in order to avoid suggesting that astronomy and astrology (including astrometeorology) can or should be clearly distinguished in the medieval period, let alone distinguished along modern lines. Although medieval authors might make a conceptual distinction between the two, the terms *astronomia* and *astrologia* were not used consistently to refer to one or the other of the modern disciplines, and they were always associated in practice.

12. Cicero, *De divinatione*, 1.18.340, in Cicero, *De senectute, De amicitia, De divinatione*, trans. William Armistead Falconer, Loeb Classical Library (Cambridge, Mass.: Harvard University Press, 1923).

13. See David Wardle's commentary in *Cicero: On Divination, Book I* (Oxford: Clarendon Press, 2006), 165.

14. Cicero, *De divinatione*, 1.42.93 (translation of last sentence modified for accuracy). Here "observation" corresponds to *animadversio*, but *observatio* was used in the preceding sentence, and it is clear that the two words are synonyms.

15. Ibid., 2.56.16. See Lehoux, *Astronomy, Weather, and Calendars*, 5.

16. Pliny, *Natural History*, 18.85.132, trans. H. Rackham et al., Loeb Classical Library (Cambridge, Mass.: Harvard University Press, 1935). Other examples in 18.57.209–10 and 18.65.238, and in 8.69.172–73 (animal husbandry).

17. Ibid., 30.14.44.

18. Ibid, 7.56.194. For navigation, see 7.56.209.

19. Ibid., 18.69.284.

20. Ibid., 37.33.83 (testing of opal). See also 37.54.145 (black antipathes), 37.59.66 (hephaestitis); 2.75.183 (well at Syene).

21. Augustine, *Epistulae*, 55.15, ed. K. D. Daur, 2 vols. (Turnhout: Brepols, 2004), 1: 245–46.

22. See chapter 2 in this volume. See also Gianna Pomata, "A Word of the Empirics: The Ancient Concept of Observation and Its Recovery in Early Modern Medicine," *Annals of Science*, forthcoming.

23. Although the eighth-century encyclopedist Isidore of Seville signaled a distinction between "observation" and "observance" in his *Differentiae; or, On the Property of Words*—he defined *observatio* as "relating to study, learning, and art," and *observantia* as "relating to cult and religion"—he elsewhere used both terms in the latter sense: observation, like observance, involved obedience to religious authorities and rules. See Isidore of Seville, *Differentiae, sive de proprietate sermonum*, in Isidore, *Diferencias libro I*, ed. and trans. Carmen Codoñer (Paris: Belles Lettres, 1992), 151 (400), 162.

24. Dieter Harmening, *Superstitio: Ueberlieferungs- und theoriegeschichtliche Untersuchungen zur kirchlich-theologischen Aberglaubensliteratur im Mittelalter* (Berlin: Erich Schmidt, 1979), 76–80.

25. For example, Bede, *De ratione temporum*, chaps. 3 and 25 (observations of the new moon and of the setting sun, respectively) in *PL*, vol. 90, cols. 353A and 405A.

26. For a general introduction to these two forms of astronomical timekeeping, see Stephen C. McCluskey, *Astronomies and Cultures in Early Medieval Europe* (Cambridge: Cambridge University Press, 1998), chaps. 5–6.

27. Benedict of Norcia, *Regula Sancti Benedicti*, chaps. 8–11 and 16–17, http://www.intratext.com/X/LAT0011.HTM; English translation and http://www.ccel.org/ccel/gregory/life_rule.i.html (both accessed 14 Dec. 2008).

28. On the horological challenge posed by the timing of Vigils and Matins, see Benedict, *Regula Sancti Benedicti*, chap. 11; English translation http://www.ccel.org/ccel/gregory/life_rule.iv.i.html (both accessed 14 Dec. 2008).

29. Gregory of Tours, *De cursu stellarum*, ed. B. Krusch, in *Monumenta Germaniae historica. Scriptorum rerum Merovingicarum* (Hannover: Hahn, 1885), 1.2: 854–72, monthly calendar at 870–72; http://www.mgh.de (accessed 22 June 2008). See Stephen C. McCluskey, "Gregory of Tours, Monastic Timekeeping, and Early Christian Attitudes to Astronomy," *Isis* 81 (1990): 8–22; and McCluskey, *Astronomies and Cultures*, 101–10. According to McCluskey, Gregory's rules reflected the actual appearance of the night sky at the latitude of Tours near the end of the sixth century, which means that his system was based on contemporary observations.

30. McCluskey, "Gregory of Tours," 18 and 21.

31. Text transcribed in Giles Constable, "Horologium stellare monasticum (saec. XI)," in *Consuetudines Benedictinae variae (Saec. XI–Saec. XIV)* (Siegburg: F. Schmitt, 1975), 1–18, on 17.

32. Peter Damian, *De perfectione monachorum*, 17, in *De divina omnipotentia et altri opuscoli*, ed. Paolo Brezzi, trans. Bruno Nardi (Florence: Vallecchi, 1943), 286–88. See Faith Wallis, "Images of Order in the Medieval *Computus*," in *Ideas of Order in the Middle Ages*, ed. Warren Ginsberg (Binghamton: Center for Medieval and Early Renaissance Studies, SUNY Binghamton, 1990), 45–68.

33. See Susanne Pickert, "Jerusalem Sehen: Wahrnehmung und Andacht: Lateineuropäische Reiseberichte des 12. bis 15. Jahrhunderts als Anleitung zur geistigen Pilgerfahrt" (Ph.D. diss., Humboldt-Universituat zu Berlin, 2007), 59–74.

34. See Robert R. Newton, *Medieval Chronicles and the Rotation of the Earth* (Baltimore: Johns Hopkins University Press, 1972), chap. 2; and McCluskey, *Astronomies and Cultures*, 80–84, for the calendrical complexities involved.

35. *Chronicon S. Andreae Castri Cameracesii*, ed. Ludwig Bethmann, in *Monumenta Germaniae historica, Scriptores* (Hannover: Hahn, 1846), 7: 550. For other examples of observed eclipses interpreted as divine omens, see Newton, *Medieval Chronicles*, 32–33, 244, 344, 368–69.

36. See David A. King, "Aspects of Practical Astronomy in Mosques and Monasteries," in *The Call of the Muezzin*, vol. 1 of *In Synchrony with the Heavens: Studies in Astronomical Timekeeping and Instrumentation in Medieval Islamic Civilization*, 2 vols. (Brill: Leiden, 2004), 847–80. King emphasizes that the times of religious ritual were in practice determined not by astronomers but by legal scholars, who "advocated practices based on folk astronomy" (859), that is, observation, such as the actual sighting of the crescent moon.

37. There is surprisingly little written on observational practice in the medieval Islamic world; for an overview, see Régis Morelon, "General Survey of Arabic Astronomy," in *Encyclopedia of the History of Arabic Science*, 3 vols., ed. Roshdi Rashed, with Régis Morelon (London: Routledge, 1996), 1: 8–15. More information in George Saliba, "Solar Observations at the Maragha Observatory before 1275: A New Set of Parameters," *Journal for the History of Astronomy* 16 (1985): 113–22; S. S. Said and F. R. Stephenson, "Solar and Lunar Eclipse Measurements by Medieval Muslim Astronomers," *Journal for the History of Astronomy* 27 (1996): 259–73; and the literature cited therein. On observatories, including the Maragha Observatory, see Aydın Sayılı, *The Observatory in Islam, and Its Place in the General History of the Observatory* (Ankara: Türk Tarih Kurumu Basımevi, 1960).

38. See McCluskey, *Astronomies and Cultures*, chap. 9; Arno Borst, *Astrolab und Klosterreform an der Jahrtausendwende* (Heidelberg: Carl Winter/Universitätsverlag, 1989).

39. Lupitus of Barcelona (attr.), *Fragmentum libelli de astrolabio*, in *Assaig d'història de les idees físiques i matemàtiques a la Catalunya medieval*, ed. José María Millás Vallicrosa (Barcelona: Institució Patxot, 1931), 1: 271–75, on 274. On Lupitus, see Harriet Pratt Lattin, "Lupitus Barchinonensis," *Speculum* 7 (1932): 58–64.

40. On the representation of astronomers and astronomy in thirteenth-century France, see Katherine H. Tachau, "God's Compass and *Vana Curiositas*: Scientific Study in the Old French *Bible Moralisée*," *Art Bulletin* 80 (1998): 7–33, on 15–17. My thanks to James Evans, Owen Gingerich, Bernard Goldstein, Richard Kremer, and Michael Shank for the advice concerning this confusing image.

41. Walcher, *De experientia scriptoris*, Bodleian Library, MS Auct. F. 1. 9, fols. 86r–99r, partially edited in Charles Homer Haskins, "The Reception of Arabic Science in England," *English Historical Review* 30 (1915): 56–69. For details on Walcher's calculations, see McCluskey, *Astronomies and Cultures*, 180–84.

42. Walcher, *De experientia scriptoris*, fols. 86r–99r, in Haskins, "Reception of Arabic Science," 57. The phases of the moon were often taken into account in deciding when to engage in certain medical procedures, especially those involving bleeding.

43. Peter Alfonsi, *Epistola*, ed. José María Millás Vallicrosa, "La aportación astronómica de Pedro Alfonso," *Sefarad* 3 (1943): 65–105, on 97–105 (quotation on 99–100); on Alfonsi and his work, see Charles Burnett, "The Works of Petrus Alfonsi: Questions of Authenticity," *Medium Aevum* 66 (1997): 42–79. On Abu Ma'shar's influence in this area, see Richard Lemay,

Abu Ma'shar and Latin Aristotelianism in the Twelfth Century: The Recovery of Aristotle's Natural Philosophy through Arabic Astrology (Beirut: American University of Beirut, 1962), 146–47.

44. Raymond of Marseilles, *Liber cursuum planetarum*, quoted in Lemay, *Abu Ma'shar*, 146 n. 3: "firmamenti ipsius veritati." For the kinds of theoretical issues that required experiential verification in this period, see Joshua David Lipton, "The Rational Evaluation of Astrology in the Period of Arabo-Latin Translation, ca. 1126–1187 AD" (Ph.D. diss., University of California, Los Angeles, 1978), chap. 6.

45. Lynn Thorndike, *History of Magic and Experimental Science*, 8 vols. (New York: Columbia University Press, 1934–58), 3: 295–96, n. 16.

46. For the reception of Arabic astrology and the development of Latin astrological writing beginning in the twelfth century, see Jean-Patrice Boudet, *Entre science et nigromance: Astrologie divination et magie dans l'Occident médiéval (XIIe Xve siècle)* (Paris: Publications de la Sorbonne, 2006), esp. chap. 1.

47. Albertus Magnus, *De caelo et mundo*, e.g., 1.1.9, in *Opera omnia ad fidem codicum manuscriptorum edenda*, gen. ed., Bernhard Geyer (Monasterii Westfalorum: Aschendorff, 1951–), 5.1: 23–24; see also 2.3.13, 171.

48. Albertus Magnus, *De caelo*, 2.3.15 and 2.3.11, in *Opera omnia*, 5.1: 176–77 and 166; and Albertus Magnus, *Metaphysica* 2.2.24, in *Opera omnia*, 16.2: 514. Precession is the motion that imperceptibly carries the stars from west to east against their rapid daily progression from east to west; trepidation introduces an oscillatory perturbation in precession.

49. Albertus Magnus, *De caelo et mundo*, 2.3.15, in *Opera omnia*, 5.1: 176–77. For a detailed discussion of trepidation theory, see José Chabás and Bernard R. Goldstein, *The Alfonsine Tables of Toledo* (Dordrecht: Kluwer, 2003), 256–67.

50. Others included the order of the planetary spheres and the physical reality of Ptolemaic epicycles; see Bernard R. Goldstein, "Theory and Observation in Medieval Astronomy," *Isis* 63 (1972): 39–47, esp. 40–43.

51. On the practice and practitioners of astronomy at Paris in this period, see in general Chabás and Goldstein, *Alfonsine Tables*, 245–90, and the literature on William of Saint-Cloud and John of Murs cited below. On similar activities in other cities, see Lynn Thorndike, "Astronomical Observations at Paris from 1312 to 1315," *Isis* 38 (1948): 200–5, on 203.

52. William's *Almanach* is unedited; for the prologue, see Paris lat. 7281, fols. 141r–44v. Summary details of William's observations in Richard Irwin Harper, "The *Kalendarium Regine* of Guillaume de St.-Cloud" (Ph.D. diss., Emory University, 1966), 40–52; and J. L. Mancha, "Astronomical Use of Pinhole Images in William of Saint-Cloud's *Almanach planetarum* (1292)," *Archive for History of the Exact Sciences* 43 (1991): 275–98, on 283. See in general Emmanuel Poulle, "William of Saint Cloud," *Dictionary of Scientific Biography*, 16 vols. in 8 (New York: Charles Scribner's Sons, 1981), 14: 389–91.

53. John of Murs, *Expositio intentionis regis Alfonsii circa tabulas eius*, in "Jean de Murs et les tables Alphonsines," ed. Emmanuel Poulle, *Archives d'histoire doctrinale et littéraire du Moyen Age* 47 (1980): 261–68. On the life and work of John of Murs, see in general Lawrence Gushee, "New Sources for the Biography of Johannes de Muris," *Journal of the American Musicological Society* 22 (1969): 3–26; Emmanuel Poulle, "John of Murs," *Dictionary of Scientific Biography*, 7: 128–33; and Chabás and Goldstein, *Alfonsine Tables*, 277–81.

54. Chabás and Goldstein, *Alfonsine Tables*, 251–81, passim (quotation on 279). Goldstein contrasts this aspect of the work of late medieval Latin astronomers with that practiced by some of their contemporaries writing in Hebrew and Arabic; see Goldstein, "Theory and Observation," 39–47.

55. Gushee, "New Sources," 16.

56. Escorial, MS O.II.10, which I have not seen. Many of the relevant notes are transcribed in Gushee, "New Sources"; and Guy Beaujouan, "Observations et calculs astronomiques de Jean de Murs (1321–44)," in *Par raison de nombres: L'art du calcul et les savoirs scientifiques médiévaux* (Aldershot: Variorum, 1991), VII, 27.

57. Transcribed in Gushee, "New Sources," 14 (fol. 219v). The aim of this observation was to determine the latitude of the Abbey of Fontevrault.

58. Similarly, John observed the eclipse of 1331 in the presence of "three Franciscan friars and . . . the Queen of Navarre"; transcribed in Gushee, "New Sources," 15.

59. See note 37 above.

60. Gushee, "New Sources," 16; Beaujouan, "Observations et calculs," 30.

61. An anonymous fragment of observations of equinoxes, solstices, and planetary conjunctions made in Paris between 1312 and 1315 (Oxford, Corpus Christi College, cod. 144, fols. 97r–98v) and transcribed in Thorndike, "Astronomical Observations," 203–5, was clearly part of the project of testing planetary positions predicted on the basis of contemporary tables, although it may also have served astrological purposes.

62. For an excellent overview of this field and its Arabic sources, see Stuart Jenks, "Astrometeorology in the Middle Ages," *Isis* 74 (1983): 185–210; and Gerrit Bos and Charles Burnett, *Scientific Weather Forecasting in the Middle Ages: The Writings of Al-Kindi. Studies, Editions, and Translations of the Arabic, Hebrew and Latin Texts* (London: Kegal Paul, 2000), chap. 1. Weather conditions were understood to include a relatively broad range of phenomena, including aurorae, comets, and earthquakes, all produced by the motions and influences of the heavens.

63. The studies collected in *Scientia in Margine. Études sur les marginalia dans les manuscrits scientifiques du Moyen Âge à la Renaissance*, ed. Danielle Jacquart and Charles Burnett (Geneva: Droz, 2005), suggest that late medieval marginalia were of two main types: notes by students supplementing the text with what they had heard in a lecture, or notes from personal experience and from the practical application of the text; see Marilyn Nicoud, "Les *marginalia* dans les manuscrits latins des *Diètes* d'Isaac Israëli conservès à Paris," ibid., 191–216, esp. 208–9. The marginal *observationes* discussed here seem to have been of the latter sort.

64. London, BM, Royal 7 F VIII, fols. 176v–79v; see Lynn Thorndike, *History of Magic*, 3: 141n. This extremely interesting set of annotations, which appears to be a fair copy written in the same hand and ink as the ephemerides themselves, covers the period between August 1269 and February 1270.

65. Oxford, Bodleian Library, MS 191 (notes in an English calendar covering the years 1438–39); transcribed in G. Hellmann, *Meteorologische Beobachtungen vom XIV. bis XVII. Jahrhundert*, ed. G. Hellmann, Neudrucke von Schriften und Karten über Meteorologie und Erdmagnetismus, 13 (Berlin: A. Asher, 1901), documentary appendix, 6.

66. William Merle, *Merle's ms. Consideraciones temperiei pro 7 annis . . .*, ed. G. J. Symons, trans. Miss Parker (London: E. Stanford, 1891); this book contains a facsimile of Oxford, Bodleian Library, Digby MS 176, fols. 4r–7r. On Merle and his meteorological writings, see Hellmann, *Meteorologische Beobachtungen*, 25; and Thorndike, *History of Magic*, vol. 3, chap. 8.

67. Hellmann, *Meteorologische Beobachtungen*, documentary appendix, 2.

68. Cuno, *Iudicia de impressionibus que fiunt in aere collecta et experimentata a magistro Cunone morantem* [*sic*] *circa Sanctum Burgardum in Herbipoli*, Nüremberg, Stadtbibliothek, Cent. V. 64, fols. 92va–102rb (fifteenth century), covering events from 1331 to 1355; see Thorndike, *History of Magic*, 3: 145–46; and Jenks, "Astrometeorology," *Isis* 74 (1983): 205.

69. Jenks, "Astrometeorology," *Isis* 74 (1983): 99–201.

70. Reading *morante* for *morantem* in the colophon of *Iudicia*; see n. 68 above.

71. Lynn Thorndike, "A Weather Record for 1399–1406 A.D.," *Isis* 32 (1940): 304–23; idem, "A Daily Weather Record Continued from 1 September 1400 to 25 June 1401," *Isis* 57 (1966): 90–99; Fritz Klemm, "Über die Frage des Beobachtungsortes des Baseler Wettermanuskriptes von 1399–1406," *Meteorologische Rundschau* 22 (1969): 84.

72. Johannes Schöner, ed., *Scripta clarissimi mathematici M. Joannis Regiomontani de torqueto, astrolabio armillari, regula magna ptolemaica baculoque astronomico et observationibus cometarum, aucta necessariis Ioannis Schoneri Carolostadii. Item Observationes motuum solis, ac stellarum tam fixarum quam erraticarum* (Nuremberg: Joannis Montanus and Ulricus Neuber, 1544). The volume included Regiomontanus's and Walther's *Ad solem observationes* (covering the years 1462–1504), fols. 27r–34r; the *Eclipsium, cometarum, planetarum ac fixarum observationes* attributed to "Peurbach, Regiomontanus, Walther, and others" (1457–74), fols. 36r–43v; and Bernhard Walther's *Observationes* [of the planets and fixed stars] (1475–1504), a continuation of the preceding, fols. 44r–60v.

73. Schöner, *Scripta*, fols. 36r–37r (quotation on 36r). On *consideratio* and *observatio* in Latin translations of Ptolemy, see Pomata, "Word of the Empirics."

74. On Peurbach's and Regiomontanus's observations, see John M. Steele and F. Richard Stephenson, "Eclipse Observations Made by Regiomontanus and Walther," *Journal for the History of Astronomy* 29 (1998): 331–44, on 331–35.

75. Schöner, *Scripta*, fols. 39r–43v.

76. Ibid., fols. 27r–v. Although the original notebooks have not survived, comparison of a partial manuscript copy of the solar observations, earlier and more accurate than Schöner's edition, shows that while Schöner abbreviated some entries, rounded up values, and introduced a number of errors, his edition seems to have been largely true to the overall format and nature of the original; see Richard L. Kremer, "Walther's Solar Observations: A Reply to R. R. Newton," *Quarterly Journal of the Royal Astronomical Society* 24 (1983): 36–47, on 37–39. On Regiomontanus's theoretical aspirations, see Richard L. Kremer, "War Bernhard Walther, Nürnberger astronomischer Beobachter des 15. Jahrhunderts, auch ein Theoretiker?" in *Astronomie in Nürnberg*, ed. Gudrun Wolfschmidt, Nuncius Hamburgensis, Beiträge zur Geschichte der Naturwissenschaften, 3 (Norderstedt: Books on Demand, forthcoming), 30–55, on 37–38.

77. Although it is impossible to know whether Walther's original notes took exactly this form, this is likely, to judge by the partial manuscript version described in the preceding note; see the reproduction in Kremer, "Walther's Solar Observations," 38.

78. Schöner, *Scripta*, fols. 27v–34r. On Walther's observations in general, see Richard L. Kremer, "Bernhard Walther's Astronomical Observations," *Journal for the History of Astronomy* 11 (1980): 174–91, on 185–86; and idem, "Walther's Solar Observations."

79. Partial reproduction and transcription in Fritz Klemm, *Die Entwicklung der meteorologischen Beobachtungen in Franken und Bayern* (Offenbach am Main: Deutsche Wetterdienst, 1973), 15–18; the diary covers February–March 1487. Peurbach also kept a weather diary; see Hellmann, *Meteorologische Beobachtungen*, 16–17.

80. The same increasing concern with accuracy is visible in his 638 observations of planetary positions (performed with a Jacob's staff and later an armillary sphere), although these were more varied in their objects than the solar ones. See Schöner, *Scripta*, fols. 44r–60v.

81. See Kurt Pilz, "Bernhard Walther und seine astronomischen Beobachtungsstände," *Mitteilungen des Vereins für Geschichte der Stadt Nürnberg* 57 (1970): 176–88.

82. Schöner, *Scripta*, fol. 47r.

83. Regiomontanus's only named assistant was Conrad Heinfogl; see Kremer, "War Bernhard Walther," 37.

84. Kremer, "Walther's Solar Observations," 46 n. 6.

85. Kremer, "War Bernhard Walther," 40–52.

86. See Richard L. Kremer, "The Use of Bernhard Walther's Astronomical Observations: Theory and Observation in Early Modern Astronomy," *Journal for the History of Astronomy* 12 (1981): 124–32.

87. See Daston and Park, *Wonders*, chap. 3, esp. 137–46; Agrimi and Crisciani, "Per una ricerca"; and Chiara Crisciani, "*Experientia* e *opus* in medicina e alchimia: forme e problemi di esperienza nel tardo Medioevo," *Quaestio: Yearbook of the History of Metaphysics* 4 (2004): 149–73.

88. McVaugh, "Two Montpellier Recipe Collections," 178. For a rare example of a more focused research program growing out of practical interests of a number of late fourteenth- and early fifteenth-century physicians, see Park, "Natural Particulars."

89. Transcribed in Michael McVaugh, "The *Experimenta* of Arnald of Villanova," *Journal of Medieval and Renaissance Studies* 1 (1971): 107–18, on 116. The genre of the *experimentum* was Arabic in origin and originally had a strongly magical bent, which was tempered in later medieval medical writing by more utilitarian recipes of the sort recorded by Arnald and his colleagues. See William Eamon, *Science and the Secrets of Nature: Books of Secrets in Medieval and Early Modern Culture* (Princeton: Princeton University Press, 1994), 36–56; and McVaugh, "Two Montpellier Recipe Collections," respectively.

2

Observation Rising: Birth of an Epistemic Genre, 1500–1650

GIANNA POMATA

A Word for the Practice

In the second half of the fifteenth century, what we would now call observational practices developed at an unprecedented pace in astronomy, astrology and astrometeorology, while also emerging in fields as diverse as medicine, alchemy, natural history, physiognomy, and antiquarianism.[1] And yet throughout the fifteenth and even part of the sixteenth century, there was apparently no word firmly attached to these practices; *experientia, experimentum, contemplatio, consideratio,* and less often, *observatio* could all be used for this purpose, in an unsettled and confused way. This teeming world of practices seems at first sight to be only palely reflected in Renaissance philosophical language, the lack of a specialized term suggesting an only incipient conceptualization of the cognitive act of observing.

Medieval Aristotelianism has been defined as "empiricism without observation,"[2] and this seems certainly true in a linguistic sense, in view of the fact that the vast literature on Aristotle's natural books invariably referred to empirical knowledge by the terms *experientia/experimentum*—never *observatio. Experimentum* was the Latin rendition of the Aristotelian *empeiría* as well as of the Hippocratic *peira,*[3] so the numerous Scholastic commentaries on Aristotelian and Hippocratic texts were sure to contain a definition of *experimentum* as a specialized term of philosophical and medical language.[4] In contrast, neither the Aristotelian nor the Hippocratic language contained the Greek original of the Latin *observatio* (*tērēsis*).[5] So it is not surprising that in late medieval philosophy *observatio* in a cognitive sense was only used sporadically and with a narrow semantic focus, that is, the observation of the heavens.[6] In the late centuries of the Middle Ages, while *experimentum* became an object of discussion in both philosophy and medicine, *observatio* had as yet little or no currency as a philosophical term, and it definitely was not an epistemic category.

More surprising is to find that nothing much seems to have changed in this respect during the Renaissance. *Observatio* does not appear to have been much used by sixteenth-century philosophers. If we turn to some emblematic works of sixteenth-century critics of Scholastic Aristotelianism and advocates of empirical knowledge, such as Vives's *De disciplinis*, Ramus's *Scholae in liberales artes*, and Sanches's *Quod nihil scitur*, we find that the key words referring to empirical knowledge are still *experientia* and *experimentum*, while *observatio* seems to have been used much more rarely.[7] This relative rarity of the term is obscured by modern translations, which often interpolate "observation" and "to observe" where the Latin text does not have the noun *observatio* or the verb *observare*.[8] Even in Bacon's *Novum organum*, the crucial words are *experientia* and especially *experimentum*, not *observatio*.[9] Also in Bacon's case, if we compare the original with a modern translation, we find repeated instances of the modern penchant for introducing "observation" more often than warranted by the Latin text.[10]

What explains *observatio*'s timid debut on the sixteenth-century philosophical stage? A debut of some kind there was: *observatio* started to appear in the trail of *experimentum*, as in the expression, typically in the plural, "*experimenta et observationes*."[11] But *experimentum* remained apparently the dominant term. Why? One possible reason is obvious. For all the limits of their value in Aristotelian Scholastic epistemology, *experientia* and *experimentum* were well-entrenched philosophical concepts, with a long history of usage at their back.[12] On its first appearance in sixteenth-century intellectual life, in contrast, *observatio* was a parvenu, with little or no philosophical pedigree—or rather, with a pedigree that had been marginal or forgotten in medieval philosophy, and was fully rediscovered only in the middle years of the sixteenth century.

By which route did *observatio* enter early modern philosophical language? In ancient Greek culture, an elaboration of the concept of observation (*tērēsis*) had first emerged in the Hellenistic age with the medical sect of the Empirics, to be further developed, with wider philosophical significance, in late ancient Skepticism. Basically unknown in the Middle Ages, the Empirics' conceptualization of *tērēsis* trickled back into Western medicine in the fourteenth century, but its meaning seems to have been fully recovered by European scholars only in the 1560s, concomitantly with the first Latin translation of the works of Sextus Empiricus, which marked the advent of Skepticism on the early modern philosophical scene.[13] As a category originally associated with medical Skepticism, *observatio* was a new entry in early modern philosophy. In fact, although the term gained wide currency in general scholarly usage in the seventeenth century, its assimilation into standard philosophical language was a very slow business, fully completed only in the eighteenth century. Surprising as it may seem, *observatio* does not even appear as an entry in philosophical dictionaries until the eighteenth century—with one significant exception, the medical lexica, which featured the lemma, reporting its ancient Empiric definition, as early as 1564.[14]

Although *observatio* was nearly invisible in the philosophers' lexica, *observationes* (in the plural) emerged and proliferated in European scholarly culture beginning in the middle decades of the sixteenth century. In fields ranging from astronomy and astrology to philology and lexicography, from jurisprudence to medicine and to travel writing, scholars wrote new kinds of texts that they presented deliberately, with assertive pride, under the new title of *observationes.*[15] In spite of its limited philosophical currency, two major shifts mark the semantic history of the term *observatio* in early modern learned culture. First, the emergence of the word in the plural as a title.[16] From annotations written on the margin of other texts, as they used to be in late medieval astrometeorology,[17] *observationes* became a distinctive and autonomous form of writing, a recognized scholarly genre. Strikingly, this move from marginalia to title happened roughly in the same decades, between the 1530s and the 1570s, in a variety of disciplines—most prominently, though not exclusively, astronomy and medicine. The rise of the *observationes* in the mid-sixteenth century was not circumscribed to a few fields or to a few years. Far from being a passing phenomenon, it inaugurated a trend that would expand dramatically in the seventeenth century.

Starting in the same years (ca. 1530–1570) and again accelerating over the course of the seventeenth century, a major change took place in the core meaning of the word *observatio*, which shifted from observance to empirical observation. Since antiquity, the words denoting observation, the Greek *tērēsis* and its Latin equivalent, *observatio*, had a double meaning. They could mean either *observance* (in the sense of obedience to a rule), or *observation* in the sense of attentive watching of objects and events.[18] Throughout its long history, the semantics of *observatio* straddled the prescriptive and the descriptive, and every translator should be advised to keep this fundamental ambivalence in mind. In the ancient world, the primary meaning of *tērēsis* seems to have been "observance," as in the expression "observing the law"; and the same was true of *observatio* in classical Latin, though the empirical observational sense was clearly featured in astronomical, medical, and other scholarly contexts.[19] In medieval Latin, the prescriptive meaning of *observatio* as observance all but completely obliterated the observational meaning.[20] It was only in the mid-sixteenth century that *observatio* conspicuously came again to indicate empirical observation, as in the ancient world, although the following of a rule remained a vital part of the term's semantic core in everyday usage as in philosophical parlance. In Montaigne's *Essais* (1580), for instance, the French word *observation* appears rarely, and mostly signifying "a rule."[21] And yet in those same years when Montaigne wrote his essays, astronomers and physicians were already busily engaged in exchanging, through intense epistolary networks, texts that they called *observationes*, which were indeed reports of what they had observed—in the heavens, in weather conditions, in the interior of anatomized bodies, in the course of diseases. Gathering momentum in the last decades of the sixteenth century and proceeding at an accelerated pace in the seventeenth, this trend is so pronounced that, while it seems to be the case that *observatio* was not yet a generally recognized epistemic category, one must state that in this very same period observation became an *epistemic genre.*

Genres are standardized textual formats—textual tools, we may call them—handed down by tradition for the expression and communication of some kind of content. In the case of epistemic genres, this content is seen by authors and readers as primarily cognitive in character. As shared textual conventions, genres are intrinsically social: contributing to a genre means consciously joining a community. Indeed, some genres are eminently instruments of "community building," tools for the establishment of a collective scholarly endeavor as a social and intellectual shared space. This, we shall see, was very much the case for the early modern *observationes*, whose rise and fortune were linked to the development of horizontal networks of exchange among European scholars. While most medieval epistemic genres were attuned to teaching and the establishment of academic rank (*quaestiones disputatae*, *lectiones*, *tractatus*, *commentaria* being all subservient to the practice of university lecturing), the early modern *observationes* were fundamentally geared to the goal of exchanging and circulating information within communities wider than those identified by school training. The opening of broader horizons beyond the narrow *familia* formed by teacher and students,[22] or beyond the closed space of the guild, is a defining trend of early modern scholarly culture, and the rise of the *observationes* is very much part of its story.

This second part of our attempt to trace the outlines of the premodern history of observation will concentrate on these major early modern developments: the rise of the *observationes* as a new epistemic genre and the conceptualizing of observation as a new epistemic category. We will first examine the rise of the *observationes* in general terms and in a variety of fields, and then "zoom in" on their origin and growth in a specific discipline—late Renaissance medicine. Far from being casual, this focus on medicine is strongly warranted by the sources. Like astronomy, medicine was a major field where observational practices, codified by a new disciplinary genre called *observationes*, arose meteorically in the late sixteenth and the early seventeenth centuries. Furthermore, the epistemology of medicine provided the breeding ground for the revival of the ancient Empiric/Skeptic concept of empirical knowledge (a concept significally different from the Aristotelian/Scholastic *experientia*), which would prove decisive for the development of the new epistemic category of observation in the seventeenth and eighteenth centuries.

From Backstage to Limelight: Enter the *Observationes*

Astronomical and astrological pursuits were the cradle of the *observationes* as a disciplinary genre. In the last decades of the fifteenth century, Johannes Regiomontanus and his pupils routinely used *observatio* and *observare* to refer to their newly intensive observations of the heavens.[23] More importantly, it is in the astronomical context that we see an early example of the transformation of *observationes* from marginalia[24] or private work records, meant at most for scribal transmission from mentor to pupil, into printed book material addressed to a wider public.

In the early decades of the sixteenth century, a new interest in more accurate observation of celestial and meteorological phenomena was spurred by the attempts to purge astrology of what was considered its superstitious component, which had been scathingly attacked by Giovanni Pico della Mirandola in his *Disputations against Divinatory Astrology* (1496).[25] But in spite of this new flurry of observational practices, the *observationes*, whether of the heavens or of the weather, were not yet an established genre of astronomical/astrological writing. The major genres were *tabulae* and *canones* for astronomy, and for astrology the variously called *judicia*, *prognostica*, *tacuina*, *ephemerides* (all variants of the almanac), which predicted the positions of the celestial bodies and their influences on earthly events.[26] Often written in the margins of an astronomical calendar or table, observations were not considered important per se, but as subservient to astrological predictions.[27] So when in the 1540s Johannes Schöner decided to publish the *Nachlass observationes* of four previous generations of astronomers, Georg Peurbach, his disciple Regiomontanus, Regiomontanus's pupil Bernard Walther, plus Johannes Werner's weather observations, his was an unprecedented enterprise that marks a turning point in the early history of the genre.[28]

Schöner's decision to print these records clearly suggests a new perception of their significance. In the dedicatory letter to his 1544 publication, he stressed that *observationes* were foundational for the "mathematical arts," provided they were made "*not by anybody* but by the art's eminent members." For this reason, he says, he decided to add to his volume, which contained primarily Regiomontanus's treatise on astronomical instruments, "a not unworthy treasury of observations [*Thesaurum Observationum*] which I found most conscientiously stowed away and diligently safeguarded in a certain small chest."[29]

This is an early example of the use of the *observationes* as a tool to promote some kind of collective empiricism (the *thesaurus* metaphor is revealing in this respect, and it will keep coming back in the history of the genre). But it is a collective empiricism of quite a different sort from that of the millennial *observationes* of farmers and seafarers described by Pliny, or of the astrologers that Giovanni Pico had strongly criticized on the ground of historical chronology.[30] The Plinian *observationes* were typically anonymous, as are proverbs and folktales.[31] What we have here, in contrast, is one of the first attempts to transform *observatio* from an agglomerate accumulation of experience, built up over the centuries by anonymous or dimly identified observers, into a specific product of just as specific an author/observer. It was in fact a deliberate effort to stamp observation with the mark of an author, and a model author at that. It is important to notice that his predecessors' *observationes* were not for Schöner what they look like to us—data pure and simple. For him, they were examples of the way observations *should be made*, because they had been made by the masters of the craft. His main motivation in publishing them was pedagogic: Schöner thought that the *observationes* made within Regiomontanus's scholarly *familia* should be used beyond that limited sphere.[32]

The *observationes* emerge here as a bridge from the small circles formed by generational chains of teachers and pupils to a wider astronomical community. Within this community, the *observationes* of the best authors should circulate to provide a standard and a model. We see here at work the humanist pedagogical preference for teaching by means of *exempla*—the virtues exemplified being in this case not moral but epistemic, the virtues of the diligent observer. We can also hear, even in this clearly observational context, the prescriptive ring of the word *observatio*—not simply an act of observing but an act guided by a rule, protocol, or code of behavior, and therefore different from *experientia*, which can be of *anybody*. In other words, the *observationes* emerge not simply as a genre, but as a genre with a canonical author—the diligent observer, whose primary feature is that he checks firsthand the observations received from tradition in order to improve their accuracy. So Gemma Frisius, *mathematicus*, physician, cartographer, and a leading spirit of the attempt to reform astrology in Louvain in the middle years of the sixteenth century, compared unfavorably the compilers of the Alfonsine Tables, who "did not use observations (*observationes*) they themselves made, but rather followed those of Ptolemy and others," with Copernicus who, in contrast, "emended many things by comparing what he had observed himself (*sua observata*) with what observed by his predecessors."[33]

When the astronomers were newly publishing *observationes* in the middle decades of the sixteenth century, they were by no means the only ones to do so. Indeed, in this period the title *observationes* shows up in a surprisingly variegated ensemble of disciplines—philology, jurisprudence, medicine, natural history, and travelogues.[34] It may seem that all that *observationes* meant in these titles was simply "notes," work-in-progress for semiprivate use, and it may have been so occasionally—but typically no, it goes deeper than that. Take for instance the case of philology. Already by the late fifteenth century, humanist scholars called their philological works on ancient texts *observationes*—a title that was going to persist in classical scholarship into the nineteenth century.[35] We may surmise that in this context *observationes* simply meant a collection of miscellaneous "notes." Actually, as in astronomy, the word was used to indicate the work of a competent scholar, who proceeds by careful scrutiny of his sources. So Mario Nizolio, the author of a magnificent Ciceronian lexicon, *Observationes in Marcum Tullium Ciceronem* (1535; over seventy printings between 1535 and 1630) proudly called himself an "*observator*."

> *Observatores*—he wrote—do not amass their materials cursorily and, so to speak, by chance, without discriminating between authors, like the *lexicographi* do. *Observatores*, in contrast, . . . restrict themselves to the work of the ancients, and of the most approved men. . . . They collect their *observationes* from things most carefully read and considered, and they credit them to their diverse authors.[36]

In Nizolio's use of *observator* we can hear a clear echo of the ancient meaning of *observare* as an action performed according to a high standard, with attentive, quasi-religious care. As in classical Latin, where *observator* meant primarily "he who obeys the law or custom," Nizolio used it as the appropriate epithet for an author who upheld the rules of true scholarship.[37] But at the same time, quite possibly, the word also meant for him "he who observes what is actually there in a text," focusing on the text's lexical elements.[38]

This reference to the observation of discrete particulars is also to be found in the use of *observator* in other contexts. An interesting example comes from quite a different field—physiognomy. In an early sixteenth-century physiognomical text that presents impressive evidence of the rise of observational practices in this area, the physician Bartolomeo della Rocca Cocles included himself among the "learned *physici observatores* of the course of nature," while bragging that none of them had observed such a great number of individuals as he had himself.[39] Here, to be sure, *observator* has primarily an observational meaning, as Cocles is referring to his vast experience in minutely observing people to trace the correlation between their physical and moral characteristics. And yet throughout the text Cocles indicates the act of observing for physiognomic purposes by a cloud of verbs—*notare, experiri, videre, aspicere, inspicere, conspicere,* and only occasionally *observare*—thus confirming our general contention that the act of observing was not yet primarily linked to a specific word. And why, we may wonder, did Cocles call himself an *observator* rather than an *experimentator,* though he often uses *experiri* to indicate the act of observing, as innumerable authors had done before him? Possibly, because the term *experimentator* was associated with *empiricus*—definitely not a flattering association, especially in medicine, where the collections of *experimenta* often included remedies drawn from the experience of illiterate peasants and old women.[40] In whatever context the words *observator* and *observatio* come up, in contrast, they invariably carry a strong connotation of high-status learning and assertive professional pride—not at all the stuff of just anybody's experience, as Johannes Schöner would have said.

This is confirmed by the fact that in the mid-sixteenth century the genre of *observationes* spread to two very status-conscious liberal arts, the law and medicine, in both cases referring to works that contained specialized professional knowledge on how to handle specific cases. The compilations of *observationes legales* or *forenses,* which start to come out in the 1530s and 1540s, had a strong link to judicial practice. They were initially compendia of solutions to hypothetical (fictive) legal cases, based on the jurists' "common opinion."[41] But they soon evolved into reports of real cases as discussed and decided by the courts.[42] Here also the term *observationes* straddled the prescriptive and the descriptive: in describing real cases, the goal was to provide guidelines, based on precedent, on how to solve similar lawsuits in the future.[43] The title *observationes* was used for collections of cases drawn from professional practice also in the case of medicine, as we shall see. In the law as in medicine, the history of the *observationes* suggests that professional practice was a factor that fostered the belief in the value of descriptive observational knowledge per se, even without a direct link to generalization and theory.[44]

Clear evidence in this sense comes from the medical *observationes.* The adoption of the title *observationes* in a descriptive sense is particularly interesting in the case of medicine, because in medical language the word had traditionally had a strong normative meaning. Throughout the Middle Ages, from Anthimus in the sixth century to Arnald of Villanova in the fourteenth, *observatio* meant an observance or regimen—a series of medical rules and prescriptions.[45] And sure enough, in the very first occurrences of *observatio* as a medical title at the end of the fifteenth century, the word is still used in the sense of observance. The title of Alessandro Benedetti's *De observatione in pestilentia* (1493) does not refer to what he observed during an epidemic but to the rules to be kept for the preservation from pestilence.[46]

Around the 1560s, however, we start seeing examples of medical works called *observationes* in an unequivocally descriptive, observational sense. An outstanding specimen is Gabriele Falloppio's *Observationes anatomicae* (1561), which inaugurated the subgenre of anatomical *observationes*, destined to flourish in the seventeenth century.[47] Falloppio's text is written in the form of a letter to a colleague, in an informal and colloquial tone, quite different, in style and format, from a systematic treatise like Vesalius's *Fabrica*. In fact, his goal was to vet the *Fabrica* for observational mistakes, in the acknowledged conviction that, like Homer, even Vesalius may have napped at times. Throughout this text, Falloppio constantly uses the verb *observare* when referring to his own anatomical observations.[48] This seems remarkable, as traditionally the verbs used for anatomical practice were various and unspecialized, such as *videre*, *perscrutari*, *inspicere*, among others. The insistent use of *observare*—often in the first person of the past tense, *observavi*—suggests a new self-consciousness on the part of the anatomical observer.

Anatomy was by no means the only part of medicine where the *observationes* developed. In fact, the genre is primarily exemplified by a new form of writing devoted, as I have already mentioned, to the observation of cases. In medicine as in the law, the collection of cases already existed as a genre, the *consilia*, which had flourished since the Middle Ages.[49] What was the need then for a new genre apparently serving the same purpose? In the case of the law the answer is plain. The medieval legal *consilia* were the consultations of a single jurist; the early modern *observationes*, in contrast, were reports of cases as decided by a specific court of law.[50] In the case of medicine, however, no such difference exists: both the medieval *consilia* and the early modern *observationes* collect the medical consultations of a single doctor—the volume's author. In medicine, moreover, the report of cases, together with the treatment administered inclusive of recipes, featured also in another traditional genre, the *experimenta*. In other words, there were already not one but two medical genres that dealt with cases. What explains the emergence of a new genre in addition to the *consilia* and the *experimenta?* A close look at the origin of the medical *observationes* will allow us to answer this question and to survey, at the same time, a new observational trend in medical practice.

How to Invent an Epistemic Genre

If the attention to individual cases was not unprecedented in medicine, the creation of an epistemic form for their description was an intellectual novelty of the Renaissance. Narrative accounts of the treatment of single patients had been inserted, as *exempla* or *casus*, in the medieval textbooks of the *practica* genre from Archimatthaeus in the twelfth century to Michele Savonarola in the fifteenth.[51] But in medieval medicine, case histories were to be found in the folds of the text, so to speak: they did not emerge as a genre on their own. Even in the *consilium*, the genre apparently devoted to the discussion of individual cases, the goal was not describing the individual case per se.[52] Though starting from a case, the medieval *consilium* dealt typically with a disease, not with a sick person, and the description of the symptoms was usually minimal, dwarfed by the heavy apparatus of references to the authorities.[53]

A decisive break from this traditional approach is signalled by the collections of *curationes* and *observationes* (the two words are often joined in the titles)[54] that start to appear in the second half of the sixteenth century. For the first time, accounts of cases were presented no longer semi-hidden in the doctrinal framework of a text, but prominently displayed as freestanding on their own, loosely organized by numerical order, often in groups of hundreds, as the *Centuriae curationum* that the great Jewish physician Amatus Lusitanus published in seven instalments between 1551 and 1566.[55] Here also, as in the case of the astronomical *observationes*, we see material that had been considered marginal or ancillary move to the forefront of attention.

The new interest in cases may have developed in medicine through the influence of the legal *observationes*,[56] or possibly through the concerns of medical astrology. In the years 1549–1554 the French physician and astrologer Thomas Bodier drew for each of his patients the chart of the astrological dispositions at the onset of illness, and subsequently carefully compiled a history of the case until recovery or death. In 1555 he published a collection of such charts and case narratives, referring to fifty-five patients, with the object of testing the medical theory of the critical days of illness.[57] As in the case of weather observations recorded with the object of checking astrological weather predictions, Bodier's idea was to compare the patients' astrological dispositions with the outcome of their diseases. Observation was for him subservient to prognostication. The astrologically motivated routine of observing the stars and the weather may have led some physicians to keep regular notes of their cases as well. Cornelius Gemma, son of the astronomer, astrologer, and physician Gemma Frisius, reported in 1561 that his father had kept two observational journals: one was "a report of everything that pertained to the positions of the stars, as well as to the daily mutations in the atmosphere," while the other contained "his experience in the whole medical art."[58] It is clear that in the middle decades of the sixteenth century an interest in observation developed at the interface of astronomy and medicine. The author of what is arguably the most important late sixteenth-century collection of medical *observationes*, the Dutch physician Pieter van Foreest, reminisced about moving in his youth from astronomical to medical observation.

> I was once engaged with my teacher Ophusius, the eminent *mathematicus*, in making on the heavens of Harlem those observations of which Johannes Regiomontanus wrote and to which additions were made by Johannes Schöner of Karlstadt. . . . Since that time, being of weak health and unable to stand the harmful exposure to the nocturnal air while observing the stars, I decided, mindful of my profession, namely, medicine, . . . that I would make observations of the microcosm rather than of the heavens.[59]

And he certainly did, publishing up to thirty-two volumes of *observationes* of his patients between the years 1584 and 1619.

But the development of the medical *observationes* was also made possible by textual resources available inside the medical tradition, namely, by the adaptation and transformation of old genres. Though certainly a novelty, *curationes* and *observationes* share some basic traits, in fact, with the older genre of *experimenta*, and in several ways seem to derive from it. Medieval collections of *experimenta* recorded remedies that had proved successful but whose efficacy could not be justified on doctrinal grounds.[60] The format for storing and transmitting this empirical knowledge was the recipe, usually prefaced by the name of the disease to which it should be applied, but sometimes also by a brief narrative of the case in which it had given good results.[61] Like the medieval *experimenta*, the late Renaissance *curationes* and *observationes* were presented as therapies legitimized mainly by efficacy. Like the *experimenta*, moreover, they contained recipes, but with an important novelty—the case narrative had now become the main object of attention.

A focus on the case narrative was introduced by Amatus's *Curationes*. In organizing his text as a case collection, Amatus departed creatively from the *experimenta*, where the focus was on the recipe, and from the *consilium*, where the cumbersome references to doctrine had overshadowed the factual details of the case. Amatus brought the case to the fore by separating each *curatio* (the

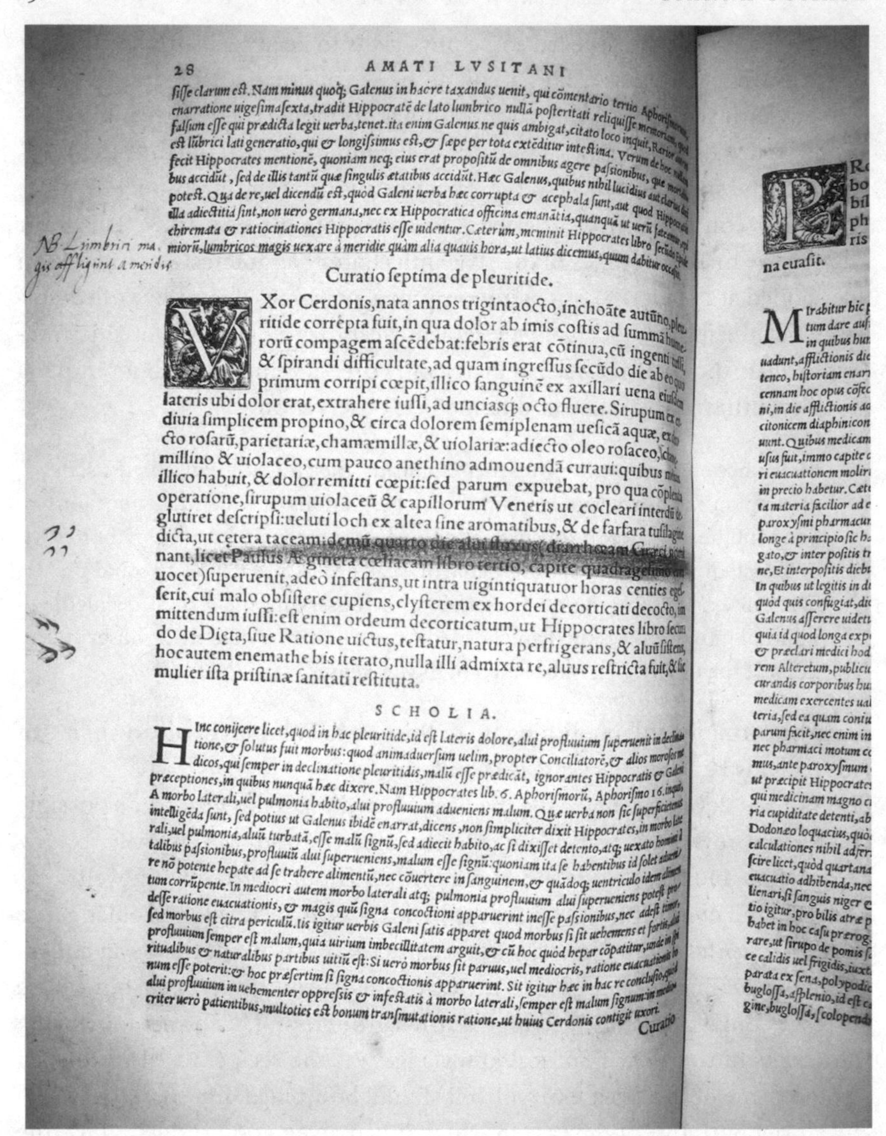

28 AMATI LVSITANI

ſiſſe clarum eſt. Nam minus quoq; Galenus in hac re taxandus uenit, qui cõmentario tertio Aphoriſmorum, enarratione uigeſimaſexta, tradit Hippocratẽ de lato lumbrico nullã poſteritati reliquiſſe memoriam, quod falſum eſſe qui prædicta legit uerba, tenet. ita enim Galenus ne quis ambigat, citato loco inquit, Rarior autem eſt lũbrici lati generatio, qui & longiſsimus eſt, & ſæpe per tota extẽditur inteſtina. Verum de hoc nullam fecit Hippocrates mentionẽ, quoniam neq; eius erat propoſitũ de omnibus agere paſsionibus, quæ mortalibus accidũt, ſed de illis tantũ quæ ſingulis ætatibus accidũt. Hæc Galenus, quibus nihil lucidius aut clarius dici poteſt. Qua de re, uel dicendũ eſt, quòd Galeni uerba hæc corrupta & acephala ſunt, aut quod Hippocratis illa adiectitia ſint, non uerò germana, nec ex Hippocratica officina emanãtia, quanquã ut uerũ fateamur, epichiremata & ratiocinationes Hippocratis eſſe uidentur. Cæterùm, meminit Hippocrates libro ſecũdo Epidemiorũ, lumbricos magis uexare à meridie quàm alia quauis hora, ut latius dicemus, quum dabitur occaſio.

Curatio ſeptima de pleuritide.

VXor Cerdonis, nata annos trigintaocto, inchoãte autũno, pleuritide correpta fuit, in qua dolor ab imis coſtis ad ſummã humerorũ compagem aſcẽdebat: febris erat cõtinua, cũ ingenti tuſſi, & ſpirandi difficultate, ad quam ingreſſus ſecũdo die ab eo quo primum corripi cœpit, illico ſanguinẽ ex axillari uena eiuſdem lateris ubi dolor erat, extrahere iuſſi, ad unciasq; octo fluere. Sirupum ex endiuia ſimplicem propino, & circa dolorem ſemiplenam ueſicã aquæ, ex decocto roſarũ, parietariæ, chamæmillæ, & uiolariæ: adiecto oleo roſaceo, chamæmillino & uiolaceo, cum pauco anethino admouendã curaui: quibus ... illico habuit, & dolor remitti cœpit: ſed parum expuebat, pro qua cõplenda operatione, ſirupum uiolaceũ & capillorum Veneris ut cocleari interdũ deglutiret deſcripſi: ueluti loch ex altea ſine aromatibus, & de farfara tuſſilagine dicta, ut cętera taceam: demũ quarto die alui fluxus (diarrhœam Græci nominant, licet Paulus Aegineta cœliacam libro tertio, capite quadrageſimo ... uocet) ſuperuenit, adeò infeſtans, ut intra uigintiquatuor horas centies egeſſerit, cui malo obſiſtere cupiens, clyſterem ex hordei decorticati decocto, immittendum iuſſi: eſt enim ordeum decorticatum, ut Hippocrates libro ſecundo de Dięta, ſiue Ratione uictus, teſtatur, natura perfrigerans, & aluũ ſiſtens, hoc autem enemathe bis iterato, nulla illi admixta re, aluus reſtricta fuit, & ſic mulier iſta priſtinæ ſanitati reſtituta.

SCHOLIA.

HInc conijcere licet, quod in hac pleuritide, id eſt lateris dolore, alui profluuium ſuperuenit in declinatione, & ſolutus fuit morbus: quod animaduerſum uelim, propter Conciliatorẽ, & alios moroſos medicos, qui ſemper in declinatione pleuritidis, malũ eſſe prædicãt, ignorantes Hippocratis & Galeni præceptiones, in quibus nunquã hæc dixere. Nam Hippocrates lib. 6. Aphoriſmorũ, Aphoriſmo 16. inquit, A morbo laterali, uel pulmonia habito, alui profluuium aduenięns malum. Quæ uerba non ſic ſuperficietenus intelligẽda ſunt, ſed potius ut Galenus ibidẽ enarrat, dicens, non ſimpliciter dixit Hippocrates, in morbo laterali, uel pulmonia, aluũ turbatã, eſſe malũ ſignũ, ſed adiecit habito, ac ſi dixiſſet detento, atq; uexato homini ... talibus paſsionibus, profluuiũ alui ſuperueniens, malum eſſe ſignũ: quoniam ita ſe habentibus id ſolet ... re nõ potente hepate ad ſe trahere alimentũ, nec cõuertere in ſanguinem, & quãdoq; uentriculo idem alimentum corrũpente. In mediocri autem morbo laterali atq; pulmonia profluuium alui ſuperueniens poteſt prodeſſe ratione euacuationis, & magis quũ ſigna concoctioni apparuerint ineſſe paſsionibus, nec adeſt timor, ſed morbus eſt citra periculũ. Iis igitur uerbis Galeni ſatis apparet quod morbus ſi ſit uehemens et fortis, alui profluuium ſemper eſt malum, quia uirium imbecillitatem arguit, & cũ hoc quòd hepar cõpatitur, unde in ſpiritualibus & naturalibus partibus uitiũ eſt: Si uerò morbus ſit paruus, uel mediocris, ratione euacuationis bonum eſſe poterit: & hoc præſertim ſi ſigna concoctionis apparuerint. Sit igitur hæc in hac re concluſio, quod alui profluuium in uehementer oppreſsis & infeſtatis à morbo laterali, ſemper eſt malum ſignum: in mediocriter uerò patientibus, multoties eſt bonum tranſmutationis ratione, ut huius Cerdonis contigit uxori.

Curatio

FIGURE 2.1. Separation of *curatio* from scholion in Amatus Lusitanus's *Curationum medicinalium centuria prima* (1551). The image is from *Curationum medicinalium centuriae quatuor* (Basel: Froben, 1556).

case narrative) from the scholion (the learned commentary) appended to it. The separation was clearly marked even typographically, with the *curatio* set in roman type and the scholion in italic for the reader's ease in immediately distinguishing them (fig. 2.1). Amatus had already used this combination of description plus commentary in his earlier work on Dioscorides' *materia medica.*[62] In the *Curationes*, he applied the same combination of description plus commentary to his own cures; only, in this case, the description did not come from an ancient authority (Dioscorides) but from his own observation. In a daring move, he combined his description of the case with the scholarly commentary (the scholion), which had been absent in the *experimenta*, being reserved traditionally to high-status classical texts.[63] In other words, he presented his own cases as significant enough to deserve learned notice.[64] In contrast to the traditional *consilium*, it was doctrine that was now confined to the subsidiary role of footnote, while the observed case had become the primary object of attention. The cognitive hierarchy of doctrine over practice (and, I might add, of theory over observation) was subtly altered in favor of the latter. Amatus's innovation exemplifies how a change of genre can involve a change in the structure of attention.

Amatus was certainly not the only physician in those years who kept records of his cures, but he was the first to publish them in this new format.[65] We can compare Amatus's *Centuriae* with an unpublished manuscript of *curationes*[66] written at Ferrara in the 1540s, the same years in which Amatus himself practiced in that city. The manuscript's author was an unidentified student of the great humanist physician Antonio Musa Brasavola's, who was indeed a close friend of Amatus, according to the latter's testimony.[67] The form and content of these unpublished *curationes* are remarkably similar to those of Amatus, with one fundamental difference: they do not include a commentary on each case. Occasionally Brasavola's pupil did add, on the margins of the manuscript, a few references to medical texts relevant to the case in point,[68] but he never developed these notes into a full doctrinal discussion, as Amatus did in the scholia appended to his *curationes*.

The manuscript suggests that the writing of *curationes* and *observationes* may have developed out of new methods of note taking in the context of the humanist reform of medical training.[69] It shows that at Ferrara in the 1540s Brasavola was encouraging his pupils to keep records of patients in a way that paid unprecedented attention to the case history of disease. We know that around the same years Giovan Battista Da Monte was doing much the same thing at Padua, teaching his students how to construct a "simple *historia*" out of the particulars of each case.[70] And we know that some of the so-called Paris Hippocratics were keeping records of their cases in the 1570s.[71] We have thus evidence that at Ferrara, Padua, and Paris, in the middle decades of the sixteenth century, a new habit of keeping records of cases was developing in medical training and in medical practice.

Neither Brasavola, nor Da Monte, nor the Paris Hippocratics, however, ever published their own *curationes*.[72] They probably viewed them as practical knowledge of minor significance, to be transmitted orally to students—useful enough in itself, but unworthy of the dignity (and effort) of publication. Amatus did, and in so doing he created an immensely successful new genre.[73] Not only were his *Curationes* read and quoted for centuries but, more importantly, they had a paradigmatic influence on the new genre as it developed in the last decades of the sixteenth century. The *observationes* that appeared in the 1570s and 1580s—François Valleriola's *Observationes medicinales* (1573), Rembert Dodoens' s *Medicinalium observationum exempla rara* (1581), and van Foreest's *Observationes et curationes medicinales* (1584)—all followed Amatus's textual structure, not only in organizing their material as a case collection, but also in adopting the hallmark of Amatus's format, the separation of case history from commentary.[74]

What explains the success of the new genre? For one thing, Amatus's seven hundred cases, told as stories one after another, strongly appealed to the huge appetite for *varietas*, which was a marked trait of Renaissance intellectual taste.[75] Some readers may have even perceived epistemic implications in the new genre's exuberant wealth of particulars. So the physician and philosopher Francisco Sanches used Amatus's account of human variability as grist for the mill of his skepticism. Addressing an imaginary Aristotelian reader, he wrote:

> You say that there is no science of individuals, because they are infinite. But species are either nothing or something imagined. Only individuals exist, only they can be perceived, it is only of them that knowledge can be gained, snatched from them. If it is not so, show me your universals in nature. You will show them to me in the particulars themselves. Yet in those particulars I do not see any universal—they are all particulars. And how much variety can be seen in them? A truly marvelous amount. . . . One man falls into a swoon at the scent, or the sight, of a rose. Yon fellow dislikes women. This woman feeds on hemlock . . . (see Amatus, *Centuria* II, *curationes* 69, 36, 76)[76]

But more to the point, Amatus's intuition of the potential of the case collection met with a crucial trend of sixteenth-century medicine—the rise of neo-Hippocratism. The revival of Hippocratic medicine played a pivotal role in reorienting medical observation in this period. A primary impulse to the collecting and publishing of *observationes* came definitely from a conscious effort to emulate the case histories in books 1 and 3 of the Hippocratic *Epidemics*, which emphasized the role of the physician as the attentive observer of the natural course of disease.[77] Nor is it a coincidence that it is in a medical dictionary, compiled within the neo-Hippocratic circles in Paris in 1564, that we find the first retrieval of the ancient Empiric notion of observation, which emphasized the distinction of observation and theory.[78] By introducing a separate space for the case narrative clearly distinguished from the scholion, Amatus paved the way for a new focus on the observational aspect of the case, as in the Hippocratic *Epidemics.* After Amatus, several medical writers used the format he had created to adhere more closely to the Hippocratic model. The goal of fashioning his case records "according to Hippocrates' rules" was declared, for instance, by François Valleriola, when he published his *Observationes medicinales* in 1573 after forty years of practice.[79]

But there were also social reasons for the success of the new genre, as is apparent from the sociology of the *observationes.* In the early stage of the genre, from the late sixteenth century to the first half of the seventeenth, the authors of *observationes* were mostly practictioners—town physicians or court physicians, not university doctors: an itinerant Jewish physician, like Amatus Lusitanus, court physicians such as Dodoens, and especially town doctors, like Valleriola and van Foreest.[80] The social profile of the authors of *observationes* is that of the *practicus,* often with a leaning to medical heterodoxy, such as Paracelsianism.[81] The main impulse to the publication of *observationes* came from a more assertive attitude of self-promotion by doctors such as court and town physicians, who stressed success in practice, over and above academic learning, as a core element of their professional identity. In stark contrast to Scholastic medicine, where the source of legitimacy was doctrine, the late Renaissance *observationes* indicate the emergence of practice as a new source of validation of medical knowledge. Amatus brilliantly captured this shift by creating the *curatio/observatio* as a new form of medical writing, one that combined an emphasis on practice with scholarly credentials, while giving new visibility, significance, and circulation to the expertise of practitioners.

Circulation is the key word here. Some of the *observationes* were published by ex-students, who capitalized on the notes they had taken when doing the rounds of patients with a famous teacher. In fact, a posthumous publication by ex-students or by physicians' sons, themselves physicians, is not uncommon for the medical *observationes,*[82] just as it was not uncommon, as we have seen above, for the astronomical *observationes.* Like the astronomical *observationes,* the medical *observationes* developed originally within the narrow circle formed by a teacher's *familia* of pupils. As for the astronomical *observationes,* however, by the mid-sixteenth century a trend was under way for the publication of these records, indicating a strong interest in them beyond the purposes of medical apprenticeship.

A family resemblance connects the texts called *observationes* in fields as various as astronomy, jurisprudence, and medicine, suggesting that the rise of observation as an epistemic genre transcended to some extent early modern disciplinary boundaries. In all these fields, *observationes* indicated the knowledge of particulars collected in the daily practice of the discipline, as distinct from the discipline's doctrinal apparatus. Traditionally this knowledge would be transmitted only within the small group formed by a practitioner and his pupils. The *observationes* emerged because of a new urge to share this knowledge in a wider scholarly community, newly envisioned as a *res publica*, a commonwealth.

The *Observatio*: A Template for Scientific Communication

In 1571, in a Europe horribly torn asunder by religious conflict and in a medical community divided, as in late antiquity, among bitterly opposed sects, the Paracelsian physician Petrus Severinus idealized the old-time Hippocratic medicine as a golden age of harmonious medical cooperation:

> Great were the industry and concord of mortals in those times. Each did openly share his observations [*observationes in medium attulit*]. Indeed they thought that the brevity of life would not suffice for the completion of the art, which can only advance by collecting many observations. . . . But all that is human is prone to ruin and error. . . . As the labors of experience declined, and language instead grew in honor, the sap from the ancient roots was exhausted. Sloth did its part. Shortcutting the art and the quick facility of hypotheses are embraced much more willingly than the long and unsung practices of experience.[83]

Observationes in medium afferre, sharing observations—to use Severinus's phrase—would become the rallying cry of the period when the new genre coalesced. It was the ideal that inspired the most important collection of *observationes* of the end of the sixteenth century, Johann Schenck's *Paratērēseis, sive observationes medicae, rarae, novae, admirabiles et monstrosae* (*Paratērēseis, or Medical Observations, Rare, New, Wonderful, and Monstrous*), the text that signals the coming of age of the *observationes* as a primary form of medical writing.[84] A town doctor like many authors of *observationes*, Schenck built his seven-volume collection not on firsthand experience, as Amatus did, but by using the jack of all trades in the humanist textual toolbox—the *ars excerpendi*. His opus is a masterpiece of excerpting and rearranging; the *observationes*, both anatomical and medical, culled from hundreds of ancient and modern authors, are organized according to the conventional head-to-toe order of textbooks of practical medicine. His text is like a gigantic concordance, and in fact he used textual devices not unlike those that Nizolio used on Cicero; he excerpted and listed observations that seemed to refer to the same object and carefully gave the source for each of them.[85] Only, what he worked on was not a single author, but collective medical experience. He cast his net very wide. The list of authors from which he excerpted *observationes* included ancient Greek and Roman, medieval Jewish and Arab medical writers, and, of the moderns, also those physicians who wrote in the vernacular. He scoured the texts of the medical tradition to retrieve, so to speak, the fragments of observation scattered in a great sea of doctrine.

And he did not look only to the past. He included his own observations,[86] and he wrote near and far to his colleagues to ask that they contribute their own unpublished *observationes*. Schenck stressed that his enterprise was made possible by a wide network of correspondents (he listed seventy-one), who helped either by sending him the references to rare observations they had come across in their reading, or by allowing him to use their own unpublished notes. Thus for instance Joachim Camerarius sent him his "*sylva* of *observationes*, containing over fifty *historiae*"; Jean Bauhin sent him his private journals, containing *observationes* and *curationes* "written for his own private use."[87] Schenck conceived and presented his collection as a truly joint enterprise.

Schenck's volume provides strong evidence that the habit of keeping detailed records of cases (especially the most unusual ones)[88] was spreading fast among European physicians. We should probably assume that the published collections of *observationes* are just the tip of the iceberg.[89] The publication of manuscript *observationes* was encouraged and eagerly anticipated. Schenck urged the colleagues in the medical faculty at the University of Louvain to publish the *historiae medicae* that Cornelius Gemma had promised but failed to publish before his death. We know that Cornelius Gemma was the author of astrological *Ephemerides* devoted to weather observations.[90] Evidently, like his father before him, he also kept records of his medical cases.

Schenck's work illustrates the extension that medical epistolary networks had reached by this time.[91] The development of scholarly correspondence in the early modern period has been amply documented and investigated in the case of natural history,[92] but there was certainly a lively medical contribution to this process. A minor genre of early modern medicine was the *epistolae medicae* (medical letters), which contained reports of cases since its inception in the early Renaissance.[93] But the sharing of *observationes* set in motion a much more intensive epistolary exchange, which involved a marked trend toward collective empiricism[94] and even some embryonic forms of shared authorship. It was not unusual for a medical author to print next to his own observations those that had been communicated to him by other physicians.

The itinerant surgeon Fabricius Hildanus, author of six *centuriae* of *Observationes et curationes chirurgicae* (published 1606–1619), cited 348 colleagues who shared his observations with him by letter or in person.[95] The defining trait of the new genre was an emphasis on the collection and circulation of observational knowledge, envisioned as a joint enterprise binding together past, present, and future members of an ideal medical community—a *res publica medica*, as Theodor Zwinger called it in his prefatory letter to Schenck's book. The history of the *observationes* is also the story of the successful attempt to turn a virtual *res publica medica* into a real community, bound together by forms of shared identity and authorship. There is no mistaking the proud sense of collective identity, projected not only into the past but even more so into the future, that the collections of *observationes* convey. Their authors plainly believed what is said in one of the celebratory poems that open Schenck's volume:

> We shall bear through the centuries the name of observers,
> This name and these writings will last forever.[96]

There is ample evidence of the success of the *observationes* in the first half of the seventeenth century. At least twenty important collections were published in this period, several of which would be reprinted, translated, and excerpted in the second half of the century and even in the eighteenth.[97] By the second half of the seventeenth century, the *observationes* were fully established as the primary medium for the circulation of information in the *res publica medica.* When the members of the newly founded Academia Naturae Curiosorum (the Academy of Those Curious about Nature), the future Academia Leopoldina, created by a group of town doctors in the imperial city of Schweinfurt in southern Germany, decided in 1670 to publish their own periodical after the model of other learned societies of Europe, they adopted the form of a miscellaneous list of *observationes.*[98] It is no surprise that physicians such as the Curiosi would choose the *observationes* as the elective format of their publishing program. When needing a model of scholarly correspondence, they naturally turned to the sharing of medical observation through epistolary networks that doctors had been practicing all over Europe for more than a century.

Animated by the dream of retrieving every particle of the "*thesaurus* of medical experience,"[99] some of the Curiosi collected for the purpose of publication all the manuscript *observationes* left by physicians in the past. So Georg Hieronymus Welsch published in 1668, together with his own *observationes*, those of five other physicians whose manuscripts he had acquired (fig. 2.2).[100] In the preface to this volume, he listed all the "not yet published *observationes*,

FIGURE 2.2. Multiple authorship. Frontispiece of G. H. Welsch, *Sylloge curationum et observationum medicinalium* (Augsburg, 1668). The six medallions, connected by a ribbon held by the *puttini*, contain the name of the medical authors whose *observationes* Welsch collected in this work, together with his own.

curationes, epistolae" he knew of, indicating the location of the manuscripts. This effort to unearth the *bibliotheca latens*, the "hidden library," of what was observed in the past went hand in hand with the activation of networks for the circulation of observations in the present.[101]

At the turn of the seventeenth century, the community-building role of the *observationes* that we have examined in the case of medicine was unfolding simultaneously in other disciplines, such as natural history and astronomy.[102] In fact, it cannot be stressed enough that the ideal of collective empiricism expressed by the sharing of *observationes* cut across disciplinary boundaries, not least because early modern disciplinary boundaries were much more porous than we tend to perceive them in retrospect. Many early modern physicians also engaged in astronomical and astrological pursuits, and conversely, some full-time astronomers like Tycho Brahe had medical and alchemical interests.[103] In some cases, and especially for medicine and natural history, drawing a dividing line between the medical and the natural historical community would completely distort the picture of what was in fact the joint pursuit of scholars who were both physicians and natural historians throughout their lives.[104] The networks of the medical *observationes* largely overlapped with those of natural history, and the links between the two fields became, if anything, even more pronounced in the first half of the seventeenth century.

The *observatio* was ideally suited to serve the polymathic interests and the nearly omnivorous pursuit of empirical knowledge that characterized many protagonists of seventeenth-century scientific life, from Peiresc and Gassendi to Boyle.[105] Typically cast in the form of a letter or a short report on a specific piece of firsthand observation (astronomical, medical, anatomical, natural historical, or natural philosophical), the *observatio* was knowledge that could easily travel. It was unencumbered by lengthy references to points of doctrine and theory. If present, these were set apart in a typographically separated scholion, which was supposed to be written "without any of the acrid salt of criticism, with the sole intent of clarifying and further explicating the narrated case with similar histories."[106] Thanks to its lightweight format and its avoidance of theoretically divisive issues, the *observatio* was well positioned to become the primary vehicle of the new "learned empiricism" that bound together European scholars across disciplinary, philosophical, and religious divisions.[107] In the second half the seventeenth century, the relatively new genre of the *observatio* would become the preferred format for intellectual exchange in the Republic of Letters, as shown by its adoption in the periodicals of the newly founded learned societies.[108]

Observation and the New Language of Experience

In the first half of the seventeenth century, *observatio* was not only established as an epistemic genre; it also started to emerge as a new cognitive category, whose use extended beyond the disciplines that had been its incubation ground, astronomy and medicine. It was in this period that the concept of *observatio* spread to natural philosophy and made its entry into mainstream philosophical language. How did this happen? A preliminary investigation suggests that it was in medicine, in the second half of the sixteenth century, that *observatio* first became a distinct epistemic category, and that from medicine the concept passed on to general philosophical language as part of a new conceptualization of experience. The category of *observatio* first emerged in neo-Hippocratic medical circles with the recovery of the ancient Empiric/Skeptic philosophical vocabulary, together with two other concepts, *autopsia* and *phainomena*, also destined to play a fundamental role in the early modern renewal of the language of experience. The combined purport of these terms, in their ancient Empiric/Skeptic acceptation, was an emphasis on the distinction between direct experience (*autopsia*) and indirect experience, the insistence on focused and repeated observation (*tērēsis*) as the foundation of empirical knowledge, and the urge to keep to the phenomena (*phainomena*, or things as they appear), avoiding useless and contentious theorization.[109]

Between the sixteenth and the seventeenth century, *phainomena*, *autopsia*, and *observatio* entered learned language and acquired an epistemic resonance that went far beyond their context of origin. *Phainomena*, which had been originally associated with astronomy, and indeed had been used mostly to refer to celestial objects, was extended to cover all natural processes, as it had been employed by the ancient Skeptics.[110] The neologism *phenomena* gradually replaced *apparentia*, which had been the medieval rendition of the Greek *phainomena* as used in astronomical texts. In his translation of Sextus Empiricus, which was one of the main sources of the Skeptical revival in the late Renaissance, Henri Estienne used *apparentia* when the Greek original referred to celestial objects, but kept *phainomena* whenever Sextus had used it with a wider philosophical meaning, thus paving the way to the introduction of *phenomena* as a neologism.[111]

The case of *autopsia* is similar and even more relevant to our purposes, since the word followed the same trajectory of *observatio*, with which it was closely associated semantically. *Autopsia* first appeared in the same medical dictionary as *tērēsis/observatio*, Jean de Gorris's *Definitiones medicae* (1564), where it is called, like *tērēsis*, "a word of the Empiric sect."[112] The term carried a strong connotation of firsthand ocular experience, or inspection. In the late Middle Ages and the early Renaissance, it had been variously translated as "*per se inspectio*" or "*intuitus proprius*."[113] *Autopsia*, the Latinized transliteration of the Greek word, is also, like *phenomena*, a late sixteenth-century neologism. Its emergence suggests a new awareness that the available philosophical terminology lacked a specific term to denote firsthand experience—and in fact no such term was available either in the Aristotelian or the Galenic conceptualization of experience.

Though explicitly identified as "a word of the ancient Empirics," *autopsia* spread rapidly into general medical usage, and by the early seventeenth century had become a commonly recognized medical category, beyond sectarian divisions between Empiricist and Rationalist schools. It was used, in fact, by physicians who certainly did not identify with an Empiric or Skeptical epistemology but were strongly committed to a program of anatomical inquiry, such as William Harvey. Harvey used the term *autopsia* repeatedly, both in *De motu cordis* and in *Exercitationes de generatione animalium*,[114] not in a narrow anatomical sense (i.e., dissection), but to indicate "*experientia propria*" (one's own experience). His use of the term was certainly deliberate, as it features in the philosophically self-conscious preface to the *Exercitationes*, where Harvey sketched "the way and order in which knowledge should be acquired," trying to reconcile his anatomical work with an Aristotelian epistemology.[115] By using *autopsia* in this context, Harvey interpolated a concept of Empiric extraction into his otherwise Aristotelian view of knowledge. The adoption of *autopsia* in mainstream medical language, irrespective of philosophical allegiance, is confirmed by a later entry in Castelli's medical dictionary (1688), which states that the term "was *once* a word belonging to the Empirical sect . . . but nowadays *autopsia*, that is, the observation [*observatio*] and memory of those things examined with one's own eyes, is eminently necessary to Rational Medicine."[116]

Like *autopsia*, *observatio* was adopted into the mainstream language of medicine in spite of its original Empiric derivation. But differently from *autopsia*, whose use remained mostly circumscribed to medicine, *observatio* was fated to gain much wider currency. A decisive step in launching the term's philosophical career was probably the fact that in the late sixteenth century, *observatio* was newly interpolated into the standard Aristotelian vocabulary of experience, of which previously it had never been part. This innovation can be traced back to Petrus Ramus's effort to revise Aristotelian epistemology by placing a stronger emphasis on experience. Extolling the Aristotelian terms for empirical knowledge, *empeiría* and *historía*, in his *Scholae in liberales artes* (1569), Ramus rendered *historía* with *observatio*.[117] This new twist to the Aristotelian vocabulary of experience spread further thanks to the philosophical eclecticism of early seventeenth-century encyclopedists, such as Johann Heinrich Alsted, who, revisiting in 1623 the Aristotelian constellation of empirical terms (*aisthēsis/sensus, historía/historia, empeiría/experientia, epagōgē/inductio*), followed Ramus in translating *historía* as *observatio.*[118] This usage seems to have been adopted even by self-conscious Aristotelians. William Harvey, for instance, added *observatio* to the classic Aristotelian account of empirical knowledge (as given in *Posterior Analytics*: from sense perception derives memory, from memory experience), stating that true knowledge is based on "one's own experience acquired through manifold memory, frequent sense perception, and *diligent observation.*"[119] By the mid-seventeenth century, the identification of *historia* and *observatio* had become commonplace, and indeed the two terms got to be practically synonymous in medical and philosophical terminology.[120] *Historia* was an epistemic category with a multifaceted pedigree, which came to signify, in the early modern period, a descriptive account of observational knowledge in any field, and as such enjoyed enormous vogue in the language of early modern "learned empiricism."[121] Associated with *historia, observatio* acquired generalized currency in scholarly language as a key term of the new vocabulary of experience, so much so that, like *historia,* it came to be identified with *experientia* itself.[122]

But *observatio* was actually quite different from the old category of *experientia.* Even when coopted into mainstream philosophical terminology, *observatio* kept a strong flavor of the Empiric/Skeptical source from which it had originated. It invariably indicated, as it did for Harvey, *experientia propria, autopsia,* authored observation—not the generic, anonymous experience of the Aristotelian *empeiría* or of the Plinian *observationes.*[123] *Observatio,* moreover, kept its ancient antagonism to hypothesis, doctrine, or theoretical speculation. A suspicion of theory, which was never part of the Aristotelian notion of experience, nor of the Galenic model of "rationalist empiricism," was at the very core of the concept of *observatio* as it gained ascendancy in early modern philosophical language. Severinus noted, as we may recall, that the decline of the medical *observationes* was directly related to the tempting "facility of hypotheses"; in the same years, Petrus Ramus and Copernicus's pupil Rheticus seriously discussed the possibility of freeing astronomy from all hypotheses and basing it exclusively on observations.[124] *Observatio* and *hypothesis* stood as two firmly separated and indeed antithetical concepts. From Amatus's *Curationes* in the mid-sixteenth century to the Curiosi's *Observationes* in the second half of the seventeenth, the distinctive hallmark of the *observatio* as a vehicle of scientific communication was the separation of the observational report from its theoretical interpretation.

FIGURE 2.3. Donato Creti, *Astronomical Observation: Jupiter*. 1711. Pinacoteca Vaticana. Photo: Vatican Museums.

Several interweaving intellectual strands combined to form the category of observation in the seventeenth century. From the astronomical tradition came the emphasis on seriality, mathematization, and the striving for more and more exact and calibrated measurement. From medicine, via the Empiric/Skeptic legacy revitalized by neo-Hippocratism, came the clear-cut distinction between direct and indirect experience, and the separation of observation from theory. From thc philological training shared by all early modern scholars came habits of accuracy in identifying sources and a whole array of textual tools, from the *ars excerpendi* to the concordance and the thesaurus, that could be used to master complexity, of texts as of things.[125] And binding all these strands together was the overarching sense of observation as an observance, a dedication of one's whole life to a rule-bound activity that had the powerful appeal of a mission. By the mid-seventeenth century, observation had become a cognitive activity with a distinct literary format, an ever-increasing barrage of instruments, an ever widening set of practices; an activity that could be pursued everywhere but had already long developed its own specialized sites, from the botanical garden to the anatomical theater to the astronomical observatory. It was an activity whose immense appeal to seventeenth-century minds was both cognitive and aesthetic, as testified by the lovely pictures in which Donato Creti used the conventions of late Baroque landscape painting to portray and celebrate astronomical observation (fig. 2.3).[126]

Notes

From conception to redaction, this essay greatly benefited from the constant advice and support of Katharine Park and Lorraine Daston, who helped me most generously with research ideas, references to sources, and interpretative insight. For their useful comments on earlier drafts, I'd like to thank all the participants in the "History of Scientific Observation" Project at the Max-Planck Institut für Wissenschaftsgeschichte in Berlin, in particular Nico Bertoloni Meli, Elizabeth Lunbeck, Mary Morgan, and Ted Porter. My thanks to Harry Marks for incisive criticism that helped me clarify my argument. All the translations are mine, unless otherwise stated. For reasons of space, the notes have been cut to the essential. I refer the reader, with apologies, to the fully referenced text at http://www.mpiwg-berlin.mpg.de/en/research/projects/DEPT1_Renn-Globalization/projects/DeptII_Da_observation/index_html.

1. See the essay by Katharine Park in this volume for the astral sciences; Anthony Grafton and Nancy Siraisi, eds., *Natural Particulars: Nature and the Disciplines in Renaissance Europe* (Cambridge, Mass.: MIT Press, 1999); Chiara Crisciani, "*Experientia* e *opus* in medicina ed alchimia: forme e problemi di esperienza nel tardo Medioevo," *Quaestio* 4 (2004): 149–73; Joseph Ziegler, "Médecine et physiognomie du XIVe au début du XVIe siècle," *Médiévales* 46 (2004): 87–105; Brian Curran and Anthony Grafton, "A Fifteenth-Century Site Report on the Vatican Obelisk," *Journal of the Warburg and Courtauld Institutes* 58 (1995): 234–48 (on antiquarianism).

2. Edward Grant, "Medieval Natural Philosophy: Empiricism without Observation," in *The Dynamics of Aristotelian Natural Philosophy from Antiquity to the Seventeenth Century*, ed. C. Leijenhorst, C. Lüthy, and J. M. M. H. Thijssen (Leiden: Brill, 2002), 141–68, esp. 141–46.

3. Aristotle, *Metaphysics,* I. 1, 980a 27–981 b13; *Posterior Analytics* 19, 100a 4–9; Hippocrates, *Aphorisms,* I.1.

4. Jole Agrimi and Chiara Crisciani, "Per una ricerca su *experimentum-experimenta*: riflessione epistemologica e tradizione medica (secc. XIII–XV)," in *Presenza del lessico greco e latino nelle lingue contemporanee,* ed. Pietro Janni and Innocenzo Mazzini (Macerata: Università degli Studi di Macerata, 1990), 9–49.

5. On Aristotle see Park, in this volume, p. 38, n. 10; on the Hippocratic Corpus, see Gianna Pomata, "A Word of the Empirics: The Ancient Concept of Observation and Its Recovery in Early Modern Medicine," *Annals of Science,* forthcoming.

6. Park, in this volume, p. 19.

7. On *experimentum* in Vives, see Carlos G. Noreña, *Juan Luis Vives* (The Hague: Nijhoff, 1970), 286; cf. for instance Vives, *De prima philosophia* in *Opera omnia* (Valencia, 1992–93), 3: 184. For Sanches, see Elaine Limbrick, introduction, in Francisco Sanches, *That Nothing Is Known,* ed. Elaine Limbrick and Douglas F. S. Thomson (Cambridge: Cambridge University Press, 1988), 25. Ramus saw *experientia* as a general cognitive category and used *observationes* in a limited astronomical sense (see below, n. 124). But he translated the Aristotelian term *historia* with *observatio,* thus giving *observatio* a wider philosophical significance. See below, n. 117.

8. To give just one example out of many: in the modern translation of Sanches's *Quod nihil scitur* by D. F. S. Thomson (ed. Limbrick, cited above, n. 7), what the translator rendered uniformly as "to observe" is almost always *videre* (to see), or less often *animum vertere* (to direct the mind to something), not *observare.* For example: "nil in rebus perpendentes" (92) is rendered as "judging nothing in terms of (observed) facts" (277). "Sed etsi plura videret, non tamen omnia posset, quod necesse est vere scienti" (162) is translated as "even if he were to observe a huge number of facts, yet even so he would not be able to observe all the facts" (287).

9. Marta Fattori, *Lessico del Novum Organum di Francesco Bacone* (Rome: Edizioni dell'Ateneo, 1980), 2: 449, 480: *experientia* appears 64 times, *experimentum* 108 times—*observatio* only 24 times. The verb *experiri* appears 25 times (plus two cases of *experimentare*), the verb *observare* 13 times.

10. I have compared the Latin text of the *Novum organum* with the nineteenth-century translation in Francis Bacon, *The Works,* ed. James Spedding et al., vol. 4 (London: Longman, 1860). For example: "ex sensu proprio" rendered as "of his own observation" (43); "particularium sylva et materies" rendered as "a store of particular observations" (94).

11. See Pomata, "Word of the Empirics," n. 113, for early examples from Giovanni Pico della Mirandola; Sebastian Fox Morcillo, *De demonstratione, eiusque necessitate ac vi* (Basel, 1556), 7–8: "artes . . . ac disciplinae fundatae et constitutae experimento, observationeque diuturna." The two words are sometimes associated in titles: Dietrich Dorsten, *Botanicon . . . additis etiam quae neotericorum observationes et experientiae* (Frankfurt, 1540).

12. On *empeiria* in Aristotle, see most recently Catherine Darbo-Peschanski, *L'Historia. Commencements grecs* (Paris: Gallimard, 2007), 112–32. On the philosophical history of the concept, see the essays collected in Marco Veneziani, ed., *Experientia* (Florence: Olschki, 2002); and the special issue of *Quaestio* 4 (2004), especially Paolo Ponzio, "The Articulation of the Idea of Experience in the Sixteenth and Seventeenth Centuries," 175–95.

13. Pomata, "Word of the Empirics."

14. Jean De Gorris, *Definitiones medicae* (Paris, 1564), s.v. "*Tērēsis, Observatio,*" 328 rv; on the philosophical dictionaries, see Pomata, "Word of the Empirics," n. 1; and the essay by Lorraine Daston in this volume, n. 1.

15. A bibliography of works titled *Observationes* for the period 1500–1800 prepared by Sebastian Gottschalk using the WorldCat, British Library, Library of Congress, and Herzog-August Bibliothek online catalogs indicates that philology, lexicography, jurisprudence, medicine, astronomy/astrology, and travel writing are the areas in which the term was used in titles in the sixteenth and seventeenth centuries. This confirms what is indicated by the entry *Observationes* in Martinus Lipenius, *Bibliotheca Realis Philosophica* (1682), which lists mostly philological works, and in Lipenius's *Bibliotheca Realis Medica* (1679) and *Bibliotheca Realis Juridica* (1679), which have long lists of medical and legal works titled *Observationes*.

16. The notion of title should be historicized, of course. In the late fifteenth and early sixteenth century, with the advent of printing, titles acquired a new significance as identifiers of books in the consciousness of authors and readers. In antiquity, in contrast, the same work could be known under several names, or have no title at all, usually substituted with the incipit. See Jean-Claude Fredouille et al., eds., *Titres et articulations du texte dans les oeuvres antiques* (Paris: Institut d'Études Augustiniennes, 1997).

17. Park, in this volume, pp. 29–32, 37.

18. On the double meaning of *tērēsis*, see David L. Blank, *Ancient Philosophy and Grammar. The Syntax of Apollonius Dyscolus* (Chico, CA: Scholars Press, 1982), 71 n. 2. For *observatio*, see the entry in Mario Nizolio, *Lexicon Ciceronianum*, 3 vols. (London, 1820).

19. Pomata, "Word of the Empirics," nn. 84, 85, 86.

20. Park, in this volume, p. 21.

21. Pierre Villey, *Lexique de la langue des Essais de Montaigne* [1933] (New York: Burt Franklin, 1973), 452, s.v. "observation." I have checked the word usage also through the digitized version of Montaigne's *Essais* (1595 edition).

22. On *familia* as the basic metaphor for the medieval teacher-student relationship, see Chiara Crisciani, "Teachers and Learners in Scholastic Medicine: Some Images and Metaphors," in *History of Universities* 15 (1997–99): 75–101.

23. Park, in this volume, pp. 33–35.

24. Park, in this volume, p. 31 and n. 63.

25. Giovanni Pico della Mirandola, *Disputationes adversus astrologiam divinatricem*, ed. Eugenio Garin, 2 vols. (Florence: Vallecchi, 1943). Steven Vanden Broecke has shown that Pico's strictures led to the attempt to reform astrology's observational basis in mid-sixteenth-century Louvain: *The Limits of Influence. Pico, Louvain, and the Crisis of Renaissance Astrology* (Leiden: Brill, 2003). This effort had strong religious motivations, as had been the case originally for Pico's work. For Protestant Reformers, see Sachiko Kusukawa, "*Aspectio divinorum operum*: Melanchton and Astrology for Lutheran Medics," in *Medicine and the Reformation*, ed. O. P. Grell and A. Cunningham (London: Routledge, 1993), 33–56.

26. Emmanuel Poulle, *Les sources astronomiques (textes, tables, instruments)* (Turnhout: Brepols, 1981), 55–64 (tables), 64–66 (almanacs).

27. Otto Brunfels's dictionary of astrological terms (1534) does not include an entry for *observatio*. It has instead the entry "*αποτέλεσμα/judicium*," defined as "sententia astrologica, latine praedictio vel judicium": *De diffinitionibus et terminis astrologiae libellus isagogicus*, in *Astronomicorum libri VIII*, ed. Nicolaus Prucknerus (Basel, 1551). The term *observationes* seems to be fairly rare among the incipits of seventeenth-century astrological manuscripts: Lynn Thorndike and Pearl Kibre, *A Catalogue of Incipits of Medieval Scientific Writings in Latin* (London: Medieval Society of America, 1963), has only one example at 455, col. a. Only one example also in Jole Agrimi, *Tecnica e scienza nella cultura medievale: inventario dei manoscritti relativi*

alla scienza e alla tecnica medievali (secc. XI–XV) nelle biblioteche di Lombardia (Milan: Angeli, 1976), on 126.

28. *Scripta Clarissimi Mathematici M. Joannis Regiomontani. . . .* (Nuremberg, 1544). See Park, in this volume, p. 33 and n. 72. Two years later Schöner published Werner's *Canones*, including also Werner's weather observations for the period 1513–20: Johannes Werner, *Canones, sicut brevissimi, ita etiam doctissimi, complectens praecepta et observationes de mutatione Aurae* (Nuremberg, 1546). See Vanden Broecke, *Limits of Influence*, 204–5.

29. Schöner, dedication to the senators of Nuremberg, in *Scripta Clarissimi Mathematici M. Joannis Regiomontani*, a iij. Emphasis added. Osiander also used the metaphor of the "*thesaurus observationum*" in his preface to Copernicus's *De Revolutionibus* (1543).

30. Pico had called a "lie of the astrologers" their claim to have "tot miliorum annorum observationes" (*Disputationes*, ed. Garin, 2: 472).

31. But it should be noticed that Pliny sometimes mentioned the authors of astrometeorological observations, as in the case of Caesar's *parapegmata* (*Natural History*, XVIII, 65. 237).

32. Indeed, Schöner passed on to Copernicus some of Walther's as yet unpublished observations, which Copernicus used in *De Revolutionibus*: see Park, in this volume, n. 86. Sharing observations appears to have been a new feature of Renaissance astronomy. Ptolemy, for instance, mentions only one contemporary scholar (Theon) who shared his observations with him. See Olaf Pedersen, *A Survey of the Almagest* (Odense: Odense University Press, 1974), 12–13 and items 56, 57, 58, and 61 in appendix A, "Dated Observations in the Almagest," 408–22, on 416–17.

33. Cited in Vanden Broecke, *Limits of Influence*, 155 n. 31, 180 n. 138.

34. For travelogues, see, for instance, Pierre Belon, *Les observations de plusieurs singularitez et choses memorables, trouuées en Grece, Asie. . . .* (Paris, 1553). On the role of the *ars apodemica* in the history of observation, see Daston, in this volume, p. 89.

35. Among early examples are Domizio Calderini's *Observationes quaedam*, in appendix to his edition of Statius's *Sylvae* (Rome, 1475), and Marino Becichemo's V*ariae Observationes*, in his *Aurea Praelectio in C. Plinium Secundum* (Brescia, 1504).

36. *Nizolius sive linguae Latinae thesaurus* (Venice, 1551), preface. See also Mario Nizolio, *De veris principiis*, ed. Quirinus Breen (Rome: Bocca, 1956), 2: 71–72. On the editions of Nizolio's Ciceronian *Observationes*, see Quirinus Breen, "The *Observationes* in M. T. Ciceronem of Marius Nizolius," *Studies in the Renaissance* 1 (1954): 49–58, on 50–51.

37. See Forcellini, *Totius Latinitatis Lexicon*: observator et custos bonorum (Seneca, *Ep.* 41.2); Catholicae legis (*Codex Theod.* 16.5.1).

38. Nizolio drew on Cicero for philosophical views that strongly leaned to nominalism and skepticism. See Charles B. Schmitt, *Cicero Scepticus* (The Hague: Nijhoff, 1972), 72 ff.; Brian P. Copenhaver and Charles B. Schmitt, *Renaissance Philosophy* (Oxford: Oxford University Press, 1992), 207–9.

39. Cocles's *Chiromantie ac physiognomie anastasis* (1504), quoted in Joseph Ziegler, "Observing Living Bodies in Pre-Modern Learned Physiognomy: The Physiognomic Portraits in Bartolomeo della Rocca Cocles's *Anastasis*," paper presented at the Max-Planck-Institut für Wissenschaftsgeschichte, Abt. II Colloquium, 12 Sept. 2008.

40. Symphorien Champier flatly equated *experimentator* with *empiricus*: *Vocabularius . . . difficilium terminorum naturalis philosophiae et medicinae*, in S. Champier, *De triplici disciplina. . . .* (Lyon, 1508), s.v. "*empiricus*." On *rustici* et *vetulae* as authors of *experimenta*, see Agrimi-Crisciani, "Experientia-experimentum," 29–30, 42.

41. Among early examples of the genre are Goffredo Lanfranco Balbi, *Observationes nonnullarum in iure decisionum* (Lyon, 1538); Marco Mantova Benavides, *Observationum legalium libri X* (Lyon, 1546). See also Bernhard Wurmser and Hartmann Hartmann, *Observationes practicae* (Basel, 1570); Alessandro Stiatici, *Praxis iuridicalis, hoc est, observationes et animadversiones quaedam* (Venice, 1580).

42. See, for instance, Andreas von Gail, *Practicarum observationum tam ad processum judiciarium praesertim Imperialis Camerae . . . libri duo* (Cologne, 1578); Polidoro Ripa, *Singulares observationes in foro responsae* (Venice, 1605). These collections are organized by *centuriae*, as were also some of the medical *observationes*. Both start from a hypothetical case but then argue on the basis of a real case, relating how it was settled. This explains why the collections of *observationes* often merge with the collections of *decisiones*, which reported the verdicts of specific law courts. On the development of this literature, see O. F. Robinson, T. D. Fergus, and W. M. Gordon, *An Introduction to European Legal History* (Abingdon: Professional Books, 1985), 325–26, 346–47.

43. As would also be the case of the medical *observationes*, the writing of legal *observationes* grew rapidly over the course of the seventeenth century, and by the end of the eighteenth century the genre had reached massive proportions. See the copious listings under the heading *observationes juris* in Martin Lipen, *Bibliotheca Realis Juridica*, with additions by F. G. Struvius and G. A. Jenichenius (Leipzig, 1757), 2: 92–95; *Supplementa* (Leipzig, 1775), 352–53; *Supplementa* (Leipzig, 1789), cols. 295–97.

44. On the epistemic significance of *practica* in the legal tradition, see Danilo Segoloni, "*Practica, practicus, practicare* in Bartolo e in Baldo," in *L'educazione giuridica, II, Profili storici* (Perugia: Consiglio Nazionale delle Ricerche, 1979), 52–103. For the Renaissance see Maximilian Herberger, *Dogmatik. Zur Geschichte von Begriff und Methode in Medizin und Jurisprudenz* (Frankfurt am Main: Klosterman, 1981), 210–31, 257–58, which amply documents the close connection between the legal and the medical view of *practica*.

45. Anthimus, *De observatione ciborum* (*On the Observance of Foods*), ed. and trans. Mark Grant (Blackawton: Prospect Books, 1996); Pomata, "Word of the Empirics," n. 89 (on Arnald).

46. Alessandro Benedetti, *De observatione in pestilentia* (Venice, 1493); the same is true of Heinrich Stromer von Auerbach, *Saluberrimae adversus pestilentiam observationes* (Leipzig, 1516).

47. Gabriele Falloppio, *Observationes anatomicae ad Petrum Mannam medicum cremonensem* [Venice, 1561] (Modena: Mucchi, 1964), 4v.

48. Ibid., examples at 17v, 36r, 37v, 45v, 51v, 57r, 183v. Falloppio was very proud of his acuity as an observer and was fond of referring autobiographically to observations he had made even as a child: see the passages quoted in Giuseppe Favaro, *Gabriele Falloppia* [*sic*] *modenese. Studio biografico* (Modena, 1928), 48–49.

49. Law and medicine shared a long tradition of similarly named and similarly conceived genres, such as the *consilia*. On the parallels between juridical and medical mental framework in the Renaissance, see Herberger, *Dogmatik*, 211–75; and Ian Maclean, *Logic, Signs and Nature in the Renaissance* (Cambridge: Cambridge University Press, 2002), 84–86.

50. See Robinson et al., *An Introduction to European Legal History*, on the medieval *consilia* (94, 113–14) and on the early modern *observationes* (325–26).

51. Jole Agrimi and Chiara Crisciani, *Edocere medicos: medicina scolastica nei secoli XII–XV* (Milan: Guerini, 1988), 216–17; Chiara Crisciani, "Histories, Stories, Exempla and Anecdotes:

Michele Savonarola from Latin to Vernacular," in *Historia: Empiricism and Erudition in Early Modern Europe*, ed. Gianna Pomata and Nancy G. Siraisi (Cambridge, Mass.: MIT Press, 2005), 297–324.

52. The medieval *consilium* dealt typically with a disease, not with a sick person, as shown by Jole Agrimi and Chiara Crisciani, *Les 'consilia' medicaux* (Turnhout: Brepols, 1994); see also Chiara Crisciani, "L'individuale nella medicina tra medioevo e umanesimo: i *consilia*," in *Umanesimo e medicina. Il problema dell'individuale*, ed. Roberto Cardini and Mariangela Regoliosi (Rome: Bulzoni, 1996), 1–20.

53. The doctrinal apparatus became even heavier in the fourteenth- and fifteenth-century collections: see Crisciani, "L'individuale nella medicina," 20 n. 33. The thirteenth-century *consilia* of Taddeo Alderotti, in contrast, were relatively less focused on doctrine; but even so, the descriptions of cases were very rare. See Taddeo Alderotti, *I consilia*, ed. G. M. Nardi (Turin: Minerva medica, 1937); Nancy Siraisi, *Taddeo Alderotti and His Pupils: Two Generations of Italian Medical Learning* (Princeton: Princeton University Press, 1981), 270–302, especially 270–73.

54. As, for instance, in van Foreest's *Observationes et curationes* (cited below, n. 59) or Paul de Reneaulme's *Ex curationibus observationes* (Paris, 1606).

55. Amatus says that he finished writing the first *centuria* in Florence in 1549: Amatus Lusitanus, *Curationum medicinalium centuria prima* (Florence, 1551), preface. The seventh *centuria* came out in Venice in 1566. On Amatus (1511–68), one of the most prominent Jewish intellectuals of the late Renaissance, see Harry Friedenwald, *The Jews and Medicine* (Baltimore: Johns Hopkins University Press, 1944), 1: 332–80.

56. Francis Bacon seems to have thought so when he noted that medical writers should not ape the men of law in this respect. If the physicians wanted to write down cases—he noted—let them follow the Hippocratic model instead of copying the jurists. Bacon, *De augmentis scientiarum*, in *Works*, ed. Spedding et al., 1: 591–92.

57. Thomas Bodier, *De ratione et usu dierum criticorum* (Paris, 1555), 17r–51r. Bodier organized his fifty-five cases in fourteen *observationes*, each *observatio* grouping a certain number of cases from which he drew a specific rule for prognostication. By *observatio*, he meant a rule based on the observation of several cases, not the description of a case. On Bodier, see Steven vanden Broecke, "Evidence and Conjecture in Cardano's Horoscope Collection," in *Horoscopes and Public Spheres*, ed. Günther Oesterman, H. Darrell Rutkin, and Kocku von Stuckrads (Berlin and New York: de Gruyter, 2005), 215–17; Nancy Siraisi, "Anatomizing the Past: Physicians and History in Renaissance Culture," *Renaissance Quarterly* 53 (2000): 8. The connection between case writing and astrology should be further explored.

58. Cited in Vanden Broecke, *Limits of Influence*, 196. On Cornelius Gemma, see Hiro Hirai, ed., *Cornelius Gemma. Cosmology, Medicine and Natural Philosophy in Renaissance Louvain* (Pisa: Fabrizio Serra, 2008).

59. Pieter van Foreest, *Observationes et curationes medicinales* (Leiden, 1584), preface, 15–16.

60. Park, in this volume, p. 36; Agrimi and Crisciani, "*Experimentum-experimenta*," 39–47; Lynn Thorndike, *History of Magic and Experimental Science*, 8 vols. (New York: Columbia University Press, 1923–58), 2: 751–808.

61. The fourteenth-century *experimenta* of Arnald of Villanova present both formats: the simple recipe and the recipe cum case narrative. See the texts reported in appendix to Michael McVaugh, "The Experimenta of Arnald of Villanova," *Journal of Medieval and Renaissance Studies* 1 (1971): 107–18.

62. Amatus Lusitanus, *Index Dioscoridis. En, candide lector, historiales Dioscoridis campi*

(Antwerp, 1536); see Gianna Pomata, "*Praxis historialis*: The Uses of *Historia* in Early Modern Medicine," in Pomata and Siraisi, *Historia*, 123–24.

63. On the history of the commentary, see Marie-Odile Guilet-Cazé, ed., *Le commentaire entre tradition et innovation* (Paris: Vrin, 2000).

64. As Brian Nance has noticed, the authors of *observations* "took their own practice seriously enough to make written accounts of their cases, rather than a classical text, the subject of their own learned commentary": "Wondrous Experience as Text: Valleriola and the *Observationes Medicinales*," in *Textual Healing: Essays on Medieval and Early Modern Medicine*, ed. Elizabeth Lane Furdell (Leiden: Brill, 2005), 101–18, on 115.

65. To my knowledge, Antonio Benivieni's *De abditis ac mirandis morborum et sanationum causis* (Florence, 1507) is the only example of a case collection published before Amatus's; but his text did not include scholia, and remained therefore closer to the traditional *experimenta*. In contrast with Amatus, his contemporary Girolamo Cardano never published his *curationes* as a separate work. He first listed them in his *De libris propriis* (1557) as a continuous autobiographical narrative about a series of patients that he had successfully cured. He then revised and included them in other publications, and finally in section 3 of his *De methodo medendi* (1565) (cf. *Opera omnia*, Lyon, 1663: 7: 253–64), where they appear as a numbered series of thirty *curationes* and seven *praedictiones*. See Ian Maclean, "A Chronology of the Composition of Cardano's Works," in Girolamo Cardano, *De libris propriis*, ed. Ian Maclean (Milan: Angeli, 2004), 97–98, 103, 109.

66. Biblioteca Comunale Ariostea, Ferrara: MS Antonelli 531, *Curationes Antonij Musae Brasavoli* (hereafter BCA, MS Antonelli 131). The manuscript is in a sixteenth-century hand and one of the cures is dated 1547 (fol. 143r). For a fuller analysis, see Gianna Pomata, "Sharing Cases: The *Observationes* in Early Modern Medicine," *Early Science and Medicine* 15 (2010): 193–236, on 208–15.

67. Antonio Musa Brasavola taught practical medicine in the Studio of Ferrara since 1541: *Dizionario biografico degli italiani* (Rome, 1972), 14: 51–52; Vivian Nutton, "The Rise of Medical Humanism: Ferrara, 1464–1555," *Renaissance Studies* 11 (1997): 11–16. Amatus lived in Ferrara from 1540 to 1547. He repeatedly referred to his friendship with Brasavola: see, for instance, Amatus Lusitanus, *Curationum medicinalium centuriae duo* (Paris, 1554), 40v; and idem, *In Dioscoridis Anazarbei de medica materia libros enarrationes eruditissimae* (Venice, 1557), 14.

68. BCA, MS Antonelli 131, marginalia to fol. 157r. The references are to Giovanni Manardi's *Epistolae medicinales* (1528), Nicolò Massa's *Epistolae medicinales* (1550), and Giovan Battista Da Monte's second centuria of *Consultationes* (1558). Given the date of publication of the last of the three works, these notes appear to have been a later addition to the manuscript. They are, however, in the same hand.

69. On the relevance of note taking for the history of cognitive practices, see Lorraine Daston, "Taking Note(s)," *Isis* 95 (2004): 443–48.

70. See Giovanni Battista Da Monte, *Consultationum medicinalium centuria secunda . . . His accesserunt Curationes febrium Montani*, ed. Johann Crato von Crafftheim (Venice, 1558), 511–605. Da Monte's *curationes febrium* are twenty-two cases, some of them fictitious. The instructions for writing the *historia* of the case are at 542–43. On Da Monte's innovative teaching, see Jerome J. Bylebyl, "Teaching 'Methodus Medendi' in the Renaissance," in *Galen's Method of Healing*, ed. Fridolf Kudlien and Richard J. Durling (Leiden: Brill, 1991), 157–89, which should be supplemented with Giuseppe Ongaro, "L'insegnamento clinico di Giovan Battista da Monte (1489–1551): una revisione critica," *Physis* n.s. 31 (1994): 357–69.

71. See Iain M. Lonie, "The Paris Hippocratics," in *The Medical Renaissance of the Sixteenth*

Century, ed. A. Wear, R. K. French, and I. M. Lonie (Cambridge: Cambridge University Press, 1985), 169–74, with special reference to Guillaume de Baillou's work.

72. Da Monte's *curationes* were published posthumously by his former student Johann Crato von Crafftheim: see above, n. 70. The cases of the "Paris Hippocratic" Guillaume de Baillou were published long after his death: G. Ballonius, *Epidemiorum et ephemeridum libri duo*, ed. M. Jacob Theuart (Paris, 1640).

73. The *Centuriae* were republished, in their entirety or in part in 1570, 1580, 1620, and 1628: see the list of editions in Maximiano Lemos, *Amato Lusitano. A sua vida e a sua obra* (Porto: E. Tavares Martins, 1907), 200–3.

74. François Valleriola, *Observationum medicinalium libri sex* (Lyon, 1573); Rembert Dodoens, *Medicinalium observationum exempla rara* (Cologne, 1581); Pieter van Foreest (= Forestus), *Observationum et curationum medicinalium libri XXXII* (Antwerp, 1584–1609). In these collections, each *observatio* is structured after the model of Amatus's *curatio*, that is, the case narrative is followed by the scholion (Valleriola calls it *explicatio*).

75. See Dominique de Courcelles, ed., *La* Varietas *à la Renaissance* (Paris: École des chartes, 2001), esp. Jean-Marc Mandosio, "La 'docte variété' chez Ange Politien," 33–42; and Marie-Dominique Couzinet, "La variété dans la philosophie de la nature: Cardano, Bodin," 105–18.

76. Sanches, *That Nothing Is Known* (ed. Limbrick, cited above, n. 7), 117–18 (Latin text). The translation is mine; cf. the translation at 213–14 in the Limbrick edition.

77. Both Brasavola and Da Monte, like other protagonists of the medical Renaissance, lectured extensively on *Epidemics*. Brasavola's lecture notes (from 1544–45 and 1550) are extant in manuscript: BCA, Ferrara, MS I, 112 n. 46: "Commentaria in libros Hippocratis de morbis popularibus." Da Monte's lectures on *Epidemics* were recorded by his students and published posthumously under his name as *In tertiam primi Epidemiorum sectionem explanationes*, ed. Valentinus Lublinus (Venice, 1554). The literature on the structure of the case history in *Epidemics* is vast. For an excellent synthesis, see Cristina Álvarez Millán, "Graeco-Roman Case Histories and Their Influence on Medieval Islamic Clinical Accounts," *Social History of Medicine* 12 (1999): 19–43. On the Hippocratic model of the case history in the Renaissance, see Pomata, "*Praxis historialis*," 124–27.

78. De Gorris, *Definitiones medicae*, cited above n. 14.

79. See François Valleriola, *Observationes medicinales* (Lyon, 1573), preface. On Valleriola's *Observationes*, see Nance, "Wondrous Experience," cited above n. 64.

80. This is also the case of most authors of *observationes* published in the first half of the seventeenth century. On the sociology and geography of the *observationes*, see Pomata, "Sharing Cases," 226–30. H. C. Erik Midelfort has pointed out that most of the collections of *observationes* were published in Germany, not in Italy or France, the sites of the most prestigious and long-established universities: *A History of Madness in Sixteenth-Century Germany* (Stanford: Stanford University Press, 1999), 165–66.

81. See Pomata, "Sharing Cases," 226–28.

82. See above, n. 72. Francisco Sanches's *Observationes in praxi* were published by his relatives: *Opera medica*, ed. Dionisio Sanches and Guillermo Sanches (Toulouse, 1636). Petrus Matthaeus Rossius's *Observationes medico-chirurgicae et practicae* (Frankfurt, 1608) were published by his son Viktor Ross. Some of Brasavola's *curationes*, recorded in Ferrara in the 1540s, were printed over half a century later in Germany by Johannes Wittich, a physician and the son of a physician, who found them among his father's manuscripts. Wittich Senior had studied in

Italy and had brought back to Germany notes of the *curationes* of Brasavola and other Italian doctors. Wittich Junior published these *curationes* with those of Elideo Padoani, who had been his father's primary teacher. See Elideo Padoani, *Processus, curationes [et] consilia in curandis particularibus morbis quae prosperos habuerunt eventus*, ed. Johannes Wittich (Leipzig, 1607), 3–24, 282–84. These *curationes* are very similar to the manuscript of Brasavola's pupil described above.

83. Petrus Severinus, *Idea medicinae philosophicae* (Basel, 1571), dedicatory letter to Frederick II of Denmark. On Severinus, see Jole Shackelford, *A Philosophical Path for Paracelsian Medicine: The Ideas, Intellectual Context and Influence of Petrus Severinus, 1540–1602* (Copenhagen: Museum Tusculanum Press, 2004).

84. This work came out in Basel and Freiburg 1584–97 and was reprinted in 1600, 1604, 1609, 1644, and 1665 (with additions by L. Strauss): see Lipenius, *Bibliotheca realis medica*, 309. The word *paratērēseis* (plural of the Greek *paratērēsis*, observation) was added to the title in 1609 and kept in the following editions. On Schenck von Grafenberg (1530–98), a town physician in Freiburg, see August Hirsch, *Biographisches Lexicon der Hervorragenden Ärzte* (Vienna and Leipzig, 1884–88), s.v. "Schenck von Grafenberg, Johann."

85. Schenck's work is a *florilegium*, that is, an anthology, but he identified and quoted the authors of the *observationes* with meticulous care, in contrast with medieval *florilegia*, which often effaced the name of the author excerpted. On this feature of medieval *florilegia*, see Faith Wallis, "The Experience of the Book: Manuscripts, Texts, and the Role of Epistemology in Early Medieval Medicine," in *Knowledge and the Scholarly Medical Traditions*, ed. Don Bates (Cambridge: Cambridge University Press, 1995), 101–26.

86. Tilman Kiehne, "Die eigenen Fallbeschreibungen des Freiburger Stadtartztes Johannes Schenk" (Ph.D. diss., University of Freiburg im Breisgau, 1994). Thanks to Lorraine Daston for referring me to this work.

87. Schenck, *Paratērēseōn, sive observationum . . . volumen* (Frankfurt, 1609), preface.

88. The emphasis on rarity is typical of the early collections of *observationes*. See Pomata, "*Praxis historialis*," 131–32.

89. Valleriola, for instance, published in 1573 only sixty *observationes* but claimed he had collected six hundred in his daily practice, which he hoped to publish at some later date (Valleriola, *Observationes*, 397).

90. See Vanden Broecke, *Limits of Influence*, 187–88. *Historia* and *observatio* were interchangeable terms in the medical language of this period: see Pomata, "*Praxis historialis*," 122–37.

91. Most of his contributors were from German and Swiss towns, but he also named correspondents from Padua and Florence, who sent him the *observationes* of the "Italian friends."

92. See Giuseppe Olmi, "Molti amici in vari luoghi: studio della natura e rapporti epistolari nel XVI secolo," *Nuncius* 6 (1991): 3–31; Candice Delisle, "The Letter: Private Text or Public Place? The Mattioli-Gesner Controversy about the *aconitum primum*," *Gesnerus* 61, nos. 3–4 (2004): 161–76; Florike Egmond, "Clusius and Friends: Cultures of Exchange in the Circles of European Naturalists," in *Carolus Clusius: Towards a Cultural History of Renaissance Naturalists*, ed. F. Egmond, P. Hoftijzer and R. Visser (Amsterdam, 2007), 9–48.

93. On this genre, see Ian Maclean, "The Medical Republic of Letters before the Thirty Years War," *Intellectual History Review* 18 (2008): 15–30.

94. Paracelsus had planned to put together an atlas of the different kinds of tartar and tartar-related diseases by collecting descriptions of each region's tartar from the local physicians. See

Paracelsus, *Das Buch von den tartarischen Krankheiten* (1537–38), in *Sämtliche Werke*, ed. K. Sudhoff and Wilhem Matthiessen (reprint, Hildesheim: Olms, 1996), 11: 26–27. Nothing came out of this project, but it suggests a new trend toward collective medical inquiry.

95. See Ellis Jones, "The Life and Works of Fabricius Hildanus," pts. 1 and 2, *Medical History* 4, no. 2 (1960): 112–34; 4, no. 3 (1960): 196–209, on part 1, p. 121. Lazare Rivière, *Observationes medicae et curationes insignes quibus accesserunt observationes ab aliis communicatae* (London and Paris, 1646), published together with his own the observations of nine of his colleagues.

96. "Observatorum feremus per secula nomen/Semper enim hoc nomen semperque haec scripta manebunt." The author of the verses was the physician Martin Holtzapfel. See Schenck, *Paratērēseōn*, prefatory *epigrammata*, n.p.

97. For a list of *observationes* published in this period, see Pomata, "Sharing Cases," appendix, 232–36.

98. *Epistola invitatoria*, in *Miscellanea curiosa, sive ephemerides medico-physicae Germanicae Academiae Naturae Curiosorum*, Annus secundus, 1671 (Frankfurt and Leipzig, 1688), 1–6. On the Academia and its publishing program, see Daston, in this volume, p. 84. Another medical journal that adopted the *observationes* format was the Danish *Acta Medica Hafniensia*, founded in 1671.

99. See Georg Hieronymus Welsch and Johan Michael Fehr, *Epistolae mutuae Argonautae ad Nestorem et Nestoris ad Argonautam de thesauro experientiae medicae* (Augsburg, 1677). In this exchange of letters, Welsch sent to Fehr (one of the founders of the Academia) a sample of his *thesaurus* of medical experience, envisioned as a medical encyclopedia under whose headings would be collected all the observations by physicians of all times. This ambitious plan was never realized.

100. Georg Hieronymus Welsch, *Sylloge curationum et observationum medicinalium* (Augsburg, 1667). The other authors were Marcellus Cumanus (second half of the fifteenth century), Jeremias Martius (d. 1585), Achilles Gasser (1505–77), Johannes Udalrich Rumler (seventeenth century), and Hieronymus Reusner (b. 1558).

101. Welsch later claimed he published two thousand *observationes* from his collection of medical manuscripts: Georg Hieronymus Welsch, *Curationum exotericarum Chiliades II . . . nunc primum ex Mss. Editae* (Ulm, 1676). His description of his manuscript collection is appended to Theodor Jansson ab Almeloveen, *Bibliotheca promissa et latens* (Gouda, 1688), 73–132. Similarly, in the first half of the eighteenth century, the founder of the journal *Commercium litterarium ad rei medicinae et scientiae naturalis incrementum*, the Nuremberg doctor Christoph Jakob Trew, amassed a huge collection of letters by physicians and scholars from the sixteenth century to his times. The collection, published in microfiche by Harald Fischer Verlag, is available online: http://www.haraldfischerverlag.de/hfv/trew_briefe_engl.php.

102. Tycho and Kepler's work bears ample traces of a growing network of exchange of astronomical *observationes*. See Tycho Brahe, *Epistolae astronomicae* (Nuremberg, 1601); and Kepler's correspondence in his *Gesammelte Werke* (Munich: Beck, 1945–49), vols. 14–17. The observations of Jeremiah Horrocks for the years 1636–40, contained in his letters to his friend and fellow astronomer William Crabtree or left in manuscript, were published after Horrocks's untimely death: Jeremiah Horrocks, *Opera posthuma* (London, 1673), 347–439. See Allan Chapman, "Jeremiah Horrocks, William Crabtree and the Lancashire Observations of the Transit of Venus of 1639," *Proceedings of the International Astronomical Union* (2004): 3–26.

103. On Tycho, see Shackelford, *A Philosophical Path*, cited above n. 83, 75–84.

104. See Olmi, "Molti amici in vari luoghi"; Pomata, "Sharing Cases," 222.

105. Peiresc's omnivorous observational interests are especially striking. See Peter N. Miller, "Description Terminable and Interminable: Looking at the Past, Nature, and Peoples in Peiresc's Archive," in Pomata and Siraisi, *Historia*, 355–99.

106. *Epistola invitatoria*, cited above n. 98, 8.

107. On early modern "learned empiricism," see Pomata and Siraisi, introduction, in *Historia*, 17–28.

108. Daston, in this volume, pp. 83–85.

109. Pomata, "Word of the Empirics."

110. On the concept of *phainomena* in ancient Skepticism, see *Historisches Wörterbuch der Philosophie* (Darmstadt: Wissenschaftliche Buchgesellschaft, 1971–2007), vol. 6, cols. 463–64, s.v. "Phänomen." The *Oxford English Dictionary* documents the progressive extension of the term from astronomical to all natural processes. The lemma is absent in the seventeenth-century philosophical dictionaries of Goclenius and Micraelius, but it appears in Goclenius's *Lexicon philosophicum graecum* (1615): "Phaenomenon est per se manifestum sensibus" [phenomenon is what is in itself manifest to the senses]. Chauvin's *Lexicon philosophicum* (1713) documents the wider acceptation of the word: "phenomenon dicitur illud omne, quod in corporibus sensu percipitur" [It is said phenomenon all that is perceived in bodies by means of the senses]. On the term's early modern semantic history, see Gabriele Baroncini, *Forme di esperienza e rivoluzione scientifica* (Florence: Olschki, 1993), 116–23.

111. Baroncini, *Forme di esperienza*, 114–15, 125. In his lexical annotations to his translation of Sextus, Estienne devoted several pages to the term *phainomena*. See Françoise Joukovsky, "Le commentaire d'Henri Estienne aux Hypotyposes de Sextus Empiricus," in *Henri Estienne* (Paris: Centre V. L. Saulnier, 1988), 129–45, on 133–37.

112. Jean De Gorris, *Definitiones medicae*, cited above, n. 14, 49v: "Vocabulum est Empiricae sectae proprium, quo significabant memoriam eorum quae proprio intuitu unusquisque inspexit" [It is a word belonging to the Empirical sect, by which they used to signify the memory of those things which someone saw with his own eyes]. The definition is reproduced verbatim in Bartolomeo Castelli, *Lexicon medicum greco-latinum* (Venice, 1642), s.v. "*observatio.*"

113. "Per se inspectio" in Niccolò da Reggio's translation of Galen's *Subfiguratio empirica* (in Karl Deichgräber, *Die griechische Empirikerschule* [Berlin and Zürich, 1965], 47); "intuitus proprius" in Brasavola's index to the Giunta edition of Galen: *Index refertissimus in omnes Galeni libros* (Venice, 1551), s.v. "*autopsia.*"

114. *De motu cordis* (Frankfurt, 1628), 6: "per autopsiam confirmassem"; preface to *Exercitationes de generatione animalium* (London, 1651): "per autopsiam . . . propriis oculis certior factus" (16); "ipsamque autopsiam amplectendo" (25).

115. On Harvey's "modified Aristotelianism," see Andrew Wear, "William Harvey and the Way of the Anatomists," *History of Science* 21 (1983): 223–49; Baroncini, *Forme di esperienza*, 145–73.

116. Castelli, *Lexicon medicum* (Nuremberg, 1688): "vocabulum erat olim Empiricae sectae proprium . . . Verum & hodieque ad Medicinam Dogmaticam, vel Rationalem summe necessaria est Autopsia." Emphasis added.

117. Ramus, *Scholae in liberales artes* (reprint, Hildesheim: Olms, 1970), cols. 257–58, 318.

118. Johann Heinrich Alsted, *Compendium logicae harmonicae* (Herborn, 1623), 1: 123 and 1: 441–42: "Aristoteli adminicula ista vocantur, αισθήσις, ιστορία, εμπειρία, επαγωγή, id est, sensus, observatio, experientia, inductio." Cf. Ramus, *Scholae*, col. 318. On Alsted, see Howard Hotson, *Johann Heinrich Alsted, 1588–1638: Between Renaissance, Reformation, and Universal Reform* (Oxford: Clarendon Press, 2000).

119. Preface to *Exercitationes*, 29: "propria experientia (ex multiplici memoria, frequenti sensatione, atque *observatione diligente* acquisita)." Emphasis added. The general value he attributes to *observatio* is apparent from the assertion that "in every discipline diligent observation (*diligens observatio*) is required" (23).

120. The physician Hermann Conring, for instance, equated *observatio* with *experientia* and with *historica cognitio*. See his *Introductio in universam artem medicam* (Helmstedt, 1654), 20. See also Pomata, "*Praxis historialis*," 122–37.

121. Arno Seifert, *Cognitio Historica: Die Geschichte als Namengeberin der frühneuzeitlichen Empirie* (Berlin: Duncker & Humblot, 1976); Pomata and Siraisi, introduction to *Historia*, 1–39.

122. De Gorris, *Definitiones medicae*, s.v. "*peira*": "idem vero est observatio atque experientia" [Observation and experience are one and the same thing]. The lemma *peira* was introduced in the second edition of de Gorris's work (Frankfurt, 1578).

123. In this sense, the emergence of *observatio* as an epistemic category is part of what Peter Dear has called "the emergence of the discrete experience" in early modern natural philosophy. See Peter Dear, "Jesuit Mathematical Science and the Reconstitution of Experience in the Early Seventeenth Century," *Studies in the History and Philosophy of Science* 18 (1987): 141–43.

124. See above, n. 83 for Severinus; Marie Delcourt, "Une lettre de Ramus à Joachim Rheticus," *Bulletin de l'Association Guillaume Budé* 44 (1934): 5–15 (the letter is from 1563). Cf. Ramus, *Scholae Mathematicae* (1569; Frankfurt, 1599), book 2, 46–48, 67. Ramus tried to persuade Tycho Brahe to attempt "an astronomy without hypotheses," as Tycho himself recounted: see Tycho Brahe, *Epistolae astronomicae* (Nuremberg, 1601), 60, letter to Christopher Rothmann, 20 January 1587.

125. On the parallel between "the observational method" as applied to texts and the same method as applied to natural phenomena see Dirk van Miert, "Philology and the Roots of Empiricism: Observation and Description in the Correspondence of Joseph Scaliger (1504–1609)," in *Observation in Early Modern Letters, 1500–1650*, ed. Dirk van Miert, Warburg Institute Colloquia, forthcoming. See also Miller, "Description," in Pomata and Siraisi, *Historia*, 355–98.

126. See Christopher M. Johns, "Art and Science in Eighteenth-Century Bologna: Donato Creti's Astronomical Landscape Paintings," *Zeitschrift für Kunstgeschichte* 55 (1992): 578–89.

3

The Empire of Observation, 1600–1800

LORRAINE DASTON

By circa 1600, as the previous essay by Gianna Pomata shows, observation had become an epistemic genre, especially among astronomers and physicians but also among jurists and philologists: an increasing number of book titles proudly announced their contents as "observations," understood as the results of empirical inquiry. Characteristic of the emergent epistemic genre of the *observationes* was, first, an emphasis on singular events, witnessed firsthand (*autopsia*) by a named author (in contrast to the accumulation of anonymous data over centuries described by Cicero and Pliny as typical of *observationes*); second, a deliberate effort to separate observation from conjecture (in contrast to the medieval Scholastic connection of observation with the conjectural sciences, such as astrology); and third, the creation of virtual communities of observers dispersed over time and space, who communicated and pooled their observations in letters and publications (in contrast to passing them down from father to son or teacher to student as rare and precious treasures). By circa 1750, observation had also become an *epistemic category*, that is, an object of reflection that had found its way into philosophical lexica and methodological treatises.[1] Observation had arrived, both as a key learned practice and as a fundamental form of knowledge. As the Genevan naturalist Charles Bonnet wrote in 1757 to his fellow observer, Bern anatomist and botanist Albrecht von Haller: "I have often revolved in my mind the plan of a work that I would have entitled *Essay on the Art of Observing*. I would have collected as in a tableau the most beautiful discoveries that had been made since the birth of philosophy. . . . I would have demonstrated that the spirit of observation is the universal spirit of the arts and sciences."[2]

The consolidation of an epistemic genre primarily linked to astronomy and medicine in the sixteenth century into an epistemic category essential for all the arts and sciences by the early eighteenth century was the result of remarkable innovations in the making, using, and conceptualizing of observation: new instruments like the telescope and microscope; new techniques for coordinating and collating the information produced by far-flung observers ranging from the questionnaire to the synoptic map; new thinking about the relationship between reason and experience—or rather, about new forms of reasoned experience, most prominently observation and experiment. As an epistemic category, "observation" took its place among a throng of other early modern innovations in the realm of disciplined experience.[3] The most important of these was "experiment," whose meaning shifted from the broad and heterogeneous sense of *experimentum* as recipe, trial, or just common experience to a concertedly artificial manipulation, often using special instruments and designed to probe hidden causes. By the early seventeenth century, "observation" and "experiment," seldom coupled in the Middle Ages, as Katharine Park notes in her essay in this volume, had become an inseparable pair, and have defined and redefined each other ever since. In the period from the early seventeenth to the mid-nineteenth century, the relationship between observation and experiment shifted not once, but several times: from rough synonyms, as in the phrase "observations and experiments" that had become current by the early seventeenth century, to complementary and interlocking parts of a single method of inquiry throughout much of the eighteenth and early nineteenth centuries, to distinct procedures opposed as "passive observation" and "active experiment" by the mid-nineteenth century. The relationship between observation and conjecture was also in motion during this period, evolving from deliberate segregation in the late sixteenth and seventeenth centuries to equally deliberate interaction by the latter half of the eighteenth century, when observation became an "art of conjecture."

The emergence of observation as a recognized form of learned experience in early modern Europe did not, however, alter a fundamental aspect of observation that had been prominent since the Middle Ages, if not earlier, and is amply documented in the other essays in part 1 of this volume: observation and observance remained tightly intertwined. Although the kinds of observances required by new contexts and modes of observation did change dramatically, observation remained a way of life, not just a technique. Indeed, so demanding did this way of life become that it threatened to disrupt the observer's other commitments to family, profession, or religion and to substitute epistolary contacts with other observers for local sociability with relatives and peers. The metaphorical "family" developed among observers in the context of the emergent epistemic genre of the *observationes* in the late sixteenth century threatened by the mid-eighteenth century to displace the observer's literal family—as when French naturalist Louis Duhamel du Monceau depleted not only his own fortune but that of his nephews on scientific investigations.[4] By the late seventeenth century, the dedicated scientific observer who lavished time and money on eccentric pursuits was a sufficiently distinctive persona in sophisticated cultural capitals like London or Paris to be ridiculed by satirists and lambasted by moralists.[5] In the course of the seventeenth and eighteenth centuries, scientific observation was theorized and practiced, disseminated and celebrated with missionary-like enthusiasm, as its adherents opened up a veritable empire of observation.

Observation and Experiment

How did the term "observation" broaden its meaning and significance to become an essential aspect of both the theory and practice of natural knowledge by the late seventeenth century? The obverse of this question is how the widely diffused, all-purpose word "experiment" during the same period narrowed its scope to denote a carefully designed human intervention into the ordinary course of nature. Although Francis Bacon's own vocabulary did not fix the fluid meanings of *observatio*, *experimentum*, and *experientia*,[6] avowed Baconians played a key role in the rise of the terminology of observation and experiment in mid-seventeenth-century scientific circles. The academies (and some private groups, such as the circle around Samuel Hartlib in London[7]) founded in northern Europe during the middle decades of the seventeenth century, in imitation of earlier Italian academies like the Roman Accademia dei Lincei (established 1603), seem to have provided the crucible that fused the Baconian program for a natural philosophy grounded in an enlarged and improved natural history with the earlier medical project of collecting *observationes*.

The earliest of these transalpine academies, the Academia Naturae Curiosorum (Academy of Those Curious about Nature; later known as the Leopoldina) established in the imperial city of Schweinfurt in 1652 by a handful of German physicians, was perhaps the clearest example of this fusion.[8] In the late 1660s the officers of the Academia Naturae Curiosorum issued an invitation to the "learned all over Europe" to submit their "observations and experiments" on anything "rare and hidden in physic or medicine" to be collected and published with the names of the contributors in an annual volume, variously known as the *Ephemerides* or the *Miscellanea curiosa*, with the academy's imprimatur.[9] The early volumes reported on the activities of sister academies in Florence, London, and Paris (which were sent copies), and Bacon's House of Salomon, an imagined institution for lavishly funded scientific research, was explicitly held up as a model.

Although the Academia Naturae Curiosorum's official focus was on medicine, it self-consciously emulated the Royal Society of London (established 1660) and the Paris Académie Royale des Sciences (established 1666) in the form and scope of its publications: short firsthand reports submitted by—as the preface to the 1669 volume of the *Philosophical Transactions* put it—"all Ingenious Men, and such as consider the importance of Cementing Philosophical Spirits, and of assembling together Ingenuities, Observations, Experiments and Inventions, scattered up and down the World;. . . ."[10] As the wording of this invitation suggests, the vocabulary of the *Philosophical Transactions* was not as influenced by the medical model of *observationes* as that of the *Miscellanea curiosa.* In a 1665 letter to Breslau physician and Academia Naturae Curiosorum member Philip Jacob Sachs von Lowenhaimb, Henry Oldenburg, editor of the *Philosophical Transactions* and secretary to the Royal Society, emphasized the Society's more sweeping ambitions: "I understand that your Academy is composed of medical men only . . . But our Society, aiming at other things, is composed of men of all ranks who are distinguished in letters or by their experience [*tum literis tum experientia*], and enrolls mathematicians, physicists, mechanicians, physicians, astronomers, opticians, etc. It is about to reconstruct philosophy, not as it pertains to medicine alone, but as it concerns all that pertains to the usefulness and convenience of human life . . . to this end it is busy with nothing so much as building up a store and treasury of observations and experiments [*Observationum et Experimentorum*]."[11]

The titles of the articles published by the *Philosophical Transactions* in its first decades reflect this broader constituency and less-specialized vocabulary: many but by no means all titles relating some event or object investigated firsthand contained the word "observation" (and variants such as "observables"); of these, only some followed the medical format of numbered items. Yet these articles nonetheless bear witness to a meaning of the term "observation" that had at once expanded and sharpened: "observations" on everything from may dew to silkworms joined examples in astronomy and medicine, but the sense of "observation" in the late seventeenth-century context was now explicitly linked to *autopsia*, as opposed to remarks upon someone else's observations or hypotheses, which were designated as "considerations" or "animadversions."[12]

A parallel consolidation of term and meaning appears to have taken place in the annals of the Paris Académie Royale des Sciences in the 1660s and

1670s. Like the Royal Society of London, the Paris Académie aimed to be more comprehensive in its membership and inquiries than the medical Academia Naturae Curiosorum. But as in the case of the Royal Society, medical men were prominently represented among its members and correspondents.[13] Several of the works, including books and especially pamphlets, published under its auspices during this period, are presented as "observations."[14] After the *Histoire et mémoires de l'Académie royale des sciences* began regular publication from 1699 on, "diverse observations," under which individual short observations were presented in numbered lists with the names of the observers, became a regular feature of the *Histoire* section.[15] The term was used often in the manuscript minutes of the Académie from the pre-1699 period, almost always where astronomical or meteorological information was presented, frequently for anatomical reports and occasionally for accounts of botanical, chemical, and physical phenomena.[16] In all cases, *observation* denoted a firsthand report in which the time and place were scrupulously noted. Even those observations that were not presented in a numbered list, after the fashion of the medical *observationes*, were of well-circumscribed objects or events, including those observations that were routinely repeated (e.g., daily thermometer and barometer readings). By the turn of the eighteenth century, "observation" had become an essential practice in almost all of the sciences, not just astronomy, meteorology, and medicine—and the complement and supplement of "experiment."

In Latin and in the vernacular, the terms *experientia/experimentum* appear to have undergone an analogous focusing in the latter half of the seventeenth century, which fixed their meanings well into the eighteenth century. In the medieval period through the early seventeenth century, these words were often used interchangeably, covered a broad range of empirical procedures ranging from experience in general to the artisanal trial or medical recipe, and occurred with considerably greater frequency than *observatio* and its variants,[17] at least in texts about natural knowledge. Probably the most celebrated seventeenth-century use of the word *experimentum*, Bacon's *experimentum crucis* that decided between rival hypotheses, was introduced in the context of a sifting and comparison of observations.[18] English natural philosopher Robert Hooke, for example, perpetuated this sense when in 1679 he described the observation of stellar parallax as the *experimentum crucis* with which to test the Copernican hypothesis.[19]

Yet in the *Novum organum* (1620) and especially in his histories of various natural phenomena, Bacon occasionally and consequentially became more specific in his usage: *experimentum* referred to a deliberate manipulation that would shed light on causes inaccessible to the unaided senses and intellect, not just produce an effect. In addition to exhorting natural philosophers to pay greater heed to "all the experiments [*experimenta*] of the mechanical arts and all the operative parts of the liberal [arts],"[20] he proposed several specific "experiments" of his own, for example, regarding the rarefaction and compression of air, described in considerable detail: "We took a glass egg, with a small hole at one end. . . ."[21] These were "artificial experiments," as opposed to those provided by the ordinary course of nature, and imitated nature's "sports and wantonings": for example, gunpowder was an "artificial experiment" that explained the cause of lightning.[22] In explicit contrast to the trials of the workshop or the marvels of nature, these Baconian operations on nature were to be first and foremost experiments of "light" rather than of "fruit": only once nature had been understood could it be commanded.[23]

What Bacon called "artificial experiments" became the model for "experiment" tout court by circa 1660. The language of artifice, intervention, manipulation, demonstration (both in the sense of proof and spectacle), and causal inquiry increasingly defined the *experimentum* (known, however, as *expérience* in French and *esperienza* in Italian, a lingering echo of the medieval twins *experientia/experimentum*).[24] By the late seventeenth century, the nice-minded were drawing distinctions between *experimenta* and *observationes* on the basis of whether one intervened in the course of nature to produce an effect or studied effects as they occurred in the course of nature: according to German natural philosopher Gottfried Wilhelm Leibniz, "there are certain experiments that would be better called observations, in which one considers rather than produces the work."[25] Other distinctions emphasized that observation examined nature as presented to the senses (with or without the aid of instruments), while experiment revealed hidden effects or causes.[26] By the mid-eighteenth century, usage in English, French, and German had crystallized around some form of this distinction.

The terms nonetheless remained intertwined, if distinct, throughout the eighteenth century, as countless titles of the form "Observations and Experiments" testify. In 1756, French mathematician and *philosophe* Jean Le Rond d'Alembert characterized the interaction between observation and experiment as a never-ending loop: "Observation, by the curiosity it inspires and the gaps that it leaves, leads to experiment; experiment returns to observation by the same curiosity that seeks to fill and close the gaps still more; thus one can regard experiment and observation as in some fashion the consequence and complement of one another."[27] The English natural philosopher Joseph Priestley, author of one of the most celebrated eighteenth-century collections of "observations and experiments," similarly emphasized how experiments ramified into observations, which in turn led to new experiments, yielding further observations, stoked by endless curiosity.[28] Although various eighteenth-century accounts valorized either one of the terms at the expense of the other, almost all viewed the two forms of inquiry as working in tandem.[29]

Coordinated Observation

Since ancient times, observation had been understood as collective, as the slow accumulation of anonymous observations over generations, centuries, even millennia. But when observation was reconceived in early modern Europe as the province of doctors, scholars, naturalists, and other literate elites, the nature of that collectivity changed radically: authored observations were systematically made and recorded, exchanged in letters, published in books, and gathered by individuals, governments, mercantile corporations, and scientific societies. Some of these new collectives of observers were informal, albeit crucial to the development of sciences like botany: adopting the epistolary habits of Renaissance humanists, learned naturalists such as Conrad Gessner in sixteenth-century Zurich or Carolus Clusius in Leiden exchanged observations (both in word and image) just as they exchanged specimens and seeds of plants.[30] By the late seventeenth century such letters were sent to and solicited by the editors of learned journals such as the *Philosophical Transactions* and the *Miscellanea curiosa*, who transformed them into the first scientific articles simply by deleting the opening greetings and concluding compliments.[31] But such publications did not replace the personal correspondence of savants, which remained an important means of collectivizing observation throughout the eighteenth century and could rival the networks of major academies in their number of correspondents and geographic reach, as in the case of the Swedish naturalist Carl Linnaeus.[32]

Other early modern observer collectives were more formal and centralized, depending on paid labor and hierarchies of command rather than voluntary contributions from self-declared citizens of the Republic of Letters. The Holy See and the Spanish Council of the Indies issued voluminous questionnaires to solicit the observations of missionaries and colonial administrators, respectively, in foreign lands; trading companies such as the Dutch East India Company instructed their functionaries to file detailed reports on their travels.[33]

Although formal and informal observer collectives were differently organized, the boundary between them was often blurred: the Royal Society resorted to questionnaires and eagerly interrogated merchants about the natural history of faraway lands; well-traveled Jesuits published accounts of their missions abroad that were reviewed in learned journals and plundered for observations; botany, imperialism, and commerce were braided together in the global trade in new pharmaceuticals; humanist travel for personal edification shaded imperceptibly into official travel in the service of the crown, deploying similar observational grids. Bacon's imagined "Merchants of Light," described in his utopian fragment *The New Atlantis* (1627), were supposed to sail the world's seas as spies in order to supply the "Interpreters of Nature" at the pinnacle of the House of Salomon with "knowledge of the affairs and state of those countries to which they were designed, and especially of the sciences, arts, manufactures, and inventions of all the world"[34]—a neat and prescient conflation of the diplomatic, mercantile, and scientific models of early modern collective observation.

The explosion of collective observational activity created a new challenge of integration: how to coordinate observers, standardize instruments and regimens, and correlate results? When observations had been rare and costly to make, as in medieval astronomy, or left uncollected and untransmitted in doctors' personal notes or individual diaries, as Katharine Park describes in her essay in this volume, or confined to local phenomena such as the weather and farming conditions, integration had posed few problems. But as observations multiplied, diversified, and diffused and the ambitions of observational programs like those of imperial powers or transcontinental trading companies swelled, ways of collecting and sorting out the results became urgently needed.

Compendia were a typically humanist response to the problem: adapting the techniques of commonplace books, erudite compilers with well-stocked libraries combed the work of ancient and modern authors to assemble thick volumes of selected, indexed observations on all manner of topics. This was the bookish method plied by medical authors such as Johann Schenck and also by naturalists such as Gessner in his *Historiae animalium* (1551–60)[35] or Bacon in his unfinished *Sylva sylvarum* (1627).[36] The collective empiricism encouraged by seventeenth-century periodicals like the *Miscellanea curiosa* and the *Philosophical Transactions* modified the humanist compendium model to solicit new observations made by named contemporaries, substituting eyewitness testimony for bookish scholarship. But the use of the library to construct series of observations, sometimes reaching back to antiquity, continued to be an important observational technique.

The limitations of compendia soon became evident, especially as methods of observation were refined and standards raised: from the standpoint of naturalists increasingly skeptical about the reliability of classical authors like Pliny,[37] observations attributed to authors of varying credibility or to no authors at all and made under diverse or unspecified conditions heaped up helter-skelter seemed unlikely to supply the solid foundation for a reformed natural philosophy.[38] Even compendia of observations freshly made by trustworthy reporters were too heterogeneous to be summed into generalizations or sifted for regularities: despite the efforts of some editors to append "scholia" or "histories" to individual observations in order to bring out their connections to other observations and larger significance, in the spirit of Bacon's "major observations,"[39] the contents of mid-seventeenth-century scientific journals remained stubbornly miscellaneous—and therefore a disappointment to those who, like Oldenburg, hoped to use them to mobilize the Republic of Letters for a program of coordinated, global observation.

Several attempts were made to counter the dispersion of observations, both before and after the fact. Since the mid-sixteenth century (and well before, in the case of Venetian ambassadors),[40] states and mercantile enterprises trained their representatives in foreign parts to observe and report according to standardized schemes: questionnaires, synoptic tables, Ramist branching charts. Observational grids ranged from curt instructions like Sir William Petty's unpublished lists ("Get the best map of the country." "The value of fruites in winter and somer.") to voluminous lists of questions like the two hundred published by the diplomat and humanist Heinrich Rantzau, which covered everything from the exact point of sunset to musical instruments to the salaries of local clergy.[41] Starting with the Swiss encyclopedist Theodor Zwinger's *Methodus apodemica* (1577), manuals aimed at scholars, young gentlemen, ambassadors, missionaries, merchants, colonial administrators, and other travelers instructed readers on what to look at and how in foreign climes.[42] By the early seventeenth century, observation had become a named practice that travelers were exhorted to cultivate, as in the revised 1630 English translation of Giovanni Botero's *Relationi universali* (1597–98), which added a section "Of Observation."[43]

The questionnaire format was adopted by the Royal Society, which eagerly sought information from travelers in order to compile its Baconian natural histories, despite the problems of verifying marvelous tales from distant lands.[44] Robert Boyle recommended the preparation of a compendium of travel reports to Oldenburg in 1666 and in the first volume of the *Philosophical Transactions* published a natural history questionnaire for any "Countrey, Great or Small."[45] In his fragmentary "The General History of the Air," Boyle had also called upon everyone "who hath leisure, opportunity, and time" to keep a diary of "his own observations of the change and alteration of the air from day to day," emphasizing the utility of such mundane "histories."[46] Instead of the questionnaire format, tables or "schemes" like that proposed by Hooke in 1663 were intended to make the weather observations sent in by correspondents all over Europe commensurable and comparable with one another.[47] In 1723, James Jurin, in his capacity as secretary to the Royal Society of London, went one step further in his Latin invitation to potential observers, offering to provide instruments and giving detailed instructions as to when, where, and how to deploy them.[48]

In such dragnet calls for observations to be sent in from far and wide, the scientific societies of the late seventeenth and early eighteenth centuries shifted the emphasis from observation as individual self-improvement, a prominent theme in earlier humanist travel guides, to observation as a collective, coordinated effort in the service of public utility. As English philosopher John Locke wrote when he published his own weather observations: "I have often thought that if such a Register as this, or one that were better contriv'd, with the help of some Instruments that for exactness might be added, were kept in every County in England, and so constantly published, many things relating to the Air, Winds, Health, Fruitfulness, & c. might by a sagacious man be collected from them, and several Rules and Observations concerning the extent of Winds and Rains, & c. be in time establish'd, to the great advantage of Mankind."[49]

Questionnaires, schemata, and instruments supplied by a central authority—princely, ecclesiastical, or scientific—aimed to press observations into a uniform grid. But these preliminary standardizing measures (which also included supervised drawings, as Daniela Bleichmar describes in her essay in this volume), even when followed scrupulously by roaming merchants, officials, missionaries, and dispersed savants, did not suffice for the smooth integration of the observations that accumulated. There were too many observations, too variously taken, and too obscurely correlated with other observations. As J. Andrew Mendelsohn documents in his essay in this volume, the predicament for the networks of seventeenth- and eighteenth-century weather watchers was dramatic: repeated efforts to sift the piles of reports in search of reliable correlations between rainfall and barometric data, wind and rain, temperature and illness, and any number of other hypotheses failed to yield the desired "Rules and Observations" of the weather.[50] Early modern statesmen were confronted with similar challenges: how to collate the stacks of reports and questionnaires sent in by ambassadors and local officials? As a late-seventeenth-century response to the problem of integrating observations by far-flung correspondents, a vogue for "synopses," "calendars," "registers," "tables," and other visual digests edged out the indices and *loci communes* devised by humanist compilers a century earlier. Tables that correlated two or more observed variables, used since ancient times in astronomy, spread in the late seventeenth and early eighteenth century to meteorology, experimental natural philosophy, and natural history.[51] In an unpublished memo probably intended for one of the rulers he served, Leibniz proposed a "State Table" that would digest all the oral and written accounts of well-traveled informants into a compact summary that the prince could "look over in a moment" and thereby grasp "the connections of things."[52]

Leibniz compared his handy table to "maps of land and sea," and one of the most successful efforts to integrate the results of collective observation was a world map showing prevailing wind patterns prepared by the English astronomer and natural philosopher Edmond Halley in 1686 (fig. 3.1). On the basis of published accounts, conversations with mariners, and his own seafaring observations, Halley discerned a few general "rules" (albeit with exceptions) in the direction of the trade winds above and below the equator and the seasons of regional storms such as Caribbean hurricanes and Indian monsoons. Like Leibniz's table, Halley's map or "Scheme" showed "at one view all the various Tracts and Courses of these Winds."[53] Although Halley's synoptic map and general explanation of global wind patterns was an all-too-rare triumph of collective observation in the seventeenth and eighteenth centuries, it can stand as an emblem for the ambitions of such programs. Inspired by Bacon's project for a "history of the winds," Halley's synthesis drew, as Bacon had hoped, on "a Multitude of Observers, to bring together the experience requisite to compose a perfect and compleat History of these Winds," including not only natural philosophers but navigators and travelers. But in contrast to Bacon's vision of a centralized, state-financed, hierarchically organized corps of observers subordinated to the "Interpreters of Nature" in Salomon's House, Halley's informants were volunteers, and he himself was a seafaring observer, a "Merchant of Light" as well as an "Interpreter of Nature." The merging of these two roles of roaming observer and discoverer of "greater observations, axioms, and aphorisms"[54] was to prove consequential for the practices of learned observation: the eye of the body and the eye of the mind had to be taught to work in harmony.

Observational Practices

By the late seventeenth century, special procedures, carried out by specially qualified people under special circumstances, distinguish the scientific observation from the all-purpose remark. At the very least, scientific observers were expected to exercise unusual care, sometimes as a group cross-checking its individual members. In his preface to the third year of the *Philosophical Transactions*, Oldenburg expressed the hope that "our Ingenious Correspondents have examin'd all circumstances of their communicated Relations,

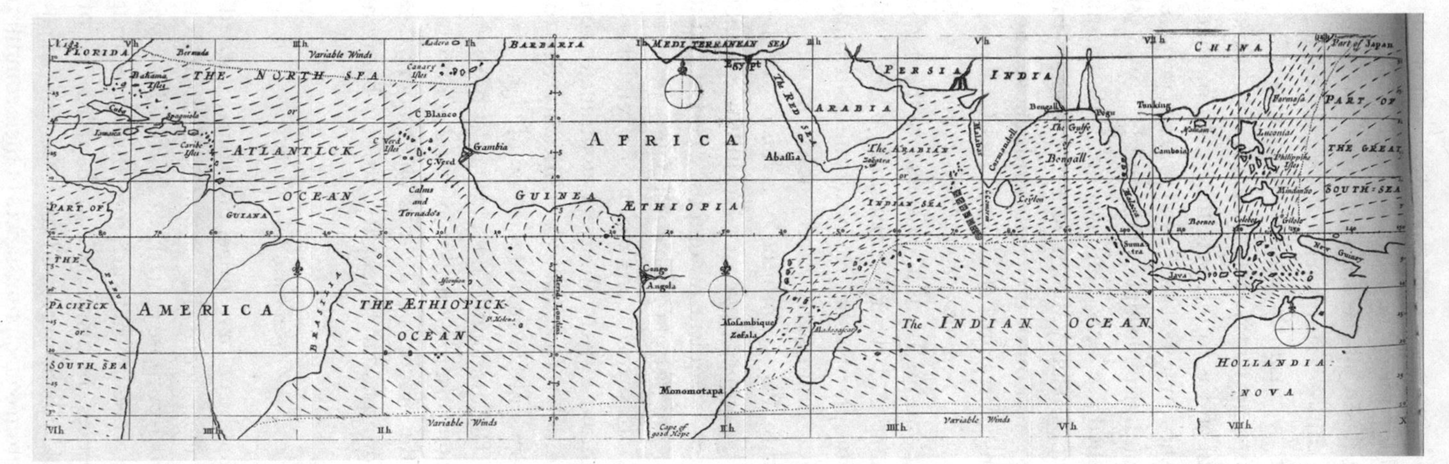

FIGURE 3.1. Map of the world winds. Edmond Halley, "An Historical Account of the Trade Winds, and Monsoons, observable in the Seas between and near the Tropicks, with an attempt to assign the Phisical cause of said Winds," *Philosophical Transactions of the Royal Society of London* 16 (1686–92): 153–68, foldout between pages 49 and 51.

with all the care and diligence necessary to be used in such Collections."[55] These sentiments were echoed in the Paris Académie's *Histoire naturelle des animaux*, which purportedly contained "no facts that have not been verified by the whole Company, composed of people who have eyes for seeing these sorts of things, in contrast to the majority of the rest of the world. . . ."[56] Scientific observers in the seventeenth and eighteenth centuries self-consciously developed novel practices that schooled perception, attention, judgment, and memory. Tools such as the notebook and the magnifying glass were enlisted in these practices. As observation became repetitive as well as collective, the challenge of synthesizing the sequence of notes made by an individual complemented that of integrating the ensemble of reports produced by a community.

REPETITION

Although sustained observation over generations had since ancient times been considered characteristic of the ways in which astronomers, farmers, sailors, and shepherds discovered regularities about the stars and the weather, regimens of repetitive observation of the same object were rare before the early modern period. The example of astronomy, as the most ancient of the observational sciences (and the one longest and most consistently associated with the term), is instructive concerning how a cumulative observational tradition became a repetitive one.

When the French astronomer Jean Picard journeyed to the Danish island of Hven in 1671 to conduct astronomical observations from the ruins of Danish astronomer Tycho Brahe's castle Uraniborg and bring back Tycho's manuscript observations to Paris, he bore witness to the strong sense of continuity that bound even the most boldly innovative early modern astronomers to their predecessors.[57] Part of the care with which late sixteenth- and seventeenth-century astronomers preserved and transposed past observations stemmed from the superhuman temporal scale along which some celestial events, such as the precession of the equinox, unfolded. Only observations carried out over centuries, and in some cases millennia, could discern and specify cycles with long periods or subtle correlations. But part of their solicitude also derived from a desire to test—not just add to—and improve upon past observations, a process that paradoxically led them first to vaunt their own advances and later to cultivate an ever more scrupulous awareness of possible sources of error. Pride in progress as well as fear of error were both tied to what was, at least in the Latin West,[58] a new practice in astronomical observation, with parallels in other early modern observational sciences: the systematic repetition of the same observation night after night, over years and decades.

The consequences of this new practice of sustained and repetitive observation (rather than at special points like quadrature or conjunction) are thrown into relief by a comparison of Tycho's methods of the 1570s and 1580s with those employed a century later. Although Tycho made his reputation with the observation of singular events such as the nova of 1572, analogous to the contemporary medical *observationes* of unusual cases, once established at his purpose-built observatory Uraniborg in the late 1570s, he began a program of sustained observation of the sun, moon, planets, and fixed stars on every clear night for over twenty-one years. His account of solar observations made clear that he was well aware of the novelty of this program: "First of all we determined the course of the sun by very careful observations during several years. We not only investigated with great care its entrance into the equinoctial points, but we also considered the position lying in between these and the solstitial points, particularly in the northern semicircle of the ecliptic since the sun there is not affected by refraction at noon. Observations were made in both cases and repeatedly confirmed, and from these I calculated mathematically both the apogee and the eccentricity corresponding to these times."[59]

Tycho's arduous, costly, decades-long regimen of observation, involving many new instruments of his own invention and of unprecedented size and accuracy, was intended to make future observations superfluous, at least in those areas to which Tycho had devoted the most time and effort. At least his mature observations, Tycho thought, were "completely valid and absolutely certain"[60] and would never need to be repeated. Yet by the 1670s, leading astronomers considered Tycho's observations insufficiently exact. As Astronomer Royal John Flamsteed wrote to Samuel Pepys in 1697 apropos of Tycho's cherished fixed-star observations, "though what he did, far excelled all that was done before him; yet it was much Short of the exactness requisite in this Business."[61] One reason why Flamsteed could pronounce Tycho's observations outdated was the introduction of telescopic sights and the micrometer, both invented circa 1640 but not put into systematic use until the 1660s. Although the telescope had been responsible for some spectacular discoveries in the hands of Galileo and others during the seventeenth century,[62] it by no means displaced sextants and quadrants; telescopic sights arguably contributed more to astronomical observations during this period than the telescope itself did, refining angular resolution to 15 seconds of arc by 1700.[63]

It was not only improvements in instrumentation that persuaded late seventeenth- and eighteenth-century astronomers that their observations were an advance on Tycho's, just as Tycho had vaunted the quality of his observations over those of all previous astronomers. The very practice of sustained, continuous observation that Tycho had institutionalized sharpened the astronomers' awareness of the possibility, perhaps the inevitability, of error. The more observations are made, the more likely it is that they will diverge from one another. Tycho's Uraniborg had been in operation for only about twenty years, but this was long enough to notice a scatter of data and to redouble vigilance to counteract the possible effects of atmospheric refraction, the sagging and stretching of heavy instruments under their own weight, jumpy clocks, and a myriad of other disturbances. But the problem did not go away, no matter how many precautions were taken. With the establishment of observatories like those in Greenwich and Paris in the late seventeenth century,[64] observations stacked up over decades and even centuries. At the Paris Observatory, Picard worried about whether the smoked glass through which the sun was observed might distort the solar diameter or whether the effects of refraction were greater in the winter than in the summer—and many other sources of minute errors.[65] By the first half of the eighteenth century, a heated debate had begun among astronomers about what to do with discordant observations. In astronomy, these were issues that were moralized, mathematized, and ultimately psychologized.[66] Despite these problems, however, by the mid-eighteenth century, all scientific observation was ideally repeated, continuous observation, in studied contrast to the singular or rare phenomena that had dominated medical and scientific collections of *observationes* a century or so earlier.

NOTE TAKING

Some form of note taking has probably since ancient times been part of taking note, of remarking, describing, and remembering—in short, of observing. But note taking itself has a history, one that was consequential for the practices of observing in the early modern period.[67] Two notebooks, one from the late seventeenth century and the other from the late eighteenth century, illustrate some of these changes.

The first was kept by Locke, from September 1666 to April 1703, and entitled "Adversaria physica," or "memoranda on physic."[68] It is a large-ish (approximately 8″ × 12″) calf-bound volume, written in ink, and continuously paginated. The entries, written in Latin, English, and French, relate mostly to medical but also to some natural philosophical matters, mingling excerpts from reading (with references), recipes for medications (e.g., Lady Chichley's eye ointment), practical tips (e.g., where to get the best French olive oil), and some of Locke's own observations, initialed "JL." At the back of the volume is a weather diary, presenting daily thermometer, barometer, hygrometer, and wind observations for a period of almost thirty-seven years (fig. 3.2) These are the only dated entries; insofar as there is another order, it is spasmodically alphabetical, with an elaborate but incomplete index at the front and back of the volume; most of the entries are flagged with a marginal keyword (e.g., "Reason," "Fulmen," "Palpitatio cordis").[69]

Locke's mélange of reading notes and observations was not exceptional in seventeenth-century commonplace books.[70] Culling facts from experience bore some resemblance to culling information and insight from books, and the commonplace books that held the latter were similar in form and aims to the lists and tables that held the former. Among the personnel in Bacon's House of Salomon there were not only "Mystery-men" who collected experiments in the mechanical arts; there were also "Depredators" who collected experiments from books.[71] The keeping of commonplace books of quotations and moral adages culled from the reading of classical authors was a pillar of early modern education in rhetoric.[72] The engrained humanist habits of excerpting, ordering, and recombining the entries of commonplace books offer a suggestive parallel for at least the recording of facts about nature, as in Locke's case. Bacon himself is alleged to have preferred the keeping of commonplace books to other forms of note taking on reading, "because they have in them a kind of Observacion."[73]

A notebook from about a century later offers a study in contrasts. On 10 July 1774 the Genevan naturalist Horace-Bénédict de Saussure began a little yellow notebook (approximately 5″ × 7″), which he labeled "Voyage autour du Mont Blanc en 1774, 10e Juil. Brouillard en crayon No.1. Extraits de l'Agenda." Each page was headed with the day of the week and the date, followed by a lettered (a, b, c, etc.) sequence of short observations, beside each of which was noted the time, often to the minute. Although Saussure recorded a terse "agenda" of the main topics to be covered by the observations on the notebook's flyleaf, he strayed from "primitive and secondary mountains" when something else caught his eye along the way: a ruined château, the strata of slate that struck him as displaced from their original position, the nickname of his local guide, barometer and thermometer readings, a terrifyingly steep mountain pass traversed in the snow in mid-July. The timed entries and the execrable handwriting suggest that the entries were made in real time, bouncing along on a bumpy mountain road. Some entries are exceptionally in ink and in a far more legible hand: "Sunday, 17 July. (a) This morning was set aside for rest or at least some observations at Cormayer. However, I was not

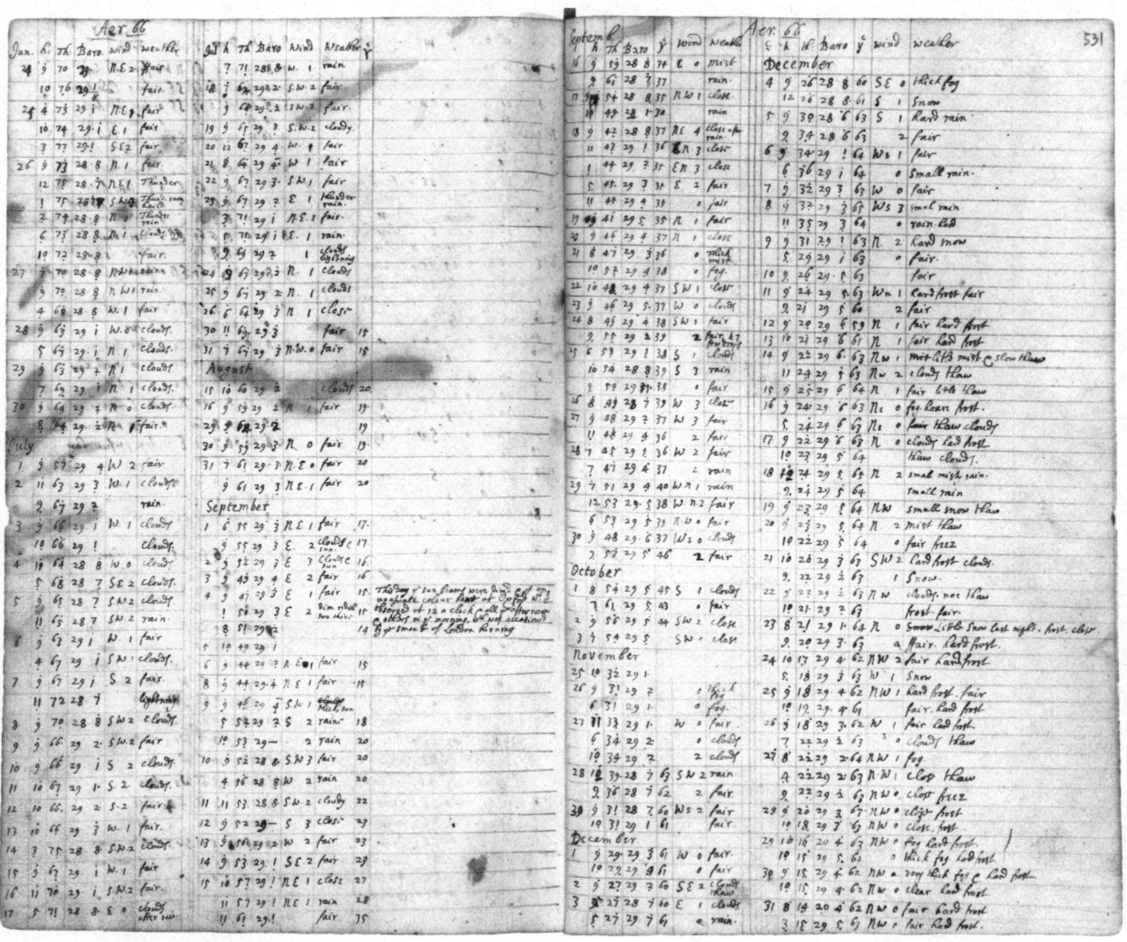

FIGURE 3.2. Weather table. John Locke, "Adversaria physica." Bodleian Library, University of Oxford, Shelfmark MS Locke d.9.

Heur. du départ | Samedi 18e. Juillet 1789 | Distances

VII 14' A de Martigny, [illegible], cotoye à d. petrosilex boisés | brisés

plus loin, cornés & porphyroïdes & petrosilex débris

58' lits horiz. d'ardoise luisante, grise, mince, non efferv. puis [illegible] A

VIII 15' Saran petit mauvais village, vignes à d, vergers & prairies | 0 | 44'

à g. en tout boisé, mais on voit [illegible] parplaces | 17

22' Saran le grand, Toujours rocs A | 7

32' à droite monticule de Gypse blanc en couche horiz. grises & blanches, Bientôt après ardoises grises, luisantes, minces, tendres, descendant un peu au Sud, *peu après quartz en parplaces, & plus dur, & sèche mais toujours finement feuilleté | 10

* à l'opposé à g. sur bras

IX Ravin calcaire, [illegible] & à droite en haut [illegible] | 28

30' Hauteur d'où l'on voit tout la vallée pré inculte | 30

50' Riddes village très couvert & boisé | 20

X 5' Pont de bois sur le Rhône | 15

26' St Pierre | 20

42' à g. pied rapide de calc. très inclinées descendant au Sud | 15'

48' Village boisé

57' [illegible] digue, torrent, [illegible] | 9'

XI 38' Pont de la Morge, Colline d'ardoise tendre | 41'

Colline semblable à d. couches très inclinées

Aussi à g. [illegible] descendant à l'Est, quartz, calcaire

mince, avec feuillets schisteux

XII 20' Sion, Repas [illegible] IV. 27' | 12'

Laissé à d. coll. calcaire de Tourbillon, couches tourmentées

[illegible] chemin entre vignes en terrasse & Rhône jaune

V 26' [illegible] Village après torrent | 39'

45. serré entre Rh. & calc. incl. desc. au Sud & couvert à l'Est

Bientôt après Glatières, Gypse blanc & gris [illegible]

& [illegible] Roche comme ferrugineuse; (Sur le chem. Chat. & vill. Granges montant. [illegible])

[illegible] ardoises & Gypses, puis schistes divers, & calcaires [illegible]

VI 30' Bas des Glatières, Sion VII. 45'

FIGURE 3.3. Pencil and ink observation notebook entries. Horace-Bénédict de Saussure, "Voyage autour du Mont Blanc en 1774, 10e. Juil. Brouillard en crayon No.1. Extraits de l'Agenda," Bibliothèque de Genève, Archives de Saussure 14/1.

at all tired from yesterday, however arduous it had been. (b) I made a trial of several rocks gathered yesterday with eau fort . . ." (fig. 3.3).[74]

This is a typical observation notebook from the latter half of the eighteenth century: pocket format, dated entries further broken down into subentries by a sequence of letters or numbers, real-time entries in pencil and retrospective entries in ink, descriptive observations interspersed with reflections, conjectures, and personal details. There are no thematic indices or reading notes. The model is the journal, more specifically, the travel journal kept en route rather than the commonplace book filled by the desk-bound scholar: Saussure's cardboard-bound notebook was small and light enough to be carried along everywhere; when Locke traveled to the Continent, he left the bulky "Adversaria physica" at home. Above all, the axis of organization has shifted from the topical to the temporal. Locke's notes were assembled with an eye to collation by subject matter; his commonplace book recycled material from old books into the stuff of new books and was itself a proper book, hefty and leather-bound; the entries (with the exception of the weather tables) are as timeless as the pages of a book. Saussure's record is in contrast driven by the calendar and his pocket watch. Time was almost always the vertical dimension of eighteenth-century tables of observation, whether the object of observation was lunar perturbations, the temperature, the incidence of smallpox, or the reproduction of aphids.

PAYING ATTENTION

For Enlightenment naturalists like Saussure and his uncle Bonnet, observing was first and foremost an exercise of attention. As Swiss Protestant minister and naturalist Jean Senebier wrote in his influential 1775 treatise *L'art d'observer*, "[a]ttention alone renders the observer master of the subjects he studies, in uniting all the forces of his soul, in making him carefully discard all that could distract him, and in regarding the object as the only one that exists for it [i.e., attention] at that moment." The peculiar economy of attention cultivated by the Enlightenment naturalists was pointillist, magnifying, and therefore deliberately repetitive. Visually and intellectually, the observer pulverized the object into a mosaic of details, focusing first on one, then another. Senebier directed the fledgling observer to compensate for the "feebleness of his soul and senses in fragmenting [*morcelant*] the subject of his observations and in studying each of its parts separately." Only the narrowness of focus could sufficiently concentrate attention to the level of intensity required for exact observations.

So pencil-thin and intense was the beam of attention that it could hardly be sustained over long periods. Hence the observer must return over and over again to the same object, picking out different details, different aspects each time, and multiply confirming what had already been observed.[75] Still better was the repetition of observations by several observers, not because the veracity of the initial observations was in doubt, but rather to widen the panorama of different perspectives on the same object. In this spirit, Bonnet urged Italian naturalist Lazzaro Spallanzani to repeat the observations of others, including his own: "Nature is so varied that we can hardly vary our attempts too much."[76] The use of microscopes and, especially, the more portable and versatile magnifying glass also tended literally to focus and circumscribe the observer's attention.[77]

SYNTHESIS AND DESCRIPTION

The result of these practices was an avalanche of descriptive detail, both visual and, especially, verbal. It was a byword among the naturalists that it was by the detail with which observations were reported that one could separate the novice from the old hand, the artisan from the savant, the bumbler from the "genius of observation."[78] The most ingenious efforts of observers were directed toward the discernment of the most fleeting details, the finest nuances. Saussure invented an instrument called the cyanometer to measure the shades of blue of the sky, ranging over fifty-three graduations, from milky white to midnight blue.[79]

No study of natural particulars could afford to become permanently mired in particulars. Bacon had feared naturalists might drown in them; Enlightenment observers gladly wallowed in them—but no one deemed them an end in themselves. The practices of taking notes and paying attention as they were cultivated during the mid- and late-eighteenth century tended to fragment the object of inquiry: numbered, dated notebook entries chopped up time into slices; narrowly focused attention dissolved wholes into tiny parts. Whereas collective observation posed the problem of the coordination of many individuals, the challenge of the practices of synthesis confronted the individual observer: how to glue all these fragments back together again into a coherent mosaic—but not in order to reconstitute the actual object of observation. Instead, the result of the synthesis was a general object—variously described by Enlightenment astronomers, anatomists, and naturalists as an archetype, an ideal, an average, or a pure phenomenon—that was more regular, more stable, more universal, more *real* than any actually existing object.[80]

Although observers were sometimes struck by singular phenomena such as an aurora borealis or a monstrous birth, by the mid-eighteenth century they attempted whenever possible to situate individual objects and events in a series. This practice had its antecedents in the longstanding astronomical practice, common since the late sixteenth century, of creating long baselines of multiple observations of the same star or planet. In other sciences of the eye, observers repeated observations of the same or similar objects in order to establish a series. Linnaeus prided himself upon having examined thousands of plant specimens, many supplied by former students dispatched to distant lands.[81] Johann Wolfgang Goethe, reflecting in 1798 on his researches in morphology and optics, described the quest for the "pure phenomenon," which can be discerned only in a sequence of observations, never in an isolated instance.[82] If such a sequence was not readily available to direct observation because of the rarity of the phenomenon, it was compiled from past records: French astronomer Louis Godin began his report to the Académie Royale des Sciences on the October 1726 aurora with a compilation of all previous such sightings, starting with Flavius Josephus in Roman times and concluding with a summary of the features common to all such cases.[83] Ideally, not only the naturalists but also their artists were supposed to be familiar with a broad range of exemplars, so that images as well as descriptions would be the distillation of not one but many individuals carefully observed.[84]

The process of how particulars were forged into generalities is most graphically displayed in the observation notebooks. Under the rubrics of "Reflections," "Results," or "Remarks" (or—in the case of Saussure—simply the shift from pencil scribblings to inky fairhand) were recorded the digestion of first impressions into second (and sometimes third) impressions. These were observations upon observations, the refinement and distillation of raw materials into what Bacon had called "vintages"—or, in his histories, "major observations."[85] Here the older Renaissance practices of humanist note taking were preserved in spirit if not in substance: what sixteenth-century scholars had done for the writings of Cicero and Livy, eighteenth-century naturalists did for oysters and aphids. A first round of observations selected the noteworthy; a second round winnowed these further by comparisons and cross-correlations, seeking patterns and regularities; a third synthesized the features now understood to be the most significant or essential into the general observation.

Observation as a Way of Life

"Never has so much been observed, as in our century."[86] By the mid-eighteenth century, observation was practiced, theorized, and celebrated in almost all sciences. But because observation and observance remained conjoined, the very success of observation, demanding ever more time and dedication from its practitioners, made it controversial as a way of life, an observance too absorbing to be easily compatible with other social, professional, and religious commitments.

Although it never ceased to be arduous, and was recognized to be dangerous at times, early modern scientific observation was seldom described as work (except perhaps by the astronomers: Flamsteed at least complained that it was "labour harder than *thrashing*"[87]). On the contrary, its delights had become so intoxicating they verged on obsession. In a time when travel was fraught with hardship and peril, Clusius could write of the "great pleasure" his wanderings to "observe plants" had provided.[88] Observation obliterated fear and even pain: when in 1770 the Genevan savant André Deluc, armed with thermometer, hygrometer, and barometer, set out to explore the peaks and glaciers of the Alps, he had a foolproof remedy for vertigo: "There is not the slightest danger for those who do not perceive the increase in height, except by a sort of pleasant sensation, which occurs when one is not afraid, and by the pleasure of continually discovering new objects."[89] Leaves dismembered under the microscope, an aurora borealis spotted after many nights' vigil, thermometer readings faithfully registered in the chill dawn, every day for decades on end—these were pleasures of "discovering new objects" evidently so intense that they tempted Enlightenment naturalists to defy parental counsel, neglect civic duties, and deplete family fortunes.

Although moralists were critical of naturalists who sacrificed their families and their health to a demanding regimen of observation,[90] the naturalists themselves could at least count upon the sympathy of their colleagues, with whom they were in constant, copious, and often commiserating correspondence. Although the publication of observations had become increasingly common by the late seventeenth century, the format by which even printed observations were first communicated was the letter. The Sicilian naturalist Paolo Boccone, for example, chose to publish his observations on coral in 1674 as a series of letters to named correspondents scattered across Europe: the Avignon doctor Pierre Guisony, the Pisa professor of mathematics Alexander Marchetti, the London Fellows of the Royal Society Hooke and Nehemiah Grew.[91] Naturalists had been exchanging observations and specimens among themselves since the sixteenth century, a practice that by the early seventeenth century had cemented a strong sense of community among them.[92]

It was in correspondence that fledgling naturalists apprenticed themselves to recognized masters, as the young Bonnet did to the French naturalist René-Antoine Ferchault de Réaumur and Spallanzani in his turn did to Bonnet, presenting their most precious observations for comment and approval. Observations presented in correspondence were also the way in which one naturalist took the measure of distant colleagues: an insufficiently circumstantial or detailed observation report reflected badly on its author—and conversely. Réaumur, for example, accepted La Hague naturalist Pierre Lyonet's corrections to his own observations on the generation of aphids, bowing to the talents of a master observer and draughtsman: "The figures you have sent me are drawn with such a great air of truth that I believe them to be very perfectly conformable to nature." And in their letters naturalists cheered each other on, comrades in a fellowship scorned by outsiders: when Lyonet was too downhearted to continue to observe insects after his proposal of marriage had been thwarted at the last minute by "a most strange caprice" of the lady's mother, Réaumur remonstrated with him not to give up on insects: "[I]t would be a great pity if you became indifferent to them [insects]; they will not fail to repay the attention you have given them with new marvels that they will make you see. I plead for my good friends."[93]

The sociability of specialized correspondence substituted for the more usual sort, since the demands of strict regimens of observations, like those of religious observances, clashed with those of friends and family. The astronomer Picard, for example, rose at 5:30 a.m. and observed with at most a break of an hour or two until midnight, beginning anew at 5:30 the next morning.[94] By the mid-eighteenth century, observant gentlemen all over Europe were interrupting their daily routines to take thermometer and barometer readings to record in diaries and journals.[95] Weather watching, especially if pursued at fixed times of day, could become a way of life, a regimen that set schedules, shooed guests to the door, and fostered clock consciousness. Tycho contemplated a move to Basel because there, close to France, Germany, and Italy, "it would be possible by correspondence to form friendships with distinguished and learned men in different places," whereas on his property in Knudstrup "a continuous stream of noblemen and friends would disturb the scientific work and impede this kind of study."[96] Réaumur moved out of central Paris to have more room for his beehives and fewer visitors—and where, as Mary Terrall shows in her essay in this volume, he could interweave observation regimens with household routines.[97] For the dedicated observer, normal social life became all but impossible. In his *Traité de météorologie* (1774), the Oratorian and corresponding member of the Académie Royale des Sciences Louis Cotte admitted that the perfect weather observer would have to "renounce almost all other business and every pleasure. Not only would he have to live for years on end in the same place; he would have to be home regularly every day for the hours of his observations."[98]

A shared commitment to observation could, however, forge as well as sever social bonds, even surmounting other barriers to friendly contact. When a group of French Jesuits landed at the Cape of Good Hope on their way to Siam in June 1685, they were greeted with suspicion by the Reformed Dutch colonists, who suspected one of the Jesuits' microscopes (draped with an ornamental cover) of being an outlawed Eucharist chalice "because . . . you are the greatest enemies of our religion." Yet their Dutch hosts were pleased to "lead the life of an observer with us" when the Jesuits measured the longitude by following the satellites of Jupiter, and both parties parted on the warmest of terms, the Jesuits presenting the Dutch with a microscope and a small burning mirror in exchange for gifts of tea and wine.[99] Despite the criticisms of moralists and the warnings of physicians, observers were not so much antisocial as highly selective about the company they kept: although they went to considerable lengths to evade conventional social obligations, they craved contact with other observers, if only by letter. Observation was a solitary and obsessive but also communal pursuit.

Conclusion: Observation as a Way of Reasoning

By the late eighteenth century, the relationship between observation and conjecture had taken yet another turn. As we have seen in chapters 1 and 2, medieval natural philosophers associated observation with conjecture because its results were uncertain, confined to particular instances, and mute concerning causes, while early modern physicians had prized observations just because they were allegedly divorced from foolhardy conjecture and system spinning. In this spirit the Roman professor of medicine Giorgio Baglivi recommended observation as an antidote to "the ardent and eager pursuit of new Hypotheses."[100] But in the course of the eighteenth century, observation became a tool of conjecture, a way of excluding some explanatory hypotheses and hatching new ones, which could in turn be submitted to a new round of observation and often experiment as well. In contrast to late seventeenth-century injunctions to segregate observation and conjecture strictly, mid-eighteenth-century manuals of scientific observation insisted that observation was a way of reasoning about, not just collecting experience: while it was deplorable to observe with prejudice for or against a system, it was utter folly to observe without ideas.[101]

The work of the French naturalist Georges Cuvier illustrates how powerful, sophisticated, and deliberate observation had become by the turn of the nineteenth century and serves as a conclusion to the long story of how observation became an essential way of reasoning in the sciences. Cuvier was celebrated among his contemporaries for his anatomical comparisons of extant and fossil organisms, a form of research that depended on the full armamentarium of techniques and resources developed by scientific observers since the sixteenth century.

In his pioneering monograph on the relation of contemporary African and Indian elephants to fossil pachyderm remains, Cuvier mobilized the well-stocked library and museum in order to construct series, in the sense of both long timelines and arrays of gradually differing specimens. From ancient sources on domesticated elephants to the latest fossil discoveries in Russia and the Americas, Cuvier marshaled an exhaustive list of all previous relevant observations in order to establish the geographic distribution of the species. He attributed particular observations to named individuals, with places, times, and stringent evaluation of reliability, discerning progress in the quality of the more recent observations: in 1577 the Swiss savant Felix Platter had mistakenly identified fossil bones as those of a giant, but Cuvier's esteemed Göttingen colleague Johann Friedrich Blumenbach had recently pronounced them as definitely elephant.

While depending heavily on a community of observers dispersed in space and time, Cuvier voiced his preference for firsthand observation wherever possible, dissecting three elephants himself and having a large drawing made "under my eyes, with much care." Drawings and measurements now counted as essential parts of an observation and were also subject to critical scrutiny. Focusing literally with a magnifying glass trained on fossil teeth, Cuvier inspected minute differences of size, shape, and wear as a function of stage of life. Pages of tables displayed the results of his observations of all the elephant molars he had observed, arranged by minutely noted features such as the length, width, and number of lamia. Amid this elephantine mass of information, ancient and modern, first- and secondhand, literary and visual, qualitative and quantitative, descriptive and tabular, Cuvier sought general, constant features that withstood thousands of comparative observations: "However the size alone of the [fossil elephant's] molars suffices in order to recognize them, because it is much more constant."[102]

For Cuvier and his contemporaries, observation had become a tool to think with, a genuine logic of discovery and proof. It was still collective and *longue durée*, but its practitioners were no longer anonymous nor were its results summarized in proverbs and rules of thumb. The work of observation consisted in collating and comparing the observations of others as well as making one's own. The store of observations burgeoned, repeated by individuals and multiplied by communities. The mission to reveal unsuspected correlations among phenomena persisted, but methods of repetition, note taking, establishing series, and inventing synoptic devices such as tables and maps had replaced what Cicero had called "natural divination."

More than ever before, observation was also an observance, regulating waking and sleeping, looking and overlooking, attention and memory, solitude and sociability. When von Haller, perhaps the most celebrated scientific observer of the Enlightenment, fell gravely ill in 1772, he recorded his own symptoms with the same ingrained habits of noting date and time, counting and measuring, and, above all, repeating an observation once, twice, three times:[103] "At five o'clock in the evening the room was a bit too warm, and there being several people there, I felt very ill, with an intermittent pulse after 1–2 or 3 pulsations. I took acid elixir and had the window opened: the air, although very warm, being a sirocco, had a surprising effect: the pulse immediately regularized itself. Three times I made the same experiment."[104] Observation and observance converged in the practices that remade the observer, body and soul.

Notes

1. See, for example, [Jean-Joseph Ménuret de Chambaud], "Observateur (*Gram. Physiq. Méd.*)," in Denis Diderot and Jean Le Rond d'Alembert, eds., *Encyclopédie, ou Dictionnaire raisonné des sciences, des arts et des métiers* [1751–80] (Stuttgart and Bad Cannstatt: Friedrich Fromann Verlag, 1967), 11: 310–13; also the three entries (one general, the other two referring to astronomical and medical observation specifically) on "Observation, *Observatio*," in Johann Heinrich Zedler, *Grosses Vollständiges Universal-Lexicon* [1732–50] (Graz: Akademische Druck- und Verlagsanstalt, 1994), 25: 278–87.

2. Charles Bonnet to Albrecht von Haller, Geneva, 22 July 1757, in Otto Sonntag, ed., *The Correspondence between Albrecht von Haller and Charles Bonnet* (Bern: Hans Huber Publishers, 1983), 107.

3. These included *historia, casus, observatio, experimentum, experientia, phaenomenon, factum.* See Arno Seifert, *Cognitio historica. Die Geschichte als Namengeberin der frühneuzeitlichen Empirie* (Berlin: Duncker & Humboldt, 1976); Peter Dear, *Discipline and Experience: The Mathematical Way in the Scientific Revolution* (Chicago: University of Chicago Press, 1995); "Fatti: storie dell'evidenza empirica," ed. Simona Cerutti and Gianna Pomata, special issue, *Quaderni storici* 108, no. 3 (2001); and Gianna Pomata and Nancy G. Siraisi, eds., *Historia: Empiricism and Erudition in Early Modern Europe* (Cambridge, Mass.: MIT Press, 2005).

4. [M. J. A. N. Condorcet], "Éloge de M. Du Hamel," in *Histoire et mémoires de l'Académie royale des sciences. Année 1783* (Paris: Imprimerie Royale, 1785), 131–55.

5. See, for example, Thomas Shadwell, *The Virtuoso* [1676], ed. Marjorie Hope Nicolson and David Stuart Rodes (London: Edward Arnold, 1966); and Jean de La Bruyère, *Les caractères de Théophraste traduit du grec avec les caractères ou les moeurs de ce siècle* [1688], ed. Robert Pignarre (Paris: Garnier-Flammarion, 1965), 338.

6. See the census of keyword frequencies in Marta Fattori, *Lessico del Novum organum di Francesco Bacone* (Rome: Edizioni dell'Atteneo & Bizzarri, 1980), 106–7, 168–74, 208–9, 342.

7. On the Baconian projects, several of them medical, of the Hartlib circle, see Charles Web-

ster, *The Great Instauration: Science, Medicine and Reform, 1626–1660*, 2nd ed. (New York: Peter Lang, 2002).

8. Uwe Müller, "Die Leopoldina unter den Präsidenten Bausch, Fehr und Volckamer 1652–1693," in *350 Jahre Leopoldina. Anspruch und Wirklichkeit*, ed. Benno Parthier and Dietrich von Engelhardt (Halle: Deutsche Akademie der Naturforscher Leopoldina, 2002), 45–94. Andreas Büchner, *Academia sacri romani imperii Leopoldino-Carolinae naturae curiosorum historia* (Halle and Magdeburg: Johann Gebauer, 1755), 181–97, gives the text of the original bylaws of 1652 (originally published in 1662 under the title *Salve academicum*). Laws 1–8 concern the monographs.

9. Müller, "Die Leopoldina"; Büchner, *Academia*, 194.

10. [Henry Oldenburg], "A Preface to the Third Year of These Tracts," *Philosophical Transactions of the Royal Society of London* 2 (1667): 409–15, on 414; also Mordechai Feingold, "Of Records and Grandeur: The Archive of the Royal Society," in *Archives of the Scientific Revolution: The Formation and Exchange of Ideas in Seventeenth-Century Europe*, ed. Michael Hunter (Woodbridge and Suffolk: Boydell Press, 1998), 171–84.

11. "Nr. 376. Oldenburg to Sachs, 30 May 1665," in A. Rupert Hall and Marie Boas Hall, eds., *The Correspondence of Henry Oldenburg*, 13 vols. (Madison: University of Wisconsin Press, 1965–86), vol. 7 (1670–71), 432–35.

12. Contrast, for example, "Of an Observation, not Long Since Made in England, of Saturn," *Philosophical Transactions* 1 (1665–66): 152–53; or "Observations Concerning Cochineel, Accompanied with Some Suggestions for Finding out and Preparing Such like Substances Out of Other Vegetables," *Philosophical Transactions* 3 (1668): 796–97, with "Some Considerations upon Mr. Reeds Letter . . . in what Sense the Sap May be Said to Descend. . . . ," *Philosophical Transactions* 6 (1671): 2144–49.

13. On the early membership of the Royal Society, see Michael Hunter, "Catalogue of Fellows, 1660–1700," 159–252, in Hunter, *The Royal Society and Its Fellows, 1660–1700: The Morphology of an Early Scientific Institution* (Chalfont St. Giles: British Society for the History of Science, 1982); and for the Académie Royale des Sciences, Roger Hahn, *The Anatomy of a Scientific Institution: The Paris Academy of Sciences, 1666–1803* (Berkeley and London: University of California Press, 1971), 31–49; *Index biographique de l'Académie des sciences*, 1666–1978 (Paris: Gauthier-Villars, 1979).

14. For example, *Relation d'une observation faite à la Bibliothèque du roy, à Paris, le 12 May 1667, sur les neuf heures du matin, d'un halo ou couronne à l'entour du soleil* (Paris: J. Cusson, 1667); or *Extrait d'une lettre écrite à Monsieur de la Chambre, qui contient les observations qui ont esté faites sur un grand poisson dissequé dans la Bibliothèque du roy, le vingt-quatrième Juin 1667* (Paris: F. Léonard, 1667).

15. Anne-Sylvie Guénon, "Les publications de l'Académie des sciences," in *Histoire et mémoire de l'Académie des sciences. Guide de recherches*, ed. Éric Brian and Christiane Demeulenaere-Douyère (Paris: Technique & Documentation, 1996), 107–40.

16. Procès-verbaux, Archives de l'Académie des sciences, Paris. The manuscript registers begin with the meeting of 22 December 1666.

17. Gianna Pomata, "A Word of the Empirics: The Ancient Concept of Observation and Its Recovery in Early Modern Medicine," *Annals of Science*, forthcoming.

18. Francis Bacon, *Novum organum*, II.xxxvi, in Basil Montagu, ed., *The Works of Francis Bacon*, 17 vols. (London: William Pickering, 1825–34), 9: 375–90.

19. Robert Hooke, "An Attempt to Prove the Motion of the Earth by Observations,"

Lectiones Cutlerianae [1679], reprinted in R. T. Gunther, *Early Science in Oxford* (London: Dawsons, 1968), 8: 4.

20. Bacon, "Distributio operis," *Novum organum*, in *Works*, 9: 173.

21. Bacon, *Novum organum*, II.xlv, in *Works*, 9: 416–17; cp. the table of densities, followed by "Modus experimenti" and "Observationes" in *Works*, 11: 184–92.

22. Bacon, "Phaenomena universi," *Works*, 11: 180.

23. Bacon, *Novum organum*, I.lxx, *Works*, 9: 219–21.

24. On the emergent rhetoric of *experientia*/*experimentum* in both French and English during this period, see Christian Licoppe, *La formation de la pratique scientifique: le discours de l'expérience en France et en Angleterre, 1630–1820* (Paris: Éditions de la Découverte, 1996), 19–87. The accounts of the experiments performed by the Florentine Accademia del Cimento (roughly, "Academy of Trials"; established 1657) were entitled *Saggi di naturale esperienze fatte nell'Academia del cimento* [1666], rendered as *Essayes of Natural Experiments made in the Academie del Cimento* in the 1684 English translation.

25. Quoted in Hans Poser, "Observatio, Beobachtung," in *Historisches Wörterbuch der Philosophie*, ed. Joachim Ritter and Karlfried Gründer (Darmstadt: Wissenschaftliche Buchgesellschaft, 1984), vol. 6, cols. 1072–81, on col. 1073.

26. See, for example, [Dumarsais], "Expérience," in Diderot and d'Alembert, eds., *Encyclopédie*, 6: 297–98; "Erfahrung," in Johann Georg Wald, *Philosophisches Lexicon* [1778] (Hildesheim: Georg Olms, 1968), 1082–84.

27. [Jean Le Rond d'Alembert], "Expérimental," in Diderot and d'Alembert, eds., *Encyclopédie*, 6: 298–301, on 298.

28. Joseph Priestley, *Experiments and Observations Relating to Various Branches of Natural Philosophy*, 3 vols. (London: J. Johnson, 1779–86), 2: ix. Priestley's experiments on the color of marine acid furnish a striking example of the back and forth between observation and experiment (1: 78–80).

29. One exception was vitalist medicine, whose proponents argued vehemently in favor of observation over experiment: see, for example, [Chambaud], "Observateur (*Gram. Physiq. Méd.*)," in Diderot and d'Alembert, eds., *Encyclopédie*, 11: 310–13, on 312–13.

30. Brian W. Ogilvie, *The Art of Describing: Natural History in Renaissance Europe* (Chicago: University of Chicago Press, 2006), 34–36, 46–44 and passim.

31. Charles Bazerman, *Shaping Written Knowledge: The Genre and Activity of the Experimental Article in Science* (Madison: University of Wisconsin Press, 1988).

32. Staffan Müller-Wille, *Botanik und weltweiter Handel: Zur Begründung eines natürlichen Systems der Pflanzen durch Carl von Linné, 1707 78* (Berlin: Verlag für Wissenschaft und Bildung, 1999).

33. Justin Stagl, *A History of Curiosity: The Theory of Travel, 1550–1800* (Chur: Harwood, 1995), 47–52, 126–31. See also chap. 15, this volume.

34. Francis Bacon, *The Great Instauration and New Atlantis*, ed. J. Weinberger (Arlington Heights: Harlan Davidson, 1980), 57.

35. Ogilvie, *The Art of Describing*, 180; Laurent Pinon, "Gessner and the Historical Depth of Natural History," in Pomata and Siraisi, *Historia*, 241–67.

36. British Library, Additional MSS 38, 693. Graham Rees estimates that of some thousand items in the *Sylva*, only twenty-six seem to be based on Bacon's own observations: Graham Rees, "An Unpublished Manuscript by Francis Bacon: *Sylva sylvarum* Drafts and Other Working Notes," *Annals of Science* 38 (1981): 377–412.

37. Charles Nauert, Jr., "Humanists, Scientists, and Pliny: Changing Approaches to a Classical Author," *American Historical Review* 84 (1979): 72–85.

38. See Bacon's strictures against "lazy and haphazard observation," "stupid, vague, and abrupt experiments," and "frivolous and scant natural history." Bacon, "Distributio operis," *Novum organum*, in *Works*, 9: 171–72.

39. Büchner, *Academia*, Law XVII, 194; on the background and adverse reactions (interpreted by some as criticism) to the scholia, see Müller, "Die Leopoldina"; on Bacon's *observationes majores*, see Bacon *Historia vitae et mortis*, in *Works*, 10: 134–45, and *Parasceve ad historiam naturalem et experimentalem*, 11: 423.

40. On the reports of early modern Venetian ambassadors, see Giovanni Comisso, ed., *Gli ambasciatori Veneti, 1525–1792. Relazione di viaggio e di missione* (Milan: Longanesi, 1985).

41. See, for example, the questionnaires of Philip II of Spain (1577), Sir William Petty, and Heinrich Rantzau (1587), all reproduced in Mohammed Rassem and Justin Stagl, eds., *Geschichte der Stadtbeschreibung* (Berlin: Akademie Verlag, 1994), 133–43; 285–93; 157–82; also Justin Stagl, "Vom Dialog zum Fragebogen. Miszellen zur Geschichte der Umfrage," *Kölner Zeitschrift für Soziologie und Sozialpsychologie* 31 (1979): 11–631.

42. Stagl, *A History of Curiosity*, 47–94.

43. [Giovanni Botero], *Relations of the Most Famous Kingdomes and Common-wealths throwout the World*, trans. Robert Johnson, rev. ed. (London: John Haviland, 1630), 1–45. On the Venetian background to Botero's and other compilations of travel reports and their influence on later compilations, see Joan-Pau Rubiés, "Instructions for Travellers: Teaching the Eye to See," *History and Anthropology* 9 (1996): 139–90, on 152–58.

44. Bacon himself had relied heavily on José de Acosta's *Historia natural y moral de las Indies* (1590) for his *Historia ventorum*: Daniel Carey, "Compiling Nature's History: Travellers and Travel Narratives in the Early Royal Society," *Annals of Science* 54 (1997): 269–92. See also Maurizio Bossi and Claudio Greppi, eds., *Viaggi e scienza. Le istruzioni scientifiche per i viaggiatori nei secoli XVII–XIX* (Florence: Leo S. Olschki, 2005); and Daniel Carey, "Hakluyt's Instructions: The Principal Navigations and Sixteenth-Century Travel Advice," *Studies in Travel Writing* 13 (2009): 167–85.

45. Robert Boyle, "General Heads for a Natural History of a Countrey, Great or Small," *Philosophical Transactions of the Royal Society* 1 (1665–66): 186–89.

46. Robert Boyle, *The Works of the Honourable Robert Boyle* [1772], ed. Thomas Birch, 6 vols. (Hildesheim: Olms, 1966), 5: 642.

47. Thomas Sprat, *History of the Royal Society* (London: J. Martyn, 1667), 173–79.

48. James Jurin, "Invitatio ad observationes meteorologicas communi consilio instituendas," *Philosophical Transactions of the Royal Society of London* 32 (1723): 422–27. On other Royal Society observational networks, see Andrea Rusnock, "Correspondence Networks and the Royal Society, 1700–1750," *British Journal for the History of Science* 32 (1999): 155–69. Other scientific academies also launched correspondence networks of weather observers: see Aleksandr K. Khrgian, *Meteorology: A Historical Survey* [1959], trans. and ed. Ron Hardin (Jerusalem: Israel Program for Scientific Translations, 1970), 71–72; Gustav Hellmann, "Die Entwicklung der meteorologischen Beobachtungen in Deutschland, von den ersten Anfängen bis zur Einrichtung staatlicher Beobachtungsnetze," *Abhandlungen der Preussische Akademie der Wissenschaften, Physisch-Mathematische Klasse* 1 (1926): 1–25; Henry E. Lowood, *Patriotism, Profit, and the Promotion of Science in the German Enlightenment: The Economic and Scientific Societies, 1760–1815* (New York: Garland, 1991), 117–23.

49. John Locke, "A Register of the Weather for the Year 1692, Kept at Oates in Essex," *Philosophical Transactions of the Royal Society of London* 24 (1704–5): 1917–37 on 1919.

50. Lorraine Daston, "Unruly Weather: Natural Law Confronts Natural Variability," in *Natural Law and Laws of Nature in Early Modern Europe*, ed. Lorraine Daston and Michael Stolleis (Aldershot: Ashgate, 2008), 234–48.

51. M. Campbell-Kelly, M. Croarken, R. Flood, and E. Robson, *A History of Mathematical Tables: From Sumer to Spreadsheets* (Oxford: Oxford University Press, 2003).

52. Gottfried Wilhelm Leibniz, "Entwurf gewisser Staatstafeln," in Rassem and Stagl, eds., *Geschichte der Stadtbeschreibung*, 319–29, on 325. See also chap. 16, this volume.

53. Edmond Halley, "An Historical Account of the Trade Winds, and Monsoons, observable in the Seas between and near the Tropicks, with an attempt to assign the Phisical cause of said Winds," *Philosophical Transactions of the Royal Society of London* 16 (1686–92): 153–68, on 163.

54. Bacon, *The Great Instauration*, 80.

55. [Oldenburg], "A Preface to the Third Year," 410.

56. *Mémoires de mathématique et de physique. Année MDCXCII. Tirez des registres de l'Académie Royale des Sciences* (Amsterdam, 1723), "Avertissement," sig. 2r.

57. Kurt Møller Pedersen, "Une mission astronomique de Jean Picard: Le voyage d'Uraniborg," in *Jean Picard et les débuts de l'astronomie de précision au XVIIe siècle*, ed. Guy Picolet (Paris: Éditions du Centre Nationale de la Recherche Scientifique, 1987), 175–203.

58. Aydin Sayili, *The Observatory in Islam and Its Place in the General History of the Observatory*, 2nd ed. (Ankara: Türk Tarih Kurumu Basimevi, 1998); David A. King, *Astronomy in the Service of Islam* (Aldershot: Variorum, 1993); E. S. Kennedy, *Astronomy and Astrology in the Medieval Islamic World* (Aldershot: Ashgate, 1998).

59. Tycho Brahe, *Tycho Brahe's Description of his Instruments and Scientific Work* [*Astronomiae instauratae mechanica*, 1598], trans. and ed. Hans Raeder, Elis Strömgren, and Bengt Strömgren (Copenhagen: I Kommission Hos Ejnar Munksgaard, 1946), 110–11.

60. Ibid., 110.

61. Eric Forbes, ed., *The Correspondence of John Flamsteed, the First Astronomer Royal* (Bristol: Institute of Physics, 1997), 2: 635; quoted in Frances Willmoth, "Models for the Practice of Astronomy: Flamsteed, Horrock and Tycho," in *Flamsteed's Stars: New Perspectives on the Life and Work of the First Astronomer Royal, 1646–1719*, ed. Frances Willmoth (Woodbridge, Suffolk: Boydell Press, 1997), 49–75, on 63.

62. Galileo Galilei, *Sidereus nuncius* (Venice: Thomas Baglioni, 1610). See also the introduction to idem, *Sidereus Nuncius: Or the Sidereal Messenger*, trans. and ed. Albert van Helden (Chicago: University of Chicago Press, 1989).

63. Allan Chapman, "The Accuracy of Angular Measuring Instruments Used in Astronomy between 1500 and 1850," in Chapman, *Astronomical Instruments and Their Uses: Tycho Brahe to William Lassell* (Aldershot: Variorum, 1996), 133–37, especially 134–35. On the telescope, see Albert van Helden, "The Invention of the Telescope," *Transactions of the American Philosophical Society*, 67, no. 4 (1977): 1–67.

64. Olaf Pedersen, "Some Early European Observatories," *Vistas in Astronomy* 20 (1976): 17–28; and in the same volume, Eric G. Forbes, "The Origins of the Royal Observatory at Greenwich," 39–50; and René Taton, "Les origines et les débuts de l'Observatoire de Paris," 65–71.

65. Suzanne Débarbat, "La qualité des observations astronomiques de Picard," in Picolet, *Jean Picard*, 157–73, especially 160.

66. On astronomy, see O. B. Sheynin, "Mathematical Treatment of Astronomical Observa-

tions (A Historical Essay)," *Archive for History of Exact Sciences* 11 (1973): 97–126; Stephen M. Stigler, *The History of Statistics: The Measurement of Uncertainty before 1900* (Cambridge, Mass.: Harvard University Press, 1986), 11–61; Simon Schaffer, "Astronomers Mark Time: Discipline and the Personal Equation," *Science in Context* 2 (1988): 115–45; on meteorology, Christoph Hoffmann, *Unter Beobachtung: Naturforschung in der Zeit der Sinnesapparate* (Göttingen: Wallstein, 2006), 23–88.

67. Lorraine Daston, "Taking Note(s)," *Isis* 95 (2004): 443–48.

68. "Adversaria physica," Bodleian Library, Oxford University, MS Locke d.9. "Adversaria" in classical Latin originally meant a merchant's waste-book or journal, in which items are entered as they occur, for later use.

69. Locke himself published an article on how to organize commonplace books: [John Locke], "Méthode nouvelle de dresser des recueils," *Bibliothèque universelle et historique* 2 (1686); 315–28; see also Richard Yeo, "John Locke's 'Of Study' (1677): Interpreting an Unpublished Essay," *Locke Studies* 3 (2003): 147–65.

70. Ann Blair, "Scientific Readers: An Early Modernist's Perspective," *Isis* 95 (2004): 420–30, and the bibliography given therein.

71. Bacon, *The Great Instauration*, 79.

72. Sister Joan Marie Lechner, *Renaissance Concepts of the Commonplaces* (New York: Pageant Press, 1962), 153–99.

73. Quoted in Vernon F. Snow, "Francis Bacon's Advice to Fulke Greville on Research Techniques," *Huntington Library Quarterly* 22 (1969): 369–78, on 372.

74. Horace-Bénédict de Saussure, "Voyage autour du Mont Blanc en 1774, 10e. Juil.," Bibliothèque publique et universitaire de Genève, MS Saussure 14/1. For a detailed analysis of a comparable scientific notebook, see Marie-Noëlle Bourguet, "La fabrique du savoir. Essais sur les carnets de voyage d'Alexandre von Humboldt," "Festschrift für Margo Falk," special issue, *Humboldt im Netz* 7, no. 13 (2006): 17–33.

75. Jean Senebier, *L'art d'observer*, 2 vols. (Geneva: Chez Cl. Philibert & Bart. Chirol, 1775), 1: 144, 145, 188.

76. Bonnet to Spallanzani, 27 Dec. 1765, in Charles Bonnet, *Oeuvres d'histoire naturelle* (Neuchâtel: Samuel Fauche, 1781), 5: 10.

77. On the seventeenth-century vogue for—and ultimate disillusionment with—the microscope, see, respectively, David Freedberg, *The Eye of the Lynx: Galileo, His Friends, and the Beginnings of Modern Natural History* (Chicago: University of Chicago Press, 2002), 151–53, 179–94; and Catherine Wilson, *The Invisible World: Early Modern Philosophy and the Invention of the Microscope* (Princeton: Princeton University Press, 1995), 225–50. References to the use of magnifying glasses are scattered but frequent in the notes and publications of eighteenth-century naturalists, perhaps because such instruments were always ready to hand for reading purposes: see Giuliana Biavati, "Gli occhiali: una storia attraverso l'otticadelle ambivaenze iconografiche," in Museo Civico di Storia Naturale "G. Doria," *La lente. Storia, scienza, curiosità attraverso la collezione Fritz Rathschilder* (Genoa: Editioni Culturali Internazionali Genova, 1988), 11–31, and also the illustrations of eighteenth-century reading glasses, 109–10.

78. See, for example, Bonnet to Réaumur, 27 July 1739, Dossier Charles Bonnet, Archives de l'Académie des Sciences, Paris.

79. Jean Senebier, "Mémoire historique sur la vie et les écrits de Horace Bénédict Desaussure," in Horace-Bénédict de Saussure, *Voyages dans les Alpes* [1779–96], 4 vols. (Geneva: Éditions Slatkine, 1978), 1: 28.

80. See Lorraine Daston and Peter Galison, *Objectivity* (New York: Zone Books, 2007), 63–82; also Patrick Singy, "Huber's Eyes: The Art of Scientific Observation before the Emergence of Positivism," *Representations* 95 (2006): 54–75. See also chap. 16, this volume.

81. Gunnar Eriksson, "Linnaeus the Botanist," in *Linnaeus: The Man and His Work*, ed. Tore Frangsmyr, rev. ed. (Canton, Mass.: Science History Publications, 1994), 63–109.

82. Johann Wolfgang Goethe, "Erfahrung und Wissenschaft" [1798, publ. 1893], in Dorothea Kuhn and Rike Wankmüller, eds., *Goethes Werke*, 14 vols., 7th ed. (Munich: C. H. Beck, 1975), 13: 25.

83. [Louis] Godin, "Sur le météore qui a paru le 19 octobre de cette année," *Mémoires de l'Académie royale des sciences. Année 1726* (Paris: Imprimerie Royale, 1728), 287–302, on 297–98.

84. Daston and Galison, *Objectivity*, 84–98. See also chap. 15, this volume.

85. Bacon, *Historia vitae et mortis*, *Works*, 10: 134–45, and *Parasceve*, *Works*, 11: 423.

86. Carrard, *Essai*, 1. Carrard won the Haarlem Academy of Science's prize for the best work on observation: Hans Poser, "Die Kunst der Beobachtung: Zur Preisfrage der Holländischen Akademie von 1768," in *Erfahrung und Beobachtung: Erkenntnistheoretische Untersuchungen zur Erkenntnisbegründung*, ed. Hans Poser (Berlin: Technische Universität Berlin, 1992), 99–119; J. G. Bruijn, *Inventaris van de Prijsvragen uitgeschreven door de Hollandsche Maatschappij der Wettenschappen 1753–1917* (Haarlem: Hollandsche Maatschappij der Wettenschappen, 1977). Senebier received an honorable mention.

87. William J. Ashworth, "'Labour harder than thrashing': John Flamsteed, Property and Intellectual Labour in Nineteenth-Century England," in Willmoth, *Flamsteed's Stars*, 199–216.

88. Carolus Clusius, *Rariorum plantarum historiae* (Antwerp: Moretus, 1601), f. *3r.

89. [André Deluc and Jean Dentand], *Relation de différents voyages dans les Alpes de Faucigny* (Maestricht: Chez J. E. Dufour & Ph. Roux, 1776), 29.

90. Lorraine Daston, "Attention and the Values of Nature," in *The Moral Authority of Nature*, ed. Lorraine Daston and Fernando Vidal (Chicago: University of Chicago Press, 2004), 100–26. The medical dangers of the observer's way of life were described, using the examples of Bonnet, Swammerdam, von Haller, and others, by Samuel Auguste Tissot, *De la santé des gens de lettres* [1768], ed. Christophe Calame (Paris: Éditions de la Différence, 1991), 30, 63, 77.

91. Paolo Boccone, *Recherches et observations naturelles* (Amsterdam: Jean Janson, 1674).

92. Ogilvie, *The Science of Describing*, 53. Compare the sixteenth-century epistolary networks of physicians described in the essay by Gianna Pomata in this volume.

93. Réamur to Lyonet, 19 March 1743; Lyonet to Réaumur, 11 May 1742; Réaumur to Lyonet, 22 Dec. 1742, Bibliothèque publique et universitaire de Genève, Fonds Trembley 5, Correspondance entre tiers.

94. Solange Grillot, "Picard observateur," in Picolet, *Jean Picard*, 143–56, on 152.

95. Jan Golinski, "Barometers of Change: Meteorological Instruments as Machines of Enlightenment," in *The Sciences in Enlightened Europe*, ed. William Clark, Jan Golinski, and Simon Schaffer (Chicago: University of Chicago Press, 1999), 69–93; Vladimir Jankovic, *Reading the Skies: A Cultural History of the English Weather* (Manchester: Manchester University Press, 2000); Jan Golinski, *British Weather and the Climate of the Enlightenment* (Chicago: University of Chicago Press, 2007).

96. Tycho Brahe, "On that which we have hitherto accomplished in astronomy with God's help, and on that which with his gracious aid has yet to be completed in the future," in Tycho Brahe, *Tycho Brahe's Descriptions*, 106–18, on 108–9.

97. Réaumur to Jean-François Séguier, 25 April 1743, in Académie des Belles-Lettres, Sciences et Arts de La Rochelle, *Lettres inédites de Réaumur* (La Rochelle: Veuve Mareschal & Martin, 1886), 15.

98. Louis Cotte, *Traité de météorologie* (Paris: Imprimerie Royale, 1774), 519.

99. Guy Tachard, *Voyage de Siam des pères Jesuites, envoyez par le roi aux Indes et à la Chine, avec leurs observations astronomiques, et leurs remarques de physique, de géographie, d'hydrographie, et d'histoire* (Paris: Arnould Seneuze & D. Horthemels, 1686), 82–87.

100. Giorgio Baglivi, *The Practice of Physick, Reduc'd to the Ancient Way of Observation* [1696] (London: Andrew Bell, Ralph Smith, and Daniel Midwinter, 1704), I.i.4, 6.

101. Carrard, *Essai*, 242.

102. Georges Cuvier, *Recherches sur les ossemens fossiles, ou l'on rétablit les caractères de plusieurs animaux dont les révolutions du globe ont détruit les espèces*, 2nd ed., 5 vols. (Paris: Chez G. Dufour et E. d'Ocagne, 1821–24), 1: 104, 113, 9, 168–71, 165. On the context of Cuvier's investigations of pachyderms, see Martin J. S. Rudwick, *Georges Cuvier, Fossil Bones, and Geological Catastrophes* (Chicago: University of Chicago Press, 1997), 13–41; Claudine Cohen, *The Fate of the Mammoth: Fossils, Myth, and History* [1994], trans. William Rodarmor (Chicago: University of Chicago Press, 2002), 105–24.

103. Von Haller's observational notebooks preserve traces of many repeated observations, especially on controversial matters such as embryological development: Albrecht von Haller Sammlung, Burgerbibliothek Bern, N AvH 5, MS Hall. 1617, partially transcribed in Albrecht von Haller, *Commentarius de formatione cordis in ovo*, ed. Maria Teresa Monti (Basel: Schwabe, 2000). See also Albrecht von Haller to Charles Bonnet, 1 Sept. 1757, in Sonntag, *Correspondence*, 109.

104. Albrecht von Haller to Samuel Auguste Tissot, 1 March 1772, in Erich Hintzsche, ed., *Albrecht Hallers Briefe an Auguste Tissot 1754–1777* (Bern: Verlag Hans Huber, 1977), 344.

PART TWO

Observing and Believing: Evidence

All scientific observation ultimately aims to provide evidence: for the bare existence of phenomena, for or against a hypothesis, for the significance of this or that detail in the broader context of inquiry. But in order to achieve any of these evidentiary goals, observation must first be conceptualized as a distinctive way of acquiring knowledge, with its own methods, guarantees of reliability, and functions vis-à-vis other modes of investigation. The essays in this part explore all of these aspects of observation as evidence, from the mid-seventeenth to the early twentieth centuries, in medicine, natural history, and physics. In each of these essays, the evidentiary weight of the observation is intertwined with the personal credibility and skills of the observer: across centuries and disciplines, the persona of the virtuoso observer, open-eyed and open-minded, attentive, and preternaturally patient, persists.

Domenico Bertoloni Meli's essay on the observation of color in seventeenth-century physiology, chemistry, and natural history is situated in the context of early modern medicine. To embrace observation as a source of evidence at first overwhelmed naturalists with an embarrassment of riches: out of the myriad phenomena, what was worth observing, as evidence of what? This was as much a philosophical as a practical problem: new doctrines of primary and secondary qualities espoused by Galileo and others, as well as chemical and medical experiments, sowed doubts in the minds of Pisan physicians Giovanni Borelli, Carlo Fracassati, and Marcello Malpighi as to the value of color as evidence of the essential properties of substance, despite a venerable medical tradition of using color to diagnose disease and ascertain the composition of blood. Observations on caked blood indicated that color could be all too easily altered by exposure to air or water. Yet the very same observations that had discounted the evidentiary import of color for the Pisans were eagerly received by their colleagues in Oxford and London as sterling evidence for the role of air in the lungs. Converted from his earlier skepticism about the evidentiary value of color, Malpighi went on to observe colors with exquisite attention to variety and nuance in his study of silkworms.

Michael D. Gordin raises the question of how the evidence of observation commands belief in starkest form in his essay on the exuberant nineteenth-century Russian naturalist and man of letters Nikolai Petrovich Vagner. Vagner made a career and forged a persona as the indefatigable, attentive observer of highly implausible phenomena. Who would have believed that some insect larvae could contain yet another generation of larvae, without fertilization or even maturation? Certainly not the editor of the German scientific journal to whom Vagner submitted his sensational article, at least not until he had himself replicated the observations and the leading authority on embryology had endorsed them. Flush with victory, Vagner tested his observational prowess on still more improbable effects: the apparitions from the beyond reported in Spiritualist séances. Armed with his scientific reputation as an ace observer and a razor-sharp polemical pen, Vagner was undaunted by the skepticism of colleagues like Mendeleev and the jeers of literati like Tolstoy. This was evidentiary observation simultaneously at its most humdrum and most radical: unbiased patience and perspicacity would, Vagner was convinced, reveal brave new worlds.

Charlotte Bigg's essay also presents a case of virtuoso observation: French physical chemist Jean Perrin's attempts to observe the ultimate unobservable, the atom, in his study of Brownian motion. Although Perrin's apparatus and pencil-and-paper methods were low-tech in comparison to those used by other laboratories to test Einstein's claim that Brownian motion could prove the reality of atoms, they were extraordinarily sophisticated in their use of statistical ideas to minimize observational error. Perrin's observational strategies also highlight the often blurred boundaries between experiment, observation, and theory in the early twentieth century: his preparations of colloid solutions were manipulative enough to qualify as "experiments," but the actual tracking of the paths of the suspended granules counted as "observation." Einstein's 1905 theoretical paper on the relationship between atomism and the measurement of Brownian motion directly inspired Perrin's efforts. All of these elements contributed to the evidentiary weight of Perrin's iconic drawing of Brownian motion.

4

The Color of Blood: Between Sensory Experience and Epistemic Significance

DOMENICO BERTOLONI MELI

Introduction: Between Anatomy and Philosophy

In his monumental *Canon*, the Persian philosopher and physician Ibn Sīnā—known in the West as Avicenna—discussed the nature and composition of blood with regard to its role in nutrition. In an important passage, he argued that blood is a humor consisting of four components, as can be ascertained by pouring the blood drawn from a patient into a vessel and observing its separation into a foamy "colera rubea," a turbid "fex" or "melancolia," a portion resembling egg white, and lastly a watery part. The first three parts are themselves humors, namely, bile, melancholia, and phlegm, whereas the last part is that which expells its excess as urine. Avicenna relied on a range of features, including color, for the identification of the blood's components. This passage attracted the attention of physicians and alchemists alike because of the importance of blood, the compound nature of the humors, and the problem of their separation. In this paper I will discuss some implications of Avicenna's passage for the nature of color and blood in the seventeenth century.[1]

In his 1651 *De generatione*, for example, William Harvey argued that blood is heterogeneous and is composed of different humors, but while the animal is alive "it is a homogeneous animate part, compounded out of soul and body"; this unity disappears in death when the soul fades away and blood decomposes into its constituents and becomes corrupted. Harvey also noticed that blood found in the lungs was especially florid, but he believed that the difference in the color of blood from arteries and veins depended on accidental circumstances, such as the size of the openings: blood squirting from a tight opening, like that in an artery, was brighter, whereas blood from a wider opening, like that in a vein, was darker. He added that between venous and arterial blood there were no physical differences and that arterial blood collected in a bowl would soon look venous.[2]

These observations and reflections on blood and its components call into question the nature of color as a tool of investigation in a number of areas ranging from chemistry to philosophy. Color is one of the most immediate sensory experiences and at the same time one of the most complex philosophical and physiological problems in sense perception. The seventeenth century was a particularly remarkable period in this regard, one that saw the crystallization of the notions of primary and secondary qualities and the publication of a number of celebrated studies and experiments on the nature of light and colors, as well as the investigation of the significance of color change in blood. This essay moves across a varied terrain conceptually and geographically: it starts by providing a brief synopsis of physical-philosophical stances on color in a few decades around midcentury, beginning with Galileo's *Assayer* (Rome, 1623) and Descartes' *Dioptrique* (Leiden, 1637). Moreover, Robert Boyle and Robert Hooke joined a chemical with a mechanistic standpoint in *Experiments and Considerations Touching Colours* (London, 1664), *The Origine of Formes and Qualities* (Oxford, 1666), and *Micrographia* (London, 1665).[3]

Anatomical investigations are particularly relevant because color enters the description of important structures and processes in the body. For this reason I will focus on a key episode, the study of color change in blood between 1659 and 1669, with special emphasis on the group around Giovanni Alfonso Borelli in Pisa, including Marcello Malpighi and Carlo Fracassati, and some scholars moving between Oxford and London, including Thomas Willis, Boyle, and Hooke. Later in his career Malpighi changed his philosophical stance on color in dramatic fashion; therefore a study of his work promises to shed light on a broad range of epistemological positions. Briefly put, at an early stage, relying on his own philosophical views and the experiments of the Cimento Academy, Borelli explained to Malpighi that color was not a useful way to explore the properties of substances. The *Saggi*, or samples of experiments of the Cimento Academy (Florence, 1667), tackled the problem of the nature of color change experimentally, discussing tests with color indicators leading Borelli to believe that colors could easily be changed and were therefore unreliable indicators of the true nature of a substance—a much more radical stance than Boyle's. As a result, in his study on lungs and respiration, Malpighi ignored color change in blood. As reported in print by Fracassati, Malpighi observed that air—among other factors, to be sure—was responsible for color change in blood, but he did not consider this change to be indicative of a corresponding transformation in its substance and therefore as a meaningful feature of respiration.

It is especially useful to contrast the works by the Pisa anatomists with the *Tractatus de corde item de motu & colore sanguinis et chili in eum transitu* [*Treatise on the Heart as well as on the Motion and Color of Blood and on the Transit of Chyle through it*] (London, 1669) by the physician Richard Lower, a treatise examining respiration—among other topics—in which color change in blood is prominently included in the title. Both Boyle and Hooke were engaged not only in philosophical and experimental reflections on color, but in anatomical investigations as well: Hooke offered a decisive contribution to Lower's work, one that Lower chose to acknowledge in print. Lower was a student and follower of Thomas Willis, a physician, anatomist, and chemist whose reflections on the nature of blood and the site where its color changed in the body proved quite influential.

In a concluding section I show that at a later stage, after having broken with Borelli and having become associated with the Royal Society, Malpighi gave increasing attention to color: not only did his description of the silkworm display a stunning sensitivity to color, but he also attributed an epistemic significance to it, since the change of color of the silkworm's eggs indicated whether they had been fertilized. I suspect that Boyle's work joined forces with Malpighi's medical background and artistic sensibility in effecting this remarkable transformation, both in the style of description and in its philosophical underpinning. This episode provides material for reflection on the nature of observation and its epistemic presuppositions and consequences.

The issue of color in philosophy, anatomy, or medicine in the seventeenth century is a huge one that cannot possibly be exhausted within the compass of a short paper, even one confined to the study of blood; therefore my aim here is limited to raising some questions and stimulating further investigations through the lens of a particularly significant episode rather than offering a comprehensive examination of the issues at stake.

Color in the Mid-seventeenth Century

I wish to open this short section by discussing Galileo's celebrated passages from the *Assayer* in which he introduced the distinction between what we can call "objective" and "subjective" qualities, later called primary and secondary. In section 48 Galileo discussed the nature of heat and then went on to argue that some qualities—such as size, motion, spatial relation to other bodies, and number—are inseparable in our mind from corporeal substance. Other qualities, however, such as tastes, odors, and colors reside only in the sensorium of the perceiving animal; if this were removed, they would disappear. Heat, according to Galileo, was one of those qualities; heat would consist in a multitude of tiny particles—the *ignicoli*—moving at great speed, which are the only entities existing independently of the perceiver; consequently there is no such thing in nature as heat independently of those who perceive it. Looking more closely at Galileo's text one notices a significant difference among the purely subjective qualities: in some cases, as with tastes, odors, and sounds, Galileo provided an explanation of their origin, such that tastes and odors are associated with the shape, size, and speed of particles entering the pores of the tongue or the nostrils. Galileo had already discussed sounds in the celebrated and much discussed fable of the cicada; whatever their specific forms of production, however, such as the vibration of a string, they stemmed from the motion of air. By contrast, Galileo left the issue of light and especially color open, arguing first that he understood very little about it and then that it would require a long time to explain the little he knew.[4]

Probably Borelli had Galileo's *Assayer* in mind when in the 1649 *Delle cagioni de le febbri maligne* he applied a similar reasoning to medical matters, pointing out that neither tastes, nor smells, nor colors are reliable or indeed viable ways to distinguish poisons from healthy foods: as we shall see, for Borelli those qualities could be changed without a change in substance and therefore all we can do to find a substance's properties—medical or otherwise—is to study its effects.[5]

I believe that Descartes too was familiar with Galileo's *Assayer*, which was published in Rome just before his arrival in the eternal city; several passages from the 1644 *Principia philosophiae* echo quite closely Galileo's text. Descartes retained Galileo's dichotomy between "objective" and "subjective" qualities, namely, qualities like size, shape, and motion on the one hand, and colors, tastes, and odors on the other, arguing that there is nothing in nature that corresponds to color as such independently of the perceiving subject. Light played a major part in Descartes' natural philosophy, so much so that his treatise *Le monde* was originally conceived as a treatise on light: although he treated the problem in different ways depending on the problem he was addressing, overall he understood light in terms of pressure from particles of a fluid. Already in the *Dioptrique* Descartes moved one step further than Galileo in providing a mechanical account of the corresponding quality of colored particles, namely, their spin. According to his view, the different rotational speed of light particles makes us see color: red for the greatest spin and blue for the smallest. Descartes, too, dealt with the color of blood, framing his study in a neo-Galenic fashion to try to explain how white chyle is transformed into red blood in the liver: his answer was that just as the white juice of black grapes is turned into red wine, so chyle passing through the pores of the liver "takes on the color, and acquires the form, of blood," a comparison borrowed from Galen.[6]

Moving to England, we find writings of Boyle and Hooke especially pertinent to the philosophy of color: both were engaged in anatomical experiments on respiration and the reasons for the change of color of blood. Several documents from that period testify to Boyle's interests in the matter: for example, he wished to investigate the differences between arterial and venous blood, as well as their color, taste, odor, and specific gravity. Boyle's *Experiments Touching Colour* claim that color and color changes are due to the change of the mechanical texture of bodies, especially their surfaces. Boyle's argument that color is related to the roughness of surfaces led him to accept the report by Sir John Finch that John Vermaasen, a blind man in the Netherlands, was able to distinguish colors by touch, a report savagely lampooned by Jonathan Swift in *Gulliver's Travels.* According to Vermaasen, black and white had the roughest surfaces or the "most asperous," while red and blue were the least rough or "asperous," the full range going from black to blue. It is of particular interest that Boyle reported several experiments with color indicators, much like the Cimento Academy, though he did not reach Borelli's radical conclusion that color is ultimately unrelated to the nature of substance. Rather, he showed a typically restrained attitude to formulating a general theory. Boyle, however, did surmise that colored bodies appear opaque but may in fact consist of transparent corpuscles. In *The Origine of Formes and Qualities* he argued that colors are not inherent qualities of a body due to its substantial form; rather, they derive from the mechanical texture of its minute parts and can be easily changed by changing that texture. The very first of ten experiments in his book involves the dissolution of camphor into oil of vitriol, producing a deep yellow-red color in striking contrast to the colorless ingredients; add water to the solution, however, and the solution turns colorless and camphor regains its piercing odor that it had lost in its dissolution. Boyle's *Memoirs for the Natural History of Humane Blood* (London, 1684), published twenty years later, testifies to his lasting interest in blood. The book was dedicated to John Locke, who in the mid-1660s was interested in the color change of blood and believed it was due to the niter in the air.[7]

Robert Hooke, too, indulged in speculations on light and color in several bodies, such as Muscovy glass—a mineral composed of tiny flakes with varying optical properties as they got smaller—and a diamond presented by Mr. Clayton to the Royal Society, which produced light when rubbed, struck, or beaten in the dark, a matter discussed by Boyle too. Hooke concluded from painstaking examination of the behavior of Clayton's diamond that light resulted from a very short vibratory motion. While examining the color of bodies, Hooke argued that even those appearing opaque are composed of tiny transparent elements, hence the importance of his study of Muscovy glass, which thus appeared not as a peculiar exception but as exemplary of the structure of bodies.[8]

Whatever the specific view about the color of bodies, both Boyle and Hooke, unlike Borelli and his followers, did not dismiss the significance of color and color change altogether. Rather, they adopted a more flexible approach whereby color did have some correlations to bodies, if not strictly to their material substance, at least to its arrangement in the texture and especially to the surface of bodies.

The Pisa Scene: Borelli, Malpighi, and Fracassati

Between 1656 and 1667 Borelli held the chair of mathematics at Pisa University. Although traditionally this position was not especially highly remunerated or of very high rank, Borelli's close contacts with the Medici rulers and their academy enabled him to enjoy an unusually high standing at the university, where he was the "philosophical" and "political" leader of a group that included at different times the professors of medicine Malpighi (1656–59) and Fracassati (1663–68). Let us focus on Malpighi first: his position was especially interesting because, besides being an anatomist, he was also a professor of the practice of medicine and a physician, and this adds another dimension to the issue of color. Although the venerated practice of uroscopy—involving the careful inspection of urine, including its color—may have fallen into disuse, color remained a key feature of medical diagnosis as a meaningful indicator of health and disease: jaundice, for example, relied on the observation of a yellow tinge in some solid and fluid parts of patients. Therefore it was natural for Malpighi to pay attention to the color of body parts, as he did in a letter to Borelli in 1660 in which he commented on the changing color of some callous particles—possibly of blood—extracted from a patient and friend afflicted by pain in the articulations; he reported that those particles turned from white into a rotten color, *color di marcia*, or the color of rotten or putrid matter. In a revealing reply, Borelli stated that the change of color of those callous particles was not a matter of great interest, "knowing that the colors of things can be very easily changed."[9]

In a later letter Borelli discussed the issue of color at greater length in a medical and therapeutic context: the topic of discussion was the nature of some fevers afflicting Pisa and the search for the best therapy. Since postmortems revealed an excess of bile in the victims' cadavers, bile played a major role in his and Malpighi's reflections. Malpighi had argued that no fever arises in those cases in which bile is mixed with blood, as the example of jaundice shows. In his reply Borelli questioned whether bile is truly to be found in the arteries and veins of patients; he recalled having tested by means of a piece of paper the urine of a patient with exceedingly yellow face and eyes and found that the paper did not turn yellow. He pointed out that since nature can change colors very easily, it would be conceivable that jaundice could be due to causes other than bile. Thus in this instance Borelli still considered color as a valid symptom in that the yellow face and eyes of the patient indicated jaundice, yet he questioned the traditional causal mechanism linking the appearances to bile. At this point he embarked on a chemical-philosophical excursus on color—and taste too—arguing that colors can be changed without a corresponding change of "substance," by which he meant the constituent matter of the body. He mentioned the experiments performed at the Cimento Academy and later published in the *Saggi.* It is to these experiments that we now turn.[10]

Study of color change occupied a small part of the agenda of the Cimento Academy around 1660. Its activities aimed to promote experimental philosophy without an explicit philosophical agenda for or against novelties in order to present irrefutable experimental results and avoid sterile philosophical disputes. As in many other cases, however, it seems likely that the experiments on color indicators did follow a philosophical agenda in challenging the view that colors were related to substantial forms, in that colors could be easily changed without changing the substance generating them in any meaningful way. This way of proceeding by allusions or coded messages was standard at the Cimento. The *Saggi* of the Cimento states that, the academy truly did not wish to meddle with color changes studied by the chemists, but the members investigated some of those changes in connection with their study of the properties of mineral waters. The third experiment offers an example:[11]

> Tincture of red roses extracted with spirit of vitriol becomes a very beautiful green when mixed with oil of tartar. A few drops of spirit of sulphur make it all bubble up into a bright red foam, and it finally returns to a rose color without ever losing its scent and can no longer be changed by oil of tartar poured into it.

The text specifies that ten or twelve drops of oil of tartar and of spirit of sulphur in half an ounce of tincture of roses are enough to achieve the desired result. Although at first sight this and other similar experiments seem like neutral factual reports, Borelli's correspondence reveals a different side of the story. Borelli drew the conclusion that there was no fixed relation between color and the substance generating it: a few drops of oil of tartar or spirit of sulphur could turn a much larger amount of liquid obtained from red roses and spirit of vitriol from red to green and back to red. The red liquids, however, had such different properties that the first had a pleasant taste and was innocuous, whereas the final product could have proved lethal. Similarly, tastes too could be deceptive, as he had just experienced by noticing the similarity between two fluids with different properties, such as the brine in which olives macerate and that found in the stomach of fishes, or milk and the liquid found in the stomach of hawks: whereas the digestive fluids were very corrosive, the others were innocuous. Hence nature could easily change colors and tastes without changing a body's substance; conversely, it could make very different substances look and taste similar: changes in color or taste were unrelated to substantive transformations.[12] As we are going to see, in this tradition joining subtle philosophical thinking with the latest experimental results, the far less dramatic color change in blood from dark to bright red and back to dark seemed unworthy of serious investigation; the change may be attributed to the rearrangement of blood components in the lungs, but the investigators in Borelli's circle did not test where and in what circumstances it occurred.

These observations about color had an anatomical counterpart in the study of blood carried out by Borelli and Malpighi when they overlapped as professors at Pisa University between 1656 and 1659. At the time Malpighi planned a dialogue in Galilean form dealing with medical and anatomical issues; although in the end that work was not published and is now lost, in 1665 Malpighi incorporated portions of it into a *Risposta* he drafted against some traditionalist Galenist physicians at Messina, which was published only posthumously in 1697. The Galenists argued that even barbers know that blood contains bile, phlegm, and melancholia or black bile, as can be seen in blood let from a healthy person, a reference dating back to Avicenna's *Canon.*[13]

Malpighi disagreed with the Galenists and challenged their interpretation: both taste and odor of the various parts are intermingled and therefore they cannot be easily distinguished, so much so that even the bitterness of bile is overshadowed by the sweetness of blood. Thus color turns out to have a crucial role: the components of blood could allegedly be detected by visual inspection by taking some congealed blood, showing a bright red portion at the top and a darker, heavier portion at the bottom: the former could be identified as rich in bile—Avicenna's "colera rubea"—that is yellow and also lighter in weight and therefore rises to the top; the latter could be identified as melancholy—Avicenna's turbid "fex." It is at this point that Malpighi deployed his philosophy of color derived from Borelli, arguing that being dark or bright are "accidents" (*accidenti*) unrelated to the change of substance or its "mixing" (*temperie*). In fact, they can be repeatedly reversed since they depend on causes that have nothing to do with what the Galenists think. Malpighi went on to report a number of experiments on the caked blood intended to show that color is not a valid indicator of the nature of a substance. He started by arguing that putting some salt on the dark portion of blood will turn it very bright red; yet the earthy nature of salt ought to have turned it dark like melancholia, according to the Galenists. The simplest experiment consisted in turning upside down the caked blood and then observing the inversion of colors, the dark portion at the bottom turns bright red once it is at the top and, vice versa, the bright portion at the top turns dark once it is at the bottom. By putting the cake under water, even the bright red portions turn entirely dark. Malpighi was fully aware of the role of air in the changing color of blood: in another passage dealing with pulmonary disease, he argued that blood spits are bright red because blood is mixed with air, whereas blood in the rest of the body can be quite different in color and texture.[14]

The blood experiments carried out at Pisa between 1656 and 1659 were first reported in print by Malpighi's friend and colleague Carlo Fracassati in his 1665 treatise on the brain, *De cerebro*, with a clear attribution to Malpighi. Fracassati's treatise is a rather disorganized work covering nearly twenty double-column folio pages in the 1699 edition from the *Bibliotheca anatomica*. His report occupies just a few lines and follows closely the style of argument we have seen above, including the challenge to the link between dark blood and melancholia; Fracassati, too, explicitly mentioned the role of air in the changing color of blood from dark to bright red.[15] Yet his acknowledgment did not imply the recognition of the anatomical significance of that transformation: Borelli and his group thought that the substance of blood remained the same whether it was mixed with air or not. Indeed, in a later passage dealing with the changing color of blood mixed with various substances Fracassati explicitly warned readers not to trust colors, "*ne crede colori*," as he put it.[16] Thus it would be erroneous and anachronistic to attach great significance to Fracassati's report, as if it had claimed that Malpighi had discovered that exposure to air turns venous blood into arterial and, conversely, privation of air turns arterial blood into venous. In fact, Borelli's correspondence and the study of the Cimento experiments offer a revealing and entirely different context to interpret Malpighi's and Fracassati's claims: color is not a valid indicator of substance and—one may add—it is therefore legitimate to ignore it in the study of nature and in anatomical investigations in particular. Moreover, mixing with air was only one among several processes that turn the color of blood bright red, besides sprinkling it with salt, for example.

We are now equipped to attempt a fresh reading of Malpighi's celebrated *Epistolae* (1661) on the lungs, in which he announced the discovery of their microscopic structure. By studying the lungs of frogs, whose microstructure is easier to detect, he showed that the lungs were not spongy as was traditionally believed, but rather consisted of a series of smaller and smaller cavities or *alveoli* delimited by membranes covered by a network of blood vessels. Malpighi was able to see the anastomoses or junctions between arteries and veins, and also venous and arterial blood flowing in opposite directions. These findings provided direct visual proof of Harvey's circulation and showed that blood always flows inside blood vessels, thus closing the missing link in Harvey's system. Malpighi, however, did not stop with structure and tried to provide an explanation of the purpose of respiration, one directly influenced by Borelli. Their account has been aptly described as purely mechanical in that they did not attribute any role to chemistry. The role of the lungs was simply to mix blood with chyle so that it could nourish all the parts of the body. The motion of inflation and deflation of the lungs allows them to mix the blood better. This account soon proved grossly inadequate, since it was shown that animals could not breathe the same air but need fresh air to enter their lungs. More significantly from our perspective, in line with Borelli's views as highlighted in the contemporary correspondence and with the experiments of the Cimento, Malpighi paid no attention whatsoever to color change of blood in respiration. The finding that air changes the color of blood from dark to bright red seemed irrelevant, since Malpighi had shown that blood is never in direct contact with air but flows always inside blood vessels. Thus Malpighi did not see a connection between the color of blood in the lungs and the purpose of their structure, or to put it another way, he did not see a connection between the color of blood and respiration.

The English Scene: Boyle, Hooke, and Lower

It may seem peculiar to start a brief account of the English scene from a review of Fracassati's experimental report of Malpighi's observation; however, since there was a fundamental shift in the way the Pisan experiment was interpreted at Oxford and London, one may well argue that the same observation played a radically different role. Even the name differed: despite Fracassati's clear attribution to Malpighi, the experiment became known in England as "Fracassati's." In a brief report in the *Philosophical Transactions*, Henry Oldenburg teased out of Fracassati's disordered work the few lines dealing with color change in blood. Oldenburg reported that when blood has turned cold in a dish, the portion at the bottom is darker than at the top. The standard explanation that this observation would reveal the presence of melancholia in blood, however, was disproved by exposing blood to air, showing that blood becomes a florid red, "An experiment as easie to try, as 'tis curious."[17] As the admirable work by Robert Frank has shown, Fracassati's report reached England in the midst of a flurry of investigations on respiration and exerted a considerable impact. In a letter to Oldenburg of 26 October 1667, Boyle gave guarded approval to the truth of the experiment and Fracassati's interpretation that air plays a role in the color of blood. They differed significantly, however, in their interpretation of the significance of this observation: by now Avicenna's original report had underwent a major reconceptualization, from a proof of the composite nature of blood to evidence of the role of air in respiration.[18]

We now take a step back in time to consider a major figure, the Oxford Sedleian Professor of Natural Philosophy Thomas Willis. Willis combined medical, chemical, and anatomical interests with a sympathetic attitude to Descartes' mechanical understanding of nature, making the notion of fermentation a hallmark of modernity. In *Diatribae duae medico-philosophicae* (London, 1659), an influential treatise dealing with fermentation and the nature of fevers, Willis had provided a chemical reason for the change of color of blood, arguing that this phenomenon resulted from the combination of the sulphurous particles of blood with those of salt and spirit. In line with Descartes and other Continental investigators, he located the site where blood changes color in the heart: thus, unlike Borelli, he attributed a significant role to the change of color of blood. Willis too referred to the stratification of blood components once blood cools in a bowl, much like milk and wine; blood separates into a purer sulphureous part at the top, which in healthy individuals is bright red, and a thicker darker part at the bottom. We encounter here exactly the classical observation Malpighi and Fracassati had reported.[19]

Following his teacher Thomas Willis, Lower attached great importance to blood, its fermentation, and color change. In a letter to Boyle of June 1664, Lower discussed the reason for the difference of color between arterial and venous blood, arguing that arterial blood is bright red, whereas blood that has circulated through the muscles and thereby lost many particles before reaching the veins is darker. Unlike Harvey, Lower could confirm that blood let from the artery of a dog and kept in a "porrenger" or a small bowl remained bright red for one or two days, whereas blood let from a vein of the same dog remained dark, except for a thin layer at the top.[20]

In October and November 1664 the Royal Society debated whether air enters the body through the lungs. The fact that during the vivisection of a dog it was possible to revive the heartbeat by blowing air into the *receptaculum chyli*, whence it reached the heart through the thoracic duct, suggested a role for air in heart pulsation. On November 7, Hooke, together with Oldenburg and Jonathan Goddard, a former student of Francis Glisson at Cambridge, inserted a pair of bellows into the trachea of a dog and inflated its lungs. Opening the thorax and cutting the diaphragm, Hooke observed the heart beating regularly for over one hour as long as air was in the lungs. Hooke could not determine whether air entered the lungs, but he could establish that the motion of the heart was related to the inflation of the lungs, even though the two were not synchronous.[21]

The English anatomists soon elaborated on this experiment and went beyond this initial result relying on Lower's skill with vivisection. Emphasis on experimentation was a hallmark of both the Royal Society and the Italian Cimento Academy, but in this case the English investigators asked questions about issues the Italians had deemed of no significance, such as the color of blood. Initially, in *De febribus vindicatio* (London, 1665), a defense of Willis's *Diatribae* against the attack by the Bristol physician Edmund Meara, Lower had claimed that blood changes color in the heart as a result of a ferment in the left ventricle and also that blood in the lungs was venous, probably also because in his early trials the animal's lungs had collapsed and were empty of air; but regardless of where the color of blood changed, the very fact that it changed was deemed significant. Lower described his experiment in the same year in which Fracassati reported Malpighi's observations on the role of air in changing the color of blood—a finding still unknown to Lower.[22] But additional experiments refuted his initial view.

On 10 October 1667 Hooke and Lower performed an experiment at the Royal Society analogous to that of 1664, but this time they relied on two pairs of bellows instead of one, producing a continuous airflow. An incision in the pleura allowed air to exit the lungs, which thus remained inflated. In this way the animal was kept alive without motion in the lungs, thus showing that their motion was not required to keep the animal alive. By cutting a portion of the lungs, they could observe the blood moving through the lungs whether they were inflated or not.[23] This experiment refuted the purely mechanical view of respiration put forward by Malpighi and Borelli and later adopted by others in England.

Lastly, Hooke and Lower performed yet another two-part experiment on a dog. First, in the initial vivisection, they closed the trachea and showed that the blood coming from the cervical artery, after the blood had gone through the left ventricle of the heart, was venous. Thus the change of color of blood did not occur in the heart. Then the animal died, and they performed the insufflation experiment with the two pairs of bellows mentioned above, managing to obtain arterial blood from the pulmonary vein. Thus it was not the motion of the lungs, or a ferment in the heart, or the animal's heat that was responsible for the change of color of blood in the lungs, but only air, in line with Fracassati's report but against the view of Malpighi, Borelli, and Fracassati himself. This experiment strikes me as being especially significant in showing that the change of color of blood was not due to the soul or one of its faculties, because the animal was dead; although Hooke and Lower followed Harvey with respect to their acceptance of the circulation and emphasis on vivisection experiments, in this respect one wonders what they would have made of Harvey's belief in the soul and its location and role in blood.[24]

Thus in Borelli's group the experimental evidence with color indicators and the anatomical evidence that blood flows always inside blood vessels joined forces in denying a meaningful role to the change of color of blood, even after the realization that air was one of the factors responsible for this change. By contrast, in the group around Willis and Boyle the medical and chemical traditions joined forces in attributing a significant role to color and the change of color of blood, a phenomenon that anatomical experiments located in the lungs.

Finale: Malpighi and the Colorful Silkworm

Matters did not end there for Malpighi. Based on a range of sources, he revised his views on respiration, eventually accepting that a portion of air enters the blood through the lungs and plays a chemical—as opposed to a purely mechanical—role in respiration. In *De polypo cordis,* for example, first published in 1668, Malpighi studied the composition of blood starting from pathology, notably the polyps found during postmortems in the heart of deceased patients. In this work he dealt with the color of blood from a different perspective: observing blood through the microscope, Malpighi noticed that the red coloration was due to a large number of "red atoms," while the rest consisted of a network of whitish fibers. Malpighi put his finger on the color dichotomy between the macroscopic and the microscopic world, whereby what appeared on unaided visual inspection as a homogeneous red humor was shown by the microscope to be quite different; he then commented on the stratification of coagulated blood. Thus he relied on microscopic observations to reassess his own 1659 experiment and observation based on Avicenna and reported by Fracassati in 1665. He attributed the black color at the bottom not to melancholia—as some had believed—but to the great abundance of those particles he had called "red atoms," which he claimed returned to purple by a mere change of position. In a later passage of *De polypo cordis*, Malpighi spelled out that the lungs filter from the air a "salt of life" that awakens—"*suscito*"—the red potion of blood: thus the changing color of blood emerged as a significant feature of respiration.[25]

At this point, rather than following Malpighi's attempts to salvage what he could from his earlier views, I wish to shift to another topic. In 1669, the same year in which Lower published *De corde*, the Royal Society published Malpighi's treatise on the silkworm, *De bombyce*, and elected its author a fellow. By that point Malpighi had broken off his friendship and correspondence with Borelli and had departed in key respects from his former mentor's philosophical stance, notably with respect to color. His publications show a growing interest in color, but it is with *De bombyce* that Malpighi let his sensitivity to color burst forward in dramatic fashion. The reader of *De bombyce* is struck by a different Malpighi from that of the *Epistolae* on the lungs: now color—including many shades of gray—takes on a significant epistemic role in the description of silkworms, revealing an author with a striking sensitivity to nuanced shades and a remarkable ability to describe color in words. Color had become an integral part of Malpighi's descriptions, not only as a source of pleasure but also as a philosophical feature of the object under investigation. The fact that Malpighi was an art collector and enthusiast is related to his ability to observe and describe nature.[26] From the first pages of *De bombyce* we read of eggs turning from *violacea* to *caerulea* or light blue, then *sulphurea* and thereafter *cinerea* or ash-colored. Nor are Malpighi's identifications of color approximate: on one single page we find him distinguishing between *cinereus* or ash-colored and *fuliginosus* or soot-colored in describing the color of the just born silkworm, a color that soon turns into *perlatus* or pearl; the head is *coracinus* or raven black; the hairs and legs are *ziziphini* or jujube-colored. Elsewhere Malpighi describes the color of the silkworm as *achatis* or agate in those parts free of folds, and *argenteum* or silvery elsewhere. The silk thread is *luteus* or *auratus*, yellow or golden, or also *subalbus* or whitish with *sulphuris tinctura* or sulphur shade. Malpighi's sheer delight in describing and his remarkable sense of color are striking. Only in this way can we explain his extraordinarily nuanced descriptions. We are also reminded of his artistic interests, in which color played a major role. In an exactly contemporary letter of 24 November 1668 to the noted Sicilian collector Antonio Ruffo, who owned paintings by Rembrandt including *Aristotle with a Bust of Homer*, now at the Metropolitan Museum in New York, and *Homer*, now at the Mauritshuis in The Hague, Malpighi provided a rich account of artistic news about recent acquisitions and prices. He regretted that a fever—probably the same that in his *Vita* he attributed to excessive work on the silkworm—had prevented him from going to Parma and Correggio to see works by "Correggio e Parmigianino"; he did go to Mirandola, however, where he saw a nude Venus by Titian with "mezze tinte di Paradiso," or heavenly halftones. These observations on Malpighi's language and artistic interests, especially about color, go hand in hand with Matthew Cobb's attribution of the watercolor of the silkworm now at the Royal Society to *De bombyce*, since the drawing and especially the color range correspond remarkably to a development stage described in the text: Cobb included the color reproduction of the watercolor in his article. It is reasonable to surmise either that Malpighi himself was responsible for the watercolor, or that the artist who executed it worked directly under his supervision. Here too the letter to Ruffo proves useful, since Malpighi states that in the summer of 1668 he had employed a young painter "per dessegnarmi alcune cosette," to draw for me a few little things, and also to make copies of paintings by members of the Carracci family—Ludovico, Agostino, and Annibale. The young painter executed a few little things for Malpighi exactly at the time of his most intense work on the silkworm; thus it seems plausible that Malpighi used the same painter to help him draw and color the silkworm and make copies of paintings by the Carraccis.[27]

Color was not just a pleasurable appendage to the treatise: Malpighi identified the significance of color differences of eggs from *violacea* or purple to *sulphurea* or pale yellow as an indication of whether fertilization has occurred. He also made an attempt at artificial insemination by sprinkling male semen on the eggs, but his experiment failed and the eggs remained sterile, as testified by the lack of color change.[28]

This brief excursus has uncovered profound links among views about color and rival philosophical, anatomical, medical, and chemical perspectives. Sense perceptions and observations were mediated by deep-rooted and radically different philosophical positions in the process of observation: Borelli and his group—notably Fracassati and, for a while, Malpighi—downplayed the role of color, while Fracassati went so far as to warn readers not to trust color, "*ne crede colori.*" By contrast, Willis, Boyle, Lower, and Hooke adopted an approach according to which color appeared related to at least some properties of a substance and was therefore worthy of attention.

Even such an apparently straightforward and simple observation as the change of color in blood has required unraveling a complex web of philosophical opinions and chemical experiments. Malpighi's stance is especially revealing because he crossed boundaries in dramatic fashion: his initial tendency—probably stemming from his medical training—was to consider color as a significant diagnostic sign; following Borelli's prodding, color was then ignored in his investigations of the structure of the lungs and respiration, only to reemerge following his break with Borelli around 1667–1668. Malpighi's attention to color burst forth in all its esthetic nuances and philosophical significance in the study of the silkworm and the fertilization of its eggs, published by the Royal Society in 1669, and remained a feature of his views on nature until the end of his life as pontifical archiater.

Notes

1. Avicenna, *Canon* (Venice: In edibus Luce Antonij Junta, 1527), 7r; William R. Newman, "An Overview of Roger Bacon's Alchemy," in *Roger Bacon and the Sciences*, ed. Jeremiah Hackett (Leiden: Brill, 1997), 317–36.

2. William Harvey, *Exercitationes de generatione animalium* (London: Typis Du Gardianis, impensis Octaviani Pulleyn, 1651), trans. with an intro. and notes by Gweneth Whitteridge as *Disputations Touching the Generation of Animals* (Oxford: Blackwell, 1981), 254–55; Robert G. Frank, *Harvey and the Oxford Physiologists* (Berkeley: University of California Press, 1980), 40–41, 205–6.

3. Alistair C. Crombie, "Le proprietá primarie e le qualitá secondarie nella filosofia naturale di Galileo," in *Galileo*, ed. Adriano Carugo and Paul Tannery (Milan: ISEDI, 1978), 207–37. Surprisingly, William F. Bynum and Roy Porter, eds., *Medicine and the Five Senses* (Cambridge: Cambridge University Press, 1993), ignores color.

4. Galileo Galilei, *The Assayer*, sections 21 and 48, translated by Stillman Drake and Charles D. O'Malley in *The Controversy on the Comets of 1618* (Philadelphia: University of Pennsylvania Press, 1960), 234–48 and 308–14. Susana Gómez López, "Marcello Malpighi and Atomism," in *Marcello Malpighi, Anatomist and Physician*, ed. Domenico Bertoloni Meli (Florence: Olschki, 1997), 175–89; Pietro Redondi, *Galileo Heretic*, trans. Raymond Rosenthal (Princeton: Princeton University Press, 1987).

5. Giovanni Alfonso Borelli, *Delle cagioni delle febbri maligne della Sicilia. Negli anni 1647 e 1648* (Cosenza: Gio. Battista Rosso, 1649), 143.

6. René Descartes, *Treatise on Man* (Cambridge, Mass.: Harvard University Press, 1972), 9, see also n.19; it is noteworthy that Descartes uses the Aristotelian notion of "form" here. Galen, *On the Usefulness of the Parts of the Body*, 2 vols., trans. and intro. by Margaret Tallmadge May (Ithaca: Cornell University Press, 1968), 1: 205–6; A. I. Sabra, *Theories of Light from Descartes to Newton*, 2nd ed. (Cambridge: Cambridge University Press, 1981), 65–68; John Cottingham, "Descartes on Color," *Proceedings of the Aristotelian Society* 90 (1989–90): 231–46; Stephen Gaukroger, *Descartes: An Intellectual Biography* (Oxford: Oxford University Press, 1995), 158–64, 262–69, 345–46.

7. Robert Boyle, *Works*, ed. Michael Hunter and Edward B. Davis, 14 vols. (London and Brookfield, Vt.: Pickering & Chatto, 1999–2000), 4: 40–42, 50–51, 150, 5: 395–96; Laura Keating, "Un-Locke-ing Boyle: Boyle on Primary and Secondary Qualities," *History of Philosophy Quarterly* 10 (1993): 305–23; Frank, *Harvey and the Oxford Physiologists*, chap. 7, especially 184–88; Boyle, *Memoirs for the Natural History of Humane Blood*, in *Works*, vol. 10; Alan E. Shapiro, *Fits, Passions, and Paroxisms* (Cambridge: Cambridge University Press, 1993), 99–105; Harriet Knight and Michael Hunter, "Robert Boyle's *Memoirs for the Natural History of Humane Blood* (1684): Print, Manuscript and the Impact of Baconianism in Seventeenth-Century Medical Science," *Medical History* 51 (2007): 145–64; William R. Newman and Lawrence Principe, *Alchemy Tried*

in the Fire: Starkey, Boyle, and the Fate of Helmontian Chymistry (Chicago: University of Chicago Press, 2002), 276–77; William R. Newman, *Atoms and Alchemy* (Chicago: University of Chicago Press, 2006), 182–85.

8. Robert Hooke, *Micrographia* (London: John Martyn and James Allestry, 1665), 47–79. Sabra, *Theories of Light from Descartes to Newton*, 187–95; Shapiro, *Fits, Passions, and Paroxisms*, 99–105.

9. Borelli to Malpighi, Pisa, 5 March 1659 [*more pisano* = 1660], cited in Domenico Bertoloni Meli, "Additions to the Correspondence of Marcello Malpighi," in Bertoloni Meli, *Marcello Malpighi, Anatomist and Physician*, 275–308, on 281–82. On the role of the senses in medical diagnosis, see Bynum and Porter, eds., *Medicine and the Five Senses*; Tobias Heinrich Duncker, "'Wie nämlish könnten diese einander gleich sein . . . ?' Zur Hermeneutik farblicher codierung in der antiken Medizin," *Farbe, Erkenntnis, Wissenschaft. Zur epistemischen Bedeutung von Farbe in der Medizin*, ed. Dominik Gross and Tobias Heinrich Duncker (Berlin: Lit Verlag, 2006), 29–38.

10. Borelli to Malpighi, Pisa, 20 Dec. 1661, in Marcello Malpighi, *Correspondence*, ed. Howard B. Adelmann, 5 vols. (Ithaca: Cornell University Press, 1975), 1: 105–9, on 107–8; in reply to Malpighi's undated letter, 1: 104–5. See also Nicolaus Steno, *Observationes anatomicae* (Leiden: Apud Jacobum Chouët, 1662), paragraphs 33–34.

11. W. E. Knowles Middleton, *The Experimenters* (Baltimore: Johns Hopkins University Press, 1971), 234–37, on 234–35; see also 362, 364. As to Aristotle's views on color, see *Categories*, 9b10–32; *Metaphysics*, 1007a31–3; Boyle, *Works*, 4: 150, 152; Domenico Bertoloni Meli, "Authorship and Teamwork around the Cimento Academy," *Early Science and Medicine* 6 (2001): 65–95; William Eamon, "Robert Boyle and the Discovery of Chemical Indicators," *Ambix* 27 (1980): 204–9.

12. Borelli to Malpighi, Pisa, 20 Dec. 1661, in *Correspondence*, 1: 105–9, on 107–8.

13. The author of the tract by the Galenists, *Galenistarum triumphus* (Cosenza: Apud Io. Baptistam Russo, 1665), is Michele Lipari. The text is reproduced from a manuscript by Corrado Dollo, *Modelli scientifici e filosoofici nella Sicilia spagnola* (Naples: Guida, 1984), 290–304, on 296; the unique copy of the printed version is described in Rosario Moscheo, "The *Galenistarum triumphus* by Michele Lipari (1665): A Real Edition, Not Merely a Bibliographical Illusion," in Bertoloni Meli, *Marcello Malpighi, Anatomist and Physician*, 331–35.

14. Marcello Malpighi, *Risposta*, in *Opera posthuma* (London: A. & J. Churchill, 1697), 32, 40–41.

15. Carlo Fracassati, *De cerebro*, in Daniel Le Clerc and Jean-Jacques Manget, eds, *Bibliotheca anatomica*, 2 vols., 2nd ed. (Geneva: Johan. Anthon. Chouët & David Ritter, 1699), 2: 76b: "sed tenes etiam quam male ad oculorum fidem provocent, nam inter alia color saturatus, & nigricans in sanguine, quia fundum scyphi occupavit, & ideo sanguis melancholicus habetur statim ac in lancem projicitur, & aëri inde miscetur, mutatur; scis, quomodo confertim melancholia abeat, & debeant spectatores huic insulsae operationi, quam superciliose aggrediuntur, ludibrium, si parum morentur; etenim color non idem manet, clarior, ac nitentior redditur; nonnes Achilles hic in Thersitem degenerat? Sed de his alias, dum (ne me plagii arguas) fateor tuum hoc esse inventum, & te praeunte hoc didicisse; de natura tamen sanguinis, si aliquid ab obsitis jam situ suo sententiis expectamus, decipimur."

16. Fracassati, *De cerebro*, 2: 79a, emphasis in the original: "Si arbitrarer ex colore nos posse in indicia naturae sanguinis mutuari, non omitterem, quomodo purpureus color ex digestione salium volatilium cum oleis emicet sola etiam coctione, ut in succis symphyti, pyrorum etc. quomodo ab acido hoc in sanguine contingat; sed hic recte admoneor, *ne crede colori*."

17. [Henry Oldenburg], "An Experiment of Signior Fracassati upon Bloud Grown Cold," *Philosophical Transactions* 2 (1667): 492; Frank, *Harvey and the Oxford Physiologists*, 205–6.

18. Boyle to Oldenburg, 26 October 1667, in Robert Boyle, *Correspondence*, ed. Michael Hunter, Antonio Clericuzio, and Lawrence M. Principe, 6 vols. (London and Brookfield, Vt.: Pickering & Chatto, 2001), 3: 357–59, on 357.

19. Thomas Willis, *Diatribae duae medico-philosophicae* (London: Tho. Roycroft, impensis Jo. Martin, Ja. Allestry, & Tho. Dicas, 1659), separate pagination: *De fermentatione*, 1, 10; *De febribus*, 13–15, 20; Frank, *Harvey and the Oxford Physiologists*, 165–68.

20. Lower to Boyle, 24 June 1664, in Boyle, *Correspondence*, 2: 282–91, on 288–89.

21. Frank, *Harvey and the Oxford Physiologists*, 157–59. The insufflation experiment whereby air is blown into the heart was already known to Galen and is mentioned by Harvey and Malpighi.

22. Richard Lower, *De febribus vindicatio* (London: Jo. Martyn & Ja. Allestry, 1665), 117–18; Frank, *Harvey and the Oxford Physiologists*, 188–92 (at 190), and 206.

23. Robert Hooke, "An Account of an Experiment of Preserving Animals Alive by Blowing through Their Lungs with Bellows," *Philosophical Transactions* 2 (1667): 509–16; Frank, *Harvey and the Oxford Physiologists*, 330–31.

24. Richard Lower, *Tractatus de corde* (London: Typis Jo. Redmayne impensis Jacobi Allestry, 1669), 165–67; Frank, *Harvey and the Oxford Physiologists*, 189, on Willis and Lower on Harvey and the soul, 214–15.

25. Marcello Malpighi, *Vita*, in *Opera posthuma*, 16, idem, *Opere scelte* (Turin: UTET, 1967), 193, 200–1, 212–14, 533; Domenico Bertoloni Meli, "Blood, Monsters, and Necessity in Malpighi's *De polypo cordis*," *Medical History* 45 (2001): 511–22; John M. Forrester, "Marcello Malpighi's *De polypo cordis*: An Annotated Translation," *Medical History* 39 (1995): 477–92, on 483–84; Howard B. Adelmann, *Marcello Malpighi and the Evolution of Embryology*, 5 vols. (Ithaca: Cornell University Press, 1966), 1: 196–97.

26. On physicians and art collecting in this period, see Pamela H. Smith, *The Body of the Artisan* (Chicago: University of Chicago Press, 2004), chap. 6 on Sylvius de le Boë, and "Science and Taste: Painting, the Passions, and the New Philosophy in Seventeenth-Century Leiden," *Isis* 90 (1999): 420–61.

27. Marcello Malpighi, *De bombyce*, in *Opera omnia*, 2 vols. (London: Thomas Sawbridge and others, 1686–87), vol. 2, 2 (wrongly numbered 66), 7, 20; Matthew Cobb, "Malpighi, Swammerdam, and the Colorful Silkworm: Replication and Visual Representation in Early Modern Science," *Annals of Science* 59 (2002): 111–47, on 119–21; Malpighi to Antonio Ruffo, 24 Nov. 1668, in Malpighi, *Correspondence*, 1: 388–89; Martin Kemp, *The Science of Art* (New Haven: Yale University Press, 1990), part 3; Trevor Lamb and Janine Bourriau, eds., *Colour: Art and Science* (Cambridge: Cambridge University Press, 1995); John Gage, *Color and Meaning: Art, Science, and Symbolism* (Berkeley: University of California Press, 1999); and idem, *Color and Culture: Practice and Meaning from Antiquity to Abstraction* (Berkeley: University of California Press, 1993).

28. Marcello Malpighi, *De bombyce*, in *Opera omnia*, 2 vols. (London: Thomas Sawbridge and others, 1686–87), 2: 37, 42–43; idem, *Dissertazione epistolare sul baco da seta*, in *Il bacofilo italiano*, vol. 2 (1860), 1–90, on 77–78, 87–88; Howard B. Adelmann, *Marcello Malpighi and the Evolution of Embryology*, 5 vols. (Ithaca: Cornell University Press, 1966), 2: 856–58.

5

Seeing Is Believing: Professor Vagner's Wonderful World

MICHAEL D. GORDIN

> In a word, he was a very original Cat, although he didn't like any kind of originality and persecuted it: first of all, because he couldn't differentiate at all the original from the fashionable, and chiefly because everything original, in his opinion, shielded from us everything ordinary, simple, that we should study and that demands our help.
>
> NIKOLAI VAGNER, "Who Was Kot-Murlyka?"[1]

Russian zoologist Nikolai Petrovich Vagner (1829–1909) enjoyed going for walks in the countryside armed with a clear head, a magnifying glass, and an avid sense of curiosity. You never knew what you might observe. For example, as he wrote in a German scientific article in 1863, describing one of his walks:

> In the environs of Kazan' I found on 12 August 1861 under the bark of a dead elm a group of little white worms that didn't move. Under the microscope these little worms revealed themselves to be larvae of arthropods with antennae and tracheas, in a word, insect larvae. Each of them was filled with other larvae.
>
> I believed at first that what I was dealing with was a case of parasitism, which is so common among insects. The similarity of the enclosed larvae with the enclosing ones, a similarity that extended to the chief external identifying marks, however, soon led me to the thought that I was dealing with a normal formation, not with a pathological occurrence. On the other hand this was something too unusual, that in an insect larva a second generation of larvae could develop, and only after much to-ing and fro-ing and after many investigations did I come to the conviction, supported by evidence, that I had finally found the truth.[2]

Vagner wielded his prose with agility: he moved from observation, to description, to investigation, to conviction—all in the span of a few sentences. His finding, later dubbed pædogenesis, sparked intense disbelief and some sustained controversy. In the end, Vagner was vindicated, merely by sticking to his observations, invoking distinguished authorities among eminent scientists, and publishing early and often. This was a lesson of perseverance that he would long remember, much to the detriment of his reputation in later years.

He had made a high-stakes discovery against the objections of critics, and he would continue to stalk ever more elusive and unbelievable prey.

Vagner spent his life committed to the observation of animals in all their varieties, but especially two: invertebrates and humans. There are three major ways to consider the observation of people and bugs together. The first is to observe the bugs closely and then convince people of your observations (and the implicit interpretation that goes along with them). I shall call this "observation as persuasion." The second claims that one can observe humans as part of the same natural processes as their spineless counterparts. This approach flattens distinctions in the natural world in favor of comprehensive and general laws of biology—which the naturalist of course *also* claims to "observe": "observation as generalization." The third tactic capitalizes on the "observational authority" gained from observing insects—a tricky and painstaking business—to build credit for controversial observations of human phenomena, such as paranormal occurrences at séances. Vagner never did anything by halves. He would employ all three.

This essay follows several strands in Vagner's highly idiosyncratic career: professor of zoology and comparative anatomy first at Kazan' and then at St. Petersburg University, and also an author of classic children's stories published under the persona of Kot-Murlyka (lit. "Cat-Purr"). Although the narrative is biographical in structure, its purpose is to excavate the category of "observation" for naturalists in Imperial Russia and thus add to the taxonomy introduced by Katharine Park, Gianna Pomata, and Lorraine Daston in part 1 of this volume. Among the several dominant Russian terms to describe the investigation of nature—*opyt* ("attempt," "experiment," and "experience"), *issledovanie* ("research," with the definite connotation of literally following), and *ispytanie* ("probe," bearing whiffs of putting something to trial)—only "observation" (*nabliudenie*) carries the specific sense of the *visual*, a sense of being passive, of letting nature wash over one. That is as far as the etymology takes us, and it is not very far at all.

The reason to focus on Vagner, to explore how he connected both his practices of observation and his observational narratives of the insect and the human world for his intended audiences, is that he repeatedly made explicit many of the cultural assumptions that lay behind "observation," and that at times put the category in tension with contemporary interpretations of "science." Vagner did not perceive the three observational strategies mentioned earlier as distinct; each was simply part of what it meant to "observe" in a hostile universe. For him, what linked both strategies and observations together was the ubiquity of "struggle" (*bor'ba*) in the inorganic, organic, and social worlds. It is commonplace now to emphasize the tropes of *control* in understanding experimental science, which prides itself on manipulation of an environment to heighten a single effect. Vagner, however, saw struggle as the defining feature of *observational* science as well: not only did one have to struggle to maintain clarity and focus on surprising phenomena, struggle to free oneself of preconceptions and biases, and then struggle to persuade people of the controversial reality; one also literally *observed* struggle. It was both the content and the form of what the naturalist-observer did. *Nabliudenie* was about as active as you could get.

Observation as Persuasion: Weird Sex

Even specialists on the history of the life sciences in Russia may not recall Nikolai Vagner. Although his name lightly veils his family's German ancestry, Vagner was Russian through and through. He was born in 1829 at Bogoslovskii Zavod in the region of Perm, but he moved as a child to Kazan', where his father, Pëtr Ivanovich Vagner, had obtained the chair in zoology at the local university. Nikolai received his secondary education at the second Kazan' *gymnasium* and entered the university in 1845. In 1849 he received his candidate degree and one gold medal for his thesis, "On the Best Characteristic Signs for the Classification of Insects." He then took a post as teacher of natural history and agriculture at the Nobleman's Institute in Nizhnii Novgorod, where he stayed until 1851, returning to Kazan' University for a master's degree.

From a career that began in the provinces, Vagner began to push to the metropoles. In 1855 he defended his doctoral dissertation, "A General View on Arachnids and a Particular Description of One of the Forms (*Androctoceus occitans*) Belonging to Them," in Moscow (it was later translated into Dutch).[3] In 1858 he went abroad for postdoctoral work in Giessen, and then returned to Moscow to edit the *Journal of the Moscow Agricultural Society*. He returned to Kazan' in 1861 as an adjunct in comparative anatomy and physiology and became an extraordinary professor the same year. On 9 June 1862 he became ordinary professor of zoology. He edited the *Scholarly Notes*, the official academic journal of the university, from 1862 to 1864, and from 12 May 1869 was the first president of the Society of Naturalists at Kazan' University. Vagner traveled to Europe for another academic trip in 1870–1871 and returned as professor of zoology and comparative anatomy at St. Petersburg University.[4]

To hear Vagner tell it, this training had only limited impact on his abilities as an observer. As he imagined his own student training from the perspective of the late 1880s, he and his schoolmates were not adept at observing from nature:

> We were all patriots and unconditional monarchists and were not troubled by any doubts or questions. This absence of ideal interests was expressed also in the interests of science and life. We related to science completely superficially, not from the philosophical side. We studied lectures from their formal and factual sides. We wrote everything down accurately and constantly and, it stands to reason, considered it a sin to skip a lecture of the main subject. These written lectures were almost the only sources of our knowledge.[5]

Not that we should necessarily take all this too seriously. Vagner was prone to exaggeration in all his writings, and he was very fond of dramatic revelation.

We know, for example, something more about his practices of entomological observation—practices that informed all of his graduate work at Kazan'—from the publications themselves. Consider, for example, that walk in the environs of Kazan' that yielded those observations of insect larvae. Observation was a simple affair. In the context of the vastly understudied Russian steppe and Volga regions, there were plenty of publishable zoological finds to be obtained simply by walking around with a magnifying glass. This was fortunate, because at this stage in his career, Vagner did not have access to substantial state support for expeditions or even the less-elaborate makeshift sets arranged by René-Antoine Ferchault de Réaumur in the essay by Mary Terrall in this volume. For Vagner, one needed only to be alert and to take one's eyes to where they would do the observing for you. And on that particular walk, he found something worth publishing.

He translated his original Russian article into German and sent it to the *Zeitschrift für wissenschaftliche Zoologie*, published in Munich by Karl Theodor Ernst von Siebold. Von Siebold chose to sit on the manuscript, considering the findings too implausible to publish. St. Petersburg naturalist Karl Ernst von Baer (1792–1876) later articulated this feeling of disbelief:

> One could expect that this discovery of Vagner's would create a great sensation, but also many doubts, before it was to achieve complete confirmation or refutation. That in a formed insect larva a brood of new larvae of the same sort could develop had until then never been observed; likewise, many deviations from the usual methods of reproduction in the higher animals also had never been observed in the lower orders.[6]

Only after von Siebold had found analogous phenomena (replicating Vagner's observations) outside Munich *and* received living samples from Vagner did he agree to publish the piece in 1863—with an introductory footnote explaining his delay.[7]

The findings remained controversial even as additional observations of the phenomenon poured in.[8] The dispute over the existence of these unusual *Diptera* larvae was resolved not through Vagner's persistence, or additional eyewitness evidence, but mainly through the support of the grand old man of Russian biology, Academician Karl Ernst von Baer. Von Baer published two articles supporting the discovery in the *Bulletin* of the St. Petersburg Academy of Sciences. Not only did he add his own observations of Vagner's specimens, his tremendous authority in European zoology as the world's greatest living embryologist led the Petersburg Academy to award Vagner the coveted Demidov Prize in 1863.

On the one hand, von Baer sealed the credit for the discovery for Vagner, and on the other, he completely eclipsed it. Von Baer endowed the phenomenon of sexually immature larvae reproducing parthenogenetically with its still-current moniker of pædogenesis: "Provisionally only one difference from *parthenogenesis* has revealed itself, that the newly emergent individual, from an immature and for that matter unfertilized egg, emerges as a sexually mature individual."[9] Although nationalist Soviet textbooks highlighted Vagner's role, most later accounts of the history of the phenomenon elided the Kazan' entomologist completely. Stephen Jay Gould, in the standard reference on the history of theories of reproduction, devotes much attention to pædogenesis, but never mentions Vagner. Instead, he attributes the discovery to its namer, von Baer, even though the title of the source Gould cites references Vagner prominently.[10]

Vagner, of course, had no sense of his impending marginality, and in 1865 was ready to gloat, retelling his discovery as a victory of observation:

> If each day one watches the larvae carefully with naked [*unbewaffneten*] eyes, one sees clearly that new larvae grow from these, and after 7–10 days again bring forth new larvae just as the former did.
>
> Such observations must surely be proof enough for any skeptic, as long as there arises no suspicion that the observer himself has intentionally meddled with the facts.[11]

We should note Vagner's stress on the *absence* of intervening instruments, on the importance of the trained *naked* eye as the conduit of (passive?) observations. Indeed, by the early twentieth century, it was considered so easy to find pædogenesis in the wild in various species of *Diptera* that it was touted as an excellent pedagogical tool to introduce students to fieldwork.[12] Pædogenesis had become entirely domesticated *qua* observation.

In his later entomological studies, such as his attempts to explain the coloring of butterfly wings by subjecting larvae to electric currents, Vagner again returned to the tropes of observation explicitly—even though in this instance he was dealing with a more obviously *experimental* setup, a baby step toward Perrin's devices described in the essay by Charlotte Bigg in this volume.[13] Vagner would later relate that in his sole meeting with famed French physiologist Claude Bernard, the latter expressed admiration for precisely this kind of experiment *cum* observation developed by the Russian with phrases like "zoology will only stand on real scientific ground when physiology becomes its leader."[14] (Bernard's emphatic writings on the inferiority of observation to experiment lead one to doubt the true extent of agreement between the two.)[15]

By this time, however, Vagner was less interested in being a narrow observer of insects. Instead, he shifted to grander schemes, hoping to observe even the abstract, the extremely general. As he commented in his St. Petersburg University lectures in 1879: "In the world there exists one phenomenon, which comprises in itself all of the rest. This world phenomenon, in which are assimilated all the facts observed in nature, is the gradual complexification (*oslozhnenie*) or development of everything that exists."[16] He wanted here to convince humans that they fit into the same patterns as insects. And this was not going to be just a struggle. It was to be a struggle *about* struggle.

Observation as Generalization: Where's Wallace?

Consider these three quotations by Vagner, the first drawn from a textbook for children, the second from one for adults, and the third from an editorial from his interdisciplinary magazine, *Svet*:

> Thus, you see, in nature there is a constant battle. Here one can't be weak, clumsy, unaware, clueless and lazy. . . . Couldn't one truly call this battle a *battle for life* or *for existence?* And from this battle, in the very end, there constantly emerge more perfect, more solid, and stronger animals.[17]
>
> [Nature's] goal is achieved by the battle for existence, the battle among the elements of organs, the battle among the organs themselves, the battle among organisms, finally, the battle among entire groups of organisms. Everything battles so that it can destroy everything weak, ugly, which doesn't harmonize with the environment and in general with surrounding conditions. In the strata of our planet, and maybe on other planets, there are buried many unharmonious, ugly forms, which were at their time and place made so that development would further new, more complex, harmonious forms via impossible anachronisms.[18]
>
> In the life of the world, struggle continues as an endless, central theme, diversifying itself in billions of different forms. It began from the first steps,

> from the first germs of the planetary system and, developing, moved wider, further—into the endless distance of the future.[19]

Struggle metaphors in the life sciences in the second half of the nineteenth century evoke natural selection: Darwinian evolution through Malthusian conflict. But there was a bit of a twist here: Vagner was the single prominent advocate in late imperial Russia of natural selection . . . in the form expressed by Alfred Russel Wallace, *not* Charles Darwin. This was a peculiar position, and one first needs to understand Vagner's motivations for advocating evolution, and then Russians' resistance to the same, in order to make sense of it.

After the acceptance of pædogenesis, Vagner set his sights higher. He had a dim opinion of "those scientists who do not want to see in natural history anything besides naked data and facts."[20] He exhorted an interdisciplinary meeting of Russian scientists to seek the whole picture:

> Open a book of any scientific journal and you will see that fruitful scientific works, those which really push science forward, those which expand our worldview or give results directly applicable to life, stand out like bright oases.—Meanwhile, there follows a plethora of works that comprise the material for the future growth of science. Finally, there appears a large contingent of scientific workers who, with true pleasure, gather various petty facts with the firm faith that even these will one day prove useful. One should admit that this faith is sometimes justified, but how much labor in this blind, impassioned work falls in vain![21]

In the small hive of Russian naturalists, Vagner did not want to be a worker bee. He wanted to throw his weight behind a monumental general law, recognizing that "[s]uch general laws are explicated slowly."[22] These laws were not the fruits of speculation; they were *observed realities* just like the multiplying larvae in the decayed elm stump (or like the economies observed in the essays by Harro Maas, Theodore M. Porter, and Mary S. Morgan in this volume). You just needed to know *how* to observe.

This is where natural selection comes in. The tortured history of the reception of Darwin's theory in Russia has been recounted many times. Russian naturalists were generally supportive of the idea of common descent. This can be easily observed in the publication history of the classic works: *The Origin of Species* was translated into Russian in 1864 by S. Rachinskii, with a second edition in 1865; noted physiologist I. M. Sechenov translated *The Descent of Man* in 1871, the same year as the English edition, with a second edition in 1874; *Variation of Animals and Plants, The Expression of Emotions,* and *The Voyage of the Beagle* all appeared in the 1870s; and between 1907 and

1909 botanist K. A. Timiriazev ("Darwin's Russian bulldog") oversaw an edition of eight volumes of Darwiniana in Russian.

Among the general enthusiasm, criticism came from two fronts. Conservative intellectuals, such as Nikolai Strakhov and Nikolai Danilevskii, attacked the theory for being irreligious, materialist, and corrosive of morals.[23] The main source of scientific objections, as detailed by Daniel Todes, was the perception among Russians that Darwin's reliance on the thinking of Thomas Robert Malthus was incorrect.[24] Alfred Russel Wallace, Darwin's rival for priority in discovering natural selection, is oddly absent in all of this scholarship.[25] And yet, if one wanted to test the resistance to Malthusianism in Russian culture, Wallace was the more Malthusian and human-directed of the two evolutionists.[26]

There were not one but three Russian editions of Wallace's *Contributions to the Theory of Natural Selection* (1870). The first appeared in 1876, edited by a man named Lindeman, who mistranslated, arbitrarily reordered chapters, and cut out the theistic conclusions.[27] The second edition came out from a certain G. B. Our Vagner did not care for this version at all, mostly because G. B. "directed toward me his astonishing grumbles that I had the insolence to demand respect for the views of scientists such as Wallace."[28] Vagner had had enough; he would put out his own translation, complete and unexpurgated, with updated footnotes, new illustrations, and an appendix that included his own views about what Wallace had gotten right and wrong.[29] In 1879 a man named Gusev published an entire book devoted to these translations and interpretations of Wallace, and he found that "only in the translation under N. P. Vagner's editorship does Wallace appear before the Russian public with his present and full form of thought on the question of the origin of man."[30]

Why so much attention to Wallace? The first reason was his strict adherence to natural selection, rejecting all vestiges of Lamarckian adaptationism, which Darwin's "pangenesis" theory of heredity in part preserved. Vagner saw pædogenesis as a consequence of a Malthusian pressure: there were not enough males to go around fertilizing eggs, and asexual reproduction by adult insects (parthenogenesis) would require resources to bring the females through gestation, so there was a selective pressure in favor of immature asexual reproduction. The second reason Vagner backed Wallace over Darwin concerns the most famous conflict between the two British naturalists: the adequacy of natural selection to explain human consciousness. Here is Wallace's judicious account of the difference:

> My view, on the other hand, was, and is, that there is a difference in kind, intellectually and morally, between man and other animals; and that while

> his body was undoubtedly developed by the continuous modification of some ancestral animal form, some different agency, analogous to that which first produced organic *life*, and then originated *consciousness*, came into play in order to develop the higher intellectual and spiritual nature of man.[31]

Wallace and Vagner both thought there were limits to natural selection, and they both found evidence for it in the same place: Spiritualism. Vagner's adherence to the doctrines of modern Spiritualism—table turning, table rapping, spirit materialization, automatic writing, and so on—will be addressed in the next section. Vagner's conversion was independent of Wallace's views on the subject, but it surely only enhanced Vagner's belief in natural selection as the best candidate for a universal law of development that the most prominent British scientific advocate of both natural selection and Spiritualism was Wallace.

For Wallace, belief in Spiritualism was crucially about *observation*:

> Each fresh observation, confirming previous evidence, is treated [by critics] as though it were now put forth for the *first* time; and fresh confirmation is asked of it. And when this fresh and independent confirmation comes, yet more confirmation is asked for, and so on without end. This is a very clever way to ignore and stifle a new truth; but the facts of Spiritualism are ubiquitous in their occurrence and of so indisputable a nature, as to compel conviction in every earnest inquirer.[32]

Wallace's observations of Spiritualism that led to his own conversion took place at séances in 1865–1866. Malcolm Jay Kottler and others have argued convincingly that this experience of supposed communication beyond the borders of death persuaded Wallace that more must be involved in the evolution of man than natural selection alone.[33] The supposed contradiction between Wallace the rigorous naturalist and Wallace the devoted mystical spiritualist has spawned the plethora of recent biographies of Wallace.[34] The same thing sparked some of Vagner's interest, and he threw himself into Spiritualism *à la* Wallace: first piggybacking on his observational authority from his entomology to argue for the validity of his findings, and then invoking Wallace (in counterpoint to von Baer's earlier intervention on his behalf) as an even *better* observer with a stronger reputation to justify his own claims.

Observational Authority: Psychic Polemic

Nikolai Vagner was an avid polemicist, but not an unfair one; he recognized that Spiritualism was particularly difficult for the arch-rationalist (and here the scientist was the prime exemplar) to swallow. In 1902, Vagner published an appropriately titled monograph, *Observations on Spiritualism*, which attempted to explain how difficult these phenomena were to observe:

> Here the conditions of observation are so capricious (i.e., varied and elusive) that it is almost never possible to predict that such and such a phenomena will appear and such and such an experiment will work. . . . The main reason for this is 1) the unusual complexity of these phenomena and 2) the impossibility of knowing and studying them as a consequence of the inadequacy and limitations of our own organism.[35]

Anyone interested in Spiritualism required special instruction in how to conduct inquiries into the relevant phenomena: "In all these cases [of dark or dim rooms] I advise people who want to be convinced of the reality of these phenomena to maintain composure, patience, and not to arrive at a definitive conclusion after the first séances in which they have participated. Only after long and careful pondering can the observer arrive at the true conclusion."[36] Fundamentally, the observer *was supposed to resist* the existence of these phenomena at first: "I willingly concede that these facts are in the highest degree improbable, that they sharply contradict all contemporary psychological and natural-historical data. They unexpectedly open before us that quasi-fantastic world, in the existence of which we are unaccustomed to believe to the extent our consciousness has developed, developed apparently quite firmly, thanks to exact, experimental research. But nevertheless these are facts."[37]

From his first public writings on Spiritualism in 1875, Vagner always stressed that Spiritualism should not be understood as a mystical belief system, but as a collection of "mediumistic phenomena"—raw observations—that needed to be investigated using the standard methods of science. (In this, his position foreshadows that of the stroboscopists in the essay by Jimena Canales in this volume.) The crucial ingredient for a Russian séance, exactly parallel to contemporary British standards, was a "medium," someone purported to channel phenomena between the two worlds. To the extent that Vagner thought a séance was an "experiment," it was an experiment to test whether the medium's claims were accurate. The way this was done was through "observation": one eliminated all interference that might disturb the medium (bright lights, intrusive experimental devices), and then simply observed. This is similar to Vagner's entomological research in its emphasis on mere presence and attention, for those were his proven mechanisms. One needed to be skeptical, to be sure, but if one were *too* skeptical, then this might distort the phenomena, scare the medium, and ruin the observational setup.

This form of open inquiry to establish the phenomena was where Vagner had started in the 1870s, although soon he began to consider Spiritualism a quasi-religious system.[38] The stakes were quite high: "For me, spiritualist phenomena explain the life of the entire visible and invisible world. They connect the physical world and the transcendental, science with religion."[39] But one should not assume Vagner to be antirational or dogmatic from the beginning. By his own lights, this *profession de foi* was no different from his advocacy of the universal laws of biology. Wallace was not only an interlocutor and a source; he was an *exemplar* that Vagner observed and hoped to emulate.

When he returned from Europe in 1871 and began to teach at St. Petersburg University, Vagner had at first scoffed at his schoolmate and now colleague, chemist Aleksandr Butlerov, for advocating scientific investigation of mediumistic phenomena. But Butlerov insisted that Vagner at least attend a few séances with the noted European medium Camille Bredif, organized by himself and his cousin Aleksandr N. Aksakov. Vagner would later describe his conversion experience in terms analogous to his account of the discovery of pædogenesis in the early 1860s:

> In the fall of 1874 in Petersburg Bredif arrived, and Butlerov invited me, together with A. Ia. Danilevskii and A. I. Iakobi, to participate in his séances. The latter, however, could only participate in two séances. The whole array of strong visible phenomena convinced me finally of the existence of mediumistic facts and pushed me to print a letter in the *Messenger of Europe*. Butlerov and A. N. Aksakov took a lively part in this publication. They in no way believed that my letter would convince anyone of the existence of mediumistic facts. But in the end it turned out differently. My scientific authority and my firm conviction awakened the entire intelligentsia. From all sides Butlerov and I began to receive letters, with a request for admission to séances with Bredif. Meanwhile, the party of a priori skeptics did not dither and began to print articles in refutation of those facts that they had not seen.[40]

The reference to pædogenesis is not tendentious on my part. Consider the April 1875 article referred to above, the one that sparked the Spiritualist controversy of 1875–1876 by triggering chemist Dmitrii Mendeleev's campaign against Spiritualism.[41] Vagner began this article with the oft-repeated quotation from *Hamlet* (I.v) when the doomed prince tells his friend, "There are more things in heaven and earth, Horatio, / Than are dreamt of in your philosophy." This quotation was very much in vogue among British Spiritualists, but Vagner was not interested in invoking them (yet). Instead, he recalled that the permanent secretary of the Academy of Sciences, Konstantin S. Veselovskii, had intoned these words in reference to Vagner's entomological discoveries.[42] Vagner tried to bootstrap from his acknowledged reputation as a skillful observer in the realm of the small and wriggly to the dead and immaterial. In a footnote to a German article on Spiritualism by Vagner, A. N. Aksakov cemented the connection:

> Among the main scientific works of Prof. Vagner that are well known to specialists an entirely special sensation was aroused by his remarkable discovery of a particular form of asexual reproduction (pædogenesis). . . . This discovery of Prof. Vagner's was first met generally with distrust, the reported facts were seen as something unbelievable and impossible; new observations, however, soon set them firmly as facts.[43]

Vagner's counterparts in the polemic understood what he was doing and called him on it.[44] S. Rachinskii (also the translator of Charles Darwin's *Origin of Species*) conceded that *something* must have occurred in the room if Vagner and Butlerov said it had; he just thought the medium produced these phenomena himself (or herself).[45] The hypothesis of spirit action from the beyond was simply unnecessary.

Perhaps no one agreed with Vagner more on the link between observational authority and Spiritualism than Mendeleev, who targeted Vagner (among others) with his supposedly impartial investigative commission.[46] Vagner maintained a constant back channel of correspondence with Mendeleev during the commission's activity (fall 1875–spring 1876), and he repeatedly insisted that the chemist was observing irresponsibly. First, observation was fundamentally an *individual* affair and did not require (or might even be harmed by) the presence of too many persons: "If you sincerely strove to convince yourself that mediumistic phenomena exist, then the form of your action would have been entirely different.—For this you didn't need to gather a commission of scholarly physicists and mechanics. You yourself are the authority and judge for yourself."[47] The second problem was the prejudice induced by excessive skepticism on Mendeleev's part: "The black worm [of suspicion] drove you further. He showed you things that any reasonable observer would have associated with the sphere of subjective sensations and hallucinations. And you?! You came forth with these sensations as with proofs against the medium, accusing him of charlatanry."[48] (Here also was an echo of entomology; the skepticism of von Siebert and others had hindered the acceptance of Vagner's findings.)

Mendeleev, for his part, emphasized two major problems with Spiritualists as observers of nature. The first was that they were convinced in advance that the phenomena existed, and hence were likely to conflate the very subjective and objective events that Vagner wanted to distinguish. Characteristically for a physical scientist, who demanded *experimental* confirmation of controversial claims, Mendeleev believed that predisposition could seriously distort *observations* conducted in a purely natural-historical vein of nonintervention. The desire to observe would in turn yield the observation of what one desired.[49] The second problem was the refusal of many Spiritualists to recognize fraud. In this manner, Vagner was very similar to Wallace, who claimed that no medium had ever committed fraud in his presence. (Even fraudulent mediums produced mediumistic phenomena *despite themselves* when he was present.)[50]

None of the criticism prevented Vagner from defending the faith. Long after Mendeleev had ceased to care about the propagation of Spiritualism by scientists, Vagner continued to explore new methods of observing the phenomena, such as the difficult craft of spirit photography, a significant concession on his part to the potential of instruments to aid naked-eye observations (and with resonances in the essay by Kelley Wilder on Henri Becquerel in this volume).[51] Much like William James and Oliver Lodge in parallel contexts, Vagner continued to push a program of psychical research that would help uncover (via observation) the general laws of these phenomena. In fact, Vagner considered observation via automatic writing a superlative exemplar of observation for any scientist: "This phenomenon, by its objectivity, especially affords facility for observation, and deserves full attention and investigation from competent persons and institutions."[52] He also never stopped engaging in "real science" during this period. He conducted a series of expeditions to the White Sea on Russia's Arctic coast in 1876, 1877, 1880, 1882, and 1887, and in 1881 helped to establish the first Russian marine biological station, which he directed until it closed down. His research on Arctic marine invertebrates, published in the 1880s, was a major collective achievement.[53]

Vagner's visibility in the popular press sagged after Mendeleev's assaults, intensified censorship against mediumism in general, and direct attacks on his character by rationalist writers—and even some by pronounced antirationalists, such as Fedor Dostoevskii.[54] And then, after a decade, Vagner's polemic reemerged in a rather curious episode pitting him against the most famous of his contemporaries, writer Lev Nikolaevich Tolstoi. The conflict between Vagner and Tolstoi broke out in response to Tolstoi's four-act play, *The Fruits of Enlightenment*, first drafted in 1889 and finished in 1890. It had begun as a quick sketch in November 1886 but was then shelved until, as Tolstoi would have it, "my daughters asked to be able to play it, I started to correct it, in no way thinking that it would ever go further than our home, and it ended with its being distributed."[55] *The Fruits of Enlightenment* is essentially a classic farce, in which a smart peasant girl (Tania) manages to win her hapless lover (Semën) and persuade her noble master, Leonid Fedorovich Zvezdintsev, to sell land to Semën's village family at reasonable rates. The rub is how

Tania pulls this off. Zvezdintsev was an avid Spiritualist, and Tania rigs a séance by persuading him that Semën, while napping, has mediumistic powers, and then "materializing" the deed of sale during the séance.

Vagner was not amused, and he wrote an irate letter to Tolstoi after attending a reading of the play in Petersburg, horrified at what he saw as a "satire on professors, on scientists!"[56] In particular, he was upset with the learned professor, Aleksei Vladimirovich Krugosvetlov (meaning "surrounded by light," which may have been a barb at Vagner's crypto-Spiritualist magazine venture from the late 1870s, *Svet* [*Light*]). Tolstoi described Krugosvetlov thus: "Scientist, about 50 years old, with calm, nicely self-confident manners and a similar slow, singing speech. Speaks carefully. To those who don't agree with him he relates curtly, contemptuously. Smokes a great deal. A thin, mobile person."[57] This is a pretty good description of Vagner. The tone of the professor as he explains away other theories and defends his doctrines is also strongly Vagnerian. To pick one Krugosvetlov monologue: "The same thing here also. The phenomenon is repeated, and we subject it to research. And that's just the start, we subject the researched phenomena to laws general to other phenomena. The phenomena, after all, appear supernatural only because the reasons for the phenomena are attributed to the medium himself."[58] Perhaps most Vagnerian of all, Krugosvetlov refuses to admit the fraud after it is exposed by Semën's rival for Tania's affections, the valet Grigorii.[59]

Whatever Vagner's personal wounded feelings, his objective ire was sparked by Tolstoi's refusal to accept mediumistic facts as they stood. Criticizing scientists for hubris is one thing—and Tolstoi had repeatedly attacked physicians (and Darwinists)—but refusing to accept the sincerity of Spiritualists was quite another. It was not simply insulting, it was uncharitable—and hence un-Christian, a criticism he felt sure would prick Tolstoi. Here, Vagner's Spiritualism was bound together with his corporate identification with scientists, his insistence on observation, his orientation toward struggle, and finally, his literary persona. It is time, at last, to turn to Kot-Murlyka.

Observers Observed: Endless Childhood

As early critics of Spiritualism noted, one of the reasons why Vagner needed to be stopped from propagating his dangerous creed was that he was a gifted literary stylist: "The uncontrived sincerity of tone, the energy and literary achievements of exposition, finally, the adoration of scientific authority—all of these are very important tools, found in the hands of Mr. Vagner, and which he uses to the fullest."[60] Perhaps the best comparison is with the writer who gave him the name for his literary persona: E. T. A. Hoffmann, whose

"Kater Murr" exhibited just the range of nonchalance, mockery, and sincerity that characterized Vagner's writings. One of the foremost critics of classic Russian literature, D. S. Mirsky, lauded Vagner as the only author of his age to write well outside of the canons of the "natural school."[61] His anti-Spiritualist critics did not shy away from approaching Vagner as primarily a writer about nature; neither, I contend, should we.

I conclude with Vagner's position in Russian literature for several reasons. First, separating off his scientific work from his literary activity implies an artificial separation denied by the subject himself and absent from the sources. More importantly, belles lettres in late Imperial Russia held a position of authority most closely reflected by science in contemporary Western culture. My goal is to show that the practices and theories of observation Vagner at first developed in the field and séance room became tools to develop a prominent literary status, and thus the history of Russian culture is incomplete without attention to "observation" as a category, scientific and otherwise. For Vagner treated his literary efforts like his Spiritualist publications and scientific articles: as tools to train Russian children to *observe*—not investigate, not experiment on—the world around them. Literature, and especially literature for children, was thus the highest stakes game of all.

According to an 1892 autobiographical piece, Vagner's interest in literature emerged from folktales told to him by his family's nanny, Natalia S. Aksënova. A precocious youth, he memorized extended tracts of various stories that pleased him with their rhythm and cadence, and performed impromptu one-boy shows for his relatives, but he did not endeavor to write fiction until he reached adulthood. The trigger for his own ventures was the 1868 publication of Hans Christian Andersen's tales:

> Reading the praiseworthy and even rapturous reviews of these tales in our magazines and newspapers, I bought them and read them through. Many of them I enjoyed as well, but I was also dissatisfied with many of them; I found them weak and asked myself the question: couldn't I perhaps write something like this or even better? Thus the task was posed, and in three years I had written about a dozen stories, which comprised the first edition of the "tales" of "Kot-Murlyka."[62]

Andersen may have been the immediate trigger, but Kot-Murlyka took on further tasks. His writing for children assumed two forms: nonfictional guides to induce children to observe outdoor nature, and fantastical tales to induce children to observe their inner natures.[63]

Although Vagner's literary reputation hangs entirely on the second group, the first was more important to him. He produced several popular texts of natural history for children of various ages, including two editions of a translation of Paul Bert's French textbook. As he indicated in his preface to the first edition, the purpose of having such a book available in Russian was to encourage children: "For them [children] it [nature] is an open book, in which they can, unwittingly, learn a great deal if only they had a guide near them. It would show them how one can observe, look into things, and chiefly, think about subjects and phenomena."[64] Children, more than adults, were natural observers, and must observe *Russia*: "Let's stop, although not for long, on still another country, which lies both in Europe and in Asia, but which for every Russian lies closer than any other, because it lies in his own heart. You, of course, have guessed, that I want to speak to you about the population of your mother country, with which you are connected through your birth, language, your faith, customs, habits, finally, your character."[65] Vagner was once again pushing for a repeat of his pædogenetic discovery. Accomplished with a magnifying glass, a pair of walking shoes, and some natural curiosity, it had been a victory for Russian natural history. If children simply indulged the childlike in themselves, they would produce wonderful science.

Then one had the tales of Kot-Murlyka, which appeared in nine editions between 1872 and 1913, and in 1923 were issued in their first Soviet edition. This was also the last edition until 1990, right before the collapse of the Soviet Union.[66] The corpus ranges widely among morality tales, children's heroic narratives, just-so stories, and prose poems about nature.[67] Interestingly, some of his peers considered these stories deeply unsuitable for children:

> They [the stories] are undesirable, first, because they are almost all unbearably heavy tales, full of woe, suffering, unilluminated human gloom. All of this acts too strongly on the reader, acts oppressively, even frighteningly in places. Second, these stories for children are not desirable even for those who are older because the philosophy of many of them can lead an impressionable, sensitive child to despair, to a total lack of desire to live.
>
> One of the main themes of the tales of Kot-Murlyka is the inevitability of woe and sufferings, the naturalness of woe and suffering, the battle of knowledge and blind naïve faith in the wonderful [*chudesnoe*], in which the sympathies of the author are always on the side of the victor—knowledge. Next comes a description of the muddle of human society, in which poverty is completely unavoidable, and in which even the battle with poverty and human unhappiness is completely fruitless. Finally—the heavy battle for existence, the extinction of everything weak, incomplete, or not suitable for life.[68]

Even here, in the realm of fictional tales for children, the same themes emerged: his opponents' resistance to careful observation, and Vagner's imperviousness to opposition. The man kept writing until the end of his life (although his later works abandoned some of the fantastic features and took a regrettable slide into anti-Semitism),[69] just as he held on to his Malthusian Darwinism (or should we say Wallaceism?), as he had persisted with pædogenesis, and as he continued to defend Spiritualism. All of these were marked with struggle, by struggle, and through struggle: in the discovery of pædogenesis, it was the struggle of observation as persuasion, convincing recalcitrant entomologists; in his attachment to Wallace's natural selection, observation as generalization found struggle woven in the fabric of nature; and that struggle needed itself to be struggled against in order to achieve the expectant calm of the Spiritualist observer. It was not so much that Vagner believed his observational authority sufficed to bear him along against the slings and arrows of outrageous fortune; rather, it was the very nature of observation, properly done, to exhibit and elicit these kinds of attacks. The fact that his critics continued to attack him suited him just fine. He knew he would eventually be vindicated. All his observations supported it.

Notes

1. N. P. Vagner, "Kto byl Kot-Murlyka?" in *Skazki Kota-Murlyki* (Rostov-on-Don: Prof-Press, 2001), 5–10, on 5–6. Russian dates are given according to the old-style Julian calendar, European dates in the new-style Gregorian; transliterations follow the modified Library of Congress standard. All unattributed translations are my own.

2. Nicolas Wagner, "Beitrag zur Lehre von der Fortpflanzung der Insectenlarven," *Zeitschrift für wissenschaftliche Zoologie* 13 (1863): 513–27, on 514.

3. N. P. Vagner, *Obshchii vzgliad na klass zhivotnykh paukoobraznykh* (Arachnidae) *i chastnoe opisanie odnoi iz form, k nemu prinadlezhashchikh* (Kazan': Universitetskaia tip., 1854).

4. Biographical facts are drawn from G. A. Kliuge, "Vagner, Nikolai Petrovich," in N. P. Zagoskin, ed., *Biograficheskii slovar' professorov i prepodavatelei Imperatorskago kazanskago universiteta (1804–1904)* 2 vols. (Kazan': Tip. Imperatorskago Universiteta, 1904), 1: 286–89; N. Knipovich, "Vagner (Nikolai Petrovich)," *Entsiklopedicheskii slovar' Brokgauza i Efrona*, v. 5 (St. Petersburg: I. A. Efron, 1891), 343–44; and V. I. Mil'don, "Vagner, Nikolai Petrovich," in P. A. Nikolaev, ed., *Russkie pisateli, 1800–1917* (Moscow: Sovetskaia entsiklopediia, 1989), 1: 385–86.

5. N. P. Vagner, "Vospominan'e ob Aleksandre Mikhailoviche Butlerove," in A. M. Butlerov, *Stat'i po mediumizmu* (St. Petersburg: A. N. Aksakov, 1889), iii–lxvii, on vi.

6. K. E. von Baer, "Über Prof. Nic. Wagner's Entdeckung von Larven, die sich fortpflanzen, Herrn Ganin's verwandte und ergänzende Beobachtungen und über die Paedogenesis überhaupt," *Mélanges biologiques tirés du Bulletin de l'Académie impériale des sciences de St. Pétersbourg* 9 (1865): 64–137, on 66.

7. Wagner, "Beitrag zur Lehre von der Fortpflanzung der Insectenlarven," 513n.

8. M. Ganin, "Neue Beobachtungen über die Fortpflanzung der viviparen Dipterenlarven," *Zeitschrift für wissenschaftliche Zoologie* 15 (1865): 375–90; Fr. Meinert, "Weitere Erläuterungen über die von Prof. Nic. Wagner beschriebene Insectenlarve, welche sich durch Sprossenbildung vermehrt," trans. and ed. C. Th. V. Siebold, *Zeitschrift für wissenschaftliche Zoologie* 14 (1864):

394–99, on 397; N. Wagner, Meinert, Pagenstecher, and Ganin, "La reproduction parthénogénésique chez quelques larves d'insects diptères," *Annales des sciences naturelles, Série zoologique* 5, no. 4 (1865): 259–91; and K. E. von Baer, "Bericht über eine neue von Prof. Wagner in Kasan an Dipteren beobachtete abweichende Propagationsform," *Mélanges biologiques tirés du Bulletin de l'Académie impériale des sciences de St. Pétersbourg* 6 (1863): 239–41, on 240.

9. Von Baer, "Über Prof. Nic. Wagner's Entdeckung von Larven," 97. Emphasis in original.

10. Stephen Jay Gould, *Ontogeny and Phylogeny* (Cambridge, Mass.: Belknap Press, 1977); B. N. Shvanvich, *Kurs obshchei entomologii: Vvedenie i izuchenie stroeniia i funktsii tela nasekomykh* (Moscow: Sovetskaia Nauka, 1949), 698–99. This elision had already begun in the nineteenth century: Otto Taschenberg, "Historische Entwicklung der Lehre von Parthenogenesis," *Abhandlungen der Naturforschenden Gesellschaft zu Halle* 17 (1892): 366–453, on 398–400.

11. Nicolas Wagner, "Ueber die viviparen Gallmückenlarven," ed. C. Th. v. Siebold, *Zeitschrift für wissenschaftliche Zoologie* 15 (1865): 105–17, on 107. Emphasis in the original.

12. E. P. Felt, "Miastor and Embryology," *Science*, n.s. 33, no. 843 (24 Feb. 1911): 302–3, on 302.

13. Nicolas Wagner, "Influence de l'électricité sur la formation des pigments et sur la forme des ailes chez les Papillons," *Comptes rendus* 61 (1865): 170–72, on 172.

14. N. P. Vagner, "Vospominanie o Klod-Bernar," *Svet* 2, no. 2 (Feb. 1878): 64–65, on 65.

15. Claude Bernard, *An Introduction to the Study of Experimental Medicine*, trans. Henry Copley Greene (New York: Henry Schuman, Inc., 1949), 196–226.

16. N. P. Vagner, "Zoologiia," stenograph from lectures [1879], Rossiiskaia Natsional'naia Biblioteka (hereafter RNB), St. Petersburg, Russia, 18.167.6.46.

17. Paul Bert and Nikolai Vagner, *Ocherk zoologii* (St. Petersburg: N. P. Vagner, 1883), 166. Emphasis in the original. Vagner became enmeshed in a publication dispute over this translation, since publisher F. Pavlenkov offered a rival translation at the same time. Vagner framed his case even in this domain in terms of battle metaphors: "In all of these conditions, I hope that the public will understand and value that *here there isn't a simple competition, but a battle of scientific work with publishing capital and knowledge with its all too shameless exploitation.*" "Dokumenty po delu N. Vagnera s F. Pavlenkovym," [1883], RNB, p. 4. Emphasis in the original.

18. N. P. Vagner, *Istoriia razvitiia tsarstva zhivotnykh: Kurs filogeneticheskoi zoologii* (St. Petersburg: N. A. Lebedev, 1887), 33.

19. N. P. Vagner, "Bor'ba," *Svet* 1, no. 11 (Nov. 1877): 241–47, on 241. This was the first in a series of editorials on struggle: idem, "Prisposobliaemost' i tselesoobraznost'," *Svet* 2, no. 3 (March 1878): 69–72; idem, "Bor'ba za sushchestvovanie," *Svet* 2, no. 4 (April 1878): 105–10; and idem, "Organizm i obshchestvo," *Svet* 2, no. 7 (July 1878): 105–10.

20. Vagner, *Istoriia razvitiia tsarstva zhivotnykh*, 4.

21. N. P. Vagner, "O sredstvakh dlia resheniia slozhnykh nauchnykh voprosov," in *Rechi i protokoly VI-go s"ezda Russkikh Estestvoispytatelei i Vrachei v S.-Peterburge s 20-go po 30-e dekabria 1879 g.* (St. Petersburg: Tip. Imp. Akademii nauk, 1880), otd. 1: 39–47, on 41.

22. Vagner, *Istoriia razvitiia tsarstva zhivotnykh*, i. See also N. P. Vagner, *Zhorzh Kiuv'e i Et'en Zhoffrua-Sent-Iler (Fiziologicheskii ocherk)* (Kazan': Universitetskaia tip., 1860), esp. 55.

23. N. Ia. Danilevskii, *Darvinizm: Kriticheskoe izsledovanie*, vol. 1, pt. 1 (St. Petersburg: Merkurii Eleazarovich Komarov, 1885); and N. N. Strakhov, "Durnye priznaki," in *Kriticheskiia stat'i (1861–1894)*, vol. 2 (Kiev: I. P. Matchenko, 1902), 379–97.

24. Daniel P. Todes, *Darwin without Malthus: The Struggle for Existence in Russian Evolutionary Thought* (New York: Oxford University Press, 1989).

25. Alexander Vucinich's ostensibly comprehensive history of Darwinism in Russia—

Alexander Vucinich, *Darwin in Russian Thought* (Berkeley: University of California Press, 1988)—ignores Wallace completely and only mentions Vagner twice in passing. Even Todes minimizes our hero: he only receives three mentions, and two of them in the context of the work of Vagner's student, Modest Bogdanov.

26. H. Lewis McKinney, *Wallace and Natural Selection* (New Haven: Yale University Press, 1972), 81.

27. See A. Gusev, *Naturalist Uolles, ego russkie perevodchiki i kritki (K voprosu o proiskhozhdenii cheloveka): Po povodu perevodov knigi Uollesa: Estestvennyi podbor* (Moscow: M. Katkov, 1879), 4–5.

28. A. R. Wallace, *Estestvennyi podbor*, trans. G. Glazenap, ed. N. P. Vagner (St. Petersburg: F. Shushchinskii, 1878), ii.

29. He was in contact with Wallace during the production of this translation, since Vagner mentioned that Wallace had sent him some illustrations for inclusion (Wallace, *Estestvennyi podbor*, iv). I can find no mention of Vagner by Wallace, although the latter recalled having met Aleksandr Butlerov, a "biologist and also a spiritualist." Alfred Russel Wallace, *My Life: A Record of Events and Opinions*, 2 vols. (London: Champman & Hall, 1905), 2: 93. This misidentification of the *chemist* Butlerov might mean that Wallace confused him with Vagner.

30. Gusev, *Naturalist Uolles*, 8. Vagner himself reviewed Gusev: N. P. Vagner, "Otvet g. Gusevu na ego knigu: Naturalist Uolles i ego russkie perevodchiki," *Svet* 3, no. 4 (April 1879): 200.

31. Wallace, *My Life*, 2: 17. Emphasis in the original.

32. Alfred Russel Wallace, *A Defense of Modern Spiritualism* (Boston: Colby and Rich, 1874), 33.

33. Malcolm Jay Kottler, "Alfred Russel Wallace, the Origin of Man, and Spiritualism," *Isis* 65 (1974): 144–92.

34. Especially Ross A. Slotten, *The Heretic in Darwin's Court: The Life of Alfred Russel Wallace* (New York: Columbia University Press, 2004); and Martin Fichman, *An Elusive Victorian: The Evolution of Alfred Russel Wallace* (Chicago: University of Chicago Press, 2004).

35. N. P. Vagner, *Nabliudeniia nad mediumizmom* (St. Petersburg: F. Vaisberg and P. Gershunin, 1902), iv, ellipses added.

36. Ibid., xviii. See also N. P. Vagner to A. D. Butovskii, 2 Dec. (n.s.) 1883, Naples zoological station, Pushkinskii Dom, Institute of the History of Russian Literature (hereafter PD), f. 1, op. 2, d. 174, ll. 3–4ob., on 11. 3–3ob.

37. N. P. Vagner, "Mediumizm," *Russkii Vestnik* 119 (Oct. 1875): 866–951, on 878, ellipses added.

38. Vagner, *Nabliudeniia nad mediumizmom*, 121.

39. N. P. Vagner to L. N. Tolstoi, 10 April 1890, PD f. 231, d. 279, ll. 4–9ob., on 1. 8–8ob.

40. Vagner, "Vospominan'e ob Aleksandre Mikhailoviche Butlerove," xxxix–xl. (This Danilevskii is *not* the anti-Darwinist referred to above.)

41. See Michael D. Gordin, *A Well-Ordered Thing: Dmitrii Mendeleev and the Shadow of the Periodic Table* (New York: Basic Books, 2004), chap. 4; and also the suggestive interpretation of Vagner in Ilya Vinitsky, *Ghostly Paradoxes: Modern Spiritualism and Russian Culture in the Age of Realism* (Toronto: University of Toronto Press, 2009), chap. 4.

42. N. Vagner, "Pis'mo k redaktoru: Po povodu spiritizma," *Vestnik Evropy* (April 1875): 855–75, on 855.

43. Nicolas Wagner [Vagner], "Ueber die psychodynamischen Erscheinungen," *Psychische Studien* 2 (1875): 97–106, on 97n. Ellipses added.

44. A. Shkliarevskii, "Chto dumat' o spiritizme?: Po povodu pis'ma prof. Vagnera," *Vestnik Evropy* (June–July 1875): 906–18, 409–18, on 906.

45. S. Rachinskii, "Po povodu spiriticheskikh soobshchenii g. Vagnera," *Russkii Vestnik* (May 1875): 380–99, on 380.

46. The chief target was Akaskov, not Vagner, as can be seen in Mendeleev's detailed footnotes to the protocols of his commission, published as D. I. Mendeleev, *Materialy dlia suzhdeniia o spiritizme* (St. Petersburg: D. Mendeleev, 1876).

47. N. P. Vagner to D. I. Mendeleev, 1 January 1875 [*sic*: 1876], Archive-Museum of D. I. Mendeleev (hereafter ADIM), St. Petersburg, Russia, Alb. 4/52.

48. N. P. Vagner to D. I. Mendeleev, 19 Feb. 1876, ADIM Alb. 4/56.

49. This argument is quite similar to Fedor Dostoevskii's writings on Spiritualism in his 1876 *Writer's Diary*. This dispute of tone between Mendeleev and Dostoevskii is addressed in Michael D. Gordin, "Loose and Baggy Spirits: Reading Dostoevskii and Mendeleev," *Slavic Review* 60 (2001): 756–80.

50. Fichman, *An Elusive Victorian*, 183. The Vagnerian equivalent was the Russian's attack against the self-proclaimed fraud (and Mendeleev ally) I. Livchak. See the account in N. Lerner, "Tainstvennye uzelki: Sluchai s Dostoevskim," *Literaturno-khudozhestvennyi sbornik "Krasnoi panoramy"* (Oct. 1928): 36–42.

51. V. Pribytkov, "O fotografiiakh, predstavlennykh professorom N. P. Vagnerom v Tekhnicheskoe Obshchestvo," *Rebus* no. 13 (1894): 131; and N. P. Vagner, "Sine ira et studio. (Po povodu mediumicheskikh fotografii)," *Rebus* nos. 15–16 (1894): 151–52, 164–66; and A. N. Aksakov, *Animizm i spiritizm: Kriticheskoe issledovanie* (1900; Moscow: Agraf, 2001), 80–85.

52. Nicholas Wagner, A. Boutlerof, and A. Dobroslavin, "Seance for Autographic Writing with Mr. Eglinton," *Journal of Society for Psychical Research* (June 1886): 329–31, on 331.

53. The results of the White Sea research were published preliminarily as N. P. Vagner, *Predvaritel'noe soobshchenie o meduzakh i gidroidakh Belago moria* (St. Petersburg: V. Demakov, 1881); and then in the gorgeous volume of idem, *Bezpovochnyia Belago moria: Zoologicheskiia izsledovaniia proizvedennyia, na beregakh Solovetskago zaliva, v letnie mesiatsy 1876, 1877, 1879 i 1882 goda* (St. Petersburg: M. M. Stasiulevich, 1885).

54. Dostoevskii had brutally mocked Vagner's Spiritualism (and in print, too!) in his very funny 1878 feuilleton "From the Dacha Strolls of Kuz'ma Prutkov and His Friends, I: Triton" (F. M. Dostoevskii, *Polnoe sobranie sochinenii* [Leningrad: Nauka, 1972–90], 21: 250–51).

55. L. N. Tolstoi to N. P. Vagner, April 1890, PD f. 231, d. 287, ll. 1–4, on 1. 1ob.

56. N. P. Vagner to L. N. Tolstoi, 22 March 1890, PD f. 231, d. 279, ll. 1–3, on 1. 1

57. L. N. Tolstoi, "Plody prosveshcheniia: Komediia v chetyrekh deistviiakh," in Tolstoi, *Sobranie sochinenii* (Moscow: Khudozhestvennaia literatura, 1978–85), ll: 101–94, on 101.

58. Ibid., ll: 158.

59. Ibid., ll: 192.

60. A. Shkliarevskii, "Kritiki togo berega," *Russkii Vestnik* 121 (Jan. 1876): 470–95, on 470.

61. D. S. Mirsky, *A History of Russian Literature: From Its Beginnings to 1900* [1926], ed. Francis J. Whitfield (Evanston, IL: Northwestern University Press, 1999), 296–97.

62. N. P. Vagner, "Kak ia sdelalsia pistalem? (Nechto v rode ispovedi)," *Russkaia shkola* 1 (Jan. 1892): 26–38, on 37.

63. I am uncertain how to classify Vagner's retelling of the Synoptic Gospels for young adults: N. P. Vagner, *Razskaz o zemnoi zhizni Iisusa Khrista po Sv. Evangeliiam, narodnym predaniiam i ucheniiam Sv. Tserkvi* (St. Petersburg: M. M. Stasiulevich, 1908).

64. Bert and Vagner, *Ocherk zoologii*, i. The same is true of the wonderful N. P. Vagner, *Kartiny iz zhizni zhivotnykh: Ocherki i razskazy* (St. Petersburg: A. F. Marks, 1901).

65. Bert and Vagner, *Ocherk zoologii*, 157. Chapter 9 was added to cover Russian nature.

66. Viktor Shirokov, "Russkii Andersen," in N. P. Vagner, *Skazki Kota-Murlyki* (Moscow: Pravda, 1991), 5–14, on 5.

67. For the complete collection, see N. P. Vagner, *Romany, povesti i razskazy Kota-Murlyki*, 4th ed., 7 vols. (St. Petersburg: Obshchestvennaia Pol'za, 1905–14).

68. Evgenii Elachich, *Sbornik statei po voprosam detskogo chteniia* (St. Petersburg: Khudozhestvennaia pechat', 1914), 132–33.

69. The relevant text here is Vagner's novel of the Crimean War, *Temnyi put'*, published as a book in 1890, although partially serialized from 1881 to 1884 in the Spiritualist journal *Rebus*. For an extensive discussion of its anti-Semitic tropes, see Savelii Dudakov, *Istoriia odnogo mifa: Ocherki russkoi literatury XIX–XX vv.* (Moscow: Nauka, 1993), 242–60.

6

A Visual History of Jean Perrin's Brownian Motion Curves

CHARLOTTE BIGG

> In sum the science of drawing consists in instituting relations between curves and straight lines. A painting containing only curves or straight lines would not express existence. (En some la science du dessin consiste à instituer des rapports entre les courbes et les droites. Un tableau qui ne contiendrait que des droites ou des courbes n'exprimerait pas l'existence.)
>
> ALBERT GLEIZES AND JEAN METZINGER, *Du "Cubisme"* (1912)

A sheet of squared paper on which three broken lines have been drawn. A connect-the-dots game gone slightly awry, with no pattern obviously recognizable. No scale is inscribed that might provide clues about the size and nature of the object or phenomenon represented here. No indications on the procedure involved in the production of this two-dimensional abstraction. No numbers, letters, or symbols to tell the viewer how to hold the figure, or in what direction the lines run; indeed, its author (or perhaps was it the publisher's initiative?) occasionally published it sideways (fig. 6.1).[1]

Yet show this image to a physicist or a mathematician and the response will be immediate: *this is Brownian motion.*

This image, published for the first time in September 1909 by French physical chemist Jean Perrin (1870–1942),[2] has acquired iconic status in the physical sciences. It was and is still perceived as an experimental confirmation and a visual equivalent of Albert Einstein's theoretical demonstration, in a paper of 1905, of "the reality of atoms and molecules, of the kinetic theory of heat, and of the fundamental part of probability in the natural laws."[3]

Yet Einstein's publications on Brownian motion do not feature any images, nor did he suggest that the phenomenon should be represented in this way. In fact, until proven wrong, Einstein even doubted that the methods he suggested for measuring Brownian motion could be realized experimentally: "I would have thought such a precise study of Brownian motion impossible to realize," he wrote in admiration to Perrin in November 1909.[4]

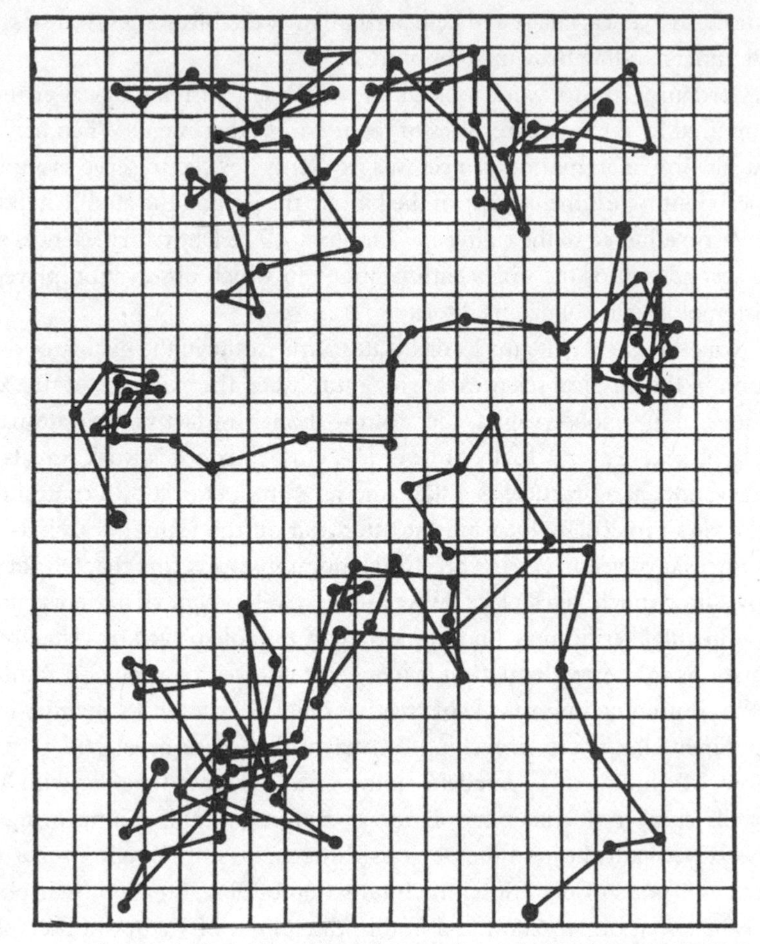

FIGURE 6.1. Jean Perrin, "Mouvement brownien et realité moléculaire," *Annales de chimie et de physique* ser. 8, 18 (1909): 81.

Einstein's and Perrin's Brownian motion work is justly famous for raising a number of issues central to the epistemology and historiography of the physical sciences, in particular, related to the nature of evidence, the relationship between theory and experiment, and realism.[5]

Rather than investigating the detailed ways in which a perfect fit between Perrin's experiments and Einstein's theory was realized, this paper explores the gap between Einstein's formulas and Perrin's image, a gap that stretched across four years and different scientific cultures but was subsequently erased when the image collapsed onto the formula it represents. What happens when we pry apart the formula from its representation? How did this image come both to encapsulate and help permanently establish a new way of seeing and understanding Brownian motion?

Certainly, Perrin's work took place within an epistemological economy structured by the twin categories of theory and experiment. When he wrote that his Brownian motion project was explicitly devised to serve as a crucial experiment to test the validity of the kinetic theory and the atomic hypothesis, Perrin adhered to the ruling epistemology of the laboratory sciences since the second half of the nineteenth century, in which observation played an epistemologically subordinate role.

Nonetheless, Perrin's work constitutes a milestone in the history of observation in the physical sciences, having established the existence of that most famous of all unobservables, the atom—though in fact neither atoms nor molecules were ever actually observed. A close look at his work reveals that observation, as a practice, as a skill, and as a product occupy a central place in Perrin's project. By focusing the attention on the techniques, skills, and resources involved in Perrin's practical laboratory work, this chapter not only shows how much takes place between the development of a theory and its experimental verification, and in particular, the role played by visual representations in the production of evidence; it also serves as a good reminder of the continuing importance of creative skill and technical ingenuity in the experimental sciences, broaching several of the issues developed in part 3 below, "Techniques." A parallel reading of this chapter together with Mary Terrall's chapter on Réaumur's observation of frogs in this volume brings out especially vividly, beyond the obvious differences, surprisingly similar concerns with ways of presenting the interaction of observers and their objects and the social organization and bodily disciplines of virtuoso observation. Finally, by uncovering the interdisciplinary dialogue that was determinant in the elaboration and subsequent appropriations of this image, this chapter shows how Perrin's work was embedded in the cultural and scientific fabric of his times. Tracing the history of this image, it shows how a virtual community was created through the making of Perrin as an observer and the making and the reception of his observations of Brownian motion.

"Mise en Observation"

Figure 6.1 appeared for the first time in the September 1909 issue of the *Annales de chimie et de physique*. As Perrin's laboratory notebooks of the time testify (fig. 6.2), this image was a composite picture of three drawings made earlier that year by Perrin during a series of experiments carried out together

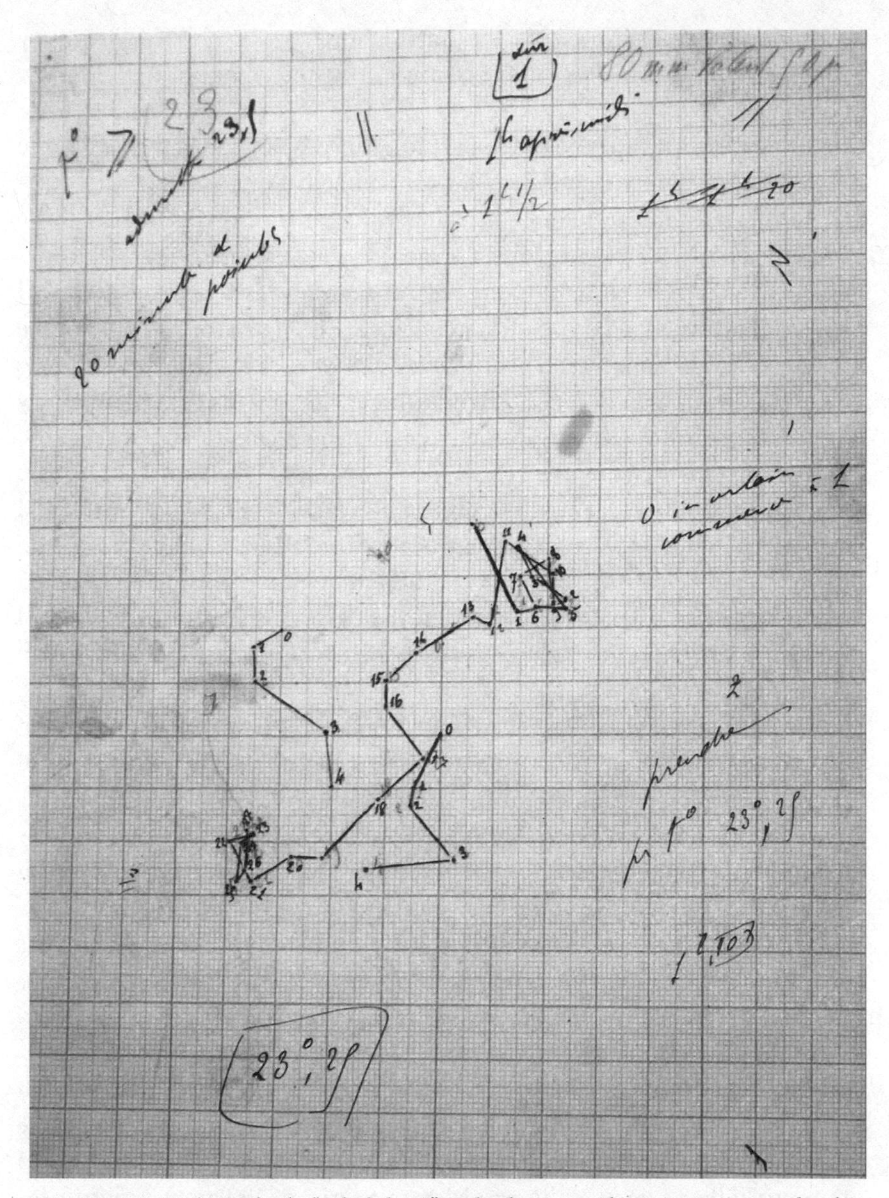

FIGURE 6.2. Perrin notebook, "Calcul de *N*," undated, c. 1909. This trajectory is reproduced in figure 6.1 (*middle*), with slight modifications in the segment angles and omitting the numbers. Dossier Jean Perrin, Archives of the Académie des Sciences, Paris, with permission.

with his student Dabrowski in the small laboratory for physical chemistry Perrin had set up at the turn of the century in an attic of one of the Sorbonne buildings.

In 1898, shortly after completing studies in physics and chemistry at the École Normale Supérieure with a Ph.D. showing that cathode rays were negatively charged (and therefore consisting of streams of particles, later to be named electrons), Perrin had begun teaching physical chemistry at the Sorbonne. He had been entrusted by the recruiting commission, in the person of Henri Poincaré, with the task of "naturalizing" on French soil the work of the mainly German pioneers and of bringing Gallic precision to the dynamic new field.[6] Taking up the challenge, Perrin recast in his own words contemporary discussions on the relationship between thermodynamics and mechanics, propagating his views in a textbook, *Traité de chimie physique. Les principes* that appeared in 1903.[7] There, but also in his teaching, popular lectures, and articles,[8] he stood up as a staunch advocate of atomism, a minority position in the French scientific community—though one that was well represented at the École Normale Supérieure and in the circle of scientists he interacted with on a daily basis, including his teacher Aimé Cotton, Marie and Pierre Curie, and Paul Langevin.

In 1903, once his laboratory had been fitted out, Perrin launched a series of experiments on the electrical properties of colloid solutions, suspensions of submicroscopic particles, that increasingly attracted the interest of chemists and biologists in these years, and on which Cotton was working with a colleague at the Institut Pasteur, Henri Mouton.[9] From there, Perrin moved on in early 1908 to the study of the Brownian motion of colloid particles.

Figure 6.1 was a product of one of the very last experiments within this new project. Perrin and Dabrowski had prepared what they referred to as their emulsion by bringing mastic, extracted from the bark of *Pistacia lentiscus* from the island of Chios and commonly used in the production of varnishes, into contact with methyl alcohol, obtaining a solution floating above a sticky insoluble residue. When diluted extensively, the solution became white as milk, in fact a suspension of spherical granules of varying sizes. Perrin and Dabrowski then subjected this emulsion to a series of "fractioned centrifugations" to obtain a suspension of grains of identical sizes. For this experiment they selected grains of a radius of 0.52 micrometers.[10]

In order to observe his emulsions, Perrin usually placed a drop of the suspension in a cavity about one-tenth of a millimeter deep, created by gluing a glass plate in which a wide hole had been bored onto an object slide. The cavity was then covered with another glass plate and sealed. This preparation could be used for several days or weeks.[11] For this particular experiment, the stability of the liquid's viscosity (and hence its temperature) was essential, so Perrin and Dabrowski immersed both the cell containing the drop and the microscope objective in a water-filled tank. The temperature was regularly measured by dipping a thermometer close to the microscope objective. On the notebook page shown in figure 6.2 the temperature is indicated, "23°,25" (23,25° C).

These procedures—the selection of the grains, enclosing the emulsion, and setting up the microscope for optimal conditions of visibility—constituted, with slight variations involving different types of microscopes and different sizes and types of grains, the starting point for all of Perrin's Brownian motion experiments. They were described by him as the "mise en observation" of his emulsions, the setting up of the conditions under which the Brownian motion of colloid particles was to be observed and produce fruitful insights.[12] Perrin relied mostly on techniques learned or developed in previous years. As he later wrote, "the study of colloids had familiarized me with the observation of Brownian motion."[13] It is noteworthy that many of these techniques were borrowed or adapted from the biologist's arsenal, from centrifugation, "as one does to separate the red cells from blood serum," to the use of dissection needles and plates engraved with a grid to help count cells in solutions, not to mention the microscope and camera lucida, standard microbiologist's devices.[14] The cheapness and ready availability of these instruments were certainly an advantage in a modestly endowed laboratory such as Perrin's. Moreover, they testify to Perrin's close acquaintance, perhaps through Mouton, of biological techniques and their creative adaptation for physical-chemical investigation.

The majority of Perrin's observations concerned the behavior of large numbers of particles, for instance, when they reached equilibrium in the emulsion, their concentration decreasing with altitude. In the particular experiment discussed here, Perrin and Dabrowski were interested instead in measuring the motion of *individual* particles.

The broken line at the center of figure 6.2 represents the motion of a single particle over a period of 20 minutes ("20 minutes de pointés"). In the published version (fig. 6.1), 16 divisions of the grid correspond to 50 micrometers. This is a simplified representation of the more complex trajectory followed by the particle, obtained by marking its position at regular intervals of time, here every 30 seconds. The trajectory is also simplified in the sense that it is a projection on a two-dimensional plane of the three-dimensional motion of the particles in the liquid. For this, Perrin attached a camera lucida to his microscope, enabling the simultaneous visualization of the particles swimming in the liquid and of his sheet of paper. While one man stationed in front of a chronometer called out the signals every 30 seconds, the other stood at the microscope eyepiece, following the motion of a particle and noting its position on paper when the signal was called. Perrin and Dabrowski regularly swapped positions, presumably to compensate for personal idiosyncrasies in each scientist's observing technique. The dots on the sheet were then numbered successively and joined by straight lines to produce trajectories as shown in figures 6.1 and 6.2. The notebook page featured in figure 6.2 indicates that this particular set of observations took place one undated afternoon between 1:00 p.m. and 1:30 p.m. Altogether, 950 observations were made using this particular emulsion, with each observation corresponding to one curve segment.[15]

Theory and Experiment

Brownian motion, the irregular and perpetual motion of small particles suspended in a liquid or a gas, varies in intensity as a function of the viscosity and temperature of the medium, but also of the size of the particles: the smaller the particle, the greater the motion. It accordingly affects microscopic and submicroscopic particles particularly strongly (as such it is well known to microscopists for interfering with microorganisms' proper motions). Brownian motion presented an opening for Perrin as a phenomenon that might be enrolled in his advocacy of atomism. In previous years several scientists, including Aimé Cotton and Pierre Curie's close friend the physicist Georges Gouy, had suggested that the Brownian motion of microscopic particles was a perceptible consequence of molecular agitation in fluids, and therefore that the phenomenon could be interpreted as empirical evidence in favor of atomism and the kinetic theory (which supposed liquids and gases to be made up of very small, hard spheres, or atoms).[16] However, as even the supporters of atomism recognized, existing measurements of the Brownian motion of particles did not correspond, by far, to the values predicted by the kinetic theory, in turn casting doubt on the atomic hypothesis.

Perrin's project explicitly aimed at producing experimental evidence in favor of the molecular-kinetic interpretation of Brownian motion. For this, he relied on new methods for measuring this motion that offered hope to bring observation in line with the values predicted by the kinetic theory. The temporal development of Perrin's Brownian motion work, as recorded in his laboratory notebooks, makes clear that Einstein's theory was a primary resource for developing his own project. The first entry in his first notebook, begun around March–April 1908, begins: "Langevin-Einstein hypothesis: each granule is assimilable to molecule (same $\sqrt{mv^2}$)." Further, Perrin consigned "requires experimental verification," outlining first ideas for producing emulsions of same-sized grains by centrifugation and measuring experimentally Avogadro's number, N the number of molecules in a mole. Then Perrin noted: "Thereafter, nothing comes in the way of verifying Einstein's formula

$$\sqrt{\overline{\Delta_x^2}} = \sqrt{t}\sqrt{\frac{RT}{N}\frac{1}{3\pi\mu a}}$$

(a radius of the granule μ viscosity), after which the application of this formula enables a to be obtained for *any* arbitrary granule, followed from moment to moment."[17]

The reference to his close friend the physicist Paul Langevin on the first page of the notebook is indicative of the latter's role in attracting Perrin's attention to Einstein's work. On 9 March 1908, Langevin had presented a paper to the Académie des Sciences entitled "Sur la théorie du mouvement brownien" that gave Einstein's formula in a form very close to that appearing in Perrin's notebook. Langevin also discussed Maryan von Smoluchowski's publications, and he assessed critically the first attempt at an experimental verification of Einstein's methods by Swedish physicist The Svedberg.[18] Perrin himself claimed in 1911 that Langevin had first brought Einstein's investigations to his attention.[19]

In his Brownian motion paper of 1905, Einstein had proposed new quantitative methods of measuring Brownian motion and of determining the dimensions of the particles, thus offering novel tools for testing the validity of the kinetic theory. He argued in particular that it was meaningless to measure the instantaneous velocity of individual particles, as previous researchers had done. Instead he proposed to measure their *mean displacement*, suggesting that the mean displacement of a particle on the x axis during an interval of time t should be proportional to the square root of t.[20] In Perrin's formulation of Einstein's formula above, the mean displacement over a given interval of time $\sqrt{\overline{\Delta_x^2}}$ can be calculated when R (gas constant), N (Avogadro's number), T (absolute temperature), μ (the viscosity of the fluid), and a (the radius of the particle) are known; conversely, N or a can be obtained when mean displacement and the other factors are known.[21] Perrin proposed first to provide an experimental confirmation of Einstein's formula by calculating the value of N based on the experimental determination of the other factors in the equation. If this N corresponded to the values of N obtained by other methods, the formula could be considered reliable and could in turn be used for determining the size of the suspended grains.

Perrin's broken lines, then, aimed at determining experimentally the mean displacement of particles of known radius in a liquid of known viscosity. The squared paper enabled a quick measurement of the length of each segment of his trajectories;[22] from there the mean displacement of a particle during successive time intervals could be calculated. Factoring in the constants *R* and *T*, Perrin could obtain, for each series of measurements, a value for Avogadro's number *N*.

Perrin found that the values of *N* obtained on the basis of 3,000 measured displacements agreed well with other determinations of *N* he and other scientists, from Lord Rayleigh and J. J. Thomson to the Curies, had made on the basis of different emulsions or different phenomena entirely. This "miracle of concordances" constituted for Perrin decisive evidence of the validity of Einstein's formula, of his method of measuring Brownian motion, and beyond, of the kinetic theory and of molecular reality.[23]

It is worth remarking that Perrin's argumentation followed the conventional epistemology of the laboratory sciences of his times in that he put forward his experiments as testing hypotheses derived from theories. As he wrote in 1912: "To this end I searched for a crucial experiment that, by approaching the molecular scale, might give a solid experimental basis to attack or defend the kinetic theories."[24] His 1909 *Annales* paper begins with a theoretical discussion before his experimental setup and results are brought up. One should be wary, however, of concluding from these claims that this case illustrates the progressive division of labor between theoreticians and experimentalists that emerged in the early twentieth century, most evidently among the German-speaking physicists. In the French context and in particular in the scientific circles in which Perrin lived and worked, other disciplinary faultlines prevailed. Perrin saw himself primarily as a physical chemist and by no means as an experimental physicist who left theory to more competent colleagues.

Of Photography versus Drawing and the Uses of Statistics in Perfecting Observation

A comparison with contemporary representations of Brownian motion shows that figure 6.1 was by no means the only way in which Brownian motion could be observed and represented, and in fact was quite unusual. Figures 6.3 and 6.4 were, for instance, published in the same year. Maurice de Broglie and Henry Siedentopf, respectively, captured the trajectories of Brownian particles on photographic plates using long exposures. Their images superficially resemble many of the photographs taken by physicists investigating subatomic entities (ions, electrons, α particles) circa 1900, such as photographs of particle tracks in a cloud chamber.

FIGURE 6.3. Long-exposure photographic recording of the Brownian motion of ultramicroscopic tobacco smoke particles. Maurice de Broglie, "Enregistrement photographique des trajectoires browniennes dans les gaz," *Comptes rendus hebdomadaires des séances de l'Académie des sciences* 148 (1909): 1164.

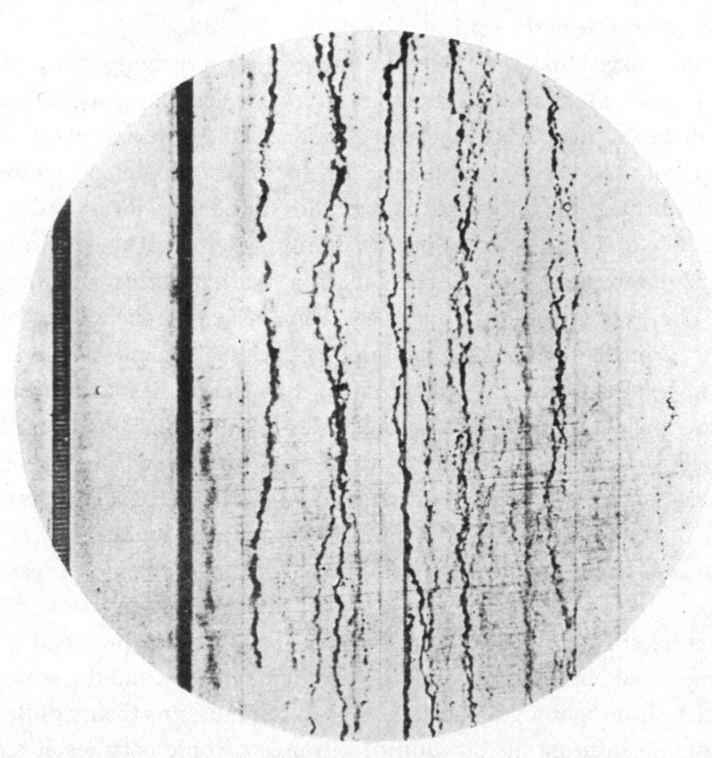

FIGURE 6.4. "Brownsche Molekularbewegung. Momentaufnahme auf fallender Platte mit aplanatischem Dunkelfeld-kondensor von Zeiss." Henry Siedentopf, "Über ultramikroskopische Abbildung," *Zeitschrift für Wissenschaftliche Mikroskopie* 26 (1909): plate.

The use of photography for observing and representing Brownian motion brought with it a whole new set of challenges. Perrin occasionally used the technique but found that though photography was less time-consuming and tiring to the eye than a camera lucida, "the eye is more sensitive than the photographic plate with regard to the visibility of very small, pale grains on a background that is nearly as pale."[25] And even when photographic observations could be made, the images obtained did not always give satisfactory results in print, as was especially the case in the observation of the smallest particles, visible only in a particular type of darkfield instrument, the ultramicroscope.[26] Maurice de Broglie apologized for the poor quality of figure 6.3: "The imperfection of the typographical reproduction accompanying this note scarcely gives an exact idea of the photographs obtained."[27] Worse still, photographs of Brownian motion were not so easily legible as experimental evidence. In figures 6.3 and 6.4, the particles' displacements could only be measured by magnifying the photographs, and the Brownian motion had to be distinguished from the overall motion of the gas or solution. In figure 6.4, the Brownian motion of falling particles can be identified as the small deviations on either side of the vertical.

Perrin's image looks very different. Figure 6.1 is a drawing, not a photograph. It looks more abstract. Even though it depicts the specific trajectories of three specific particles measured at a specific time and place, Perrin stripped from his drawing all elements that might have pointed to a specific experiment. Only the bare essentials remain: three trajectories and a grid. The absence of any indications on the figure suggests that these three lines are simply examples, perhaps even chosen at random, of Perrin's measurements. They stand for all his other measurements and all the measurements that can be made following his method. The strong presence of the grid in figure 6.1 helps emphasize the quantitative character of the observation. The reproduction of three exemplary trajectories in the publication thus served both to illustrate Perrin's own technique of measuring Brownian motion and his experimental verification of Einstein's theory. The unusual appearance of the trajectories as broken lines, underscored by the abstraction of the rendering, was a strong visual marker of the novelty of the method of measuring motion.

An interesting counterpoint to Perrin's approach was provided by the Marburg-based physicist Max Seddig, who developed around the same time a complex chronophotographic apparatus to test Einstein's theory of Brownian motion, capturing the position of ultramicroscopic particles at regular intervals of time. To avoid heating the liquid and thereby changing its viscosity during the experiment, Seddig used a stroboscopic technique to illuminate the solution intermittently, timing his photographic camera to open the shutters at exactly the same time. Seddig explicitly put forward his results as being superior to drawings because they were "objectively obtained": "Following the uncertain results of the subjective methods to date, an objective one should be attempted. As such, only the photographic process could come into question." Seddig referred here to Felix Exner's camera lucida drawings on a smoked glass plate of the Brownian motion of particles observed through a microscope (Perrin had not yet published at this point). Such drawings were necessarily unreliable for Seddig because the drawing hand was too slow to follow the particles' exceedingly rapid motions. His own photographic method, by contrast, supplied a "direct, experimental confirmation" of the kinetic hypothesis.[28] Yet Seddig, for reasons unspecified, did not publish any of his photographs, giving only the numerical values he had obtained in the form of a table. Paradoxically, his objective method did not yield images that could be shared with readers, leaving them no other choice than to trust in Seddig's skills and apparatus.

Perrin used a similar rhetoric, putting forward his own observations of Brownian motion as supplying direct evidence of the existence of atoms. But he diverged from Seddig in arguing that drawing was a perfectly legitimate technique. He admitted freely that marking the positions of the particles by hand introduced a measure of uncertainty: "each time that a grain's position is marked, a small error is made, analogous to that made when shooting at a target, which itself obeys the law of chance and which has the same effect on the readings as if one overlaid a second Brownian motion over the one under observation."[29] However, this inconvenience was largely compensated by the large number of observations made by Perrin and his collaborators. The striking agreement in the determinations of *N* made with a range of solutions and colloid particles and in different circumstances showed that small observing errors did not compromise the overall result.

There was also a more fundamental reason for Perrin to trust that the accumulation of measurements compensated for any approximation due to the lack of high-precision recording technology, hinted at in his mention above that observational error could be assimilated to a secondary form of Brownian motion: the fact that Brownian motion could only be investigated using a statistical approach because it was essentially stochastic in character:

> Thus appears a profound, eternal property of what we call a *liquid in a state of equilibrium.* This equilibrium only exists in an average manner and for great masses: it is a statistical equilibrium. In reality, the whole fluid is indefinitely and *spontaneously* agitated in movements that are all the more violent the

> smaller the portions they concern; the static notion of equilibrium is completely illusory.[30]

After investigations of radioactive decay (Exner, Curie) and the emission of α particles, Perrin's Brownian motion work constituted one of the early experimental attempts to apply statistics and probability theory to physical systems, and here Perrin followed closely the lead of James Clerk Maxwell and Ludwig Boltzmann.[31] Brownian motion, considered macroscopically, at the level of the solution, consisted in the minute and random fluctuations around the average state of the fluid. And like the radioactivity experimentalists, Perrin adopted a pragmatic attitude to fluctuations, using knowledge of their existence to develop better methods of measurement.[32] His whole strategy aimed at developing experimental and theoretical tools for smoothing out these fluctuations and obtaining average numbers that correspond to the equilibrium state of his emulsions. This statistical approach offered Perrin a means of achieving almost unlimited precision in his observations:

> Once this point is well established, one finds in this very equation, to determine the constant *N* and the constants depending upon it, a method that seems *capable of achieving an unlimited precision.* The preparation of a uniform emulsion and the determination of the values other than *N* that figure in the equation can indeed be pushed to the desired degree of precision. It is a simple question of patience and of time: nothing limits a priori the exactness of the results, and we can obtain, if we wish, the mass of an atom with the same precision as the mass of the Earth.[33]

Once experimental errors were excluded, the greater the number of observations, the closer the average of these measurements would be to the statistical equilibrium and the true value of *N*. In an experiment that involved measuring the concentration of grains at different levels of his solutions, Perrin made six series of measurements using different solutions and grain sizes. He noted a series of numbers whose average value reached a limit that corresponded to the average frequency of the grains at the level under investigation, remarking that "several thousand readings are necessary if one wants a little precision."[34] Just the sixth series of measurements involved counting no less than 11,000 grains using one method, and 13,000 grains using another.

For his Brownian motion work, Perrin was awarded the La Caze Prize in 1914. In their laudatio, the commissioners picked up on this aspect, noting that "Mr. Perrin's method is capable of achieving an infinite precision. It is only a question of patience; it comes down to a numbering of grains and a statistical calculation of averages whose accuracy increases proportionally with the square root of the total number of observations."[35] It was this aspect of Perrin's work, his patience, which most impressed some of his contemporaries—especially given that he was quite forgetful as a person. Much later, in an obituary address Louis de Broglie insisted again on the "tenacity, patience and meticulous attention required by these series of measurements," which stood in contrast to Perrin's personality, by nature "quite distracted, of a rather impulsive character, such that one might have thought him little suited to carry out a task requiring so much attention and perseverance."[36]

With the measurement of the motion of individual particles, this was accomplished by measuring mean displacement. Each displacement, represented in figure 6.1 by a straight segment, was already the result of an averaging process. The three curves of figure 6.1 thus also represented a new statistical approach to the observation of motion. The broken lines stood in stark contrast to the familiar continuous trajectory curves studied since the early days of mechanics. The zig-zag line, even devoid of any indications on the size and nature of the particles or of the interval of time chosen, stood for a new way of conceiving and measuring the motion of individual particles. It was evidence in favor of the statistical approach in the study of the phenomenon of Brownian motion and of events at the molecular level.

Techniques of Observing Motion

But where does figure 6.1 come from? The origin of this form of representation can only be recovered when the broader context in which Perrin worked and the detailed chronology of his investigations are taken into consideration.

Oddly, given that Einstein's mean displacement formula figures in the very early pages of his first notebook, Perrin only undertook experiments to measure the motion of individual particles over a year later, spending the intervening time studying the behavior of systems of particles. Even then, he did not publish figure 6.1 in any of the successive *Comptes rendus* of his work. His first mention of displacement measurements appears in a *Compte rendu* published on 6 September 1909. In the *Annales* paper published just a few weeks later, the image does not appear until page 78, and the discussion of Einstein's theory appears as somewhat of an afterthought.

If, as we have seen, the theoretical work of Einstein and Langevin played an important role in Perrin's project, these do not supply any clues to explain why Perrin turned to observe the motion of individual particles at the time he did and how he came to observe and represent this motion the way he did. For this, we need to turn to another set of resources upon which Perrin drew.

It seems that Perrin's interest in taking up the measurement of the motion of individual particles had at least one source other than Langevin's work of March 1908. Perrin's very first *Compte rendu* on Brownian motion is dated 11 May 1908. This three-page paper must have struck a chord, since at the following meeting of the academy a week later, no less than two papers were presented that dealt with Brownian motion: the one by Maurice de Broglie discussed above and one by Victor Henri, with the latter explicitly mentioning Perrin and Einstein.[37] Henri's report appeared under a *physique biologique* heading, which can be explained by the fact that he worked at the time in the physiology laboratory at the Collège de France (he was himself trained in psychology). Henri expressed doubts about Einstein's formula based on measurements using a sophisticated cinematographical apparatus combined with a microscope.

> The preparation is placed in a position precisely horizontal under the microscope. The photographs were taken using a 2mm Zeiss apochromatic objective, with a projection ocular n. 4 and a distance of 24 cm, which gives a magnification of approximately 600 diameters. The light source is a 30-ampere arc lamp; the cinematograph is placed directly above the microscope. The resulting films have twenty images per second and the exposure for each image is 1/320 of a second; consequently, the interval of time separating two images is equal to 1/20 of a second.
>
> The selected emulsion was sufficiently diluted such that only about 20 grains appeared in the field of view; in this way they can be located with accuracy and, by determining the position of a grain on a series of successive photographs, the projection of the trajectory described by each grain can be drawn. The figure represents these trajectories for five grains, the successive points corresponding to intervals of 1/20 of a second, the scale indicating the size of the μ.[38]

Henri then inserted the image shown in fig. 6.5.

Convinced of the validity of Einstein's formula and keen to disprove Henri, Perrin set one of his students, Chaudesaigues, the task of repeating Henri's experiments. Six months later, on 30 November 1908, Chaudesaigues reported to the academy that his own experiments confirmed Einstein's formula.[39] Chaudesaigues had used Perrin's emulsions and a setup combining a microscope and camera lucida. This article featured no images.

The mean displacement measurements presented in Perrin's long article of September 1909, which include both Chaudesaigues' work and the observations carried out with Dabrowski discussed above, had therefore been inspired, at least in part, by Henri's investigations; and so was, clearly, the visualization of Brownian motion that Perrin published at this point for the first time.

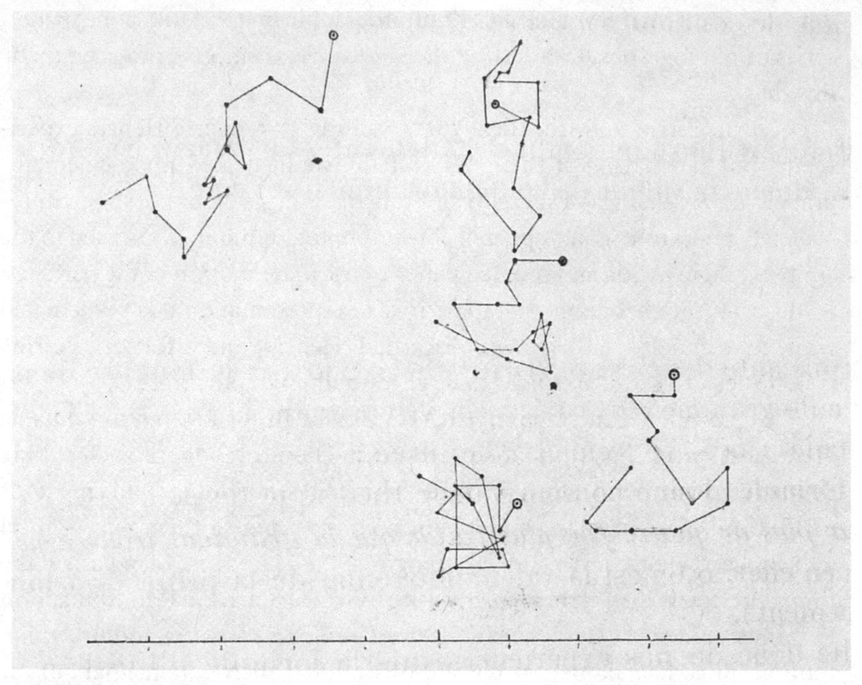

FIGURE 6.5. Victor Henri, "Étude cinématographique des mouvements browniens," *Comptes rendus hebdomadaires des séances de l'Académie des sciences* 146 (1908): 1025.

Henri, as mentioned, had made his measurements in the physiology laboratory of the Collège de France. This was Etienne-Jules Marey's laboratory, taken over after his death in 1904 by his former assistant, Charles Albert François-Franck. In 1907, a sophisticated microcinematographical apparatus had been constructed there by Victor Henri and Lucienne Chevroton (assistant and, later, wife of François-Frank) after they visited a Zeiss holiday course on microscopy, ultramicroscopy, and microphotography in Jena.[40] Upon his return from Jena, Victor Henri promptly integrated this knowledge into his lectures. Chevroton and Henri's apparatus was also made widely available to researchers for a variety of pursuits,[41] making the years between 1908 and 1910 "an extremely productive one for the development of microcinematography in biology," according to cinema historian Virgilio Tosi.[42] Henri's Brownian motion work was one side investigation using the new technology.

In a direct sense, therefore, Henri's curves were an extension of the chronophotographic investigation of the motion of humans and animals, applied now to microscopic particles in a move that might have pleased Marey, convinced as he was that his technique was a "universal graphic language" as incontrovertible as geometers' demonstrations;[43] though of course Henri departed somewhat from the chronophotographic method in that he made a series of photographs, transferring the successive positions onto a synthetic drawing.[44]

In turn, Perrin and Chaudesaigues, wanting to replicate Henri's experiments but not having at their disposal the sophisticated microcinematographic device that was a specialty of François-Frank's laboratory (Perrin revealingly refers to it as an "appareil chronophotographique"), resorted to the simple camera lucida and squared paper to track the motion of the particles. In this way, the chronophotographic *style* of representation was perpetuated while its technological basis was discarded. Henri's, and a fortiori, Perrin's trajectories in a sense *pretended* to be chronophotographic traces.

That no other contemporary investigator of Brownian motion outside Henri and Perrin chose this mode of representation (even those, such as Max Seddig, who independently developed cinema/chronophotographic methods to measure mean displacements) is of course related to their geographical and intellectual proximity and the strength of the Marey school in Paris at the time.[45] In later years Perrin regularly asked Victor Henri and Jean Comandon (a former student of Henri's who, inspired by Henri's lectures and Brownian motion films, had pioneered the use of the ultramicro-cinematograph for the study of living cells before becoming a professional science filmmaker for the film company Pathé) to prepare films of Brownian motion for showing in public lectures, such as the lecture Perrin gave in Stockholm in 1926 upon receiving the Nobel Prize for physics for his work on Brownian motion.[46]

If Perrin's image was explicitly put forward as an investigation belonging to the physical tradition of studying motion, for example, the trajectories of objects in motion using the laws of mechanics, it could thus equally be considered as belonging to the French physiological tradition of studying animal motion using the graphic method. Almost a century after Robert Brown had identified Brownian motion as distinct from the vital motions of microorganisms, Jean Perrin's visualization of Brownian motion was inspired by physiological techniques of studying the motion of human and animal organisms. Perrin's image brought together the chronophotographic style of representing motion with Einstein's displacement formula to create a new type of image in the physical sciences.

But the application of cinematographic technique to the study of Brownian motion put a new twist on the study of motion though its decomposition. While Marey's technique aimed at decomposing continuous movements into discontinuous snapshots, in the study of Brownian motion, the camera's inherent discontinuity of perception corresponded to the discontinuity of Brownian motion—or rather, to Einstein's way of measuring it, as Scott

Curtis has argued.[47] In order to measure displacement, it was necessary to measure the position of a given particle at regular intervals of time, ignoring all the intervening motions. For measuring displacements, the camera's "stroboscoping" powers, as one of Perrin's colleagues put it, were perfectly suited (this characterization gives yet another dimension to this episode, connecting it as it does to the history of the strobe discussed in the essay by Jimena Canales in this volume).

Perrin's image of Brownian motion arose out of the convergence in his work of theoretical considerations put forward by Einstein and Langevin, but also of his experience with colloid solutions, learned in part with his teachers Aimé Cotton and Henri Mouton, as well as the chronophotographic techniques of observing and representing motion. The making of Perrin's image and the sources for his novel way of observing Brownian motion can only be recovered by paying close attention to the development in Perrin's own thinking and observing practices, as documented along the way in his laboratory notebooks and the *Comptes rendus* in which he regularly presented his findings, as well as to the immediate intellectual and physical environment in which he lived and worked.

Epilogue

Perrin's work was well received among physicists and chemists in the years following the publication of his *Annales* article of 1909. It was extensively commented upon in chemical and physical journals. German and English translations appeared as booklets in 1910.[48] Perrin's image was reprinted in longer discussions of his work, the earliest reprint so far identified being a book on the experimental foundations of atomism published in German in 1910 by Werner Mecklenburg, *Die experimentelle Grundlegung der Atomistik*. This is a reprint from the popular journal *Naturwissenschaftlichen Wochenschrift* of 1909–10. Only two images feature in this 143-page booklet, Perrin's figure 6.1 and Henry Siedentopf's figure 6.4. The legend to Perrin's image reads: "The motion of colloid particles according to Perrin."[49] And of course figure 6.1 features in the publications that appeared on the occasion of Perrin's receiving the Nobel Prize in 1926 and in subsequent commemorative and obituary publications. To this day, Einstein and Perrin's Brownian motion work is taught to physics students and represented in the way Perrin suggested.[50]

In physics, Perrin's image now has only historical relevance as a particular moment in the history of the field, a significant achievement. This is not true of the *theory* of Brownian motion, which was developed in the following decades by physicists and engineers into a more general theory of fluctuations. These investigators studied Brownian motion from the macroscopic perspective, for example, in the form of electronic noise, and using very different techniques of observation and representation.[51]

In mathematics also, the theory of Brownian motion remained an ongoing concern and so did, in part, Perrin's representation, in studies that focused on the Brownian motion of individual particles. The American mathematician Norbert Wiener is a central figure in this development, whose Wiener Process is sometimes used as a synonym for Brownian motion. Wiener published a landmark article in 1923 entitled "Differential-Space," in which he developed the mathematical theory of the idealized Brownian motion of a single particle. Wiener begins his paper by noting that mathematicians and physicists are sometimes confronted with similar objects: functions or curves in systems of infinite dimensions. Wiener gives an example from statistical mechanics, the density of a gas obtained from the coordinates and velocities of its molecules. The case he chooses to discuss at length, however, is Brownian motion as theorized by Einstein.[52]

Wiener writes that the inspiration for this article came from a conversation with French mathematician Paul Lévy. But the choice of Brownian motion as a physical starting point had deeper origins. In his memoirs, Wiener later wrote:

> It was at MIT too that my ever-growing interest in the physical aspects of mathematics began to take definite shape. The school buildings overlook the River Charles and command an ever changing skyline of much beauty. The moods of the waters of the river were always delightful to watch. To me as a mathematician and as a physicist they had another meaning as well. How could one bring to a mathematical regularity the study of the mass or ever shifting ripples and waves, for was not the highest destiny of mathematics the discovery of order among disorder? . . . What descriptive language could I use that would portray these clearly visible facts without involving me in the inextricable complexity of a complete description of the water surface? This problem of the waves was clearly one of averaging and statistics, and in this way closely related to the Lebesgue integral which I was studying at the time. Thus, I came to see that the mathematical tool for which I was seeking was one suitable to the description of nature, and I grew ever more aware that it was within nature itself that I must seek the language and the problems of my mathematical investigations.[53]

Given this interest in mathematical tools that could describe nature, and more specifically the motion of liquids, it is perhaps unsurprising that Wiener read Soddy's translation of Perrin's *Annales* paper. He was particularly struck by and quotes in "Differential-Space" a passage in which Perrin commented on his Brownian motion curves:

> This motion is of such an irregular nature that Perrin says of it: "One realizes from such examples how near the mathematicians are to the truth in refusing, by a logical instinct, to admit the pretended geometrical demonstrations, which are regarded as experimental evidence for the existence of a tangent at each point of a curve." It hence becomes a matter of interest to the mathematician to discover what are the defining conditions and properties of these particle-paths.[54]

In the same passage Perrin had pointed out that his curves only gave an approximate idea of the "prodigious entanglement" of the real trajectory of individual particles: should the interval of time chosen to mark the particle's positions be reduced to a second, then each straight line would turn into a curve as complex as the initial one.[55] Perrin also compared these trajectories to the coast of Brittany, where he spent his holidays every year with, among others, the Langevins, Curies, and Borels. Unlike a map of Britanny, but like its real coast, he pointed out in *Les atomes*, if one zoomed onto the line, each straight segment turned into a broken line and each segment of that broken line also turned into a broken line, and so on ad infinitum. The variation in direction and velocity of the particles was practically infinite. Perrin added that such curves were accordingly continuous but devoid of tangents, corresponding to the nondifferentiable, "pathological" functions that his friend, mathematician Emile Borel, and others had studied since the nineteenth century: "Of course, one cannot either trace a tangent, even approximately, at any point on the trajectory. It is one of those cases where we are reminded of these continuous, nowhere differentiable functions that were wrongly seen as mathematical curiosities, since nature can suggest them just as well as differentiable functions."[56]

It is very likely that it was Borel himself, with whom Perrin was in daily contact, who inspired these remarks, as they corresponded more closely to Borel's than to Perrin's expertise and interests. In the same period, Borel published several articles exploring the relationship between mathematics and statistical mechanics and between physical and mathematical conceptions of infinity, and he frequently brought in Perrin's Brownian motion experiments.[57]

The mathematicians' conceptualization of Brownian motion and its relationship to the physical investigations discussed above is another story. But it is already apparent that not only Einstein's theory but also Perrin's new way of observing Brownian motion prompted mathematicians to recognize in his jagged curves a visual counterpart to their nondifferentiable functions, and led them to launch investigations that took into account these functions' newly revealed connection with the natural world. Conversely, the conventions of representing Brownian motion developed by Perrin were transferred to the mathematical representation of idealized Brownian motion and other "random walks." The history of Jean Perrin's Brownian motion work testifies not only to the creative skill involved in making scientific observations, but also to the transformative power of observations once they exist, to their ability to permanently change ways of seeing. This is why to this day Perrin's way of seeing Brownian motion, as encapsulated in figure 6.1, provides a blueprint for the visual representation of Wiener processes and random walks.

Notes

1. For example, Jean Perrin, "La réalité des molécules," *Revue scientifique* 49, n. 2 (1911): 774–84, on 783; idem, *Notice sur les titres et travaux scientifiques* (Paris: Gauthier-Villars, 1918), 77.

2. Jean Perrin, "Mouvement brownien et realité moléculaire," *Annales de chimie et de physique* ser. 8, 18 (1909): 5–114.

3. Max Born, "Einstein's Statistical Theories," in *Albert Einstein: Philosopher-Scientist*, ed. Paul A. Schilpp (New York: Tudor, 1949), 161–77, on 166; quoted in Stephen G. Brush, "A History of Random Processes. I. Brownian Motion from Brown to Perrin," *Archive for the History of Exact Sciences* 5 (1968): 1–36, on 14–15.

4. Albert Einstein to Jean Perrin, 11 Nov. 1909, in *The Swiss Years: Correspondence, 1902–1914*, vol. 5 of *The Collected Papers of Albert Einstein*, ed. Martin J. Klein, Anne J. Kox, and Robert Schulmann (Princeton: Princeton University Press, 1993), 216.

5. On Einstein's and Perrin's work on Brownian motion, see especially Brush, "A History of Random Processes"; Mary-Jo Nye, *Molecular Reality* (Amsterdam: Elsevier, 1972); John Stachel et al., eds., *The Swiss Years: Writings, 1900–1909*, vol. 2 of *The Collected Papers of Albert Einstein*, ed. Martin J. Klein, Anne J. Kox, and Robert Schulmann (Princeton: Princeton University Press, 1989), esp. 206–22; Roberto Maiocchi, "The Case of Brownian Motion," *British Journal for the History of Science* 23 (1990): 257–83.

6. Henri Poincaré, "Rapport sur la candidature de M. Perrin à la chaire de chimie physique," file F/17/24822, Archives Nationales, Paris.

7. Jean Perrin, *Traité de chimie physique. Les principes* (Paris: Gauthier-Villars, 1903).

8. Jean Perrin, "Les hypothèses moléculaires," *Revue scientifique* 15 (1901): 449–61 (lecture delivered on 16 Feb. 1901 to the students and friends of the University of Paris); see also idem, "La discontinuité de la matière," *Revue du mois* 1 (1906): 323–44, idem, *Notice sur les travaux scientifiques* (Toulouse: Edouard Privat, 1923), 20.

9. See especially Jean Perrin, "Mécanisme de l'électrisation de contact et solutions colloïdales," *Journal de chimie physique* 2 (1904): 601–51, and 3 (1905): 50–110. On Cotton and Mouton's work, see *Notice sur les travaux scientifiques de M. A. Cotton, chargé de cours à l'université de Paris (École Normale Supérieure)* (Paris, 1908); Aimé Cotton, laboratory notebooks 1902–7, box 13, Fonds Aimé Cotton, Physics Library of the École Normale Supérieure, Paris.

10. Perrin, "Mouvement brownien et realité moléculaire," 79.

11. Ibid., 41.

12. Perrin, *Notice sur les titres et travaux scientifiques*, 63.

13. Ibid., 56.

14. Ibid., 62; Perrin, "Mouvement brownien et realité moléculaire," 46.

15. Perrin referred to the procedure of viewing the motion of an individual particle and marking its successive positions as "pointés" or "lectures," the resulting segments constituting "observations," that then yielded measurements ("mesures"). Perrin, "Mouvement brownien et realité moléculaire," 76–80; idem, "La réalité des molécules," 783.

16. See, for example, Aimé Cotton, laboratory notebook 1904–5, box 13, Physics Library of the École Normale Supérieure, Paris; Georges Gouy, "Le mouvement brownien et les mouvements moléculaires," *Revue générale des sciences* 6, no. 1 (15 Jan. 1895): 1–7.

17. Jean Perrin, "tube de crookes," laboratory notebook, pages 1–3. Jean Perrin file, Archives of the Académie des Sciences, Paris.

18. Paul Langevin, "Sur la théorie du mouvement brownien," *Comptes rendus hebdomadaires des séances de l'Académie des sciences* (hereafter *CRAS*) 146 (1908): 530–33, on 533.

19. Perrin, "La réalité des molécules," 783.

20. Albert Einstein, *Investigations on the Theory of the Brownian Movement*, trans. A. D. Cooper (London: Methuen, 1926).

21. Albert Einstein, "Über die von der molekularkinetischen Theorie der Wärme geforderte Bewegung von in ruhenden Flüssigkeiten suspendierten Teilchen," *Annalen der Physik* 17 (1905): 549–60.

22. "The grains I had prepared were traced using the camera lucida, which gives horizontal projections of the displacements. Operating with squared paper, one directly obtains the projections on two rectangular axes of the different segments already obtained, but their measurement is unnecessary since the sum of the squares of these projections is equal to the sum of the square of the segments, such that to obtain the average square of the projection on an axis, it is enough to measure these segments one by one, to calculate their square, and to take half of the average of these squares." Perrin, "Mouvement brownien et réalité moléculaire," 76–77.

23. Jean Perrin, *Les atomes* (Paris: Gauthier-Villars, 1913), 289.

24. Jean Perrin, "Rapport sur les preuves de la réalité moléculaire," in *La théorie du rayonnement et des quanta*, ed. Paul Langevin and Maurice de Broglie (Paris: Gauthier-Villars, 1912), 153–253, on 157.

25. Perrin, "Mouvement brownien et realité moléculaire," 49.

26. On Perrin's use of the ultramicroscope, see Charlotte Bigg, "Evident Atoms: Visuality in Jean Perrin's Brownian Motion Research," *Studies in the History and Philosophy of Science* 39 (2008): 312–22.

27. Maurice de Broglie, "Sur des mesures de mouvements browniens dans les gaz et la charge des particules en suspension," *CRAS* 148 (1909): 1315–18, on 1315.

28. Max Seddig, *Messung der Temperatur-Abhängigkeit der Brown'schen Bewegung*, Habilitationsschrift zur Erlangung der Venia legendi der Akademie in Frankfurt a. M. (Frankfurt, 1909), 19, 10, 12 and 32. This argument is echoed by cinema historian Scott Curtis, who claims that Seddig's use of photography "gave substantial evidentiary support to [his] scientific theories." Scott Curtis, "Die kinematographische Methode. Das 'Bewegte Bild' und die Brownsche Bewegung," *montage/av, Zeitschrift für Theorie und Geschichte audiovisueller Kommunikation* 14, no. 2 (2005): 23–43, on 24.

29. Perrin, "Mouvement brownien et realité moléculaire," 79.

30. Ibid., 10–11.

31. On the history and context of this statistical approach, see Stephen Brush, *The Kind of Motion We Call Heat: A History of the Kinetic Theory of Gases in the Nineteenth Century* (Oxford: North Holland, 1976); Theodore Porter, *The Rise of Statistical Thinking, 1820–1900* (Princeton: Princeton University Press, 1986); Judy Klein, *Statistical Visions in Time: A History of Time Series Analysis, 1662–1938* (Cambridge: Cambridge University Press, 1997), esp. 161–93.

32. Deborah Coen, "Scientists' Errors, Nature's Fluctuations, and the Law of Radioactive Decay, 1899–1926," *Historical Studies in the Physical Sciences* 32 (2000): 179–205. See also Paul Hanle, "Indeterminacy before Heisenberg: The Case of Franz Exner and Erwin Schrödinger," *Historical Studies in the Physical Sciences* 10 (1980): 225–69.

33. Perrin, "Mouvement brownien et realité moléculaire," 62–63.

34. Ibid., 43.

35. Anon, "Prix L. La Caze," *CRAS* 159 (1914): 867.

36. Louis de Broglie, "La réalité des molécules et l'oeuvre de Jean Perrin," speech read to the Academy of Sciences on 17 December 1945, 13–14. Dossier Jean Perrin, Archives of the Académie des Sciences, Paris.

37. Maurice de Broglie, "Sur l'examen ultramicroscopique des centres chargés en suspension dans les gaz," *CRAS* 146 (1908): 1010–11.

38. Victor Henri, "Étude cinématographique des mouvements browniens," *CRAS* 146 (1908): 1024–26, on 1025.

39. Chaudesaigues, "Le mouvement brownien et la formule d'Einstein," *CRAS* 147 (1908): 1044–46.

40. Lucienne Chevroton, Victor Henri, and André Mayer, "Rapport sur une mission scientifique à Iéna, en vue d'étudier, à l'institut de microscopie et dans les laboratoires de la fondation Carl Zeiss les progrès réalisés dans la technique de la microscopie, de la microphotographie et de l'ultramicroscope," *Nouvelles archives des missions scientifiques* 17 (1908): 1–73. Note that experiments in microcinematography had previously been made at the laboratory, notably by François-Franck, whose setup has much in common with Chevroton and Henri's apparatus. See Thierry Lefebvre, "Contribution à l'histoire de la microcinématographie: De François-Franck à Comandon," *1895* 14 (1993): 35–46.

41. For example, Lucienne Chevroton and Fred Vlès, "La cinématique de la segmentation de l'oeuf et la chronophotographie du mouvement de l'oursin," *CRAS* 149 (1909): 806–9; Lucienne Chevroton, "Dispositif pour les instantanés et la chronophotographie microscopiques. Techniques des prises de vue," *Comptes rendus des séances de la société de biologie* 66 (1909): 340.

42. Virgilio Tosi, *Cinema before Cinema: The Origins of Scientific Cinematography* (London: British Univ. Film & Video, 2005), 179. On Chevroton, Vlès, Comandon, and French biological microcinematography, see Hannah Landecker, "Cellular Features: Microcinematography and Early Film Theory," *Critical Inquiry* 31 (2005): 903–37.

43. Etienne-Jules Marey, *La méthode graphique* (Paris: Masson, 1878), iii. Marey's method was inspired by the German physiologists' self-inscription techniques. See Andreas Mayer, "Autographen des Ganges. Repräsentation und Redressement bewegter Körper im neunzehnten Jahrhundert," in *Kunstmaschinen. Spielräume des Sehens zwischen Wissenschaft und Ästhetik*, ed. Andreas Mayer and Alexandre Métraux (Frankfurt am Main: Fischer, 2005), 101–38; Frederic L. Holmes and Kathryn M. Olesko, "The Images of Precision: Helmholtz and the Graphical

Method in Physiology," in *The Values of Precision*, ed. M. Norton Wise (Princeton: Princeton University Press, 1995), 198–221.

44. Henri, "Étude cinématographique des mouvements browniens," 1025.

45. And perhaps more broadly, to the tradition of graphical and geometrical methods developed by engineers in France after the Revolution, which Marey admired. See Theodore Porter, *Karl Pearson. The Scientific Life in a Statistical Age* (Princeton: Princeton University Press, 2004), 215–48.

46. Perrin first mentioned such a film in 1911. Perrin, "La réalité des molécules," 776. It was perhaps the film made by Comandon and Siedentopf when the latter visited Comandon in 1911, prior to setting up a similar laboratory for ultramicro-cinematography in the Zeiss factory. Jean Comandon, "L'ultramicroscope et la cinématographie," *La presse médicale* 94 (25 Nov. 1909): 841–43. The only extant Brownian motion film by Comandon that could be found is kept in the Musée Albert Kahn in Boulogne near Paris. This film was likely shown by Perrin, as suggested by the film still reproduced in Georges Urbain and Marcel Boll, eds, *La science: ses progrès, ses applications* (Paris: Larousse, 1933–34), 2: 368.

47. Scott Curtis has made this argument about Max Seddig's experiments using cinematography and ultramicroscopy. Curtis, "Die kinematographische Methode."

48. Jean Perrin, *Brownian Movement and Molecular Reality*, trans. Frederick Soddy (London: Taylor and Francis, 1910); and Jean Perrin, *Die Brownsche Bewegung und die wahre Existenz der Moleküle*, trans. J. Donau (Dresden: Theodor Steinkopff, 1910), the latter being a reprint from the *Kolloidchemische Beihefte*.

49. Werner Mecklenburg, *Die experimentelle Grundlegung der Atomistik* (Jena: Gustav Fischer, 1910).

50. See, for example, Thomas William Körner, *Fourier Analysis* (Cambridge: Cambridge University Press, 1989), where figure 1 appears sideways on 44.

51. See Leon Cohen, "The History of Noise, on the 100th Anniversary of Its Birth," *IEEE Signal Processing Magazine* (Nov. 2005): 20–45; Günter Dorferl and Dieter Hoffmann, "Von Albert Einstein bis Norbert Wiener—frühe Ansichten und späte Einsichten zum Phänomen des elektronischen Rauschens," Max Planck Institute for the History of Science preprint 301.

52. Norbert Wiener, "Differential-Space," *Journal of Mathematics and Physics* 58 (1923): 131–74.

53. Wiener quoted in Steve Heims, *John von Neumann and Norbert Wiener* (Cambridge, Mass.: MIT Press, 1980), 68–69.

54. Wiener, "Differential-Space," 133. The quotation is in Perrin, *Brownian Motion*, 64.

55. Perrin, "Mouvement brownien et realité moléculaire," 80–81.

56. Ibid., 31.

57. Emile Borel, "Le continu mathématique et le continu physique," *Scientia* 6 (1909): 21–25; idem, "Les théories moléculaires et les mathématiques," *Revue générale des sciences pures et appliquées* 23 (1912): 842–53; idem, "L'infini mathématique et la réalité," *Revue du mois* 18 (1914): 71–84.

PART THREE

Observing in New Ways: Techniques

Scientific observers continually honed the techniques and technologies they employed to observe the external and internal world, at some times discarding the old as they developed the new and at other times simply redefining in what observation consisted. The essays in this part—which examine the evolution of observational practices in natural science, economics, stroboscopy, and psychoanalysis—underscore how historically various and contingent such practices have been. In Terrall's and Canales's essays, we see a new mode of observation emerge at the hands of resourceful observers; in Maas's and Lunbeck's, we see beleaguered researchers mounting arguments in support of the scientific credentials of their chosen mode. All of our protagonists adroitly parried the challenge of devising ways to see the ephemeral and the invisible.

Mary Terrall plumbs unpublished notes on the reproductive habits of frogs left by the eighteenth-century French naturalist René-Antoine Ferchault de Réaumur, a scientist with wide-ranging interests and expertise known as an exemplary observer, to recreate in vivid detail the ups and downs of an observational program as it unfolded over several years in the 1730s. Réaumur's quarry was the remarkably elusive mating frog. Ensconced in his country home, supported by assistants—among them the talented and indispensible artist Hélène Dumoustier, whose drawings illustrate his magnum opus on insects—and numerous servants, Réaumur adopted a range of approaches in his dogged attempt to ascertain how in fact the male's semen fertilized the female's eggs.

Frustration—at his frogs, at other naturalists, at visitors who would interrupt his carefully staged observations—runs like a leitmotif through Réaumur's record of his experimentation. Observation demanded contrivance and ingenuity, endless patience and luck. Chronicled in the unfolding of real time, not after the fact, it appears less systematic and more haphazard than historians have appreciated, less linear and more as often a matter of chance as of design. This was heightened in the domestic setting in which Réaumur carried out his program; the rhythms of a bustling household, the shape of a multifaceted life in science, and observational demands for sustained and uninterrupted attention could come into sharp conflict.

We leave this bustling domestic setting for the solitude of the gentleman-scholar's private study in Harro Maas's essay. Focusing on the "powerful instrument" of the armchair, Maas traces the fortunes of observation in economic thought from its eighteenth-century heyday to its eclipse in the midst of the marginalist revolution in late nineteenth-century economics. Observation was a capacious category in classical political economy, not yet juxtaposed to theory but interwoven with it, associated both with the labor of assembling incontrovertible and measureable facts—of population growth, of nations' comparative wealth—and with the hard work of reflection and speculation. Receptivity to the economic in the everyday—the nature of transactions with tradesmen, for example—supplemented with a taste for paradox characterized the perfect observer, a standing to which even the cloistered academic, ensconced in his armchair, might aspire. With the late nineteenth-century quantitative turn in economics came a narrowing of observation's compass. The solitary thinking and reasoning that had constituted so much of the economist's work was abjured as mere introspection and the deploying of statistics in the service of mathematical modeling correspondingly extolled as properly observational. The armchair economist increasingly became a target of opprobrium, a lazy atavism indifferent to the real world, observing nothing but the contours of his mind.

Yet neither the armchair nor the observational activities associated with it could be so readily repudiated. Maas gives us a John Maynard Keynes—pictured contemplatively in his armchair on the cover of his magnum opus, *The General Theory*—disdainful of his fellow economists' "mazes of arithmetic." In contrast, Keynes adeptly gathers, assesses, and synthesizes vast quantities of economic and literary data as well as firsthand observations into a compelling whole, much as his now-spurned disciplinary predecessors had done. Maas suggests we need a more robust conception of observation than twentieth-century economists have given us to fully understand Keynes's way of working, one that takes better account of economists' heterogeneous practices.

Jimena Canales's paper on the short scientific career of the stroboscope takes up the question of what it might mean to see something that corresponded to no physical reality. Scientists had long been interested in so-called flicker effects, visions produced by intermittent flashes of light, but they shied from experimentally investigating them. Canales shows how the development and dissemination of the electronic stroboscope in the 1950s spawned a flurry of research into the brain mechanisms underlying these effects as well as into the specific visual imagery that they produced. Strobe-induced hallucinations were widely reported in the literature, with subjects' fantastic productions duly reproduced, sorted, and classified. For a brief moment in the late 1950s, stroboscopic research enjoyed a legitimacy that later faded as artists, Beat poets, and novelists joined both mainstream researchers as well as those working at the edges of scientific respectability in experimenting with mescaline and the newly synthesized hallucinogenic LSD and in celebrating the drugs' potent effects. As the strobe migrated from laboratory to disco, it trailed its observational legacy: the possibility of seeing what did not literally exist.

The psychoanalyst as observer of the other's interiority is the subject of Elizabeth Lunbeck's essay, which focuses on the analyst Heinz Kohut's career-long brief for empathy as the discipline's defining mode of observation. Empathy—from the German *Einfühlung*, literally, feeling into—is now common coin, the capacity for which divides those who rightfully participate in our common life from those benighted souls who stand outside it, but the concept is of surprisingly recent vintage, having been adumbrated only at the end of the nineteenth century to be conscripted into psychology early in the twentieth. Freud thought it necessary, the only possible way to understand the mental life of another, but he would famously have the analyst coolly deploying evenly suspended attention in the analytic encounter, using his own unconscious as an instrument receptive to what the patient's unconscious transmitted. Analysts since have struggled with the austerity of Freud's prescribed technique, meant in part to insure his discipline's scientific status against claims it was but suggestion under another name. His contemporary Sándor Ferenczi offered that only the analyst's empathy was adequate to the therapeutic task, and we see here, in a recounting of the fierce disagreements between the two pioneering analysts over the nature of the analytic persona and transaction, how quickly the term was burdened with associations of gratification, indulgence, and impropriety—associations that constantly threatened to overwhelm Kohut's attempt, thirty years later, to resuscitate empathy in the service of analytic science.

Empathy in Kohut's hands was manifestly an observational tool, a specific and distinctively analytic cognitive process, but, as Lunbeck shows, from the start he was unable to delimit its usages so strictly. Hostile commentators charged him with an unanalytic mysticism; those more sympathetic threatened to mobilize empathy as mere unanalytic sympathy. To the end of his life, Kohut found himself exhorting his fellow analysts to attend to the technical dimension of his signature concept, but his pleas fell on deaf ears. He nonetheless managed to effect a successful psychoanalytic revolution around it, challenging Freud's primacy while staying within the analytic fold. In the end, the defining ambiguity of empathy proved essential to this remarkable achievement.

1. Astronomers at work. Psalter of Blanche of Castille. Paris, Bibliothèque de l'Arsénal, MS 1186, fol. 1r (Paris, ca. 1226–32). The central figure is making an observation using an astrolabe. The figure on the left appears to be recording the result of the observation, while the one on the right consults a text, possibly a very early ephemerides or set of astronomical tables. By permission of Bibliothèque nationale de France.

2. Donato Creti, *Astronomical Observation: Jupiter*. 1711. Pinacoteca Vaticana. Photo: Vatican Museums.

3. Salvador Rizo? *José Celestino Mutis*, circa 1800, oil on canvas, 48.8 × 36.2 inches (124 × 92 cm). Real Academia Nacional de Medicina, Madrid.

4. José Guío (Malaspina expedition), *Rubus radicans* Cav., watercolor, 1790, 11.8 × 19.3 inches (30 × 49 cm). A manuscript annotation from botanist Luis Née indicates, "The fruit should be green, the painter is mistaken" ("El fruto ha de ser verde, se equivoca el pintor"). Real Jardín Botánico, Madrid, ARJB VI/40.

5. Salvador Rizo? *Antonio José Cavanilles*, circa. 1800, oil on canvas, 33.9 × 26 inches (86 × 66 cm). Museo Nacional, Bogotá, Colombia. Collección del Museo Nacional de Colombia, Bogotá, Colombia. Photograph for Museo Nacional de Colombia by Juan Camilo Segura.

7

Frogs on the Mantelpiece: The Practice of Observation in Daily Life

MARY TERRALL

> I separated a pair of frogs that had been mating for five or six days. I observed with attention the place on the female where the male's hands press, and I noticed nothing that looked like an aperture. It remains to observe the same places on frogs closer to being ready to lay their eggs. . . . The fingers either do not touch at all or barely touch the skin of the female. . . . The hand is turned in such a way that the shagreen portion molds itself and ensconces itself in the flesh of the [female] frog. I observed this shagreen body [i.e., nuptial pad] of the male at the moment when it had just been separated [from the female]. The grains seemed to me more distinct, bigger, more inflated than those of males who are not yet mating. It seems that it is an assemblage of an infinite number of glands. If one then pays attention to the strong and constant pressure holding these glands against the skin of the female frog, one will conclude that these glands are almost joined to the skin of the female. Since these glands are swollen, the pressure must cause some liquid to come out, and it could not do so without penetrating the skin of the female. Is it not therefore possible that the liquid trying to escape from these glands penetrates the skin of the [female] frog? . . . Why should we not suspect that this liquid is necessary for the fertilization of the eggs? (fig. 1)[1]

At his death in 1757, the French naturalist René-Antoine Ferchault de Réaumur left a vast collection of natural history specimens, books, and manuscripts, the material remains of a lifetime of scientific observation. Some of the papers—hundreds of letters and thousands of sheets of notes, drawings, early versions, and final drafts of manuscripts—ended up at the Paris Academy of Sciences, as their owner had directed in his will. Others, including many folio sheets of drawings of insects, amphibians, and marine animals, were appropriated by various academicians. Some of these papers eventu-

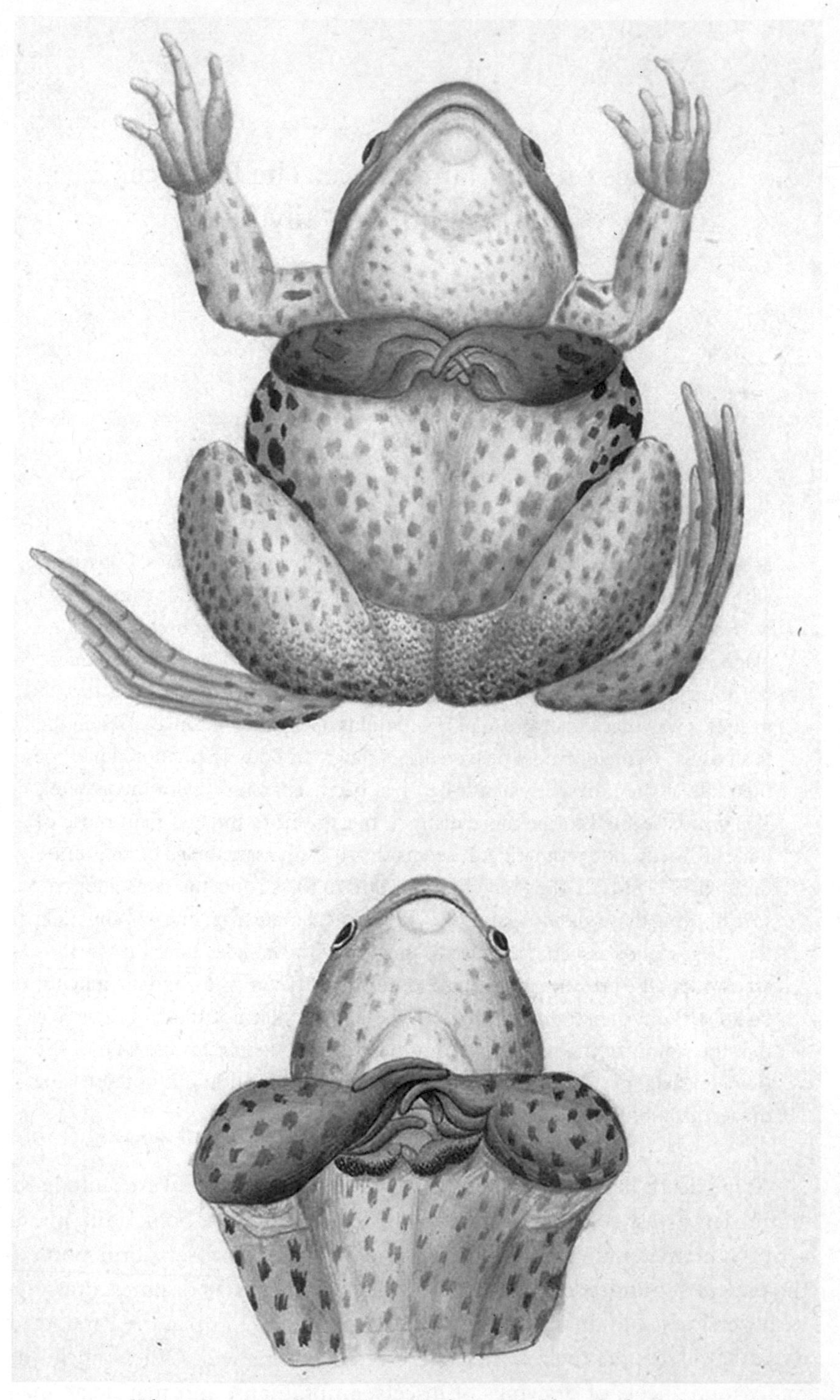

FIGURE 7.1. Hélène Dumoustier de Marsilly, mating frogs, viewed from the front, showing the male's hands reaching around from behind and gripping the female's chest. Lower image shows nuptial pads on thumbs. Drawings made for Réaumur. By permission of Bibliothèque centrale du Muséum national d'histoire naturelle, Paris 2009.

ally came to rest in the library of the Museum of Natural History in Paris. Among the manuscripts preserved today in the archives of the Academy of Sciences are a handful of unbound sheets of notes on the mating behavior of frogs.[2] These eclectic and inconclusive notes record an observational program, never published, that engaged Réaumur and several associates over the course of several springtime mating seasons in the 1730s. Their observations raised as many questions as they answered; in his notes, the naturalist's frustration is palpable at times. For the historian, likewise, the documents leave unresolved many questions about exactly what happened and what was seen by the observers. Straightforward descriptions of simple events or phenomena share the page haphazardly with queries, comments on reports by other observers, and plans for future experiments. Nevertheless, the unpolished notes, with all their confusion and spontaneity, allow us to spy on scientific observation in action, much as Réaumur and his collaborators spied on the frogs in their jars.

The scene opens, in the earliest of the dated pages, on the grounds of Réaumur's country house just outside Paris. "Although frogs jump, I don't think it could be more than two or three feet high. On this supposition, I circled my little pond in Charenton with a plank enclosure high enough so that the frogs could not jump over it, and into this enclosure I put a large number of frogs to watch them mating."[3] The fence marked off a space where Réaumur could collect and keep track of the frogs, concentrated in one area. Once confined, they could be plucked out of the pond and brought into the house. On 12 March 1736, he found two frogs coupled together in his enclosure. "They allowed themselves to be picked up without separating from each other. They stayed coupled together in my hand. I put them in a large bell jar with a bit of earth."[4] With the dirt partially above the surface of the water, the pair remained attached to each other, moving around in the jar without separating. Day after day he found them in the same attitude. Over the course of the next week, with the weather warming up, he found several more pairs of mating frogs in the pond, and moved them indoors for more convenient viewing. "I took two [of the three pairs]. One pair fell to the ground, I picked them up. The female ran onto the mantelpiece of my hearth; I caught her, and in spite of all that, they did not separate."[5] The energetic female made her brief escape from the observer's hands, but she and her mate were soon recaptured and enclosed in a jar near the hearth or on the writing table, where they became the objects of Réaumur's daily scrutiny. The chase across the mantelpiece was a rare burst of activity punctuating the long wait for the immobile pair to lay and fertilize their eggs. On 26 March, he noted, "Today the first pair of my frogs is still coupled; they have been like this for at least fifteen days."[6] Although he writes in the first person (as the master of the house and the orchestrator of the experiments and observations done there), we need not imagine that Réaumur built the wooden enclosure in the pond himself; he very likely also had assistance in catching the frogs he planned to confine there. However, direct and declarative statements about his own actions show him intimately involved in handling the frogs, moving them around, confining them in glass receptacles, placing them on his desk while he works, and so on. The frogs, meanwhile, took their time, and did not always cooperate with the projects being formulated on the desk where they sat motionless in their jars.

Réaumur approached his observations with little knowledge about how frogs went about the process of reproduction. Above all, he did not understand how the eggs were fertilized, and he kept his mind open to every imaginable possibility. Were the eggs already fertilized when they emerged from the female? Did the hands of the male, as described in the passage that opens this essay, have anything to do with fertilization or with the extrusion of the eggs? What does the male emit, if anything? Does the male even have a sex organ as such? "Could not the real copulation of frogs take place through the fingers? But if there is no opening in the place [on the female's chest] that is squeezed by the shagreen of the thumbs, might one not find something in the anatomy of the legs to convince us that seminal matter is transported to the shagreen of the thumbs, that it is filtered there and that it filters through the skin of the [female] frog?"[7] Such conjectures led to experimental intervention to clarify the means of fertilization and, in particular, to determine whether the gripping thumbs of the male were somehow implicated.

With touching optimism, Réaumur conceived a series of experiments: "The truth of all this will be easy to discover through the experiments I plan to carry out:

> exp. 1: I will put pants [*culottes*] made of bladder on a male frog.
>
> exp. 2: If I can manage to surprise a frog at the time she makes the eggs, I will pull the male off her before all the eggs have emerged; I will separate the eggs that came out first from those that came out last, and I will see if those are fertile and if the others are not.
>
> exp. It would be a good idea to separate the frogs that will have been mating for 12 or 15 days or more to see if they will produce eggs, and if their eggs will be fertile.[8]

A pair of little pants was duly constructed, though the notes do not specify exactly who did the cutting and stitching. The impersonal pronoun "*on*" (translated here as "we") signals the participation of someone else in the delicate business of constructing the garment out of stretchy pieces of animal bladder. "On the 21st of March we put a pair of pants made of bladder, very tight pants that, above all, seal off the hind end, onto a male frog coupled since the night before. Its mating was not disturbed by this; it continued to be attached to the female. If the lacing [on the sides of the pants] holds, this experiment should teach me some very curious facts."[9] But the bladder material became too soft and floppy in the water and started to shred, so that he could not be sure that the frog was adequately covered. In detailing his troubles with getting the pants to stay on the frogs, Réaumur effaced his helpers through the first-person singular pronoun. "I made numerous attempts before figuring out how to give the male frogs pants I could be satisfied with. . . . Waxed taffeta seemed a better choice, but after having made the pants and put them on, the frogs abandoned them in front of me."[10] The leg holes were too big—pulling their legs up inside the garment, the frogs pushed free of them. Human ingenuity overcame this difficulty with the custom tailoring of the waxed taffeta garment: "I managed to give them pants that they could not take off 1) by piercing the two holes such that there was no more space between them than the size of the rear end [of the frog] or just a bit more. 2) by making these holes only as big as the diameter of the thigh. 3) by sewing a few stitches on the sides and near the thighs after the pants had been put on. But what ensured the whole thing was that I put suspenders [*bretelles*] on these pants. I passed them over the arms of the male frog, under the head, between his body and that of the female."[11]

Improbable as this experimental design may sound, it was documented in finely rendered drawings recently discovered in the library of the Museum of Natural History in Paris. One of these clearly shows the stitches joining the taffeta along the sides and the straps passing over the frog's shoulders to secure the garment (fig. 7.2).[12] In spite of the successful outfitting of the male, the results were disappointingly inconclusive, a problem that plagued Réaumur's frog observations for the next few years. He just could not determine to his own satisfaction whether the pants had interfered with fertility or not. Indeed, his notes do not mention looking for seminal fluid in the pants, although the waterproof waxed taffeta held out some hope for such a discovery. (Frogs normally remain partially submerged when laying eggs.)

Two females laid eggs while mating with males who kept their pants on for four or five days before separating from the females. Oddly, Réaumur does not record whether these eggs developed; the notes do not even make clear whether he was keeping close track of the eggs, though he indicated that he intended to do so. It is quite possible, given the unsystematic nature of the notes, made on loose sheets, that some pages are missing. He does remark that the next year he planned to get the males into their pants earlier in the

FIGURE 7.2. Hélène Dumoustier de Marsilly, drawing of mating frogs. The male is encased in waxed taffeta pants with shoulder straps. By permission of Bibliothèque centrale du Muséum national d'histoire naturelle, Paris 2009.

process: "It will be good to get pants on several males before they are coupled [with the females]."[13] As frustrating as these observations may have been for those on the spot, they are similarly so for the historical observer. Like the frogs copulating in their jars, the behavior of the naturalists does not necessarily conform to our expectations—proposed experiments go undone or unrecorded, and suggestive lines of investigation seem to be dropped. These fragmentary sources can nevertheless help to retrace at least a few of the paths taken by our observers, who no doubt never imagined themselves being observed across a gap of several hundred years.

Observing Nature in Charenton

In 1736, Réaumur was at the height of a long career as a prolific and powerful member of the Paris Academy of Sciences. A moderately wealthy member of the landed gentry, Réaumur lived an active intellectual and social life in a comfortable house in the city, where he had an extensive laboratory and a constantly expanding natural history collection. Perhaps most famous for his invention of the thermometric scale that bore his name, he was also a consummate observer, especially of insects. His great work on the subject was published in six volumes between 1734 and 1742, overlapping with the period of his frog observations.[14] In addition to the Paris residence, he rented a country house on the river at Charenton, where the Marne flows into the Seine.[15] The attractive situation of the house, with a garden extending right down to the river, lent itself to many investigations in the spring and summer, especially on aquatic insects and amphibians. It was also close enough to Paris to allow him to go to afternoon meetings of the Academy of Sciences even when living out of town; Réaumur kept a carriage and enough servants and assistants of various sorts to staff both residences.[16] Here he observed frogs, in a small-scale and decidedly low-tech operation suited to the country-house setting.

The post of laboratory assistant or manager of collections, a kind of apprenticeship in observing, served as a stepping stone to a place in the academy for several young men over the years. One of these was Jean-Antoine Nollet, not yet launched on his spectacular career as instrument maker and physics demonstrator to the upper classes. Nollet performed all sorts of experiments at Réaumur's behest, with air pumps, thermometers, electrical machines, chemical apparatus, and microscopes. Though he probably did not live in Réaumur's house, he certainly did some of his work there, and was well known to the extended household. Years later, he recalled working with the frogs at Charenton. Writing to Lazzaro Spallanzani in 1768, Nollet remarked, "About thirty years ago, M. de Réaumur and I did considerable research on [the reproduction of frogs]. We followed, with a great deal of care and patience, those embraces and copulations for weeks; I remember having put pants [*caleçons*] of waxed taffeta on these little animals, and having watched them for some time, without ever being able to see anything that indicated an act of fertilization."[17] Note that Nollet recollected his own essential role in dressing the frogs, as well as the inconclusive results. (As we shall see, Nollet's remark inspired Spallanzani to repeat the experiment.)

Also part of Réaumur's extended household was Hélène Dumoustier, a lady with a considerable talent for drawing and observation. From the early

1730s, she lived, along with her sister and widowed mother, either in his house or nearby.[18] Over the course of many years, Dumoustier made hundreds of drawings to illustrate Réaumur's work on insects. She was not working for pay, and modestly refused to be named in his works. However, elliptical mentions in print and explicit references in letters and manuscript notes document her active participation in many different kinds of observations, indoors and out. She accompanied Réaumur on rambles and collecting expeditions, including his annual trip to his ancestral estate in Poitou, where he spent the two-month academic vacation in intensive natural historical work. She was often at Charenton, where she observed the mating frogs and made the drawings reproduced above. She must have worked alongside Nollet and Réaumur as they made and altered the pants for the frogs, though neither of them mentions her explicitly in this connection.

In springtime by the riverbank at Charenton, frogs would have been obvious subjects for observation, just as mayflies would be later in the summer.[19] In fact, Réaumur included frogs in the broad category of "insect," alongside salamanders, worms, and many other things. Like insects, frogs were both mundane and mysterious, and both had long been the focus of inquiry into the problem of generation, especially the vexed questions of spontaneous generation and the preexistence of germs. Contemporary literature on frogs, which Réaumur read carefully, left key details of their reproductive physiology unresolved. In particular, he mentioned the drawings of mating frogs by the renowned Dutch anatomist Jan Swammerdam, who had also passed many hours in the company of frogs and tadpoles. Réaumur noted, with evident annoyance, that he did not see some things as Swammerdam had represented them.

Jan Swammerdam's Observations of Frogs

Aristotelian natural philosophy attributed the origin of frogs, insects, and other lowly creatures to the prolific powers of nature. In the eternal cycle of corruption and generation, these animals emerged spontaneously from rotting inanimate matter.[20] Seventeenth-century microscopists, including Francesco Redi, Marcello Malpighi, and Swammerdam himself, made careful studies of the creatures that hatched in pond slime or rotting meat, with the explicit aim of undermining such claims about spontaneous generation. They refuted "equivocal" generation with observations of previously unseen structures unfolding according to predictable and consistent patterns of development. Swammerdam, an accomplished and innovative anatomist who turned his gaze and instruments on insects with passionate intensity, made it his mission to uncover hidden structures, thereby denying the possibility of order arising by chance, or spontaneously. His *Historia insectorum generalis*, based on years of systematic observation and dissection, showed that plant seeds, insect larvae, and frog eggs all contain the parts of the future adult.[21]

Swammerdam set out to reveal truths about God and nature, but also to demonstrate the qualities of the expert observer. Those clumsy observers who saw only what their prejudices told them to see had encrusted nature with "soil [*ordures*]," which he promised to clean away with his meticulous deployment of lenses and dissecting scissors.[22] Cutting through the skins and membranes hiding internal structures, he pictured himself cutting through the grime of what others had said. Seeing the unseen and making it visible to others through demonstrations, text, and images became the hallmark of Swammerdam's observational practice, emulated by many others in subsequent generations.

Swammerdam condensed a vast number of observations into a foldout table at the end of his *Historia insectorum*, where he arranged different categories of insects alongside frogs (representing "animals with blood") and plants to display analogies between the different types of "changes or growth of parts" in all these organisms.[23] Swammerdam's more extensive treatise on the frog, part of his magnum opus on insects, remained unpublished at his untimely death in 1680. The manuscript, including fifty-two plates engraved from his own drawings, changed hands several times over the years, ending up in the library of the Paris anatomist Joseph Duverney, a friend of Réaumur. In 1727, Hermann Boerhaave bought the manuscript and spent years shepherding it through the publication process in Utrecht. The two volumes came out under the Dutch title *Bybel der natur* in 1737 and 1738 (with a parallel Latin translation).[24]

Since we do not have Swammerdam's observational notes, we cannot see into the minutiae of his daily practice, as we have done for the observations in Charenton, though he does emphasize the large number of observations behind his generalized conclusions. In addition to dissecting frogs and experimenting with the eggs, he raised the young and charted their development. One section of *Bybel der natur*, illustrated with three new plates, lays out the anatomy of the frog's reproductive organs and the growth of the tadpole from the egg. He also observed the mating habits of mature frogs. "To carry on the intercourse of the sexes . . . the male Frog leaps upon the female, and when seated on her back, he fastens himself to her very firmly. For this reason, the Dutch country boors, with great propriety, tho' in their vulgar way, call this manner of copulation, the riding season of the Frogs, as the male is carried about, riding, as it were, by the female." The grip of the male was so tight

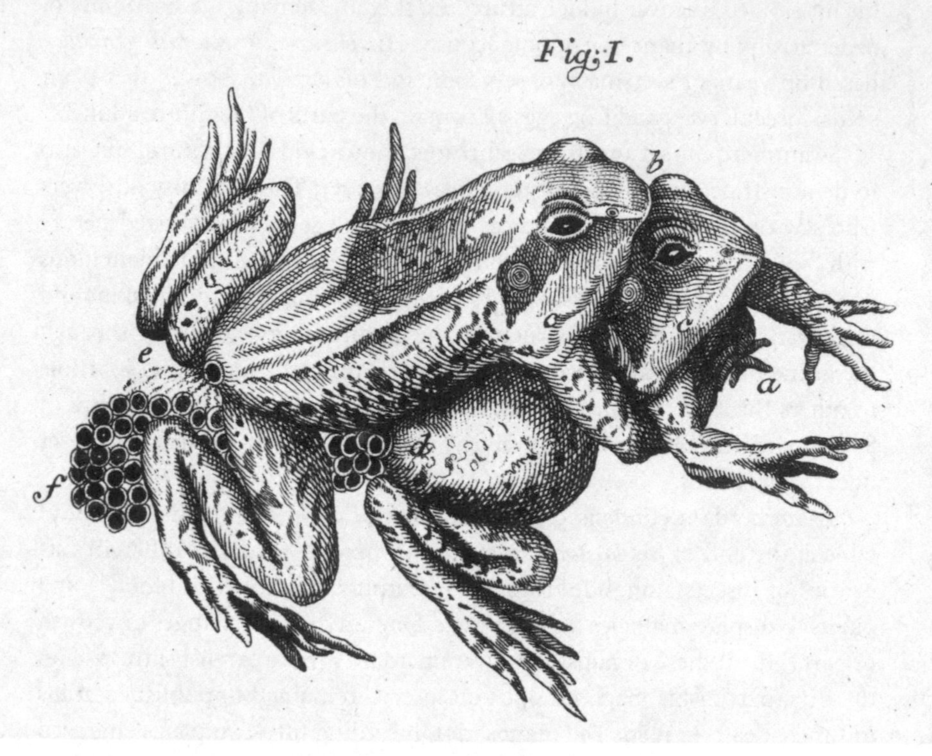

FIGURE 7.3. Jan Swammerdam, *Bybel der natur* (1737–38), plate 48 (detail).

that Swammerdam could not separate them with his hands alone; he had to use a metal spatula to pry them apart.[25] He depicted the male frog's hands squeezing the female's chest, the emission of the eggs by the female, and the fertilization of the egg mass by the male (fig. 7.3). This latter event, which became a stumbling block for Réaumur when he could not manage to see it, Swammerdam represented as entirely unproblematic.

Duverney had printed at least some of Swammerdam's plates in the 1720s, well before the publication of the book; these were known to Réaumur and his friend the anatomist Jacques Winslow, another Paris academician.[26] Réaumur's earliest notes on frogs (1736) specify that he did not observe the male's hands joining as depicted by Swammerdam. The position and action of the hands then became a focus of the investigation, as we have seen; one of Dumoustier's drawings clearly shows the hands gripping the chest without meeting, an alternative to Swammerdam's representation. Like Swammerdam, Réaumur thought the male might actively squeeze the eggs out of the female's body. If the eggs were sprayed by the male when they emerged, that event should be visible too. Inspired by Swammerdam's image, Winslow had looked for the spraying of the eggs, without success. Their difficulties made Réaumur skeptical about what his Dutch predecessor had actually seen. "If he was not simply saying this according to what he believed it should be, if he saw it, he really should have told us how he managed to see it, whether he saw it distinctly and how many times."[27] Réaumur suggested that Swammerdam might have casually assumed, without seeing it happen, that frogs spray their eggs. In other words, he might have made a drawing of a presumption rather than an observation. This petulant critique of his predecessor's evasion of his responsibility to other observers reflects Réaumur's frustration with his frogs as well as with his sources. Observers ought to reveal more than their bare results, or their conjectures, and images need narratives to enable others to see the same phenomena for themselves. The details Réaumur supplied about his own procedure, including his many setbacks, tended to attest to the credibility of the observations—even, or perhaps especially, when they failed.

Watching and Waiting in Charenton

Watching patiently, looking attentively, noticing—these are the most salient activities of observation, as many of the essays in this volume attest. Without imaginative questioning and conjecturing, however, no amount of patient attention would unravel the mysteries of something like the fertilization of frog eggs. Réaumur's notes shift continually between proposed experiments, observed events and experimental interventions, possible interpretations, and comments on what other authors had said on the subject. Observation and experiment were inextricably connected in this enterprise; Réaumur used the notations "Obs." and "Exp." more or less interchangeably in his notes to designate future plans, introduced by phrases such as "I will make . . ." or "We must see if. . . ." In contemporary usage, as we would expect, experiment often implied more active intervention than observation, but the two could hardly be separated, since the results of experiment had to be carefully observed. When Nollet, well trained by his apprenticeship with Réaumur, gave his inaugural speech as the first professor of experimental physics at the College de Navarre, he defined his field by invoking the rare but necessary qualities of the good observer. "The Observer must have unwearying patience, an attention that no circumstance escapes, a prompt and lively penetration, a wise and moderate imagination, a great deal of caution and circumspection in his judgments."[28] Nollet was talking about physics but cited Réaumur's work on insects as a prime example of this kind of attentive observation, an activity that embraced experiment as one of its methods. The adept experimenter was really nothing other than a wise observer, who could invent and adapt techniques appropriate to the subject at hand.

Réaumur never transformed his notes into a consistent narrative of discovery. The jumble of queries, experiments, details of the frogs' behavior, and possible explanations betrays the confusion and the open-endedness of the inquiry. He might start a chain of experiment, observation, and reasoning, but often the chains of inference break down when he cannot determine exact outcomes. While he notices some things as part of purposive experimental interventions, many observations follow from unplanned events. When a male died unexpectedly two or three days after the start of copulation, for example, he dissected the testicles, where he found "a great quantity of liquid as white as milk and thicker."[29] This led him to wonder why he could never see traces of the milky substance in the water around fertilized eggs. It seems obvious to us that the male fertilizes eggs by spraying semen over the eggs, and indeed, this was what Réaumur expected to see, based on analogies with other animals and on Swammerdam's depiction of the event. But he could not see it, even when he looked carefully, and so he remained undecided.

Even the act of writing notes, though a crucial part of the whole enterprise, was more haphazard than we might expect. In some cases, Réaumur seems to have reflected on what he had seen several days after a series of observations. Sometimes he carefully noted dates and times, but other times he collapsed multiple observations into a single summary statement. The fragmentary and even unfinished nature of the observations is characteristic of Réaumur's mode of operation. Like many naturalists in this period, he had many different programs and projects going on simultaneously, and only some of them resulted in finished texts. Declarative statements about the actions or anatomy of the frogs ("My first pair of mating frogs have not changed their position at all since yesterday morning and evidently they will not move at all now.") alternate with questions for further investigation ("Is it possible that the male emits from his body only a liquid reduced to a vapor?"). Sometimes he proposes an interpretation of what he has seen: "If the female makes eggs, and these eggs are fertile, it will prove that they were fertilized during the copulation [i.e., before being emitted]. But if none of the eggs is fertile, it seems very likely that their fertilization depends on the presence of the liquid [from the male]." And sometimes he noted experiments to be done at a later time. After setting down his detailed observations of the nuptial pads of the male, with his conjectures about their possible function in fertilization (quoted at the beginning of this essay), he thought of another use for waxed taffeta. "To verify these conjectures, which, in truth, are quite singular, I should put gloves on some male frogs just as I put pants on others, or rather place a piece of waxed taffeta in such a way that the male could not press the female except through this taffeta."[30] In this way, he imagined interfering with any possible transmission of the liquid from the glands of the thumb. He does not seem to have pursued this avenue of investigation, however, and by the next season, he had shifted his attention from the hands to the anus of the frog.

In the second year, Réaumur apparently did not devote as much effort to the frogs, although he did collect pairs of frogs in an attempt to see the eggs emerging from the female. His notes are brief, and retrospective, probably from the end of the mating season. "No matter what I did this year, I have not yet been able to surprise the frogs at the time of making their eggs. From the 3rd or 4th of April until the 6th I had twelve or thirteen pairs in a bell jar resting on my desk; the females laid their eggs there and some even laid them while I was working next to them on my desk, without my having succeeded in seeing them come out. What I learned at least was that their laying [*ponte*] is done in less than a quarter of an hour, and perhaps in much less than this."[31] The bell jar on the naturalist's desk, with the coupled frogs keeping him company as he writes, nicely illustrates the integration of the objects of observation into the mundane work routines of the household. Even in such close proximity, the frogs eluded his wandering gaze. We might say that the rhythm and pace of the frogs' mating did not match very well the working habits of the naturalist, who, unlike the frogs, was usually doing more than one thing at once.

At the beginning of the 1740 season, Réaumur started off with a concerted effort to catch the frogs in the act of laying eggs and perhaps of fertilizing them.[32] He wanted to see the eggs emerging from the female and witness the response of the male; since the frog lays her eggs quickly, and underwater, this presented some practical difficulties. "Although the female lays a great number of eggs and they form, with the viscous matter that envelopes them, a considerable mass, it is an operation that lasts a very short time, less than a minute. In vain have I tried for several years to seize the moment of the egg laying of the frogs that I kept in glass jars; I have not succeeded at it."[33] He decided to attempt a more systematic approach. He put twelve pairs of frogs in twelve glass jars, one pair in each jar, and distributed some of them to additional observers. "Of these twelve pairs, I kept eight for myself, and I gave two to Mlle Dumoustier, and two to M. Guettard." He clearly recognized Dumoustier as a skilled and reliable observer, and in the present instance, she proved herself worthy of this confidence. (Jean-Etienne Guettard was a young protégé of Réaumur, who often assisted in observations in this period.[34]) The lady made the key observation in this case. After her first pair of frogs frustrated her by emitting their eggs when she was not looking, she redoubled her efforts to watch the remaining pair. "Her attention to examining them was repaid at the end of a quarter of an hour; . . . she noticed eggs beginning to come out of the hind end of the female; on the instant, as I had recommended, she turned her eyes toward the hind end of the male and fixed them on it. Hardly had she focused on it when she saw a jet come out of it. She did not know what else to compare it to other than what it resembled most, a jet of pipe smoke. In leaving the male's rear end it was as big as the quill of a feather, and a bit farther out it divided into a great number of jets of finer filaments, like those into which a jet of smoke divides. This only lasted an instant and that was all she could see." Finally! Not only did she see the eggs emerging, but she saw the male's emission through the water. Her own words come through in Réaumur's summary of her description, when he reports her vivid image of the jet of pipe smoke almost verbatim. This observation opened up a new line of speculation: could it be that the liquid emerging from the male had been vaporized? "What I know and what my jars have given me the opportunity to examine several times, is that it does not seem that the male spreads a detectable quantity of milky matter over the eggs." He had found this milky substance in the testicles, but when he looked closely for traces of it in the water and on the eggs, he saw nothing. "The water where the eggs had just been deposited did not look milkier to me, and the strongest magnifying glass did not permit me to perceive any milky filaments on the eggs."[35]

After Dumoustier's success, Réaumur continued to pursue the elusive prize himself. Although he fully trusted her observation, his own skill was in question if he could not succeed in seeing it, especially given that he had started with four times as many jars as she. As his mating pairs continued one by one to lay their eggs unobserved, it became a kind of contest between frog and observer. He describes placing six or eight of his jars in a semicircle on his desk; when he got down to the last three, without having observed the event, he decided to try to control the timing by heating the water in the jars. (He had recently perfected this technique for forcing insect chrysalises to hatch ahead of their natural schedule.) His notes narrate the scene in the study, where he moved the jars, one by one, closer to the fire and back to the desk. In the first case, everything was going well when he was interrupted by a visit from two of his academic colleagues. The frogs were restless on account of the warm water; although he kept an eye on them, he missed the big moment while conversing with his visitors. The second time, encouraged by the evident effect of raising the temperature of the water, he went too far and the frogs got agitated by the heat and separated for good. Ultimately—on his last remaining pair—Réaumur saw the eggs emerge, as the male "gave out a croak, he arched his back a bit and pulled in his legs at the same time, then he stretched out and retracted his flanks. He repeated these movements three times in a row. I had only my [naked] eyes, but well positioned to see; I could perceive nothing coming out of his hind end."[36] This story sounds almost too good to be true. The first chance was ruined by inattention caused by visitors; the second by his overenthusiastic use of the fire to speed things along; and the third and last leading to the climactic moment of these weeks of observations. Even so, he saw no sign of any fluid, not even the transparent jet observed by Dumoustier, and although the frogs stayed connected for several more hours, through another bout of egg laying, none of the eggs developed.

And so the mystery of fertilization remained unresolved, in spite of careful and inventive observation. The seasons of living so intimately with frogs never culminated in the "history" of the creature analogous to those Réaumur produced for many insects. We can only conclude that he was not satisfied that he had really understood the mechanics of fertilization, even after seeing the eggs deposited in the water. Mysteries remain for the historical observer as well. Why does Réaumur not record more often whether the eggs developed? Why does he mention only once, in the final case, the croaking that accompanies mating? Why does he not follow up on his ideas about separating eggs into several batches under different conditions? Why was the experiment with the taffeta pants not more conclusive? Why was he so unsystematic about the whole operation, including his note taking? The scrawled sheets of notes are but a partial record of scientific observation in the genteel surroundings of the country house at Charenton. Réaumur, Nollet, and Dumoustier were all engaged in other work at the same time; they traveled back and forth to Paris; visitors came and went; the business of the regular academy meetings continued; letters were written; manuscripts were corrected. Observation was part of all this, taking place in the interstices of daily life, bringing together shifting sets of people and raising a steady stream of questions.

Spallanzani's Frogs

Though he knew nothing of Réaumur's amphibian investigations when he started working with frogs, the Italian naturalist Lazzaro Spallanzani pursued many of the same questions thirty years later, with more success. He started working with frog eggs to see if he could identify the structure of the tadpole before fertilization, to confirm the preexistence of germs. His experimental work with microorganisms, snails, salamanders, and frogs participated in the continuing controversy about spontaneous generation and epigenesis, a controversy kept alive especially by John Turberville Needham's work on infusoria.[37] Time and again, Spallanzani's observations led him to refute Needham with evidence for preexisting germs. Without entering into the complexities of this debate about the origin of life and organization—the cutting edge of the life sciences well into the 1780s—I look here at just a few moments in Spallanzani's experiments on fertilization in frogs that harken back to Réaumur in surprising ways. Unlike his predecessor, Spallanzani did publish his results, so we have the final version of a text to read alongside his notebooks and letters.

When he first started to collect and examine the eggs, Spallanzani had to experiment with keeping the mating pairs in tubs, much as Réaumur had done. "I did not see when the frog actually delivered herself of the eggs: my presence troubled them, and caused them to hide themselves in the water, but almost every half hour they presented me with little bits [of egg masses]."[38] Although he could not see the act of fertilization directly, he accepted Swammerdam's depiction of the event as happening at the moment when the eggs emerged from the female. Examining eggs before and after fertilization, he argued that their identical structure implied the preexistence of the tadpole in the unfertilized egg.[39] Later, he did many experiments to confirm that eggs taken from the female by dissection were infertile. "I see in my journals that, although I opened 156 mating female frogs, and although I put the eggs into water as soon as I had pulled them from the body of their mother, nevertheless all these eggs were sterile, and soon decayed."[40]

Spallanzani reported on his experiments with tadpoles and eggs in 1768 in a little book (*Prodromo di un'opera da imprimersi sopra la riproduzioni animali*) reporting preliminary results from an experimental program on generation and regeneration that he planned to pursue further. It was this book that prompted Nollet to write to his Italian colleague about helping Réaumur to put pants on frogs many years before. Spallanzani copied Nollet's letter into his research notes and quoted it verbatim in the published account of his experiments on fertilization.[41] "The idea of the pants [*calzoncini*] did not displease me," he remarked, "although apparently bizarre, and even ridiculous for those who do not enter profoundly into such matters, and I decided to do it."[42] Though Nollet and Réaumur, "in spite of their wisdom and their diligence," had been unlucky (*poco fortunati*) in their observations, Spallanzani claimed to have succeeded. He said nothing about how he had constructed the garment or whether he had had difficulty in getting it onto the frog. Mysteriously, there is no mention of the experiment with the little pants in the extant manuscript notes, although the fertilization experiments are otherwise recorded in considerable detail.[43] In the text, Spallanzani noted that the haberdashery did not interfere with the desire of the males to mount the females. The results were "as expected"—the eggs did not develop. But his success was apparently more spectacular: "I observed very visible little drops in the pants. These little drops were the real seminal liquid of the frog, since I produced with it a true artificial fertilization, as we shall see in the second memoir."[44] This tantalizing remark, found in the published text and absent in the notebooks, gives no clue as to how he managed to see and collect the drops, though it certainly reinforces the image Spallanzani constructed of himself as an expert observer, succeeding where such illustrious predecessors as Nollet and Réaumur had failed. Spallanzani went on to perform many further experiments on artificial fertilization using various techniques, usually involving dissection to acquire large enough amounts of the fluid, and on the seminal fluid itself.

Charles Bonnet, excited by the potential of this type of observation to support his own theoretical convictions, pushed Spallanzani to explore the mystery of conception by manually fertilizing frog eggs with seminal fluid taken from male frogs.[45] Bonnet was particularly interested in whether semen from one species could fertilize another. "It is well known that eggs of fish with scales are fertilized by the male more or less like those of frogs. Therefore I would like you to try to fertilize fish eggs by spreading over them the seminal fluid of frogs. Who knows whether this would not result in a singular kind of hybrid? . . . In a matter as obscure and as interesting as generation, we should be allowed to imagine experiments or combinations, even the most bizarre and the farthest from the ordinary course of nature."[46]

Spallanzani did not pursue these suggestions until some years later, in 1777, in experiments carefully documented in his notebooks. Artificial fertilization brought the mysterious process under not only the eyes but also the hands of the observer. Spallanzani referred to this as "giving life" to tadpoles and "imitating nature in the means she employs for multiplying these amphibians."[47] He claimed to have overlooked nothing in his quest to uncover the truths hidden in the frog's entrails or in the structure of the eggs, or in the behavior of the frogs themselves. "Without bragging, I have seen in my journals that I opened two thousand and twenty-seven copulating frogs or toads for this dissertation and the following."[48] Spallanzani also investigated the mechanics of fertilization, much as Réaumur had done. He described the motions and cries of the frogs in considerable detail: "I saw the female very agitated, jumping around here and there in the container, climbing up from and going back down into the water. . . . letting out a cry in a low voice; the male kept his legs circled around the body; he also made remarkable contortions, and he accompanied the female's voice with a kind of interrupted song, which it would be difficult to describe."[49] Although he saw a pointed bit of the male's anatomy, which he took to be the sex organ, at first he could not see the actual spraying of semen. Realizing that the transparency of the semen could render it invisible in the water, he decided to pluck mating pairs from the water as the eggs emerged, to see if he could witness the fertilization. Once he had taken the frogs out of their element, he saw the whole thing, including the "little jet of clear liquid." "One could observe this repeatedly; from time to time the female stopped laying her eggs, and then the male stopped shooting out the jets of this transparent liquid over them; I observed this curious scene with seven pairs of mating frogs that had been removed from the water."[50] Here Spallanzani provided exactly the kind of clear description of his procedure that Réaumur demanded.

Observations of frogs, their eggs, and their semen, from the time of Swammerdam to the publication of Spallanzani's elegant experiments with artificial insemination in the 1780s, were motivated in part by continuing controversies over spontaneous generation, epigenesis, and preexistence. But the challenge of deciphering mating habits, along with fertilization and egg laying, was more a practical than a theoretical challenge. Naturalists were universally curious about the variety and peculiarity of these aspects of animal behavior, especially in creatures far from the familiar quadrupeds. Réaumur himself observed fertilization and egg laying and mating in hundreds of species of insects, as well as birds, although he never articulated a theory of generation as such. Both Swammerdam and Spallanzani worked tirelessly to support preexistence, but their daily efforts focused on the practical problems of how to see elusive events, behavior, and mechanisms rather than how to frame theoretical explanations. In my own observations of these investigations I have left aside the ideological or theoretical convictions of my protagonists, concentrating instead on extracting mundane practices of observation, from notes, letters, and published texts to show how observing became part of the daily life of science pursued in households and gardens and ditches.

The notes and other evidence testify to the frustrations as well as the successes of this kind of work. As the case of the frogs has shown, seeing a crucial event such as the laying of the eggs or the emission of the semen took patience, but also ingenuity. The observer could not simply sit back and watch and record unfolding events. Observation required conjectures and queries as well as physical interventions. The essays by Charlotte Bigg and Anne Secord in this volume support this point. Even something as simple as laying eggs had to be staged or set up so that it could be seen and perhaps manipulated. The frogs had to be confined to glass jars, situated to suit the observer;

the water level had to be adjusted, along with the temperature; garments had to be sewn; techniques had to be devised for dissection, collection of semen, and handling of eggs. Every little variation gave rise to questions, conjectures, hypotheses. Observation (inseparable from experiment) generated a continual flow of ideas—for explanations, for interventions, and sometimes even for theories.

In a sense, the observer was never alone, even though he or she might spend long stretches closeted with immobile amphibians. Texts and images extended the community of observers through time, so that Réaumur could converse, as it were, with Swammerdam, and Spallanzani could repeat what Réaumur had done, having heard about his experiments through a personal letter. In pursuit of elusive and ephemeral phenomena like fertilization or the gripping of the female by the male nuptial pads, observers also worked in local groups, talking, looking, listening, writing, and drawing in their studies, laboratories, or gardens. Dumoustier, Réaumur's artist, and Nollet, the technical assistant, shared in the activity of observing, and shared the values of the wider community of observers as well. Other members of the household, or visitors who walked in on the observations, or local fishermen who supplied specimens were all part of the dynamic as well. The observations produced in these local settings—whether remaining in manuscripts and drawings or passing to others in letters or reaching a wider public through print—allow us to see into the lives, the methods, and the aspirations of long-dead observers.

Notes

1. René-Antoine Ferchault de Réaumur, "Grenouilles," Académie des Sciences, Fonds Réaumur, Paris (AS-FR-G; unpaginated folio sheets).

2. Excerpts from the manuscripts in R.-A. F. de Réaumur, *Morceaux choisis*, ed. Jean Torlais (Paris: Gallimard, 1939). Réaumur recorded his notes on loose sheets, often on the reverse of corrected manuscript copy of his book on insects. Only some sheets are dated; I give dates when they are indicated.

3. AS-FR-G. This sheet is evidently the first of the series, though undated.

4. AS-FR-G (1736).

5. Ibid.

6. Ibid.

7. Ibid.

8. Ibid.

9. Ibid.

10. AS-FR-G, "Grenouilles" (1736).

11. Ibid.

12. Bibliothèque centrale, Muséum national d'histoire naturelle (Paris), MS 972. The drawings of frogs are in a large volume of miscellaneous drawings, separated from the notes; almost

all the drawings in this collection are by Hélène Dumoustier. The frogs are drawn in her style, quite unlike the few drawings of frogs and tadpoles by Claude Aubriet in Réaumur's papers at the Academy of Sciences. Since Réaumur's notes place her at the scene in Charenton, it is highly probable that she made these drawings.

13. AS-FR-G, "Grenouilles" (1736).

14. R.-A. F. de Réaumur, *Mémoires pour servir à l'histoire des insectes*, 6 vols. (Paris: Imprimerie royale, 1734–42). One more large volume, on beetles and ants, was nearly completed but remained unpublished at his death.

15. Réaumur lived on the rue neuve Saint-Paul in this period; in 1740 he moved to a larger house in Faubourg Saint-Antoine and gave up the country house in Charenton.

16. For example, on 14 March 1736, Réaumur recorded an observation of mating frogs in Charenton; he was present at the academy meeting at the Louvre that afternoon. Académie des Sciences, procès-verbaux, 1736.

17. Jean-Antoine Nollet to Lazzaro Spallanzani, 24 Sept. 1768, in Pericle Di Pietro, ed., *Edizione nazionale delle opere di Lazzaro Spallanzani*, pt. 1 (*Carteggio*), vol. 6. This remark shows that Nollet's work for Réaumur went well beyond the better-known physics experiments and the construction of instruments.

18. Hélène Dumoustier added "de Marsilly" to her name in legal documents; in letters, she was called by either name. In 1735, her mother leased an apartment from Réaumur in his house (Archives nationales, Minutier central ET/CXXI/301); the family was traveling with him to Poitou at least several years before this. Her illustrations remained anonymous at her insistence, but Réaumur recognized her contributions to his insect work in his will. For the text of the will, see Maurice Caullery, introduction, *Histoire des scarabés par M. de Réaumur* (Paris: Paul Lechevalier, 1955).

19. Réaumur examined the life cycle of the mayfly in *Mémoires . . . des insectes*, 6: 457–523.

20. As Pamela Smith has shown, artists in the Renaissance perfected techniques for producing hyperrealistic representations of natural forms. For these artists, observing and imitating nature were inextricably linked. Frogs, lizards, snakes, and insects—the very creatures associated with spontaneous generation—became the subjects for artistic virtuosity in naturalistic representation. Pamela Smith, *The Body of the Artisan: Art and Experience in the Scientific Revolution* (Chicago: University of Chicago Press, 2004). On spontaneous generation, see John Farley, *The Spontaneous Generation Controversy from Descartes to Oparin* (Baltimore: Johns Hopkins University Press, 1977).

21. Jan Swammerdam, *Historia insectorum generalis, ofte algemeene verhandeling van de bloedeloose dierkens* (Utrecht: M. van Dreunen, 1669); French translation, *Histoire générale des insectes* (Utrecht: Guillaume de Walcheren, 1682). On Swammerdam, see Edward Ruestow, *The Microscope in the Dutch Republic: The Shaping of Discovery* (Cambridge: Cambridge University Press, 1996).

22. Swammerdam, *Histoire générale des insectes*, 27.

23. Ibid., 5.

24. Swammerdam, *The Book of Nature; Or, The History of Insects. . . .* (London: C. G. Seyffert, 1758), 107 (translation of *Bybel der nature*, 2 vols., 1737–38). Boerhaave's introduction included a biography of the author and a history of the manuscripts and the publication. Boerhaave portrayed Swammerdam as a tragic hero, thwarted at every turn by his father, his bad luck, and his melancholic temperament, and unable to bring his life's work to completion. Boerhaave, "The Life of Swammerdam," in *The Book of Nature*.

25. Swammerdam, *Book of Nature*, 110–11.

26. Boerhaave had seen some plates before acquiring the manuscripts from Duverney in 1727. Boerhaave, in Swammerdam, *The Book of Nature*, xiii. Swammerdam's images circulated in Paris among Duverney's colleagues.

27. AS-FR-G (8 April 1740). Réaumur mentions in his notes Winslow's unsuccessful attempt to confirm the accuracy of Swammerdam's depiction.

28. Jean-Antoine Nollet, *Discours sur les dispositions et sur les qualités qu'il faut avoir pour faire du progrès dans l'étude de la physique expérimentale* (Paris, 1751), 24–25.

29. AS-FR-G (1736).

30. Ibid.

31. AS-FR-G (1737).

32. No notes survive for 1738 or 1739.

33. Réaumur, "Mss. Grenouilles" (1740), AS-FR-G.

34. See Condorcet, "Éloge de Guettard," *Histoire de l'Académie royale des sciences* (Paris, 1787), 47–62. In the current instance, Guettard's observations of the frogs were not recorded.

35. All quotations in this paragraph, AS-FR-G (8 April 1740).

36. Ibid.

37. Marc Ratcliff, *The Quest for the Invisible: Microscopy in the Enlightenment* (Burlington, VT: Ashgate, 2009), chap. 9.

38. Lazzaro Spallanzani to Charles Bonnet, 6 June 1767, in *Opere di Lazzaro Spallanzani*, pt. 1, vol. 2, *Carteggio con Charles Bonnet*, ed. Pericle Di Pietro (Modena: Enrico Mucchi, 1984), 62.

39. Spallanzani, *Prodromo di un'opera da imprimersi sopra la riproduzioni animali* (Modena, 1768); translated as *Programme ou précis d'un ouvrage sur les reproductions animales* (Geneva, 1768). Observations of tadpole inside frog's eggs, *Prodromo*, 45–60.

40. Spallanzani, *Expériences pour servir à l'histoire de la génération des animaux et des plantes*, trans J. Senebier (Geneva, 1785), 8–9.

41. Lazzaro Spallanzani, "I giornali della generazione," in Spallanzani, *I giornali delle sperienze e osservazioni*, ed. Carlo Castellani, 6 vols. (Florence: Giunti, 1994), 4: 62–63. Spallanzani transcribed the passage from Nollet's letter in "Notizie tratte dall'opera del Roesel intitolata: *Historia naturalis ranarum*" (MS regg B 22). Castellani places this manuscript chronologically just before notes from spring 1777 on artificial insemination. Nollet called the garment "*caleçons*," translated by Spallanzani as "*calzoncini*."

42. Spallanzani, *Dissertazioni di fisica animale e vegetabile* (Modena, 1780), in *Opere*, pt. 4, vol. 4, 503. The translator omitted the remark about frog pants appearing ridiculous (*Expériences pour servir à l'histoire de la génération*, 13).

43. I thank Marco Bresadola for translating key passages from the Italian and for helping in the futile search in Spallanzani's notebooks for mention of the frog pants.

44. Spallanzani, *Dissertazioni di fisica*, 503.

45. Bonnet to Spallanzani, 8 Aug. 1767, in *Carteggio*, 67.

46. Ibid.

47. Spallanzani, *Expériences pour servir à l'histoire de la génération*, 128–29. He tried fertilizing salamander and fish eggs with frog semen as well. Spallanzani to Bonnet, 24 March 1777, in *Carteggio*, 296 ff.

48. Spallanzani, *Expériences pour servir à l'histoire de la génération*, 87.

49. Ibid., 10–11.

50. Ibid., 11.

8

Sorting Things Out: The Economist as an Armchair Observer

HARRO MAAS

In one of their encounters, Dr. Watson asks Sherlock Holmes where he acquired his extraordinary faculty of observation.[1] Much to Watson's surprise, Holmes answers that "to some extent" he thinks it is hereditary, pointing to his brother Mycroft, who "possesses it in a larger degree than I do."

> "You wonder," said my companion, "why it is that Mycroft does not use his powers for detective work. He is incapable of it."
>
> "But I thought you said—!"
>
> "I said that he was my superior in observation and deduction. If the art of the detective began and ended in reasoning from an arm-chair, my brother would be the greatest criminal agent that ever lived. But he has no ambition and no energy. He will not even go out of his way to verify his own solutions, and would rather be considered wrong than take the trouble to prove himself right."[2]

Reasoning from an armchair—is that observing? As recently as 1998, the late British economist and methodologist Terence Hutchison published a diatribe against economists who felt no "need to get up from their armchairs" to test their theories.[3] According to Hutchison, the armchair economist conjectures, speculates, theorizes, but is not engaged in the basic activities of a scientist: to observe, to experiment, and to test.

On Sherlock's assessment of his brother's observational skills, however, the distinction between observing, conjecturing, and theorizing is not so easily made. Armchair observation itself is pictured as a reasoning activity in which facts are rearranged to solve puzzles. The relation between observing and conjecturing is not a new thing.[4] Yet, with the rise of contemporary economics, observing and conjecturing were increasingly seen as different activities, if observation was considered an activity at all. Observing came to relate to field research or to the collection and composition of statistical data sets. Conjecturing became reserved for "theorizing," and testing theories against statistical data the hallmark of sound economic scientific practice. With the collection of data being relegated to statistical institutes, this effectively meant that the act of observing was squeezed from the economist's toolbox.

Those who did not fit the practice of testing theories against statistics were dismissed as "armchair theorists" (not armchair observers), and sometimes they took this label upon themselves as a badge of honor. At a roundtable in 1939 devoted to the desirability of quantification in the social sciences, famous Chicago economist Frank Knight broke up the apparent consensus between sociologists like Talcott Parsons and econometricians like Jacob Marschak and Oskar Lange: "Being just a 'damned arm-chair theorist' myself, I may be allowed to insist on the relevance of speculation to realities and real problems."[5] How did this happen? How and when did observing and conjecturing part company in political economy?

The answer lies in the history of observation in economics, especially in the eighteenth and early nineteenth centuries, when observation was conceived as a method of forming conjectural knowledge.[6]

Puzzles and Paradoxes

Eighteenth-century political economists associated observation, paradoxically enough, with puzzles and paradoxes. Traditionally, the purpose of paradox was to query convention and thus to serve as a vehicle for intellectual innovation. But observation could serve as the departure point for paradox. The very beginning of one of the founding texts of political economy, Adam Smith's *Wealth of Nations* of 1776, may serve as example. Its first chapter, "On the Division of Labour," ends with a carefully framed comparison between the wealth of a peasant and an African king:

> And yet it may be true, perhaps, that the accommodation of an European prince does not always exceed that of an industrious and frugal peasant, as the accommodation of the latter exceeds that of many an African king, the absolute master of the lives and liberties of ten thousand naked savages.[7]

In the early draft of *Wealth of Nations* the African king was an Indian prince, and tracing the roots of the comparison further back, we find Dutch physician and philosopher Bernard Mandeville explaining this paradox in his

Fable of the Bees (1723) in terms of a historical comparison between "the richest and most considerable Men" of earlier times and "the meanest and most humble Wretches" of his day.[8] Mandeville and Smith may both have been inspired by John Locke's *Two Treatises of Government* (1690), in which Locke wrote that "a king of a large and fruitful territory (in America) feeds, lodges, and is clad worse than a day-laborer in England."[9] It is likely that Locke formulated this paradoxical observation on the basis of secondhand information obtained during administrative work for the newly founded colony in Carolina in the 1660s.[10] Mandeville's and Smith's modified versions of Locke's paradox served to change contemporary views about the nature of the social order. In Smith's case, the paradox underscored the importance of the division of labor as the principle in society giving rise to the explosive growth of wealth in a market economy. Mandeville used it to show the importance of selfishness and luxury consumption as source and vehicle of economic growth and progress.

Mandeville emphasized that his use of paradox was not a matter of fanciful speculation, but a matter of firsthand observation. Born into a family of city physicians and drawing on his medical education at Leiden University as a contemporary of the famous Dutch physician Herman Boerhaave,[11] Mandeville claimed to have minutely observed the "trifling Films and little Pipes" of the human frame that were "either over-look'd, or else seem inconsiderable to Vulgar Eyes," but nevertheless determined man's actions.[12] Emphasizing his observational acumen, Mandeville identified himself in the dialogues of his *Treatise of the Hypochondriack and Hysterick Passions* (1711) with Philopirio, "a Lover of Experience," a defender of the new medical practices of Giorgio Baglivi and Thomas Sydenham, both celebrated advocates of observation in medicine.[13] Careful observation of inconsiderable details, which the speculative philosopher did not deem worth attention, was the basis of true knowledge. Transporting his medical practice to the emerging eighteenth-century science of man, Mandeville showed the paradoxes to which observations on the trifling matters of social life gave rise, culminating in the fundamental paradox that man's selfishness lay at the root of his sociability—what Kant admiringly dubbed man's "unsocial sociability."[14]

Mandeville criticized Sir William Temple's depiction of the Dutch as all too virtuous in his *Observations on the Dutch Republic* of 1673. While Temple considered the parsimony and frugality of the Dutch as the basis of their wealth, Mandeville argued that Temple ignored the extraordinary ostentation hidden behind the façades of the canal houses he knew from personal inspection. Even the austere black dress of the Dutch could not be taken as evidence of virtue, since the burghers measured status by the fabric quality.[15]

The Amsterdam burghers were well aware that good manners in public were dependent on turning a blind eye to the filth and fornication in the city's harbor taverns. Similarly, he commented on the wheeling and dealing of tradesmen at Edward Lloyd's coffeehouse and their private country houses. They were all too happy to take advantage of one another without considering this in the least an offense to good manners. Indeed, without attempts to cheat the other, no commerce would ever take place. Thus bringing private life into the open, Mandeville turned stories from the London coffeehouse and the private sphere into generic and constitutive facts of society.[16]

Such paradoxical observations (whether first- or secondhand) led to a reconsideration of the principles governing the social order. Indeed, they helped define the social order as a separate sphere in its own right. They defined the playing field for attempts to find the principles that might explain them—or might explain them away. Once the English republic of letters realized, for example, that Mandeville's paradoxical "private vices, publick benefits" was fundamentally at odds with Lord Shaftesbury's comforting thesis of the altruistic roots of man's sociability, moral philosophers from Bishop Butler to Adam Smith all strove to blunt the edges of the shocking paradoxes of this "minute observer of things."[17]

Rare Observations and Fictional Worlds

Paradoxes in the science of man emerged not only from the comparison of observations but also from reflections on fictional worlds.[18] Anglican clergyman and political economist Robert Malthus's *An Essay on the Principle of Population* (1798) provides a sterling and influential example of this strategy.[19] The book and its subsequent editions (six between 1798 and 1826) launched him on a meticulous and lifelong research program on the laws governing population growth. The *Essay* would captivate the British mind at least until the revision of the poor laws of 1834. The specter of population growth faded only when the rise in the standard of living defied Malthus's gloomy predictions and Victorian England seemed to have reached, in John Maynard Keynes's words, an "economic Eldorado, or economic Utopia, as the earlier economists would have deemed it."[20]

In the first edition of the *Essay*, Malthus wrote that "understanding how population pressure operated . . . required the constant and minute attention of an observing mind during a long life."[21] The completely reworked and vastly expanded second edition contained many additional materials, some of which he had collected himself on his travels through Scandinavia at a time when much of the rest of the Continent was inaccessible to an Englishman because of the Napoleonic Wars.[22] Yet, as we will see, his original formulation of the population principle was based on a rare observation that crucially figured in a thought experiment. Sustained observation over a long geographical and time span was a prerequisite to provide evidence for a conjecture. A particular observation could serve as a clue to a new discovery.[23]

The *Essay*'s foil was William Godwin's utopian *Enquiry Concerning Political Justice* (1793).[24] In his *Enquiry*, Godwin had not ignored the population issue. But he blamed human institutions rather than natural conditions for its rapid growth. He suggested that mankind would reach a utopian state of equality once institutions such as private property and marriage were abolished. Social interaction would no longer be based on self-interest but on benevolence, the passions between the sexes would be gradually extinguished, and life expectancy would rise to infinity. With Europe in revolutionary turmoil, Godwin's message struck a chord with the general public and, despite its high price of three guineas, his book sold extremely well.

Before he entered into his argument with Godwin, Malthus examined the present state of mankind, "assisted by what we daily see around us, by actual experience, by facts which come into the scope of everyman's observation."[25] David Hume, Robert Wallace, and Adam Smith were among his limited references; he quoted German theologian and demographer Johann Peter Süssmilch's population statistics from Richard Price's *Observations on Reversionary Payments* (1771). However, the experience in the "back settlements" of the American colonies served as the crucial case to argue that population would explode in circumstances of unrestricted availability of arable land.[26] This enabled Malthus to brilliantly play out a particular observation against Godwin's utopian vision of a society based on "benevolence" rather than "self-love."

In Godwin's utopia there was no human want. Malthus supposed that Godwin's utopia was realized "in its utmost purity." This did not mean, of course, that the "constancy of the laws of nature" could be suspended. Counterfactually relying on the American case, Malthus argued that population pressure would produce food scarcity when only limited quantities of land were available. As a result institutions like marriage and private property would quickly reemerge, and self-interest would regain prominence over benevolence. Within a very short period of time, Godwin's ideal order would relapse into the existing order. The conclusion followed inexorably: "We have supposed Mr. Godwin's system of society once completely established. But it is supposing an impossibility. The same causes in nature which would destroy it so rapidly, were it once established, would prevent the possibility of its establishment."[27]

This conclusion was, and still is, a sensation. Before Malthus, the acknowledged tensions between population growth and food production could be attributed to the imperfection of human institutions, as Godwin did. The observation on the American experience in the absence of land constraints suggested that nature, not human institutions, was the source of imperfection.[28] Malthus then could easily proceed to show the jarring contradictions into which Godwin's utopian order would run, were it realized in our world of scarcity. After Malthus's argument with Godwin, the observed tension between food production and population growth was transformed from a trifling fact of nature into a structuring principle of society. The example of Malthus shows how the principles of political economy could emerge, not from synthesizing individual observations to a regularity, but from sustained reasoning on a particular fact. At the end of the eighteenth century, to pick out the relevant observation among the many was a valid and consequential observational strategy in the science of man.

Observing and Reasoning

With the emergence of political economy as a separate discipline in the beginning of the nineteenth century, the Anglican cleric and logician Richard Whately emphasized that reasoning on particular facts was the appropriate method of political economy. According to Whately, these facts were not so much to be sought in countries far away, but in the proximity of one's daily life.[29] Whately formulated his views on the method of political economy in his introductory lectures on political economy, delivered as Drummond Professor in the University of Oxford in 1831. Responding to the Cambridge inductivists William Whewell and Richard Jones, Whately made a particularly strong case for identifying reasoning and observing, thus moving the process of making and collecting observations outside the realm of the new science. While Whewell and Jones emphasized the necessity of collecting vast amounts of data from statistics, firsthand observation, or bookish sources, Whately considered reflection on the immediate experiences of everyday life a sufficient basis for the science. Whately's move haunts all subsequent debates on the appropriate method of economics.[30]

Whately compared the political economist with a geologist. Geologists, who asked their correspondents for specimens that may give an idea of the "geological character" of a foreign country, commonly warned against the sending of "*curiosities.*" What they wanted was "the *commonest* strata." In this way, the geologists were similar to the political economist. Rather than being interested in the great events of history, the political economist was interested in the "common, and what are considered insignificant matters," because these might be most informative about the economy.[31] The political economist thus should attend to those facts that were commonly overlooked but within the reach of everyday observation. These were far more likely to reveal the true principles governing society than travel accounts and other uncontrollable sources. Whately approvingly quoted his former pupil Nassau Senior, who had argued in his own introduction to political economy that even a cloistered academic made "twenty exchanges every week" and so had sufficient experience "to enable him to understand how the human passions act in buying and selling." It was, according to Whately, "in fact as impossible to avoid being a practical Economist, as to avoid being a practical Logician."[32]

Whately also compared the economist's position with that of an "observant bystander over those actually engaged in a transaction." Just as in games a "looker-on often sees more of the game than the players," so the economist was the looker-on of economic life; the "looker-on is exactly (in Greek Θεωρὸς) the *theorist.*" Thus Whately effectively identified theory and observation. Like "*theorists*" in geology, "i.e., persons of extensive geological observation," political economists made sense of apparently confused materials.[33] Even though Whately distinguished the scientist who ordered materials from the scientist who debunked fallacies in opposing arguments, he identified the observational genius of the former with the logical virtuosity of the latter. The truly observing economist was the theorist. The result of the political economist's endeavors was, in Whately's words, the formulation of "*paradoxical* truths," of "*abstruse* and *recondite* wisdom." "They may be sense, but at least they are not *common-sense.*"[34]

Mandeville was Whately's example of a prober of such paradoxical truths. For Whately, it was not so much Mandeville's "minute observations" on the passions that were of importance. The "alarming novelty" of Mandeville's *Fable* was that he brought into "juxtaposition" notions that had long been current, "but whose *inconsistency* had escaped detection."[35] Mandeville's "object was to refute those against whom he was writing, by a reduction ad absurdum."[36] Thus, Mandeville's major contribution was a logical exercise. It is instructive that Malthus's late nineteenth-century biographer James Bonar analogously described Malthus as a logician, not as an observer, suggesting that Malthus's *Essay* should be understood as a logical dissection of Godwin from first principles rather than probing for them.[37]

The empirical paradoxes and puzzles that troubled eighteenth-century political economists were reframed by their nineteenth-century successors as mere logical inconsistencies. As a consequence, logical reasoning seemed the only concern for the political economist. Commenting on the relation of political economy to statistics, the Oxford economist Nassau Senior argued that "the truths of political economy" did not depend on "statistical facts."[38] Did Senior really want to imply that political economy was based on no facts, or did he aim to raise concerns about what the relevant facts were?

Induction or Deduction

All through the Victorian period, both positions, the political economist brooding over specific observations and the political economist as the master of logical argument, were played out against one another under the guise of the "inductive" versus the "deductive" method.

Whately's tutor and provost of Oriel College Edward Copleston maintained that "whatever is theoretically true" in political economy was "surely acted upon by the interested party, however unconscious he may be of the abstract principle."[39] In their opposite zeal to turn political economy into an inductive science, Cambridge men like William Whewell and his good friend Richard Jones accused the deductivists, the Ricardians and the Oxford economists, of driving "one jack-ass before the other" and of "jump[ing] from one or two trivial facts to the conclusion that every man will get as much money as he can, an axioma generallisimum."[40] Charles Babbage considered political economists "closet philosophers" because they were unacquainted with the relevant facts of the emerging factory system and so unable to articulate the true principles governing their field.[41] The writings of two outstanding political economists in the Victorian period, British philosopher and political economist John Stuart Mill and Irish political economist John Elliot Cairnes, reflect these oppositions.

Mill explained the relation between induction and deduction in his famous 1836 essay, "On the Definition of Political Economy; and on the Method of Investigation Proper to It," first published in the *Westminster Review*.[42] According to Mill, the appropriate method of political economy was the "a priori method," which combined induction and deduction. Starting from observation, the political economist formulated general principles from which he then deduced consequences. Because of society's complexity, and so the importance of disturbing causes, these consequences could seldom be observed in practice; political economy was a science of tendencies. But its principles were as sound as the best laws of mechanics.

Mill gave surprisingly little information about the exact way in which the first step, from observation to principles, was to be made. We learn only that political economy is a "mental science," dealing with principles of the mind that could be observed by mental experiments. Received wisdom has it that

Mill thus basically relied on introspection as the appropriate method for making observations on the economy.[43] And so it seems that Mill trivialized the process of observing the economy in ways similar to Whately.

But the actual practice of political economists gives a different image.[44] For the many editions of his *Principles of Political Economy* (first ed., 1848) Mill did not rely on mental experiments but rather on corresponding networks that supplied him with the materials he synthesized in a text that even William Whewell had little to complain about.[45] The generic practice of Victorian economists was synthesizing from different sources, sorting out the relevant materials, not some form of mental experimentation. The question was, What kinds of observations are relevant, how many do we need, and how do we derive a principle from them?

Taking Notes

As an example, let us look at Mill's request to Cairnes for information on the current state of Ireland, information Mill used for the sixth edition of his *Principles.* Cairnes carefully collected statistical and other materials in his notebooks, which he used not only in his report to Mill but also for his own articles on Ireland for the *Economist.*[46] Cairnes made three versions of his notes, of which the last was a running text that he sent off to Mill. Reproduced here is an image from one of Cairnes's original notebooks (fig. 8.1). Cairnes first made notes on the right-hand page, which he listed alphabetically, as was common note-taking practice.[47] These notes could consist of various things. They could be verbal or numerical and could include personal experiences, recorded experiences from others, or excerpts from other texts. He then added comments on the left-hand page. These additions consisted of references to other sources or personal reflections, and these also served to make connections between different topics or to comment on discrepancies that struck him in comparing his primary notes with (what he considered to be) common knowledge. Highlighting relevant facts and their possible connections, Cairnes gradually synthesized heterogeneous observations into a coherent whole. The notebook was the instrument that enabled the political economist to distinguish the relevant from the irrelevant and so to forge the governing principles of his subject of study. Observations and reflections were made in the same literary space; excerpting, observing, and reflecting were all in the same hand, on the same page. These were the tools the political economist used to infer principles from facts.

Thus, we see Cairnes under *c* on the right-hand page excerpting an article from the Irish political economist Neilson Hancock in the journal of the sta-

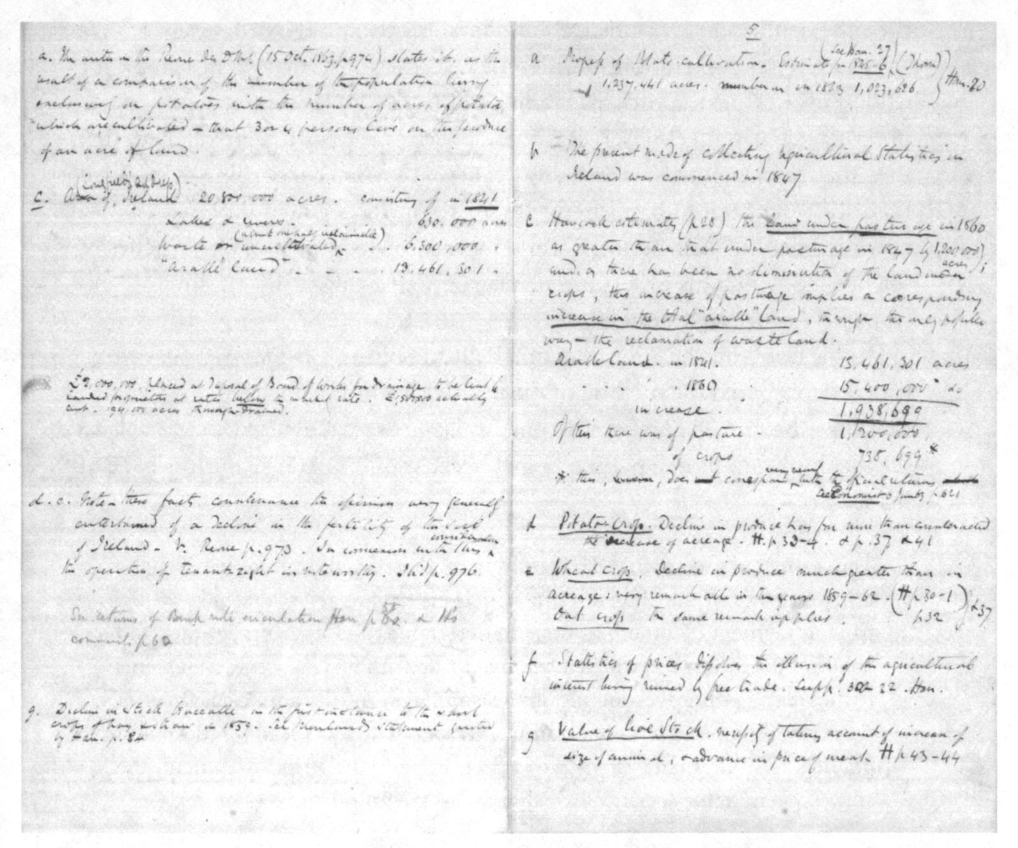

FIGURE 8.1. Fragment from John Elliot Cairnes's notebook on the state of Ireland. Cairnes papers, MS 8983, "Economic Notes on Ireland made by John E. Cairnes for John Stuart Mill," 1864. Reproduced by courtesy of the Board of the National Library of Ireland.

tistical society of Dublin, and commenting on Hancock's conclusions about the increase of land under pasturage. He made a computation of his own on the amount of arable land and, referring to the *Economist* for 6 June 1863, noted that his computation "corresponds [very nearly] with the official returns." Under *d* and *e* on the right-hand page Cairnes again took notes from Hancock's article, now on the decline of potato and wheat crops in relation to the increase in arable land. On the left-hand page Cairnes commented that this phenomenon was not in accordance with the "opinion very general entertained" about the decline of the fertility of the land in Ireland. Cairnes linked this remark to the "operation of tenant-rights" that he considered "noteworthy." For both these comments, he referred to the *Revue des deux mondes*, one of his favorite resources for statistical and other information.

In his essay in this volume, Theodore M. Porter rightly observes that in the nineteenth century statistical numbers "rarely stood by themselves." We see exactly this integration of different resources in Cairnes's notebook for Mill. It fits Cairnes's approving reading of Herschel's description of a "perfect observer." The perfect observer would "have his eyes as it were opened, that they may be struck at once with any occurrence which, according to received theories, ought not to happen; for these are the facts which serve as clews to new discoveries." The perfect observer was knowledgeable of all branches of science "to which his observations relate" or that might inform him of "extraneous disturbing causes."[48]

In his book on the American antebellum South, *The Slave Power* (1862), Cairnes compared the method of the political economist with that of a naturalist like the French comparative anatomist Georges Cuvier, who was able to see the implications of a particular fact because of his extensive knowledge of his science.

> The comparative anatomist, by reasoning on those fixed relations between the different parts of the animal frame which his science reveals to him, is able from a fragment of a tooth or bone to determine the form, dimensions, and habits of the creature to which it belonged; and with no less accuracy, it seems to me, may a political economist, by reasoning on the economic character of slavery and its peculiar connexion to the soil, deduce its leading social and political attributes, and almost construct, by way of a priori argument, the entire system of the society of which it forms the foundation.[49]

Just like the cabinet naturalist, the political economist was not involved in a merely logical exercise; the deductions from specific observational materials helped the political economist to "see" the underlying principles of society. This description is consistent with political economists arranging and reasoning upon observations to tease out their secrets, in the style of Mandeville and Malthus.

Mandeville derived the principle of selfishness not from introspection but rather from his own experiences with corrupt politics in Rotterdam before he ever entered a London coffeehouse or joined the dinner tables in Amsterdam and London.[50] In 1847 the Irish political economist James Lawson therefore wrote with some justice that political economists were not "shut up in a dark room" reasoning out their truths without recourse to observation. In contrast, the "fundamental principle" of political economy, "that man acts from self-interest . . . is the result of observation."[51] Cairnes may have agreed with Mill that the business of the political economist was done once an economic fact had been "traced to a mental principle,"[52] but this was not equivalent to the performance of an introspective mental experiment in the comfort of the armchair. Indeed, Lawson was just one among many nineteenth-century political economists who emphatically claimed they proceeded as much by observation as natural scientists, as can be witnessed from Cairnes's own notebooks.

What then did count as observation? The reflection on the published accounts of others, including statistics and travel reports, as well as on firsthand experiences (Mandeville and Malthus)? The rearrangement of observations in order to detect logical inconsistencies (Whately)? A priori inference from observational traces to an entire system (Cairnes)? These were just the kinds of observations that can be found on the reproduced page from Cairnes's notebook on the state of Ireland.

The eighteenth- and early nineteenth-century practice of forging heterogeneous observations into a coherent whole lost its credentials once observation in economics became increasingly identified with quantitative, statistical data, and the logic of inductive inference gave way to a new vocabulary of hypotheses, theories, and testing. It is only in this context that earlier practitioners began to look like the much-maligned "armchair theorists."

Plotting Numbers

In British economics, this shift coincides with the so-called marginalist revolution of the 1870s, which introduced the mathematically garbed notion of an economic agent as a utility maximizer. It is no coincidence that one of the first propagators of this theory, the British political economist and statistician William Stanley Jevons, argued that he started from "axioms" from which we can "deduce laws of supply and demand" that needed to agree with "*a posteriori* observations." These observations were for Jevons the "numerical data" in "private-account books, the great ledgers of merchants and bankers and public offices, the share lists, price lists," and so on.[53]

John Neville Keynes, Cambridge lecturer in moral sciences and father of the famous economist, was perhaps the last to defend observation in economics as a synthetic rearrangement of complex materials. As late as 1893 he described the first step of the "deductive method" as the "preliminary observation of the complex phenomena themselves, with a view to description and classification"—not just collecting statistical data. However, Neville Keynes identified this step with an "introspective investigation of human motives" without further explanation.[54] Just like Mill in his explanation of the a priori method, Neville Keynes thus completely obscured the complex practice of collecting, rearranging, and sifting of first- and secondhand observations of earlier economists.

"Introspection" was an easy target for political economists like Jevons, who championed fitting mathematical functions to statistical data, presaging the rise of econometrics in the twentieth century. Figure 8.2 shows one of Jevons's many attempts to derive a regular function from a plotting of statistical data on prices and quantities. Quantities are on the vertical axis, prices on the horizontal. The numbers next to the dots indicate years. The dotted line is Jevons's attempt to find the best adaptation of a curve to the data, much in the same way as an astronomer might go about it. In his *Principles of Science* (1874), Jevons described this procedure as an attempt to find the "rational function," that is, the function that can explain the statistical data. Jevons is probing for generality just as Cairnes was in his note-taking practice. He is very aware of the importance of the right instruments to do so and even includes the address in London where good lined paper for plotting data can be found.

But the contrast with Cairnes's method of working strikes the eye, and is indicative of the shift in economic discourse from one in which the political economist weighs different sources of evidence, to one in which statistical data collected by manufacturers, governmental agencies, and others came to count as the one and only homogenized source of observation. Cairnes sifted observations on their relevance and probed for possible connections. Jevons completely changes the relevant domain of discourse. Not knowing anything about the graph, it could be about everything, ranging from astronomical observations to laboratory measurements. The Cartesian space came to serve as a test bed for theories. From this perspective, the political economist's probing for significant facts and possible connections between them could only appear unmethodical, haphazard, and inconsequential.

Thus, Jevons aggressively ridiculed Mill's claim that his political economy was just as good as physics in terms that presage the reproach of armchair laziness: "What will our physicists say to a *strictly physical science*, which can be experimented on in the private laboratory of the philosopher's mind? What a convenient science! What a saving of expense as regard of apparatus, and materials, and specimens."[55] Treading in Jevons's footsteps, the British political economist Francis Ysidro Edgeworth contrasted the new experimental measurement practices in psychophysics favorably with Mill's recourse to "introspective marks of brain-activity."[56] Once the synthetic (re-)arrangement of a variety of different kinds of observations became identified with introspection, this venerable practice in political economy seemed a weak and outmoded alternative to gauging mathematical functions against statistical data sets.

From the end of the nineteenth century, observations became increasingly

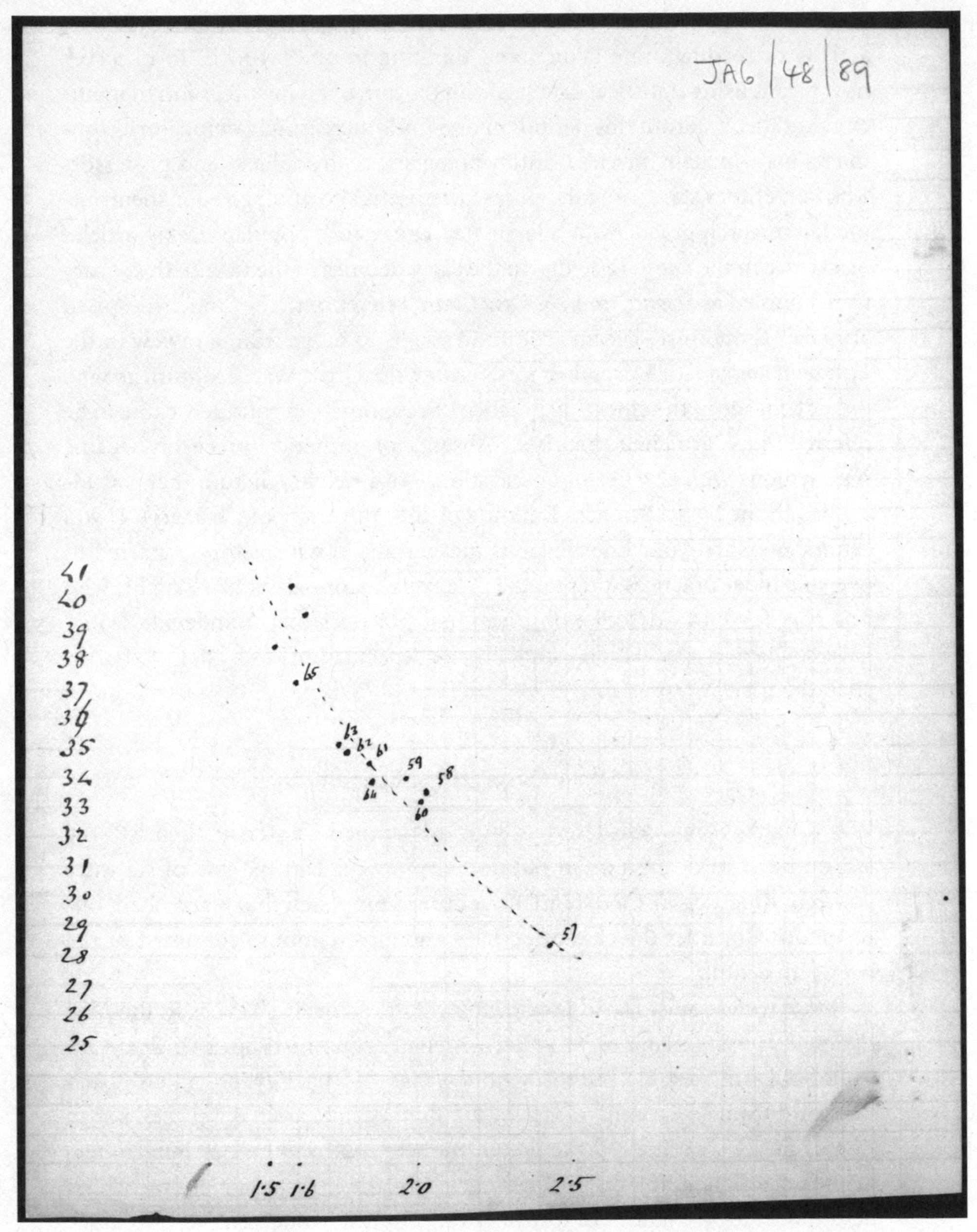

FIGURE 8.2. Alleged price-quantity graph of William Stanley Jevons. Jevons archive, item JA/48/29. Reproduced by courtesy of the University Librarian and Director, the John Rylands University Library, the University of Manchester.

identified with numerical data; observations no longer included the synthetic activity of reading, note taking, and thinking in one's study. "To observe" now meant using statistical data sets with the aim of giving precision to mathematical theory. From this period on one finds increasingly derogatory comments on armchair travelers, anthropologists, philosophers, and professors who sometimes start using the phrase themselves to apologize for their outmoded research practices. While Sir Ray Lankester's popular weekly articles on science in the *Daily Telegraph* in the early decades of the twentieth century were bundled as *Science from an Easy Chair*, reference to the "much-despised armchair economist" became common usage, to judge from a review in the *Economic Journal* of December 1915.[57] After the Great War, economists who did not engage in the emerging practice of econometrics routinely came to be referred to as "armchair theorists." Absence of numerical precision became synonymous with absence of observation. Lord Kelvin's dictum that was inscribed in the Social Sciences Building of the University of Chicago ("If you cannot measure, your knowledge is meager and unsatisfactory") started to overrule older practices, despite the protests of economists like Knight, who was now forced to describe observational practices that Mandeville would have *opposed* to speculation as merely the "speculation" of a "damned armchair theorist."[58]

Armchair Observers in Economics

Was Knight's ironic self-description as a "damned arm-chair theorist" the last outburst in defense of an outmoded practice? Did the rise of statistics and econometrics in the twentieth century completely bury armchair observation? Consider the case of perhaps the most famous economist of the twentieth century.

In 1933, cartoonist David Low pictured John Maynard Keynes in the *New Statesman* posing as the archetypical armchair economist (fig. 8.3). There are some books on the table, but the main message was that Keynes contemplated the world from an easy chair, as a logician rather than as an observer. In 1939 Keynes famously criticized the Dutch pioneer in econometrics Jan Tinbergen for his monograph on business cycles for the League of Nations, which he considered "a nightmare to live with." Even if Tinbergen were to agree with his critic's comments, Keynes feared that "his reaction will be to engage another ten computers and drown his sorrows in arithmetic."[59] Keynes was emphatically not of Tinbergen's mindset. "[Tinbergen] so clearly prefers the mazes of arithmetic to the mazes of logic, that I must ask him to forgive the criticisms of one whose tastes in statistical theory have been, beginning many

FIGURE 8.3. David Low's cartoon of John Maynard Keynes of 28 October 1933, for the *New Statesman*. Reproduced by courtesy of Solo Syndication, Associated Newspapers Ltd., London.

years ago, the other way round." Tinbergen's inductive search for regularities from statistical data sets could not deliver the "*verae causae*" the economist in fact already must have in his possession before a quantitative estimation of causes and effects could make any sense. Tinbergen's statistical analysis was "neither of discovery nor of criticism. It is a means of giving quantitative precision to what, in qualitative terms, we know already as the result of a complete theoretical analysis."[60]

Judging Keynes solely on the basis of his criticism of Tinbergen, one might be tempted to second the caricaturist's classification of Keynes as logician, the theorist or deductive economist, indifferent to real-world experience. His assertion that political economy "deals with introspection and with values" might well confirm this image.[61] It was, more or less, the image that Marschak and Lange, the representatives of the Cowles Commission at the 1939 University of Chicago roundtable on quantification in the social sciences, tried to convey in their defense of Tinbergen against Keynes. For them Tinbergen's work seemed "the road which *any* study of Business Cycles must follow if it wants to be taken seriously, and not merely be adding to the non-ending list of plausible essays."[62]

But Keynes's actual record as an economist hardly matched the image of the unworldly logician. As guest editor of the *Guardian* newspaper in the early 1920s, he incorporated statistical materials and endorsed graphical representations of the business cycle. In fact, Keynes's enthusiasm for such representations stimulated the European business cycle research that would eventually lead to publications like that of Tinbergen. Keynes also appreciated the innovations of the young Stanley Jevons, who made the use of mathematics (and utility theory) fashionable in economics. At home in the "black arts of inductive economics," Jevons was the first economist "to survey his material with the prying eyes and fertile, controlled imagination of the natural scientist."[63] This description might equally hold for Keynes himself, brooding like a naturalist over his materials. But rather than brooding over statistical data to find the optimal fit of a curve to scattered data (as Jevons did, as in fig. 8.2), Keynes had recourse to older observational practices to gather, order, and weigh statistical and literary sources in order to uncover their hidden secrets.

Keynes's work *The Economic Consequences of the Peace* (1919), for example, was written in a burst of outrage over the outcome of the peace treaty negotiations in Versailles that concluded World War One. As representative for the British Treasury, Keynes had witnessed the negotiations close-up, and his eyewitness account of the "micro-cosmos" of Versailles was an important part of his argument: as Keynes emphasized, it was his genuinely "new experience" as member of the Supreme Economic Council of the Allied Powers, far away from London, which transformed his "cares and outlooks."[64]

Keynes presented his readers with a magisterial epic that spanned the past "economic Utopia" abruptly ended by the war, the present peace treaty negotiations in Versailles, and the grim depiction of the likely future consequences of the treaty. Keynes painted vivid portraits of main characters at Versailles such as David Lloyd George and Georges Clemenceau, as well as Keynes's German counterpart, "poor M. Klotz," in perhaps the most impressive example to be found in the entire literature of economics of a general economic message built up from personal experiences and observations taken from a wide variety of resources. That was what an economist could contribute to the public from his armchair. This was not merely a game of logic. It fits the research strategy we highlighted through this essay: to resolve the incoherence of first- and secondhand observations into a synthetic picture. For the new econometricians, such a strategy was merely to produce "plausible essays."[65]

This is why Tinbergen and Keynes were talking at cross-purposes about observation in economics. In a 1987 interview, Tinbergen recounted his disappointment with Keynes:

> I had the privilege of meeting him later, just once in 1946. On that occasion I told him we had done quite a bit of research on the price elasticity of exports and that we really found that the elasticity was about 2, the figure he uses in his famous book about German reparation payments. I thought that he would be very glad that we had found that figure, and "that he had been right." But he only said: "How nice for you that you found the right figure."[66]

In contrast to Tinbergen, Keynes was not interested in digging out statistics in support of his argument. Instead, he synthesized the complex materials at his disposal to convey the essence of the situation. This form of observation was as much in evidence in his *Economic Consequences of the Peace* as it was in his *General Theory of Employment, Interest, and Money* of 1936. The cover photograph of the latter book rightly shows Keynes sitting in his armchair. But the inside story of the book is not just a theory. Just as Keynes had followed his protagonists in the microcosm of Versailles, so he now followed how the actions of the protagonists of the Great Depression played out in the macrocosm of a market economy—how the actions of the consumer, investor, banker, and speculator led to the general glut in the market that made for the daily headlines in the newspapers of the 1930s. It was an observational exercise in understanding the economic world but emphatically not one of statistics and measurement.

Sorting Things Out

The practice of observation I have examined in this essay is clearly at odds with the view that became standard after World War Two. I discussed political economists who sorted and arranged observations, including firsthand experiences, in the confines of their study in order to make sense of conflicting evidence—and sometimes to point out that it *was* conflicting.[67] With the rise of statistics and mathematical modeling in late nineteenth- and twentieth-century economics, such a practice was increasingly regarded as imprecise and insufficiently empirical. The alleged Victorian retreat to "introspection" as a resource with which to derive first principles did not help to clarify older practices. With the rise of operationalism and logical empiricism in the twentieth century, little room seemed to be left for a method of empirical research in economics that belonged neither to field research nor to econometrics. To sort things out became identified with "theorizing" rather than "observing."

A contemporary philosopher who considers observation tied to statistical data will not even recognize Keynes's research practice as "observational." But even if the complex variety of observations that can be made on society is restricted to statistical data, these data never speak for themselves. They need to be selected and interpreted. At the end of the day, it is impossible to disentangle observation from the interpretative, synthetic act of the observer. Sherlock Holmes understood this. He counted his brother's interpretative feats as proof of his extraordinary observational skills. The philosopher Kwame Anthony Appiah, in a recent contribution to the *New York Times*, underscores the point and returns us to the armchair:

> You can conduct more research to try to clarify matters, but you're left having to interpret the findings; they don't interpret themselves. There always comes a point where the clipboards and questionnaires and M.R.I. scans have to be put aside. To sort things out, it seems, another powerful instrument is needed. Let's see—there's one in the corner, over there. The springs are sagging a bit, and the cushions are worn, but never mind. That armchair will do nicely.[68]

Notes

Research for this essay was sponsored by the Netherlands Organisation for Scientific Research (NWO) grant number 276–53-004 and by the Max Planck Institute for the History of Science in Berlin. In addition to the participants of the observation project at the Max Planck, I would like to thank Marcel Boumans, Neil De Marchi, Tiago Mata, Ross Emmett and Greg Radick for their support and many comments.

1. Arthur Conan Doyle, "The Greek Interpreter," in *The Memoirs of Sherlock Holmes*, ed. and with an intro. by Christopher Roden (Oxford: Oxford University Press, 1993), 193–212.

2. Doyle, "The Greek Interpreter," 193–95.

3. Terence Hutchison, "Ultra-deductivism from Nassau Senior to Lionel Robbins and Daniel Hausman," *Journal of Economic Methodology* 5, no. 1 (1998): 43–91. Hutchison's first book, *The Significance and Basic Postulates of Economic Theory* (1938), is commonly considered the first to have made the Popperian theme of theory testing prominent in economic methodology.

4. See the essays by Lorraine Daston and Katharine Park in this volume.

5. The roundtable celebrated the ten-year existence of the Social Science Building at the University of Chicago. See *Eleven Twenty-Six: A Decade of Social Science Research*, ed. Louis Wirth (Chicago: University of Chicago Press, 1940), 167. I would like to thank Ross Emmett for drawing my attention to this quote. Oskar Lange and Jacob Marschak were, respectively, fellow and director of the so-called Cowles Commission for econometric research (that is, the integration of theory, statistics, and mathematics), which stood in an uneasy relationship with the Economics Department of Chicago, of which Knight was a prominent member. For histories of the Cowles Commission, see Carl F. Christ, "History of the Cowles Commission, 1932–1952," in *Economic Theory and Measurement: A Twenty Year Research Report, 1932–1952* (Chicago: n.p., 1952); Cliff Hildreth, *The Cowles Commission in Chicago, 1939–1955* (Berlin: Springer, 1986); Duo Qin, *The Formation of Econometrics: A Historical Perspective* (Oxford: Oxford University Press, 1993).

6. See Daston in this volume on the eighteenth-century conception of observation as the "art of conjecture."

7. Adam Smith, *An Inquiry into the Nature and Causes of the Wealth of Nations*, ed. and with an intro. by R. H. Campbell and A. S. Skinner (Oxford: Clarendon Press, 1976), 24.

8. Bernard Mandeville, *The Fable of the Bees* (Oxford: Clarendon Press, 1924), 169.

9. John Locke, *Two Treatises on Government* (Cambridge: Cambridge University Press, 1988), vol. 2, § 41.

10. J. R. Milton, "Locke, John (1632–1704)," in *Oxford Dictionary of National Biography*, ed. H. C. G. Matthew and Brian Harrison (Oxford: Oxford University Press, 2004); online ed., ed. Lawrence Goldman, May 2008, http://www.oxforddnb.com/view/article/16885 (accessed 25 Sept. 2008).

11. On Mandeville, see E. J. Hundert, *The Enlightenment's Fable: Bernard Mandeville and the Discovery of Society* (Cambridge: Cambridge University Press, 1994); Harold J. Cook, *Matters of Exchange: Commerce, Medicine, and Science in the Dutch Golden Age* (New Haven: Yale University Press, 2007), chap. 10; Neil B. De Marchi, "Exposure to Strangers and Superfluities. Mandeville's Regimen for Great Wealth and Foreign Treasure," in *Physicians and Political Economy: Six Studies of the Work of Doctor-Economists*, ed. P. Groenewegen (London: Routledge, 2001), 67–92. Mandeville published in medicine as well. See his 1706 translated and annotated version of *Riverius renovatus* of François de La Calmette, a physician he possibly met in Geneva in the early 1690s. His *Treatise of the Hypochondriack and Hysterick Passions* of 1711 appeared in an enlarged edition in 1730 with a slightly altered title. Given his family and educational background, Mandeville almost certainly knew the genre of medical observations discussed in the essay by Gianna Pomata in this volume.

12. Mandeville, *Fable of the Bees*, 3.

13. Mandeville, *Treatise of the Hypochondriack and Hysterick Passions*, xi. From the context of Mandeville's writing it is clear that "experience" referred to firsthand observation rather than experiment. E. J. Hundert consequently misspells Philopirio as Philiporo, an understandable change, because Philopirio means "lover of fire" rather than "lover of experience."

14. See Immanuel Kant, "Idee zu einer allgemeinen Geschichte in weltbürgerlicher Absicht," in *Vorkritische Schriften bis 1768, Kant-Studienausgabe*, ed. Wilhelm Weischedel (Wiesbaden, 1960), 4: 37–38.

15. Mandeville's reaction on Temple's *Observations* is discussed in Alexander Bick, "Mandeville and the Two Commercial Empires: An Interpretation of Change in the Early Modern Economy, 1690–1730" (M.S. thesis, London School of Economics and Political Science, 2004), chaps. 5–6. Mandeville's analysis of the Dutch finds its contemporary echo in Simon Schama's *The Embarrassment of Riches: An Interpretation of Dutch Culture in the Golden Age* (New York: Knopf, 1987).

16. Mandeville, *Fable of the Bees, Remark B*, 63. On the constitution of the public and private sphere in the eighteenth century, see Jürgen Habermas, *Strukturwandel der Öffentlichkeit* (Darmstadt: Luchterhand, 1962).

17. Albert O. Hirschman, *The Passions and the Interests: Political Arguments for Capitalism before Its Triumph* (Princeton: Princeton University Press, 1977), 19. The denunciation of Mandeville as a "minute observer of things" is found in the second dialog of George Berkeley, *Alciphron, or The Minute Philosopher. In Seven Dialogues. Containing an Apology for the Christian Religion, against Those Who Are Called Free-Thinkers* (London: Printed for J. Tonson, 1732), 146. For Shaftesbury's views, see especially *An Inquiry concerning VIRTUE and MERIT*, origi-

nally published 1699 and reprinted in his *Characteristicks of Men, Manners, Opinions, Times*, vol. 2 (London, Printed for J. Purser, 1737).

18. Such counterfactual reflections resemble thought experiments. For influential accounts, see Roy Sorensen, *Thought Experiments* (Oxford, 1992); Ulrich Kühne, *Die Methode des Gedanken-experiments* (Frankfurt am Main: Suhrkamp, 2005). Critical of thought experiments is Kathleen V. Wilkes, *Real People: Personal Identity without Thought Experiments* (Oxford: Clarendon Press, 1988).

19. Thomas Robert Malthus, *An Essay on the Principle of Population*, in *The Works of Thomas Robert Malthus*, ed. E. A. Wrigley and David Souden, vol. 1 (London: Pickering, 1986). On Malthus, see Patricia James, *Population Malthus: His Life and Times* (London: Routledge and Kegan Paul, 1979); Donald Winch, *Riches and Poverty: An Intellectual History of Political Economy in Britain, 1750–1834* (Cambridge: Cambridge University Press, 1996).

20. John Maynard Keynes, *The Economic Consequences of the Peace*, vol. 2 of *Collected Writings of John Maynard Keynes*, ed. Don Moggridge (London: Macmillan, 1971), 6.

21. Malthus, *Essay*, 32.

22. On Malthus's travels, see James, *Population Malthus*, 69–78; Thomas Robert Malthus, *The Travel Diaries of Thomas Robert Malthus*, ed. and with an intro. by Patricia James, with a foreword by Lord Robbins (London: Cambridge University Press for the Royal Economic Society, 1966).

23. On the historical movement between sustained observation over time versus observing rare facts, see Daston, Park, and Pomata in this volume.

24. William Godwin, *Enquiry Concerning Political Justice, and Its Influence on Modern Morals and Happiness*, ed. and with an intro. by Isaac Kramnick (Harmondsworth: Penguin, 1976).

25. Malthus, *Essay*, 53. Patricia James persuasively argues that Malthus's parish experiences at Okewood formed part of his considerations, though he mentioned them nowhere. While population in neighboring parishes remained more or less stable over the eighteenth century, population grew excessively at Okewood. Later clergymen "noted with amazement that there were so many and many pages of baptisms, and that the baptisms were so greatly in excess of the burials." See B. Stapleton, "Malthus: The Local Evidence and the Principle of Population," in *Malthus Past and Present*, ed. J. Dupâquier, A. Fauve-Chamoux, and E. Greberink (New York: Academic Press, 1983), 45–59.

26. See Richard Price, *Observations on Reversionary Payments. . . .* (London: Printed for T. Cadell, 1771), 203.

27. Malthus, *Essay*, 210.

28. This conjecture was equally at odds with the established tenets of natural religion, something that troubled Malthus himself for years to come and turned him into the bogeyman of the conservative British establishment. See Anthony M. C. Waterman, *Revolution, Economics, and Religion: Christian Political Economy, 1798–1833* (Cambridge: Cambridge University Press, 1991).

29. Richard Whately, *Introductory Lectures on Political Economy* [1832] (New York: A. M. Kelley, 1966), 71. On Richard Whately, see E. Jane Whately, *Life and Correspondence of Richard Whately, D.D., Late Archbishop of Dublin* (London: Longmans, Green and Co., 1866); William John Fitzpatrick, *Memoirs of Richard Whately* (London: R. Bentley, 1864); Norman Vance, "Improving Ireland: Richard Whately, Theology and Political Economy," in *Economy, Polity, and Society: British Intellectual History, 1750–1950*, ed. Stefan Collini, Richard Whatmore, and Brian Young (Cambridge: Cambridge University Press, 2000), 181–201.

30. On the controversy between the Cambridge men and the Oxford school, see Pietro Corsi, "The Heritage of Dugald Stewart: Oxford Philosophy and the Method of Political Economy," *Nuncius* 11, no. 2 (1987): 89–144; Harro Maas, "'A Hard Battle to Fight': Natural Theology and the Dismal Science," *History of Political Economy* 40, no. 5 (2008): 143–67; Waterman, *Christian Political Economy.*

31. Whately, *Introductory Lectures*, 233n.

32. Ibid., 229.

33. Ibid., 67–68.

34. Ibid., 72.

35. Ibid., 43.

36. Ibid., 43.

37. James Bonar, *Parson Malthus* (Glasgow: J. Maclehose, 1881), 8.

38. Senior to Quételet, Sept. 1841, quoted in M. J. Cullen, *The Statistical Movement in Early Victorian Britain* (Hassocks: Harvester Press, 1975), 84.

39. Edward Copleston, *A Letter to the Right Hon. Robert Peel, MP for the University of Oxford, on the pernicious effects of a variable standard of value, especially as it regards the lower orders and the poor laws* (Oxford: J. Murray, 1819), 36–37. Significantly, the motto of the letter was "laissez nous faire."

40. Whewell to Jones, 7 Dec. 1830 and 10 Sept. 1827, respectively. Trinity College Cambridge (TCC) Add.Ms.c51/41; Add.Ms.c51/93.

41. Charles Babbage, *On the Economy of Machinery and Manufactures* (London: C. Knight, 1832), 156.

42. John Stuart Mill, "On the Definition of Political Economy; and on the Method of Investigation Proper to It," in *Collected Works of John Stuart Mill*, vol. 4, ed. J. M. Robson (Toronto: University of Toronto Press, 1967).

43. Mill only used "introspection" in response to Auguste Comte and the commonsense philosopher Sir William Hamilton. In both cases, Mill suggested another word to better fit his ideas. To Comte, Mill responded that "retrospection" better fitted his intention than "introspection." Mill accused Hamilton for using the false "introspective method" rather than the correct "psychological method" of the association psychology of his father. See Kurt Danziger, "The History of Introspection Reconsidered," *Journal of the History of the Behavioral Sciences* 16, no. 3 (1980): 241–62.

44. Samuel Hollander and Sandra Peart, "John Stuart Mill's Method in Principle and in Practice: A Review of the Evidence," *Journal of the History of Economic Thought* 21, no. 4 (1999): 369–97; Neil De Marchi, "Putting Evidence in Its Place: John Mill's Early Struggles with 'Facts in the Concrete,'" in *Fact and Fiction in Economics: Models, Realism and Social Construction*, ed. Uskali Mäki (Cambridge: Cambridge University Press, 2002), 304–26.

45. Whewell to Jones, 30 April 1848, in *William Whewell, An Account of his Writings*, ed. Isaac Todhunter, vol. 2 (London: Macmillan, 1876), 345. On the role of corresponding networks in forging observations, see the essay by Daniela Bleichmar in this volume.

46. Tom A. Boylan and Tadhg Foley, "Notes on Ireland for John Stuart Mill: The Cairnes-Longfield Manuscript," *Hermathena* 138 (Summer 1985): 28–39; John Elliot Cairnes, "Ireland in Transition," in *John Elliot Cairnes: Collected Works*, ed. Tom Boylan and Tadhg Foley (London and New York: Routledge, 2004), 6: 221–50.

47. On note taking, see Ann Blair, "Note Taking as an Art of Transmission," *Critical Inquiry* 29, no. 5 (2004): 85–107; Richard Yeo, "Before Memex, Robert Hooke, John Locke, and

Vannevar Bush on External Memory," *Science in Context* 20, no. 1 (2007): 21–48; Daston, this volume.

48. John F. W. Herschel, *A Preliminary Discourse on the Study of Natural Philosophy* (London: Longman, Brown, Green & Longman, 1851), 132.

49. John Elliot Cairnes, *The Slave Power: Its Character, Career, and Probable Designs: Being an Attempt to Explain the Real Issues Involved in the American Contest*, 2nd ed. (New York: Carleton, 1862), 69–70.

50. For extensive descriptions of Bernard Mandeville's involvement in the so-called Costerman riots, see R. Dekker, "'Private Vices, Public Virtues' Revisited: The Dutch Background of Bernard Mandeville," *History of European Ideas* 14, no. 4 (1992): 481–98; Arne C. Jansen, "Het leven van Bernard Mandeville (1670–1733)," in Bernard Mandeville, *De wereld gaat aan deugd ten onder* [*The Fable of the Bees*, followed by *A Modest Defence of Publick Stews*], trans. and with an intro. by Arne C. Jansen (Rotterdam: Lemniscaat, 2006), 226–57.

51. James A. Lawson, "On the connexion between statistics and political economy," *Dublin: Transactions of the Dublin Statistical Society* 1, no. 1 (1847–48): 3–9.

52. Cairnes, *Character and Logical Method*, 179.

53. William Stanley Jevons, *The Theory of Political Economy* (Harmondsworth: Penguin, 1970), 88, 83.

54. John Neville Keynes, *The Scope and Method of Political Economy* [1891] (New York: Kelley, 1963).

55. William Stanley Jevons, "John Stuart Mill's Philosophy Tested," in William Stanley Jevons, *Pure Logic and Other Minor Works* [1890] (New York: B. Franklin, 1971), 215.

56. Francis Ysidro Edgeworth, *Mathematical Psychics and Further Papers on Political Economy*, ed. P. Newman (Oxford: Oxford University Press, 2003), 197.

57. Hartley Withers, review of *The Evolution of the Money Market (1385–1915)* by Ellis T. Powell, *Economic Journal* 25, no. 100 (1915): 577–79, on 578. Sir Ray Lankester's columns were bundled in 1908, 1910, and 1912.

58. Kelvin's words were famously used by Thomas Kuhn in his "The Function of Measurement in Modern Physical Science," reprinted in Thomas S. Kuhn, *The Essential Tension: Selected Studies in Scientific Tradition and Change* (Chicago: University of Chicago Press, 1977). On the use and misuse of these words, see Robert K. Merton, David L. Sills, and Stephen M. Stigler, "The Kelvin Dictum and Social Science: An Excursion into the History of an Idea," *Journal of the History of the Behavioral Sciences* 20, no. 4 (1984): 319–31.

59. John Maynard Keynes, "Professor Tinbergen's Method," in *The Foundations of Econometric Analysis*, ed. David F. Hendry and Mary S. Morgan (Cambridge: Cambridge University Press, 1995), 389.

60. Keynes, "Professor Tinbergen's Method," 382–83.

61. John B. Davis, *Keynes's Philosophical Development* (Cambridge: Cambridge University Press, 1994), 105, quoting Keynes. See also Robert Skidelsky, *John Maynard Keynes, 1883–1946: Economist, Philosopher, Statesman* (London: Macmillan, 2003).

62. Jacob Marschak and Oskar Lange, "Mr. Keynes on the Statistical Verification of Business Cycle Theories (1940)," in Hendry and Morgan, *Foundations*, 390–98. Keynes, then editor of the *Economic Journal*, refused to publish Marschak and Lange's defense of Tinbergen. For further details on the Keynes-Tinbergen debate, see Hendry and Morgan, *Foundations*, 52–60.

63. John Maynard Keynes, "William Stanley Jevons, 1835–1882: A Centenary Allocution on

His Life and Work as Economist and Statistician," reprinted in *William Stanley Jevons: Critical Assessments*, ed. J. C. Wood (London: Routledge, 1988), 66.

64. Keynes, *The Economic Consequences of the Peace*, vol. 2 of *Collected Writings of John Maynard Keynes*, ed. Don Moggridge (London: Macmillan, 1971), 2.

65. Marschak and Lange, "Mr. Keynes," 398.

66. Jan R. Magnus and Mary S. Morgan, "The ET Interview: Professor J. Tinbergen," *Econometric Theory* 3, no. 1 (April 1987): 130. See also Marcel J. Boumans, "Observations of an Expert," working paper, History of Observation in Economics, no. 4, University of Amsterdam, 2009.

67. As Mary S. Morgan shows in her essay in this volume, such conflicts could motivate economists to leave their studies and go out into the field to try to resolve them, as in the case of Phyllis Deane.

68. Kwame Anthony Appiah, *New York Times*, 9 Dec. 2007.

9

"A Number of Scenes in a Badly Cut Film": Observation in the Age of Strobe

JIMENA CANALES

In 1958, an experimental subject reported the following observation:

> It was as if I was looking into very deep water and seeing deep green coral or a coiled octopus in dark green and white—very deep. This stuff streaming out from the centre is just surface thought, but this was a deep thought. It impressed me a good deal psychologically like Jung's image—it has intrinsic significance or archetypal quality. The marble is deep; the ferns are deeper, but this was miles underneath it all. . . . A very unusual effect. It started as centrifugal motion, then the nucleus began to have an odd look and a little irregular snowflake began to form at "marble" depth and this grew and filled the whole field with salmon-pink ground bearing repeated identical irregular snowflakes. These then all melted into the impression of a THING! These small snowflakes melted into one large snowflake which became alive; it turned into a living creature—slightly eerie—like Quattermass.

The subject of the experiment was one out of thirty-five advanced psychology students or staff of the Department of Psychology at Cambridge University who were asked to stare into a strong source of intermittently flashing strobe light and to describe and draw the "visual phenomena evoked by the stimulus."[1]

The experimenter John R. Smythies reported that seven subjects undergoing the same experiment described seeing something like "bacteria seen under a microscope or pond life or powder on a liquid surface." One person saw "an aerial photo of a city with streets and blocks of houses . . . it is like looking at London from a tremendous height and seeing the whole lot swirling about," while another reported seeing "lovely tropical fish in a blue tank." Many described wallpaper—mostly Victorian—although one stood out as having "a terrific modern design" of "black lines with knobs on the end forming triangles." Some subjects saw "a continual stream of images of fully formed scenes, usually of commonplace objects and events such as trains, cars, street scenes, harbours, animals, peoples, etc." To one of them, these appeared to be "like a number of scenes in a badly cut film."[2]

Visions of this sort could appear without complicated machines. Where did these visions come from? Why were they so "extraordinarily vivid"? Why did they sometimes appear "coupled with a strong emotive sense of 'eeriness'"? Why were they frequently described as "deep," and why did they often appear to be under "clear rippling water"? When were they first seen, and when were they first described? How did they change scientists' views of what it meant to observe?

In 1823, the physiologist Jan Purkinje produced what later came to be called flicker effects simply by waving his fingers in front of one eye while staring at the sun. He drew the patterns that he saw (fig. 9.1).[3] A decade later, another scientist, David Brewster, noted that these effects sometimes appeared when "walking beside a high iron railing." What he saw "exceed[ed] any optical phenomena which I have witnessed." The visions were "so dazzling" that "the eye is soon obliged to withdraw itself from its overpowering influence," and so he prudently turned the other way.[4]

In *The Living Brain* (1953), the controversial neurophysiologist William Grey Walter suggested that intermittent flashes could appear spontaneously in the rain forest as light passed through tree leaves. He hypothesized that they could have caused important evolutionary developments, being perhaps the force that knocked the apes out of the trees and onto the ground, providing the first essential impetus for the transition to *homo erectus*. Flickering light, Walter argued, was the reason why we emerged as "sadder but wiser apes."[5]

Scientists had long compared these visions to others. Sometimes they classed them with afterimages, the images that persist in the retina after an observer has looked at bright objects or strong sources of illumination. At other times they compared them to pressure images that appeared when poking one's eyes with a moderate and long-continued uniform pressure. Some scientists classed them as entopic phenomena, that is, visualizations of the internal structures of the eye due to corneal inhomogeneities and to the shadows of the blood vessels. Walter asked whether these effects were comparable to the "peculiar responses" induced by "rhythmic stimulation" such as tickling or by listening to the beats of a tom-tom drum, which "have been endowed with mysterious and even magical properties since the dawn of consciousness."[6]

Despite the fact that many scientists claimed that these visions were

FIGURE 9.1. The first four figures starting from the top left-hand corner are "flicker" patterns. From Jan Evangelista Purkinje, *Beobachtungen und Versuche zur Physiologie der Sinne*, vol. 1, *Beiträge zur Kenntniss des Sehens in subjectiver Hinsicht*, 2nd ed. (Prague: Kupfertafel, 1823).

ubiquitous and that they easily appeared in a number of everyday situations, they rarely described them and they rarely drew them.[7] In an important nineteenth-century book on optics, the German scientist Hermann von Helmholtz ventured only briefly into "this extremely perplexing region of the most manifold phenomena" and offered brief descriptions. He mentioned a "watered silk" effect that appeared when looking at intermittent sources of light, and a central rosette figure surrounded by dots increasing in size "which may possibly be compared to a rose with many petals."[8] Up to the middle of the twentieth century, only a handful of scientists besides Helmholtz had made "brief reference to hexagonal figures, grids, radial lines and mosaics."[9]

For a few years in the late 1950s a few scientists no longer looked away and instead developed new experimental systems to study and enhance these visions. They adopted high-power electronic stroboscopes, which became available after the war and which were used in scientific, military, and industrial settings to observe fast events. But they used them in an entirely different way. Instead of illuminating the phenomena under investigation, they stared directly into the strobe, sometimes with their eyes only a few centimeters from the source of light. For two years, from 1957 to 1958, they systematically recorded their visions. From these experiments, some concluded that common understandings of observation and reality needed to be changed. Few agreed. Instead of attempting to change the meaning of observation, most continued to simply look away.

To what can we attribute this disregard? And why did they so seldom describe their visions or represent them as drawings? Brewster maintained that observing these patterns required "courage" and that drawing them was nearly impossible because of their rapidly changing nature.[10] Although he hoped that "observers who have younger eyes than mine, and who have the courage to repeat the experiments . . . will be able to obtain an accurate representation of the pattern in question," his optimism was premature; almost nobody repeated these experiments for the following hundred years.[11]

Helmholtz claimed that the reason "most observers thus far have been able to establish only a comparatively few facts and to make a few new discoveries" had to do first with fatigue and safety: "[T]hese experiments soon prove to be so trying to the eyes that severe and dangerous ocular and nervous trouble may ensue." Precautions needed to be taken, and he advised "future observers . . . not to do too many in one day."[12] These experiments required not only "practice" but at times "self-sacrifice."[13] Prohibitions against certain types of self-experimentation in the sciences had been established since at least the eighteenth century. Extremes were legendary in the literature, such as the experiments of Johann Ritter, who stared at the sun for a record twenty minutes and for the next twenty-six days was unable to see black and white, instead seeing only reversed colors.[14] Although Helmholtz admired and at times praised these kinds of experiments, he nonetheless advised against them.

The second reason Helmholtz gave for ignoring these observations was that nobody knew how to explain them using existing theories. He asked scientists to focus instead on the "great mass of relevant phenomena . . . characterized by their energy, distinctness and constancy; even if we also find isolated and more transitory phenomena for which at present there is no perfectly satisfactory explanation."[15] In short: if unexplained, better ignored.

When electronic stroboscopic technologies started to appear in the first decades of the twentieth century, a few more scientists started to note these effects. The "birth of the stroboscope" is usually traced to 1832, when the Bel-

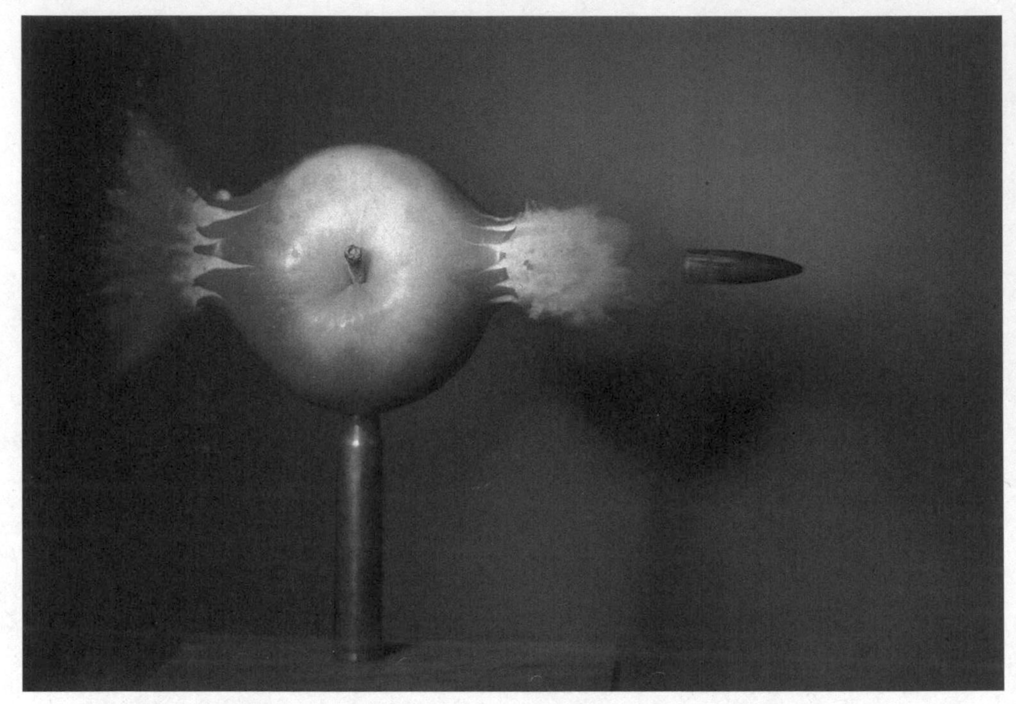

FIGURE 9.2. *Bullet through the Apple.* From Harold E. Edgerton and James R. Killian Jr., *Moments of Vision: The Stroboscopic Revolution in Photography* (Cambridge, Mass.: MIT Press, 1979), 107. Copyright Harold & Esther Edgerton Foundation, 2010, courtesy of Palm Press, Inc.

gian scientist Joseph Plateau used slotted disks turning at high speeds to provide a viewer with an illusion of movement. The mathematician and physicist Simon Stampfer built a similar apparatus and soon thereafter coined the term "stroboscope."

In the 1920s electronic stroboscopes for visualizing fast phenomena were already commercially available for industry. The development of this technology is usually credited to Harold Edgerton, professor of electrical engineering at MIT and author of well-known photographs of bullets in midflight and half-exploding balloons (fig. 9.2).

In 1942, the ability "to halt with a stroboscope and camera a bullet in flight" was hailed as one of the most important achievements of civilization, equal to the development of the telescope and microscope.[16] After World War Two, electronic strobes, such as those used and developed by Edgerton, were widely available; by 1954 there were thirty-nine different suppliers in the United States alone.[17]

A few scientists, however, used strobes in ways that strayed from conventional usage. These alternative experiments drew from a different tradition, of investigating the effects of light on humans. Such experiments contrasted sharply with those of Edgerton. Even in the rare cases when Edgerton aimed his machine at a person's eye to measure the time of a wink or to capture a delay in the iris's reaction to light, he ignored the effects of the strobe on the brain and the experience of the experimental subject.[18] He could not, however, prevent alternative uses of the technology. When he was not looking, some of his students stared directly into the machine. Edgerton never condoned these practices.[19]

Flicker before the Strobe

In the late nineteenth century, the English toymaker Charles Benham experimented with the effects that appeared when looking at black-and-white patterns twirling rapidly. Staring at a black-and-white disk, observers saw magical colors seemingly appearing from nowhere. Finding the pattern that best revealed these strange colors, he marketed a new product called the Benham top. Other spinning disks provided similar illusions. In 1928, the American psychologist Walter R. Miles picked up a five-inch paper disk for testing the speed of phonograph turntables and noticed that "if one fixates the center as steadily as he can, he observes phantom objects rippling and revolving in a most extraordinary manner." If he moved the disk in front of him he saw "grayish phantoms" and even a "reversible windmill illusion."[20] He perceived a breakdown between stasis and movement, life and inert objects: "The 5-in [Victor] disk considered as a *stationary visual stimulus* is the most *live* object of the kind that I have ever seen." He also noticed an eerie breakdown between listening and looking. Referring to the famous advertisement portraying a dog listening to a gramophone, he called attention to how the "world classic trade-mark 'His Master's Voice' shows our canine friend not only listening but looking." Yet, like others before him, he did not give further details about these visions, explaining that "the phenomena are so prominent and so bizarre that people as a rule object to looking at it for any but very brief intervals."[21]

In 1934, the Cambridge physiologist Lord E. D. Adrian and his student B. H. C. Matthews, who would later be known for their work measuring brain waves, investigated the effects of stroboscopic light on the brain. In an experiment considered foundational for electroencephalographic (EEG) research, they each stared into a 30-watt automobile headlight bulb covered with a spinning disk powered by a gramophone motor, and recorded the associated brain waves using an oscillograph. Their findings showed that, by varying the flash frequency, they could change the frequency of the recorded brain waves. Adrian and Matthews reported that "coloured patterns may be observed by looking at flickering lights" and that they had an "extremely unpleasant" feeling during the experiment. However, they did not describe these effects any further and instead tried hard to eliminate them. If their experiment was to work, the strange visions had to go: "If it is too bright the [visual] field may become filled with coloured patterns, the sensation is extremely unpleasant and no regular waves are obtained."[22]

In 1942, the noted scientist Heinrich Klüver became interested in these visions, noticing a connection between them and what he saw while experimenting with mescal "buttons" or peyote. For decades, he had been self-experimenting with the drug and extensively documenting his experience. Klüver noticed a similarity between mescal visions and those caused by flicker. "To produce flicker that is visible with open or closed eyes," he used an "alternating current of low intensity and frequency." Trying the experiment on others, he found that "when the current was on, . . . one subject, a student, suddenly saw the profiles of five faces looking to the right. These faces rapidly changed into other faces; they were seen through the 'muslin curtain' of flicker, as the subject expressed it."[23]

Many writers, poets, and artists had described drug-induced states (most famously Charles Baudelaire in *Le paradis artificiel* and Thomas de Quincey in *Dreams of an Opium Eater*), but only a few individuals had experimented with mescal.[24] Klüver was one of the few scientists to venture into this area. Although "the phenomena reported present such striking differences in appearance," he was able to find the "common elements" or "*form constants*" that "appear in almost all mescal visions."[25]

Interest in flicker continued sporadically in the following years. Two scientists, Carl R. Brown and J. W. Gebhard, analyzed these effects as part of a broader interest in "visual 'transient' phenomena."[26] They investigated the effects of intermittent light on the eye using a projector system and an episcotister that could display two to twenty flashes per second. The authors themselves were the two "observers," and their "observation of the visual field . . . disclosed a most remarkable and beautiful display of color and form perceptions."[27] The patterns and colors they observed were surprisingly limited: "all investigations indicate some regularity and constancy in the basic patterns observed."[28] For them, these were mainly a "radiating or 'windmill' pattern of yellow and blue of relative high brightness" and a "much dimmer . . . irregular mosaic of violet and yellow-green."[29] There was not a "limitless variety" in what they saw.[30]

Flicker Proved to Be a Key to Many Doors

William Grey Walter continued to experiment with the strobe in the tradition inaugurated by electroencephalographers. His research differed from previous investigations in that he used a high-power electronic stroboscope with a peak intensity comparable to that of 88,000 candles, manufactured by Scophony, Ltd., one of the earliest makers of television sets.[31] In contrast to previous experimental setups, such as Adrian and Matthews's automobile-lamp-with-gramophone system in which the length and frequency between flashes lasted much longer, his flash lasted only 10 μseconds.[32] Walter measured its effects on the brain by attaching electrodes to the subject's skull and amplifying the brain signals "ten million times or more."[33]

Walter and his coauthors found that the instrument could be used to invoke epileptic fits. It was common knowledge (at least since Roman times) that seizures could be provoked in epileptics by exposing them to flickering light (or simply by asking them to look at a potter's wheel). But Walter's results were particularly shocking because they occurred in patients who had never before suffered from epileptic attacks, and even if his subjects were under the influence of large doses of anticonvulsant drugs.[34] One alarming implication was that epileptic seizures could be induced in individuals who had "no personal record of any sort of fit."[35] While previously epilepsy had been understood as a condition affecting only a few individuals who carried the disease, Walter's research showed how it instead could be "latent" in all individuals to differing degrees.

Walter confirmed Adrian and Matthews's discovery that the flashes changed the electrical rhythmic patterns emitted by the brain. He speculated about ways in which they could be used to study (and perhaps adjust) the out-of-step cerebral rhythms associated with seizures, cerebral tumors, lesions, and other pathologies. Like Adrian and Matthews before him, Walter focused most of his investigations on what the EEG record revealed about the effects of strobe on the brain. But he also started to pay increasing attention to what the subject undergoing the experiment reported or saw. One result was clear: "As the flicker frequency is raised, the subject begins to see things which are not present in the stimulus." Walter noted that when he increased the strobe frequency, subjective sensations "of a mosaic or chessboard pattern, sometimes with a whirlpool effect superimposed," appeared. At other times these sensations were more akin to actual hallucinations, producing "impressions of bodily movement or of organized visual experiences of a bizarre and sometimes alarming nature."[36] He listed the "physiological and psychological effects" of staring into a strobe light in terms of their intensity. These varied from quite minor "visual sensations with characters not present in the stimulus" to "organized hallucinations of various types." Sometimes "the hallucinations described by some subjects were of a character so compelling that one subject was able to sketch them some weeks later."[37] The list of effects culminated with "clinical psychopathic states and epileptic seizures."[38]

Following Walter, neurophysiologists increasingly used electroencephalographic (EEG) techniques in combination with strobe stimulation. Yet they continued to focus mostly on the EEG record, ignoring the strobe visions. In 1949, during a routine investigation of a patient suffering from extreme anxiety, A. C. Mundy-Castle, a leading electroencephalographer working in South Africa, mentioned them briefly. One of his patients suffered from "vivid visual hallucinations" while he was exposed to the strobe light, informing the experimenter that his hallucinations consisted "primarily of things that happened during my life."[39] Intrigued by the "visual reconstructions from past experience" that were probably "released" from "some storage mechanism," Mundy-Castle ran more tests. In 1953 he experimented with an EEG machine and a strobe on approximately one thousand subjects who were asked to stare into a Scophony strobe at a distance of 9 centimeters. Like Walter, he noted "delusional or hallucinatory states directly evoked by flicker," which were "fortunately quiet rare."[40] He also described how many of his subjects reported visual effects "usually in the form of moving concentric rings with a faint cross or regular spoked figure radiating from the centre, together with myriads of small shadows or criss-crossed elements forming a shifting network of colours."[41] Like most researchers before him, Mundy-Castle was more interested in the EEG record than in the descriptions offered by his subjects. He stated strong reasons why he was unwilling to study these dream states: "The very nature of these responses, combined with their infrequency, renders controlled study impossible."[42] The notion of a "controlled experiment" as understood by Mundy-Castle and most other encephalographers of the time simply excluded these visions from scientific investigation.

Angiola Massuco Costa, a researcher who would become one of Italy's most important psychologists, took a slightly different perspective.[43] Experimenting on fifty subjects, she documented responses of a "fantastical-hallucinatory" nature, describing how subjects saw "horses, sails, tears, eyes, spiders," and she provided a few drawings (fig. 9.3). Some visions had "something similar to the artistic," and she speculated that they may have a "projective and symbolic value." They showed an "amazing liberty" comparable to that of artistic creativity. But how could a machine, the strobe, induce this freedom? Costa speculated that it must be a liberty of a different, restricted sort: "Obviously, I am not speaking of liberty (or spirituality) in a metaphysical sense."[44]

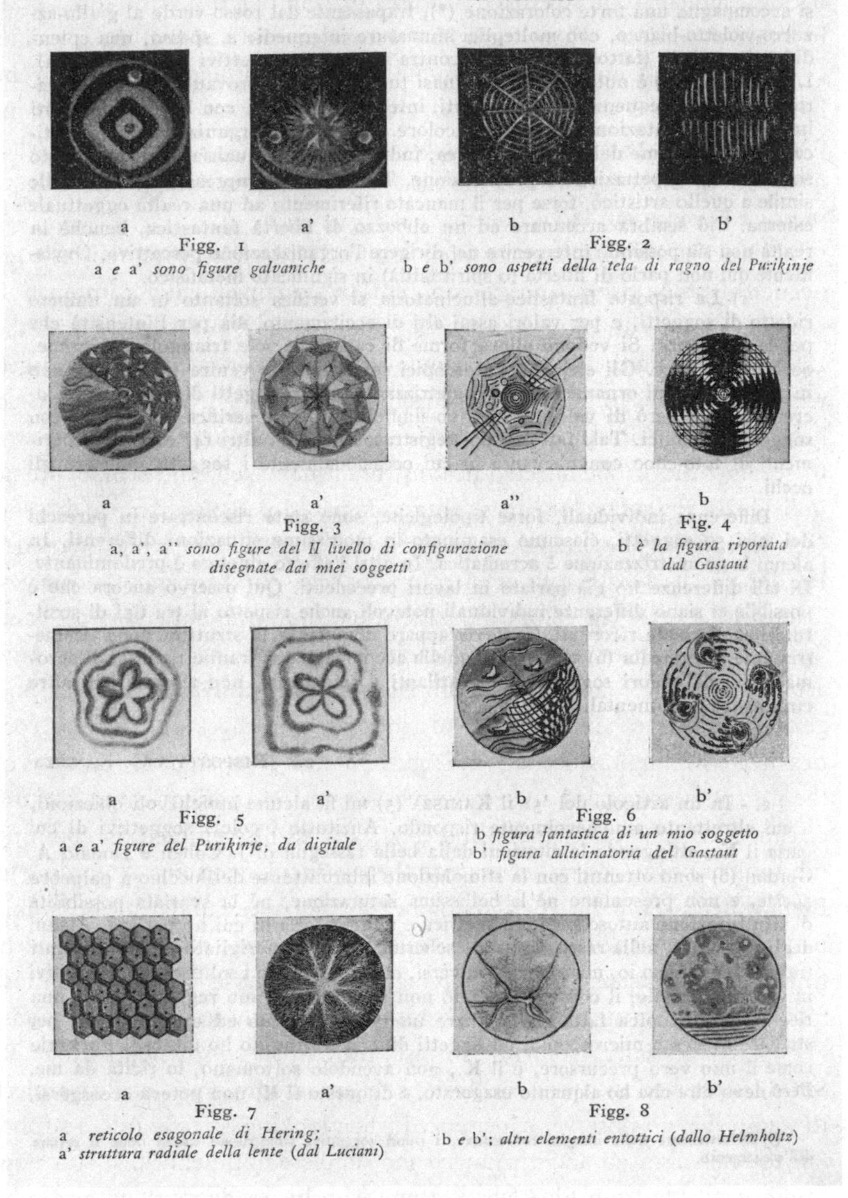
TAVOLA DOCUMENTARIA
a
a'
b
b'
Figg. 1
Figg. 2
a e a' sono figure galvaniche
b e b' sono aspetti della tela di ragno del Purikinje
a
a'
a''
b
Figg. 3
Fig. 4
a, a', a'' sono figure del II livello di configurazione disegnate dai miei soggetti
b è la figura riportata dal Gastaut
a
a'
b
b'
Figg. 5
Figg. 6
a e a' figure del Purikinje, da digitale
b figura fantastica di un mio soggetto
b' figura allucinatoria del Gastaut
a
a'
b
b'
Figg. 7
Figg. 8
a reticolo esagonale di Hering;
a' struttura radiale della lente (dal Luciani)
b e b'; altri elementi entottici (dallo Helmholtz)

FIGURE 9.3. Stroboscopic patterns. From Angiola Massuco Costa, "L'effetto geometrico-cromatico nella stimolazione intermittente della retina ad occhi chiusi," *Archivio di psicologia neurologia e psichiatria* 14 (1953): 632–35, on 634.

By the mid-1950s, most scientists considered strobe visions to be hallucinations, similar to those evoked not only by spinning disks and mescal but by other conditions. Three Canadian researchers noticed that these visions could appear not only when the subject was overstimulated with a strobe light, but *understimulated* in a sensory-deprivation chamber. In dark chambers, subjects experienced hallucinations that were "quite similar to what have been described for mescal intoxication, and to what Grey Walter has . . . produced by exposure to flickering light."[45]

Richard H. Blum, a scientist from Stanford University, explored the connection of strobe visions and schizophrenia by studying them in combination with EEG results and Rorschach tests.[46] Blum came from working with the U.S. army during the Korean War in a classified experimental unit charged with finding ways to return traumatized soldiers to the front line; he would become famous for later experiments with LSD. In 1954 his main concern was to see how the strobe affected healthy, brain-damaged, and schizophrenic individuals. Enlisting "organics" (brain-damaged patients), "normals" and "schizophrenics," he asked them to relate what they had seen while exposed.[47] He drew a chart with rubrics of colors, patterns, and meaningful images. "Normals" mostly saw colors and patterns (with movement and depth) and "schizophrenics" saw most of the meaningful images of a hallucinatory character, such as fire, waves, crabs, umbrellas, subway tunnels, dandelions, and genitalia, among others. Blum speculated that these images arose from a "cortical free-wheeling" associated with schizophrenia. Although the reactions of the patients were varied and sometimes severe (one patient entered into a catatonic-like state that lasted three days), his results centered on counting the number of images, rather than on their particular content. According to Blum, differences in the type and number of visions seen under strobe illumination were a mark of each patient's differing "ability to respond adequately to the world outside."[48] Schizophrenic patients, whose visions were most numerous and most intense, clearly responded poorly to worldly demands.

A few years later, from 1957 to 1958, the British neuroscientist John R. Smythies reached different conclusions by embarking on a detailed project to study what an observer experienced when looking directly into a strobe. Smythies "borrowed and scrounged the simple equipment" that was now readily available from EEG labs.[49] Along with his students, staff, and subjects, he stared at the strobe and recorded observations while changing the stroboscope's frequency and varying other conditions. The more Smythies worked with the stroboscope, the more complicated the patterns became.[50] Smythies published both detailed verbal descriptions and drawings. Some patterns seemed like "pond life," "bacteria," "germs," or "plankton." Others

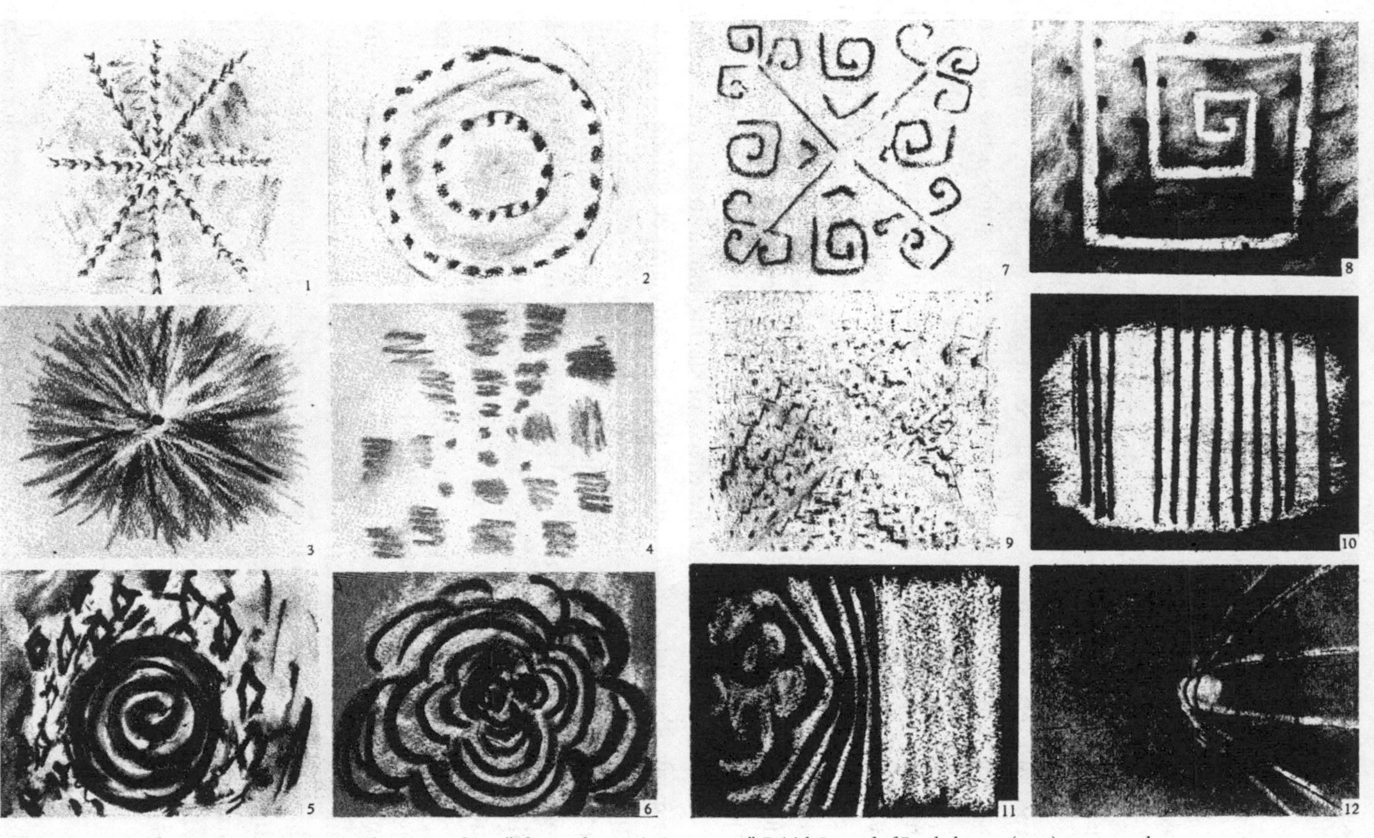

FIGURE 9.4. Stroboscopic patterns. From John R. Smythies, "The Stroboscopic Patterns: 2," *British Journal of Psychology* 50 (1959): 305–24, plates.

were "described as 'streets and houses' swirling around." Nevertheless, certain patterns (such as alphabetical symbols) *never* appeared, enabling Smythies to classify the patterns into seven main types.

While most previous researchers were interested in the EEG record, for Smythies the "stroboscopic patterns" *themselves* proved valuable. He was well versed in electroencephalographic techniques, but in this project he left EEG records aside and instead asked his experimental subjects to describe and draw their visions in pastel colors. He included numerous images in his published articles (fig. 9.4).

Smythies lamented how current research on the living brain suffered from two related problems. Scientists could either study a large number of neurons using electroencephalography, or they could study a few of them using a microelectrode, but they had no means of studying the brain at an intermediate level. In contrast to both of these options, Smythies believed that stroboscopic patterns offered a third option, possibly correlating with personality tests or with electroencephalography. Drawing from the work of others before him, including his mentor Klüver, he tried to prove that the "form constants of hallucinations represents a worthwhile field of study."[51] The stroboscopic patterns revealed the "natural history" of the brain.[52]

Smythies concluded that the effects of staring into a strobe were similar to many other visual effects, such as those appearing when staring at rapidly spinning black-and-white disks, when poking one's eyes, in hallucinations occurring when entering into or coming out of sleep, in visualizations of entopic phenomena, and in the visual phenomena of insulin hypoglycemia. He also noted that when mescaline and the stroboscope where used together, hallucinogenic effects were visibly enhanced. And both were highly addictive.[53]

Smythies had come to work on the stroboscope after studying the effects of mescaline with the controversial neurophysiologist Humphrey Osmond.[54] With his coauthor, they developed the first biochemical theory of schizophrenia by arguing that a defect in the metabolization of adrenaline could produce in the body a substance similar to mescaline (called the M-substance) that then created the effects of the disease. In doing research with mescaline and lysergic acid (a precursor to LSD), Smythies and Osmond found similarities between their effects, concluding that the mental disorder might be a chemical disorder.

"A new and sinister development"

Smythies's research had wide repercussions. The famous psychologist Carl G. Jung became interested in his work, and he invited Smythies to his home, where they delighted in some harmless Freud bashing.[55] Smythies reported that Jung was intrigued by his assertion that mescaline visions have "nothing to do with the personality having them," and that he saw in his work with Osmond a corroboration of some of his theories on the collective unconscious.[56] Aldous Huxley read a paper by Smythies on mescaline and wrote to him saying that he very much wanted to try the drug. While Smythies could not personally deliver the drug to Huxley, he put him in contact with Osmond, who gave him his first dose while on a trip to California in the spring of 1953.[57] Aldous's experiences with mescaline were recounted in *The Doors of Perception* (1954), in which he mentioned the work of Smythies and Osmond on the connection between mescaline and schizophrenia. What intrigued Huxley was the same claim that interested Jung; that these experiences were not created by the person undergoing them, but rather that they came from elsewhere: "the work of a highly differentiated mental compartment, without any apparent connection, emotional or volitional, with the aims, interests, or feelings of the person concerned."[58] Huxley next experimented with both mescaline and strobe. In *Heaven and Hell* (1956), he cited "the words used by Dr. J. R. Smythies in a recent paper in the American Journal of Psychiatry" to talk about the nature of visionary experiences: "To sit, with eyes closed, in front of a stroboscopic lamp is a very curious and fascinating experience."[59]

According to Smythies, the direction of his research changed when his colleagues and he "unfortunately" recruited Harvard professor Timothy Leary in "a plan whereby mescaline would be made available only to a carefully selected group of academics—psychologists and philosophers." But Leary instead "opened a Pandora's box with the results that we have to live with today."[60] Smythies considered the drug revolution completely sinister and condemned any "recreational use of these hallucinogens."[61] Yet he could not prevent these developments from having a "bad effect" on his research and career. Most of his peers distanced themselves from his work and from their connection to Osmond. His fellowship at Cambridge soon ran out and was not renewed.[62] Nonetheless, with the help of some of his supporters, Smythies managed to complete his clinical training at Maudsley Hospital (October 1959) and then moved to Edinburgh for the next twelve years.

But EEG and strobe research continued, often in combination with new drugs. In 1959 at the Mental Research Institute in Palo Alto, the Beat poet Allen Ginsberg was given LSD. His reaction to the drug was investigated with a stroboscope and an EEG machine.[63] The Veteran's Administration Hospital where Blum had set up his EEG strobe research a few years before became the site of government-sponsored research on the drug. Another subject, Ken Kesey, volunteered to try numerous drugs there, and later recounted his experiences in the famous novel *One Flew Over the Cuckoo's Nest* (1962). Experimenters could not keep their research completely in-house. Kesey started using LSD and the strobe in a different way, organizing the first acid drug parties illuminated by strobe light. By the end of the sixties the strobe had became essential paraphernalia of the drug revolution. It had traveled quickly from laboratories to hospitals, artists' studios, drug dens, and finally, to discos.

Ian Sommerville, William S. Burroughs's boyfriend, soon constructed a simple flicker machine, known as the "dreamachine" designed to democratize self-experimentation with flicker. Burroughs was so intrigued by flicker that he went to a lecture, talked to Walter, and publicized Walter's work. By the mid-sixties, Burroughs was advertising flicker as a way "to achieve the same results [as taking drugs] by nonchemical means."[64] He described using "flicker, music through head phones, cutups and foldins" to produce his novels, and he illustrated the technique in his films.[65]

Leary's interest in drugs was also accompanied by an interest in strobe research. In "How to Change Behavior" he wrote, "We have recently learned from W. Grey Walter and William Burroughs about photostimulation as a means of consciousness alteration. Concentrated attention to a stroboscope or flicker apparatus can produce visionary experiences."[66] The artist and poet Bryon Gysin wrote about the dreamachine in *The Process* (1969), earning for this the description by the famous punk rocker Genesis P-Orridge of being "a Dreamachine [in] human form." Sommerville, Burroughs, Leary, and Gysin all explained the effects of flicker by reference to Walter's work.

The uptake of strobe experimentation into visionary and extreme experiences often in combination with drugs by others outside of the scientific community, starting with Huxley and culminating with Leary, hurt Smythies's career. Smythies believed that some of the tasks of "Artists and Scientists" overlapped, explaining that "both have always been interested in exploring the transcendental worlds that expansion of normal consciousness leads to."[67] But few agreed.

Television

Strobe effects and visions were part and parcel of a burgeoning postmodern era. Artificial light became stroboscopic when Westinghouse's alternating current method won over Edison's direct current as a public utility. As a result, in America the glow of light started alternating sixty times per second and in Europe, fifty times. When the frequency of alternation became steadier in most households during the 1920s, everyday, artificially illuminated life started to beat with a regular, pulsating, stroboscopic rhythm. By the postwar era artificial light was only one component of a mass of other pulsating light sources. Russell J. Blattner, a leading pediatrician of the time, remarked how stroboscopic effects were all-pervasive by the early sixties: "Modern developments have increased the forms of flicker to which the seizure-prone person is subjected: fluorescent lighting, neon signs, motion pictures, and television." It was urgent, he argued, to study potential dangers lurking behind "the complexities of modern life and its attendant new forms of light reflection," particularly those of television. The doctor advised his patients to "to look away" from the set—especially if "shifting image or flicker is marked."[68]

These reactions to strobe or to an unadjusted television set led Smythies to build on a hypothesis of Walter, and to venture that visual perception functioned like television, claiming "that television uses the same mechanical principles as are used in the physiological mechanisms mediating visual perception."[69] Walter had concluded that the visual system in the brain did not work as a traditional cinematographic camera. Scientists could not produce hallucinatory effects with a strobe light and a cinematographic camera; yet these effects readily appeared if the strobe was used in combination with television. Differences between cinema and television became particularly evident when a film studio was illuminated by strobe. If the flashes coincided with moments between frames, nothing unusual appeared. If the flashes coincided with the frame frequency, the result was to have "no picture at all." In frequencies in between, a combination of these two effects appeared. But Walter noticed that "in no case will there be any 'hallucinatory' effects."

This result was completely different from the effects strobe produced on a television studio instead of on a film studio. If the strobe was directed at a scene being scanned by a television camera, strange pulses, dots, and dashes suddenly appeared across the screen, leading him to the conclusion that "the televisual system behaves very much like the neuro-visual one." The conclusion that the human visual system did not function according to traditional camera analogies was unavoidable "if we consider the [stroboscopic] effect upon the final picture of illuminating first a film studio and then a television one with a flickering light."[70]

The change in analogizing the visual mechanisms in the brain as televisual instead of cinematographic brought with it important changes in philosophy. Just as a television set does not "give us a *direct* view of the events televised," the televisual system in the brain also did not provide a direct view of reality. Smythies claimed that "current variations of naïve realism . . . in which it is believed that the physiological processes of perception mediate a *direct view* of the physical world, are wrong."[71] The "naïve realist" view that Smythies forcefully criticized informed common interpretations of what it meant to observe, including observations obtained with the stroboscope. When the stroboscope was used to illuminate fast phenomena, as Edgerton used it to produce his well-known photographs of bullets in midflight, viewers mostly explained what they saw in direct, realist terms. Edgerton's images were commonly described as "literal transcriptions" of nature, "a unique and literal transcript of that time world beyond the threshold of our eyes," furnishing "scientific records" written in a "universal language for all to appreciate."[72]

Smythies did not believe that observations were ever that simple. In subsequent publications he extended Walter's insight even further. He developed a system for finding out details about the inside of a television set *without opening it up*. The type of patterns on the television screen that appeared when a studio was illuminated by strobe depended on the type of raster mechanism inside the television. Analogously, Smythies speculated that the patterns that a person saw when staring into a strobosocope could "give us information as to details of operation of the mechanisms responsible for their production."[73] If scientists treated the brain "essentially as a 'black box'" where "the input is a temporally intermittent and spatially uniform light stimulus of the retina" and the "output is a report by the organism of the perception of geometrical patterns," a careful study of the patterns could shed light on the cerebral black box.

Walter's work, especially that which was aimed at a general readership, asked how modern bodies fit into a new postwar mass media system. The scientist lamented how modern life was "becoming more and more a one-way communication, from top and center down and out to the inert receivers."[74] Its characteristic was one of the "gaze" and the main instrumental culprit was television: "A passive solitary child gazing at the screen of a television receiver amuses only itself—the need to gaze does not promote or evoke habits of creativeness or generosity."[75] One-way mass media was "degenerating" our technologically expanded bodies into "something more like a spinal cord, able to receive instructions and implement reflex coordination" than a brain.

Television was dangerous, and so was cinema. Some dangers were immediate. Walter wrote of a case of a man who "found that when he went to the cinema he would suddenly feel an irresistible impulse to strangle the person next to him."[76] An investigator on epilepsy described two patients who "had difficulty on entering or leaving the cinema," and many others who passed out when "kneeling close to the [television] screen" or adjusting the set at close range.[77] But other dangers lurked in the future: "For Alice in Movieland the future looks drab; Tom Sawyer will have few adventures at the television set."[78] The life of the television spectator could never match a life of action.

"Observing the visionary world"

Strobe research did not fit with the widely held belief, as explained by a well-known psychologist who worked on intermittent light stimulation: "Our knowledge of the world is supposed to be built up only from the materials given by our sensory receptors."[79] Scientists had long known of exceptions to this general view, particularly those classed as illusions or hallucinations. The Swiss psychiatrist Eugen Bleuler defined hallucinations as "perceptions without corresponding stimuli from without."[80] The *Psychiatric Dictionary* (1940) defined them as the "apparent perception of an external object when no such object is present."[81] Under these definitions, strobe visions could be considered as one among many other types of hallucinations.

Interest in hallucinations was high in the late 1950s and continued into the next decade. In 1958 the American Psychiatric Association dedicated its yearly symposium to the topic, finding that it "was appropriate and timely" due to a "growing interest in the subject."[82] Investigations on LSD, mescaline, sensory deprivation, and schizophrenia dominated the conference. Research on strobe visions was notoriously absent. Why? The organizers of the conference claimed that there was "clear agreement among clinicians on the nature of the phenomena included under the term."[83] Yet strobe researchers did not agree. Smythies, the most important scientist to investigate the effects of staring into a strobe light, held a completely different view of observation, reality, and hallucinations.

Time and again what fascinated flicker researchers was how something could appear disconnected from its source of stimulation. For Costa it was "the missing reference to an objective external reality" that intrigued her.[84] Walter was fascinated by how the subject "begins to see things which are not present in the stimulus" and how even though "the stimulus source itself is white, stationary and featureless . . . all subjects report seeing coloured moving patterns."[85] Blum was enthralled by the "dissimilarity of response to similar stimuli."[86] Smythies stressed how "when we look at a uniform field under intermittent illumination, we do not merely see the stimulus but see instead these complex and interesting patterns."[87] Gebhard, one of the first and only scientists to have an interest in strobe visions, wondered why scientists were loathe to accept that there could be sensation beyond stimulation and insisted that they admit that "there are visual phenomena, although they correspond to no physical reality."[88] In consequence, he asked scholars to stop "theorizing in terms of functional 'atoms' of the form: stimulus-excitation-sensation." In this type of theorizing, he lamented, the "inevitable result is a body and soul (receptor, central projection and 'sensation') theory, logically no better than the body and soul theories of the most ancient tradition."[89]

But what was unique about strobe research was not merely its status as a form of hallucination. Rather, it revealed new fissures in a longstanding debate about what counted as reality. The debate was visible in the authors' different use of the term "observation." Klüver, for example, explained on numerous occasions that although he was clearly hallucinating under the influence of mescal, his "experiments" still lent themselves to perfectly clear scientific observation. Mescal "does not destroy the critical attitude of the observer."[90] His research was solidly based on "self-observations of qualified observers."[91] He, like most of his peers, used the terms "visions" and "observations" separately. His task was to "observe . . . the visions" and to "happily observ[e] the visionary world."[92] For most scientists who worked on strobe, to "observe" was qualitatively different than to merely see. It was more meaningful for science. "Observations" were much more than mere "visions." Brown and Gebhard, who self-experimented with a strobe light, described their work as "observations of the visual field."[93]

The word "observation" often highlighted the particular aspect of the experiment that mattered the most to the scientist. Adrian and Matthews, who focused on the EEG record and not on what the subject saw under strobe stimulation, used the word "observation" exclusively to describe the EEG record. The "observer" was the scientist in charge of the experiment, the "observation" was the EEG record, and the "subject" was the person exposed to the strobe light.[94] Blum, who focused on the relation between the EEG record and the quantity of visual imagery, also used the term "observation" in this way. Observations were the work *of* the scientist *on* his "subjects." The subjects themselves did not produce "observations" but "responses."

When Walter used the word "observation" it was again *of* patients and *of* records. The term did not refer to the testimony of the experimental subject.[95] In Walter's work "observations," although more important than "visions," were nonetheless secondary to "experiments." The introduction of flicker into EEG work was designed to "extend" an otherwise limited "field of passive observation."[96] Using strobe, "like a modern detective, we can not only tap the lines of communication, but even interject suitably phrased messages of our own and *observe* the reactions of the suspect."[97]

Smythies, in contrast, used the term "observation" to describe what the subject saw. Gesturing to his unusual use of the term, at least once he placed the word in quotation marks.[98] In other publications, he confronted the topic of "observation" directly. In his mescaline research, he ardently fought to prove that his visions counted as proper and legitimate forms of scientific observation. Mescaline was different from other drugs, he argued, because it left the subject's "observational integrity intact."[99] Although clearly hallucinating, he argued that subjects were nonetheless perfectly capable of "observing" for scientific purposes. Those who took mescaline often used visual metaphors ("They use such phrases as 'I saw,' 'As I gazed,' 'As I looked,' 'It is wonderful to see,' etc."), yet "observing" was an act that complemented the subject's experience of "looking." Smythies described his work with strobe as studying patterns that could be "observed by looking."[100] He expanded the meaning of the term "observation" and along with it of "reality." The hallucinating subject was, for him, an "observer." His view was radical, since he believed that what counted as an "observation," as a "hallucination," and—ultimately—as "reality" was culturally determined: "Thus it can be argued that the basic decision to call hallucinations 'real' or 'unreal' is a matter of convention and is determined by the rules in our language relating to the use of the word 'real' and is thus a matter of culture."[101] Eskimos, Plains Indians, and Western scientists had different views about the real, the hallucinated, and the observed.

Few scientists of the period could have agreed with the assertion that what counted as a scientific observation was cultural and conventional. In fact, this radical position highlighted the widely opposite view of observation that dominated the 1950s and its preceding centuries. For example, during the seventeenth and eighteenth centuries it was often considered to be mimetic. It was generally associated with the camera obscura, with geometrical optics, and with touch. From our perspective, the meaning of the term "observation" during that period was adequate for aristocratic societies in which bodies were clearly separated in interior and exterior parts and where art and science mixed comfortably. The view of observation from the 1950s also differed from how it was generally understood in the nineteenth century. Observation then was usually compared to photography, increasingly understood as a chemical process, and studied with physiological optics. In stark contrast to earlier centuries, it was tightly coupled with vision and sight and decoupled from touch. It was subjective and tied to the body—a concept apt for industrial societies of spectacle and surveillance, where art and science were separate disciplines and where human subjects increasingly became objects of observation.

For a very brief period during the late 1950s, a few scientists considered strobe visions a legitimate form of scientific observation. But their status as such would not last long. Perhaps some strobe researchers were asking too much. Miles claimed that spinning disks seemed to be alive, pointing to a breakdown between listening and looking. Gebhard asked his colleagues to eschew the whole stimulus-sensation language used in physiology, rewriting, in the process, theories of body and soul. Smythies not only advocated a new relation between science and art and between health and disease, but he even asked that observations be considered sometimes as wholly "disconnected" from the person experiencing them. The attempts from Beat artists to build machines and perform experiments were never strong contenders in the world of science.

By 1965 investigators used the word "by-product" to describe the strange effects of staring into a strobe light.[102] Only a few artists continued to highlight these effects. In 1966 the experimental filmmaker Tony Conrad made the film *The Flicker*, exposing the audience to stroboscopic lights in order for them to experience their hallucinogenic effects. Many left the movie theater disoriented, forgetting their bags, umbrellas and other personal belongings. What the audience underwent was described as "experimental art" or as a "countercultural experience" but hardly as a form of "scientific observation."

As flickering light (cinematographic, televisual, and other) proliferated, brightened, and became nearly all-pervasive, the controversial theorist Marshall McLuhan explained how all modern media could fit within two extremes. One extreme was revealed by "experiments in which all outer sensation is withdrawn, [and] the subject begins a furious fill-in or completion of senses." The other extreme was characterized by "the hotting-up of one sense." Both extremes produced similar responses: "The hotting-up of one sense tends to effect hypnosis, and the cooling of all senses tends to result in hallucination." McLuhan did not focus on the extremes, concentrating instead on the middle "comfort" zone of media. Yet it is these extreme cases that reveal how observation after the 1950s was increasingly described as *mediated* and studied as one node in a larger network of mass media communications.[103]

Notes

1. John R. Smythies, "The Stroboscopic Patterns: 1. The Dark Phase," *British Journal of Psychology* 50 (1958): 107.

2. Ibid., 110.

3. Jan Evangelista Purkinje, *Beobachtungen und Versuche zur Physiologie der Sinne*, vol. 1, *Beiträge zur Kenntniss des Sehens in subjectiver Hinsicht*, 2nd ed. (Prague: Kupferatel, 1823).

4. David Brewster, "On the Influence of Successive Impulses of Light Upon the Retina," *London and Edinburgh Philosophical Magazine and Journal of Science* 4 (1834): 241–42.

5. William Grey Walter, *The Living Brain* (New York: W. W. Norton, 1953), 100.

6. William Grey Walter, "Features in the Electro-Physiology of Mental Mechanisms," in *Perspectives in Neuropsychiatry*, ed. Derek Richter (London: H. K. Lewis, 1950).

7. These understudied and underobserved visions are comparable to the masses of artifacts that also escape our attention. See Bruno Latour, "Where Are the Missing Masses? The Sociology of a Few Mundane Artifacts," in *Shaping Technology/Building Society: Studies in Sociotechnical Change*, ed. W. E. Bijker and John Law (Cambridge, Mass.: MIT Press, 1992).

8. Hermann von Helmholtz, *Treatise on Physiological Optics*, 3 vols. (Mineola: Dover Publications, Inc., 1962), 2: 256.

9. John R. Smythies, "The Stroboscopic Patterns: 2. The Phenomenology of the Bright Phase and after-Images," *British Journal of Psychology* 50 (1959): 305.

10. Brewster explained: "I have never been able to draw the pattern, or to trace how the patches of interstices of the net-work spring" because "the patterns are constantly changing their colour, their intensity of light, and even their form." Brewster, "On the Influence of Successive Impulses of Light," 242.

11. Ibid., 242–43.

12. Helmholtz, *Treatise on Physiological Optics*, 2: 229.

13. Ibid., 262.

14. Stuart Strickland, "The Ideology of Self-Knowledge and the Practice of Self-Experimentation," *Eighteenth-Century Studies* 31 (1998): 453–71.

15. Helmholtz, *Treatise on Physiological Optics*, 2: 260–61.

16. Paul B. Horton, "Does History Show Long-Time Trends?" *Scientific Monthly* 55 (1942): 461–70.

17. In France, the brothers Laurent and Augustin Seguin used a machine, baptized the "stroborama," to observe and photograph rapid phenomena—mainly motors and propellers. In 1930 Westinghouse Research produced a "stroboglow," which, although not commercially available, flashed 100 times in a minute and could fit into two fifty-pound cases.

18. James R. Killian Jr., "The Meaning of the Pictures: Exploring the World of Time and Motion," in Harold E. Edgerton and James R. Killian, Jr., *Flash! Seeing the Unseen by Ultra High-Speed Photography* (Boston: Hale, 1939), 180–81.

19. Personal communication from Harold Edgerton's son, Robert Edgerton.

20. Walter Miles, "The Victor Stroboscopic Disk for Visual Experiments," *American Journal of Psychology* 40 (1928): 313.

21. Ibid., 312.

22. E. D. Adrian and B. H. C. Matthews, "The Berger Rhythm: Potential Changes from the Occipital Lobes in Man," *Brain* 57 (1934): 378.

23. Heinrich Klüver, "Mechanisms of Hallucinations," in *Studies in Personality*, ed. Q. McNemar and M. A. Merrill (New York: McGraw-Hill, 1942), 186.

24. Antonin Artaud's *Le tarahumaras* in the 1940s and Henri Michaux's *Misérable miracle* of the 1950s were two notable exceptions. Michaux considered his work both an "exploration" and an "experiment."

25. Heinrich Klüver, *Mescal: The "Divine" Plant and Its Psychological Effects* (London: Kegan Paul, 1928), 29, 36.

26. J. W. Gebhard, "Chromatic Phenomena Produced by Intermittent Stimulation of the Retina," *Journal of Experimental Psychology* 33 (1943): 401.

27. Carl R. Brown and J. W. Gebhard, "Visual Field Articulation in the Absence of Spatial Stimulus Gradients," *Journal of Experimental Psychology* 38 (1948): 189.

28. Ibid., 195.

29. Ibid., 199.

30. Ibid., 197.

31. William Grey Walter, V. J. Dovey, and H. Shipton, "Analysis of the Electrical Response of the Human Cortex to Photic Stimulation," *Nature* 158 (1946): 540.

32. It could not go at "rates higher than 25 a second as the apparatus could not be run comfortably at higher speeds." Adrian and Matthews, "The Berger Rhythm," 380.

33. Walter, *The Living Brain*, 16.

34. Walter, Dovey, and Shipton, "Analysis of the Electrical Response," 541. Walter later would restate this point: "In some epileptics, stimulation of this type . . . can induce a characteristic seizure; attacks of petit mal, grand mal, and myoclonic jerkings have been readily induced by these means." Walter, "Features in the Electro-Physiology," 74–75.

35. V. J. Walter and William Grey Walter, "The Central Effects of Rhythmic Sensory Stimulation," *EEG Journal* 1 (1949): 65.

36. Walter, "Features in the Electro-Physiology," 74.

37. Walter and Walter, "The Central Effects," 65.

38. Ibid., 63.

39. A. C. Mundy-Castle, "A Case in Which Visual Hallucinations Related to Past Experience Were Evoked by Photic Stimulation," *EEG Journal* 3 (1951): 354.

40. A. C. Mundy-Castle, "Electrical Responses of the Brain in Relation to Behaviour," *British Journal of Psychology* 44 (1953): 322.

41. Ibid., 323.

42. Ibid., 322.

43. Angiola Massuco Costa, "L'effetto geometrico-cromatico nella stimolazione intermittente della retina ad occhi chiusi," *Archivio di psicologia neurologia e psichiatria* 14 (1953): 632–35. Angiola Massucco Costa was the founder of the Istituto Superiore di Psicologia Sociale in Italy.

44. Ibid., 633.

45. W. H. Bexton, W. Heron, and T. H. Scott, "Effects of Decreased Variation in the Sensory Environment," *Canadian Journal of Psychology* 8 (1954): 70–76.

46. Richard E. Morgan, *Domestic Intelligence: Monitoring Dissent in America* (Austin, Texas: University of Texas Press, 1980), 291.

47. Richard H. Blum, "Photic Stimulation, Imagery, and Alpha Rhythm," *Journal of Mental Science* 102 (1956): 162.

48. Ibid., 166.

49. He came to Cambridge from Australia, where he had worked with the leading neurophysiologist Sir John Eccles. After working in Cambridge he moved to the Galesburg State Research Hospital in Illinois and later to the Worcester Foundation for Experimental Biology in Shrewsbury, Mass. For his experiments he used an Aldis 500-watt projector covered by an episcotister and a standard EMI electronic stroboscope. He learned to work with electroencephalography while at the National Hospital, Queen's Square, in London. John R. Smythies, *Two Coins in the Fountain: A Love Story* (n.p.: BookSurge, 2005), 41.

50. Since the effects of the strobe on both eyes was frequently different from when only a single eye was used, Smythies concluded that the observed effects were not merely retinal.

51. Smythies, "The Stroboscopic Patterns: 1," 116.

52. Smythies, "The Stroboscopic Patterns: 2," 306.

53. Smythies, "The Stroboscopic Patterns: 1," 111.

54. These studies were done at the Psychiatric Unit of St. George's Hospital.

55. Smythies, *Two Coins in the Fountain,* 35–36; John R. Smythies, "A Visit to Dr. Jung," *Alabama Journal of Medical Science* 18 (1981): 93–94.

56. Smythies, *Two Coins in the Fountain,* 36.

57. Aldous Huxley, *The Doors of Perception* (New York: Harper, 1954), 12.

58. Aldous Huxley, *Heaven and Hell* (London: Chatto, 1956), 20.

59. Ibid., 56. Huxley warned his readers of the "slight danger involved in the use of the stroboscopic lamp," particularly in epileptics: "One case in eighty may turn out badly."

60. Smythies, *Two Coins in the Fountain,* 54.

61. Ibid., 54.

62. Ibid., 34.

63. Martin A. Lee and Bruce Schlain, *Acid Dreams: The CIA, LSD, and the Sixties Rebellion* (New York: Grove, 1985).

64. William S. Burroughs, "Points of Distinction between Sedative and Consciousness-Expanding Drugs," in *LSD: The Consciousness-Expanding Drug,* ed. David Solomon (New York: G. P. Putnam's Sons, 1964), 172. He described the dreamachine in Daniel Odier, ed., *Entretiens avec William Burroughs* (Paris: Belfond, 1969).

65. Allan Ginsberg analyzed Burroughs's *The Soft Machine* (1961) in terms of the "Stroboscopic flicker-lights . . . [that] create hallucinations, and even epilepsy." Cited in John Geiger, *Chapel of Extreme Experience: A Short History of Stroboscopic Light and the Dream Machine* (Brooklyn, N.Y.: Soft Skull Press, 2003), 52. The films in "Towers Open Fire" produced by Burroughs in 1961–62 show dreamachine experiments.

66. Timothy Leary, "How to Change Behavior," in *LSD: The Consciousness-Expanding Drug,* ed. David Solomon (New York: G. P. Putnam's Sons, 1964), 105.

67. John R. Smythies, introduction to Geiger, *Chapel of Extreme Experience,* 7–8.

68. Russell J. Blattner, "Photic Seizures—Television-Induced," *Journal of Pediatrics* 58 (1961): 746.

69. John R. Smythies, "The Stroboscope as Providing Empirical Confirmation of the Representative Theory of Perception," *British Journal for the Philosophy of Science* 6 (1956): 332, n. 4.

70. Walter, "Features in the Electro-Physiology of Mental Mechanisms," 77.

71. Smythies, "The Stroboscope as Providing Empirical Confirmation," 334. Emphasis in the original.

72. Killian Jr., "The Meaning of the Pictures," 22.

73. Smythies, "The Stroboscopic Patterns: 2," 307.

74. Walter, *The Living Brain,* 267.

75. Ibid., 268.

76. Ibid., 98.

77. Blattner, "Photic Seizures—Television-Induced," 747.

78. Walter, *The Living Brain,* 268.

79. Henri Pierón, *The Sensations* (1952), cited in Sanford Goldstone, "Psychophysics, Reality and Hallucinations," in *Hallucinations,* ed. Louis J. West (New York: Grune and Stratton, 1962), 262.

80. Cited in Louis J. West, "A General Theory of Hallucinations and Dreams," in *Hallucinations,* ed. West, 276.

81. Cited in ibid.

82. Louis J. West, preface to *Hallucinations,* ed. West, vii.

83. West, "A General Theory of Hallucinations and Dreams," 276.

84. Costa, "L'effetto geometrico-cromatico," 633.

85. Walter, "Features in the Electro-Physiology of Mental Mechanisms," 74; William Grey Walter, *The Neurophysiological Aspects of Hallucinations and Illusory Experience* (London: Society for Psychical Research, 1960).

86. Blum, "Photic Stimulation, Imagery, and Alpha Rhythm," 164.

87. John R. Smythies, "The Stroboscopic Patterns: 3. Further Experiments and Discussion," *British Journal of Psychology* 51 (1960): 250.

88. Gebhard, "Chromatic Phenomena Produced by Intermittent Stimulation," 404.

89. Ibid., 387.

90. Heinrich Klüver, "Mescal Visions and Eidetic Vision," *American Journal of Psychology* 37 (1926): 513.

91. Klüver, "Mechanisms of Hallucinations," 198.

92. Klüver, "Mescal Visions and Eidetic Vision," 511.

93. Brown and Gebhard, "Visual Field Articulation," 189.

94. "The other (B.H.C.M.) [Matthews] is better in the role of observer than of subject, for in him the rhythm may not appear at all at the beginning of an examination, and seldom persists for long without intermission." Adrian and Matthews, "The Berger Rhythm," 382.

95. The fact that he considered the subject's testimony as secondary to the scientist's observations was also revealed in his choice of instruments. With a Toposcope, a machine used to evaluate various brain areas exposed to flicker, Walter claimed that the progress it brought about was comparable to the time when "a mosaic of aerial photographs" replaced "a traveler's tale." Geiger, *Chapel of Extreme Experience*, 17.

96. Walter and Walter, "The Central Effects of Rhythmic Sensory Stimulaton," 57.

97. Walter, *The Neurophysiological Aspects of Hallucinations*, 4. Emphasis added.

98. "[I]f an Archimedes spiral is rotated rather quickly and then 'observed,' the patterns for the subject can be evoked." Smythies, "The Stroboscopic Patterns: 3," 247.

99. John R. Smythies, "The Mescaline Phenomena," *British Journal for the Philosophy of Science*, 3 (1953): 339.

100. Examples of his observation language are: "[C]oloured patterns may be observed by looking" and "the sensory patterns that can be observed on looking." Smythies, "A Preliminary Analysis of the Stroboscopic Patterns," *Nature* 179 (1957): 523–24."

101. J. R. Smythies, "A Logical and Cultural Analysis of Hallucinatory Sense-Experience," *Journal of Mental Science* 102 (1956): 341.

102. Sanford J. Freedman and Patricia A. Marks, "Visual Imagery Produced by Rhythmic Photic Stimulation: Personality Correlates and Phenomenology," *British Journal of Psychology* 56 (1965): 95.

103. Marshall McLuhan, *Understanding Media: The Extensions of Man* [1964] (Cambridge, Mass.: MIT Press, 1994), 32. For the transformations of the word "media" from the late nineteenth century to its use by Marshall McLuhan, see Raymond Williams, *Keywords: A Vocabulary of Culture and Society*, rev. ed. (Oxford: Oxford University Press, 1983).

10

Empathy as a Psychoanalytic Mode of Observation: Between Sentiment and Science

ELIZABETH LUNBECK

The Chicago-based psychoanalyst Heinz Kohut entered the analytic scene in 1959 asking a deceptively simple question: in what did the analytic mode of observation consist? How was one to observe the inner processes—the thoughts and feelings—of another? Observing the external world was one thing, a matter of mobilizing the senses, employing instrumentation, and fashioning observations into theories. Observing the internal world—or, put differently, the psychological—was a dicier proposition. Observers' sensory organs on the one hand, their telescopes and microscopes on the other—these were useless in capturing and understanding individuals' interiority, their thoughts, wishes, and fantasies. Kohut argued that only introspection and empathy, introspection's vicarious counterpart, were adequate to the task of apprehending inner experience. They were, he maintained, the analyst's scientific instruments.[1]

Twenty-five years and two field-changing books later, Kohut revisited the question that had animated his first foray into print in a posthumously published paper characterized by an aggressive, bristling contentiousness. He opened with a backward glance, recalling the intense, almost violent reactions and the severe critiques that his first brief for empathy had prompted among his colleagues, referring to the offenders by name. He charged his fellow analysts with having misunderstood his simple, clear scientific message; empathy as—perforce—the echt analytic mode of observation had been his interest, and he had neither identified it with a particular emotional stance such as compassion nor had he claimed it was intuitive or even always accurate. He took aim at those who, over the course of his career, had charged him with mysticism and sentimentality in his promotion of empathy, countering that it was beyond dispute that introspection and empathy served science—as "essential constituents of psychoanalytic fact finding," they were both productive of data and results. And he then rehearsed the essentials of the critique of Freud and analytic orthodoxy that he'd developed over the course of his phenomenally successful career. The central message of his old essay—that empathy, considered epistemologically, was "a value-neutral mode of observation"—was, he claimed, like Tristam Shandy's penis, still potent despite machinations that should have severed it, and he was now prepared to expose it to all in the marketplace of ideas. This was oddly militant imagery to invoke before an audience of analysts, versed as they were in the idiom of castration and its anxieties.[2]

The question of why any of this mattered—and why Kohut would brandish an epistemologically pure empathy as his analytic phallus—lands us squarely in the middle of a psychoanalytic revolution whose reverberations are still being felt within the discipline today. Kohut's brief for empathy as a mode of observation was but the opening wedge in what over the course of his career developed into a wide-ranging assault on the foundations of the mainstream classical American analytic tradition. In a series of analytic papers published in the 1960s, at the height of Freud's influence in the United States, and then in two landmark books, *The Analysis of the Self* and *The Restoration of the Self*, which appeared in the 1970s, Kohut challenged the primacy Freud had assigned to the drives in understanding human behavior and outlined a normal narcissism that was the wellspring of human ambition, creativity, values, and ideals. By the mid-1970s, he had established, in what he and his followers called "self psychology," a thoroughgoing alternative to classicism, with a new toolkit in introspective empathy, a new metapsychology organized around developmental deficits, a new institutional apparatus of journals, conferences, foundational texts, and, in Kohut himself, a charismatic leader surrounded by an entourage of enthusiastic colleagues. Kohut shook the American analytic establishment to its core, advancing what was widely hailed as a new analytic paradigm powerful enough to rival Freud's. It was Kohut's genius to effect his revolution within, not against, psychoanalysis such that he—unlike virtually every other dissident—was not banished from the analytic fold but, rather, incorporated, even lauded by some as the new Freud. His was the first successful psychoanalytic revolution, and self psychology today is part of the analytic mainstream. Empathy, his signature concept, did yeoman work in advancing his program.

Empathy is now one of those taken-for-granted psychological concepts—like identity and authenticity, for example—that are at once ubiquitous and ambiguously capacious in the literature, meaning so much that they are ever in danger of meaning nothing at all. These terms' capaciousness is part of their appeal, of course, usefully sheltering disagreements and cementing an otherwise unattainable working consensus that allows the day-to-day practice of science to proceed. Empathy is especially exemplary in this regard, the uncertainty of its conceptual status underwritten by an array of contestations around translation practices, institutionalized forgetting, and censorship even as it gained substantiality in becoming ever more indispensable a part of the analyst's armamentarium. As the psychoanalyst Theodor Reik—a critic of the term for years—noted in 1935, the term "sounds so full of meaning that people willingly overlook its ambiguity."[3] Half a century later the same complaint was still surfacing. Did empathy refer to a cognitive or an affective capacity? Was it intuitive or ploddingly learned? Impressionistic or scientific? Capricious or disciplined? A morally tinged good or a neutral mode of observation?[4]

In this paper, I trace empathy's analytic career from its early adumbration in the work of Freud's Hungarian colleague Sándor Ferenczi in the 1920s to its near-apotheosis in Kohut's psychology in the 1960s and 1970s. Empathy was never so neutral a tool as Kohut proposed. Early on, it was a flashpoint for a series of fierce disagreements between Freud and Ferenczi that divided the psychoanalytic field between the Freudian orthodox and revisionists—many of them Ferenczi's disciplinary descendants—for much of its history. In Kohut's hands, empathy was manifestly stripped of the moral and mystical resonances it had accrued since the Freud-Ferenczi split, mobilized to serve science in the guise of an unexceptionable mode of observation. Kohut insisted that empathy was essential to analytic fact-finding and that its ally, introspection, was among the analyst's scientific instruments, yet he would prove unable to resist touting its extraepistemological dimensions—especially but not only those having to do with therapeutic efficacy. It is hardly surprising, in light of empathy's history, that Kohut's attempts to secure it for science were successful only in part.

Ferenczi's Empathy Rule

To appreciate how high the empathy stakes are, we need only peer over Ferenczi's shoulder as he puts furious pen to private paper, casting himself as an "*enfant terrible*" in revolt against his once-beloved but now irredeemably hypocritical Freud.[5] The year is 1932. Freud and Ferenczi have been colleagues for more than twenty years. They have traveled together extensively (Ferenczi having joined Carl Jung in accompanying Freud on his 1909 visit to the United States), and have exchanged more than twelve hundred remarkably intimate letters. Ferenczi has been able to sustain a relationship with

Father Freud where other teachers, collaborators, and acolytes have been cast aside—Josef Breuer in 1895, Wilhelm Fliess in 1900, Jung in 1912, Otto Rank in 1924. But he has done this at great personal cost, having learned early on that Freud—while proclaiming to want mutuality—would brook neither independence nor dissension. Ferenczi's submission has secured his position as Freud's favorite, his proclaimed crown prince and "the most perfect heir of his ideas."[6] For the last decade, however, the "wise baby" of psychoanalysis has been playing the part of unruly adolescent to Freud's coolly restrained pater familias, adopting and advocating a number of "'revolutionary' technical innovations" that pushed at the limits of psychoanalytic orthodoxy and in consequence openly strained their relationship.[7] Where Freud would famously mandate that analysis was to be carried out "in a state of frustration," Ferenczi would respond to his patients' wishes. It took a particularly disdainful, even mocking, letter from Freud—dated 13 December 1931—to push Ferenczi to the break he knew was the price of his intellectual and emotional freedom.[8] Three weeks later, in "extreme revolt," he embarked on the private *Clinical Diary*, a long-suppressed document (published only in 1985) in which he gives vent to wellsprings of creative energy tapped in this last year of his life.

Ferenczi opened his private explorations contrasting the "unfeeling and indifferent" stance of the orthodox analyst he had once been with his evolving commitment to "*natural and sincere behavior*" as best suited to establish a favorable atmosphere for analysis. The "mannered form of greeting, formal request to 'tell everything,' so-called free-floating attention" that together constituted the orthodox analytic setting were inadequate, he held, to the intensity of the analysand's suffering, the last in particular ultimately amounting "to no attention at all."[9] The formal request to "tell everything" that Ferenczi invokes here refers to the demand made on the patient to speak freely, not to self-censor, in the analyst's presence, a technique central to the development of psychoanalysis that Freud elevated to the standing of "fundamental rule" in 1912.[10] In the same year, Freud first proposed "evenly-suspended attention"[11] as the analyst's preferred stance in his "Recommendations to Physicians Practising Psycho-Analysis," one in a series of six papers published between 1911 and 1915 that together constitute his "Papers on Technique," the sacred *fons et origo* of orthodox practice. A counterpart to the recommended "free association" on the ideally compliant patient's part, the analyst's evenly-suspended attention insured that he would not subject what the patient said to unconscious censorship. Rather, he would use his unconscious as an instrument—a receptive organ, as Freud put it, much like a telephone receiver—that was oriented to receive the "transmitting unconscious of the patient." Provided the analyst had "undergone a psycho-analytic purification" in the form of a training analysis, the risk of his distorting the patient's productions would prove minimal.[12]

Freud would admit to Ferenczi in 1928 that the recommendations on technique he had made fifteen years previously were essentially negative, emphasizing "what one should not do, to demonstrate the temptations that work against analysis." Everything positive, Freud wrote, he'd left unspecified, implicitly—he now claimed—relying on the analyst's tact, his "capacity for empathy," a concept Ferenczi had recently spoken about in a lecture to his Hungarian colleagues.[13] What had happened in the intervening years, however, was that—as Ferenczi ventriloquized Freud's voice in the published version of his own address, which appeared in print as "The Elasticity of Psycho-Analytic Technique"—"the excessively docile" among analysts failed to note the elasticity required of them and "subjected themselves to Freud's 'don't's' [*sic*] as if they were taboos." In 1928, Freud did allow that his recommendations were in need of revision. And he applauded Ferenczi's advocacy of elasticity in technique, the term referring to the analyst's yielding, "like an elastic band," to the pulls of the patient while pulling back himself, a construal of the analytic encounter in the register of give-and-take that Ferenczi wrote had been suggested to him by a patient.[14] But Freud would not follow Ferenczi in what he saw as the latter's concession to an arbitrary, impossible-to-control subjectivity on the part of the analyst. Those without a capacity for empathy, Freud worried, will exploit the analytic situation, giving rein to their "own unrestrained complexes." The analytic process consisted "first and foremost" in the analyst's "quantitative assessment of the dynamic factors in the situation," not in the mystical that he worried Ferenczi was promoting.[15]

Ferenczi replied to Freud that he demanded the subjective factor be strictly controlled, explaining that the approach consisted in putting oneself in the patient's position. "One must 'empathize' [*einfühlen*]," he proclaimed.[16] Ferenczi went so far as to formulate his own psychoanalytic rule, the "empathy rule," as an alternative to Freud's "fundamental rule," and recommended that the analyst foreswear the generally adopted lofty attitude of omniscience and omnipotence in favor of a more empathic stance. Empathy, Ferenczi explained, invoking imagery borrowed from the pathological laboratory, was knowledge derived from "dissection of many minds," most notably the analyst's own, that allowed one to envision the whole range of the patient's conscious and unconscious thoughts and associations. The analyst was to be guided not by feelings but by this capacity for coolly mobilized empathy. In the consulting room, he would find his mind—here the elasticity of technique comes into play—continuously swinging from empathy to self-observation to making judgments.[17]

The concept of empathy was not native to psychology but adumbrated earlier in the field of aesthetics, with the word *Einfühlung*—literally "feeling into"—first appearing in the 1873 doctoral dissertation of the German philosopher Robert Vischer. Vischer used the term to characterize the relationship between the viewer of art and the object itself, holding that whatever aesthetic qualities the former would claim to see in the latter were not inherent to it but, rather, projected onto it by her.[18] Theodore Lipps, professor of philosophy at Munich, endowed the term with more broadly psychological meanings, and was the first—to indulge in anachronism—to situate it in an interpersonal field (as opposed to the one-person psychology of aesthetic appreciation), specifically with the publication of his *Zur Einfühlung* in 1913.[19] Freud, an avid if at times envious reader of Lipps, in whose works he admitted he'd "found the substance of my insights stated quite clearly . . . , perhaps rather more so than I would like,"[20] used the word eight times in his *Jokes and Their Relation to the Unconscious*, published in 1905, a book inspired in large part by Lipps's own 1898 *Komik und Humor*. *Einfühlung*, as Freud explained later, refers to the process, similar to identification, that allows any one person to understand another, to "take up any attitude at all towards another mental life."[21] Although he used the term twelve more times in his published writings, the word "empathy"—consensually established as the English equivalent of *Einfühlung* by around 1920—appears but three more times in the English-language *Standard Edition*, due in part to the fact that James and Alix Strachey, who supervised the translation, found the word distasteful, in Alix's estimation "a vile word, elephantine, for a subtle process."[22]

Empathy was thus not as foreign to Freud's thinking as has long been assumed. It is possible that the received wisdom (abetted by the Stracheys' idiosyncratic aversion to the word) that Freud's only sporadic invocation of the concept now so central to analytic thinking supports—namely, that warm empathy was alien to the emotionally cold and distant Freud of the consulting room—is in need of some qualification. Most notably, in one of his papers on technique, Freud advised analysts that it was imperative to the success of a psychoanalytic treatment that they approach the patient from a position of empathy or *Einfühlung*, which appears as "sympathetic understanding" in the *Standard Edition* translation, a less subjective and robust emotional stance than he actually had in mind.[23] Indeed, one analyst has argued that Freud consistently used *Einfühlung* as a clinical—neither aesthetic nor psychological—concept, and that, notably, the *Standard Edition* translation fails to render it as "empathy" in such contexts.[24] But to posit an empathic, responsive, and nimble consulting-room Freud on the basis of misguided translation practices is to go too far, for Freud was also consistent in calling primarily on the intellectual dimensions of the term, and was throughout his life suspicious of the analyst's own emotions in the analytic setting.

Informing the minor contretemps between Freud and Ferenczi was the former's urgent 1912 recommendation to his colleagues that they model themselves on the surgeon, "who puts aside all his feelings, even his human sympathy." The "emotional coldness" of Freud's enjoining stood in stark contrast to Ferenczi's recommended empathy, and it was altogether consonant with his advocacy of the analyst as mirror to the patient's psyche and, more broadly, of psychoanalysis as primarily an intellectual exercise of interpretation. Freud maintained that the analyst's coldness allowed for maximal exploration of the unconscious material produced by the analysand while protecting the analyst's "own emotional life." The analyst's own individuality and any "intimate attitude" he might want to bring to the treatment were not aids to its progress but, rather, dangers that brought the specter of suggestion into the consulting room.[25] Suggestive influences might induce patients to produce material to please the analyst, but such influences were of no utility in uncovering what was unconscious, the psychoanalyst's quarry. Only the analyst's opacity to the patient would insure that unconscious material—material of which the patient was by definition unaware—would be made available for use in the treatment. Objectivity, neutrality, and disinterestedness on the part of the analyst were the watchwords of analytic technique as explicated by Freud in his papers on technique.

Yet, Freud was well aware that emotional coldness was in many cases inadequate to the task of gaining the patient's compliance. "The cure is effected by love," Freud had written to Jung years earlier, noting that only transference, by which he then meant the patient's love for the analyst, could provide the impetus necessary for patients to engage in analysis, with its uncomfortable exhumation of troubling unconscious material.[26] Patients give up their resistances "to *please us*," Freud told his Viennese colleagues, adding, "our cures are cures of love,"[27] with these words once again underscoring the instrumentally seductive nature of the analytic encounter.[28] Freud first characterized the love for the physician—specifically, in an early case of hysteria he treated, the desire to kiss him—that he witnessed among patients in treatment as in the nature of a "false connection."[29] By 1915, when he published his paper on the phenomenon, "Observations on Transference-Love," the patient's love for the physician had been transformed into a highly explosive force and endowed with a measure of reality, a genuine phenomenon.[30]

By Ferenczi's own telling, it was Freud's indifference to the therapeutic dimension of the analytic project that prompted his apostasy. Freud's indifference is by now well documented. His correspondence is punctuated with references to the toll exacted by patients, whom he characterized as dishonest and insatiable. He was in his own words "saturated with analysis as therapy" and "fed up," eager to limit how many patients he saw, he wrote to Ferenczi, "with the clear intent of tormenting myself less."[31] He once remarked in Ferenczi's presence that "patients are a rabble," serving only to provide analysts with their livelihoods and "material to learn from"[32]—giving voice to a therapeutic nihilism that Ferenczi found especially troubling. Freud's patience with neurotics in analysis was limited, he told Ferenczi, and "in life I am inclined to intolerance toward them."[33] These sentiments were privately conveyed. But Freud also went public with his doubts, in a 1933 publication proclaiming he'd "never been a therapeutic enthusiast,"[34] and four years later, in one of the last of his works to appear in his lifetime, "Analysis Terminable and Interminable," expounding on what James Strachey as editor of the piece defensively characterized as a well-established cool, even pessimistic, attitude toward psychoanalysis's therapeutic ambitions.[35] Freud in this essay dismissively brackets the question of what eventuates in cure as "sufficiently elucidated," preferring to focus instead on obstacles in the way of such cures, then going on to settle scores in adducing as evidence Ferenczi's failed analysis with him in support of this contention.[36] Ferenczi's overweening "need to cure and to help" had led him from the path of analysis to a "boundless course of experimentation," Freud wrote, adding that he had set himself aims "altogether out of reach to-day."[37]

Indeed, it was Ferenczi's fanatical *furor sanandi*, his rage to heal, that critics would see as the Achilles heel that led him from the analytic straight and narrow. Ferenczi, who bridled against the constraints on the analyst's behavior that flowed from Freud's technical recommendations, from his conviction that patients could not be helped, saw his own need to help as the motor driving his creative explorations. "Freud no longer loves his patients," he charged in his diary. Freud was intellectually but no longer emotionally invested in psychoanalysis, disdainful of patients and in the analytic setting "levitating like some kind of divinity" above them. He sought and found the causes of therapeutic failure in the patient, not, as Ferenczi would, in the analyst. Framing it as an issue of not abusing patients' trust, Ferenczi distinguished himself from Freud in his willingness to follow their lead, to relax Freud's precepts and to "be openly a human being with feelings" both positive and negative toward the patient.[38] Where Freud hated his patients, he would love them.

Ferenczi's apostasy reached its climax in a confrontation with Freud and analytic orthodoxy in September 1932. Stopping off to visit Freud on his way to a psychoanalytic congress in Wiesbaden, Ferenczi read aloud to him the paper that would be published the next year under the title of "Confusion of Tongues between Adults and the Child." In this paper, now considered a classic, Ferenczi homed in on the professional hypocrisy he'd been worrying in his *Diary*, hypocrisy manifest in the analyst's patent politeness in the presence of difficult—angry, reproachful, critical—patients whom he in fact found hard to tolerate and often in consequence disliked. Needy patients, many of whom had as children experienced adults as duplicitous, picked up on the analyst's disdain, only imperfectly covered over by his mannered graciousness, and were unwittingly put in the position of reexperiencing, sometimes in hallucinatory, trancelike states of dissociation, the traumas of inattention, abandonment, or sexual predation that had characterized their early years. Patients exhibit "a remarkable, almost clairvoyant knowledge" concerning their analyst's thoughts and emotions, Ferenczi maintained, and in the treatment setting the most damaged and needy of them responded much like children, not to intellectual explanations but to the analyst's sincerity, "only perhaps to maternal friendliness." Patients were better served by analysts who responded honestly to criticisms than by those who hid behind their own authority. It was with the former, who abjured complacency and admitted to the possibility of error, that patients could feel the confidence and trust necessary to approach the past "as an objective memory," not as a live trauma, and with whom they could begin the process of recovery. Pay attention to the ways you speak to your patients and pupils, Ferenczi advised: "loosen, as it were, their tongues."[39]

Ferenczi's attack on the coolly detached analytic persona was heretical enough. Coupled with his focus on the traumatizing effects of incestuous seductions and "real rape,"[40] issues that Freud had long preferred to treat in the register of fantasy, Ferenczi's "errors" were serious enough to merit his banishment from the analytic fold. Freud listened "thunderstruck"[41] to Ferenczi's disquisition, warning him that he was on dangerous ground in departing so from established psychoanalytic technique[42] and begging him not to deliver the paper. Even before Freud heard Ferenczi out, he was preparing to censor him.[43] After the fact he was furious, characterizing the paper in a letter to his daughter Anna as confused, contrived, stupid, and devious[44] and Ferenczi as harmless but stupid in a telegram to a Berlin loyalist sent the day after their meeting[45]—their last, it would turn out, ending with Freud declining to shake Ferenczi's hand offered "in affectionate adieu."[46] Colleagues who wanted to forbid Ferenczi from speaking at the congress joined Freud in predicting that scandal, even sensation, would ensue were the paper to be heard. Freud tried to stand between the paper and publication, writing to Ferenczi a few days later of his hopes that the latter would recognize the impropriety of his procedure and rectify himself.[47] Although it was published the following year in German, it was not until 1949 that it appeared in English translation in a "Ferenczi Number" of the *International Journal*, Jones's promise to Ferenczi he would publish it immediately notwithstanding.[48] It was Ferenczi's fate to be branded as psychotic and enter the mainstream analytic tradition a minor figure, a once-faithful, sometimes-brilliant disciple who regrettably had lost his way.

The Burden of Empathy

Ferenczi and his explorations of empathy were largely lost to psychoanalysis until, thirty years later, Kohut reopened the conversation. The term "empathy" was used only infrequently in the analytic literature before the appearance of Kohut's landmark 1959 article; following its publication, empathy appears with increasing frequency, a regular focus of interest and debate (though it was not until 1960 that any link between Ferenczi and empathy was established).[49] Ferenczi did not explicitly conceptualize empathy as a mode of observation. But there are ways in which his treatment of it and Kohut's are quite similar.

Ferenczi—much like Kohut—maintained that empathy should be conceptualized as in the service of science. He maintained that psychoanalysis had shown it was possible to understand mental processes, to methodically investigate the mind, by means of transmissible technique—not only, as some would insist, an inexplicable "faculty called knowledge of human nature." As he saw it, the development of technique put this understanding of human nature, formerly the province of artists and psychological geniuses, within reach of anyone "of only average gifts" willing to take the time and expend the effort to learn. As it was in other sciences, so it was in the realm of the mind, with "the mystical and the miraculous" displaced by "universally valid and inevitable laws." With the establishment of the training analysis, in which the prospective analyst was himself analyzed, what Ferenczi called the "personal equation" that was at the center of the analytic relationship was diminishing. Proper training insured that an array of observers of "psychological raw material" would all reach the same objective conclusions regarding it.[50] Whatever uncertainties came up in the course of a treatment—at what precise point the patient should be told of an interpretation, for example—that could not be spelled out in advance were a matter of the analyst's tact, or empathy.

Further, like Kohut would later, Ferenczi mounted a fierce attack on analytic orthodoxy around empathy, arguing it was a far better technique for gathering data than Freud's recommended free association and evenly-suspended attention. Freud warned Ferenczi that tact—empathy—should be divested "of its mystical character for beginners" who might use it to justify "the subjective factor" in analysis.[51] Ferenczi's response was that was precisely his aim—empathy was not premised on intuition but on "the conscious assessment of the dynamic situation."[52] Ferenczi in the 1920s and Kohut in the 1960s and 1970s found themselves parrying the linked charges of mysticism, subjectivity, and maternalism—all of which defined their revisionist stances against the proclaimed scientificity, objectivity, and father-centeredness of Freud. Writing in 1975, Kohut would suggest that analysts had long been ashamed of empathy as not scientific, that the early analyst especially had been "eager to distance himself from a demimonde of sentimental fuzziness, of tenderhearted perception." Offering a scientifically valid empathy as antidote, Kohut stressed it was emphatically "not a sex-linked capacity."[53] Rather, it—along with introspection—was a tool of empirical science, an instrument with which to explore interiority.[54]

To fully grasp how burdened empathy was in the analytic domain by the time Kohut revived it, it is necessary at this point to recreate one last scene in the Freud-Ferenczi drama—an episode that saw the maternal, gratification, love, and kissing woven together into one scandalous set piece. Ferenczi's attention to the mother in psychoanalysis set him squarely against Freud, and his explorations of "the mother-role of the analyst"[55] eventually opened him to perhaps the most notorious of the many charges leveled against him, that, as Freud pointedly put it to him in the famous letter of 13 December 1931, a letter that prompted Ferenczi's diary-writing—reproduced by Jones in his Freud biography, "you kiss your patients and let them kiss you." "Why stop with a kiss?" Freud went on. "Certainly, one will achieve still more if one adds 'pawing,' which, after all, doesn't make any babies. And then bolder ones will come along who will take the further step of peeping and showing, and soon we will have accepted into the technique of psychoanalysis the whole repertoire of demiviergerie and petting parties." It was not only the kissing that irked Freud. Rather, he objected to the "technique of maternal tenderness" in toto, holding it and Ferenczi up to ridicule.[56] "He is offended because one is not delighted to hear how he plays mother and child with his female patients," Freud wrote mockingly to a colleague.[57]

Ferenczi responded by defending his asceticism of technique, but the charge of employing the "kissing technique" stuck—despite the fact no evidence supports the contention Ferenczi did any kissing.[58] Discussed by Jones in the context of Ferenczi's purportedly delusional last days, his violent paranoia and homicidal outbursts, the charge stuck, passed down among analysts as a cautionary tale of therapeutic enthusiasms run amok in the name of indulgence and love.[59] Indeed, Kohut, in one of the very few mentions of Ferenczi in his entire corpus, dredged up the unpleasant "image of the aging Ferenczi, allowing his patients to sit on his knees" as he situated his own vaunted empathy under the sign of "scientifically trained cognition," distancing it from the soft humanitarianism associated with revisionist analysts from Ferenczi onward.[60] From the start, Kohut would conceptualize empathy as "a rigorously controlled tool of observation" and do what he could to stave off its distortion by do-gooders who could see in it only "an aim-inhibited form of love."[61] This was simply too close to the unscientific, sentimental "cure-through-love" with which too many had associated psychoanalysis for too long.[62]

Empathy in the Service of Science

No one in the analytic world is now more closely associated with the concept of empathy than Kohut, who in reviving empathy, associated as it was with the outcast Ferenczi, faced the difficulty of navigating between the Scylla of the predatory healer and the Charybdis of the cold, distant, scientistic Freud.

In his hands, from the start, empathy would be classed as a scientific—not a mystical or romantic—mode of knowing. Perhaps for this reason, Kohut rarely mentioned Ferenczi, despite the fact he was a keen reader of his work. Maintaining privately that Ferenczi's "gifts were second only to Freud,"[63] Kohut was clearly well acquainted with his works, crediting him with employing the introspective and empathic method that was his own signature innovation, with this acknowledgment hinting at a disciplinary lineage rooted not in Freud but in his banished colleague.[64] Yet Kohut was well aware of the disdain heaped upon Ferenczi by the orthodox, and even as he was championing empathy he distanced himself from the overly indulgent, gratifying Ferenczi of the literature.

It is worth returning at this point to the 1959 paper in which Kohut first advanced the case for empathy as what he called "a value-neutral mode of observation."[65] Kohut's focus on observation was largely lost in the many discussions of the paper following its publication. In the analytic literature spanning 1960 through 1980, empathy was mentioned in nearly eight hundred articles while observation was almost completely ignored.[66] Observation had never had much purchase in the analytic domain, which one early analyst ascribed on the one hand to the discipline's obsession with its therapeutic efficacy and on the other to a general disinclination to probe deeply into questions of method. Doing so, he contended, would show that psychoanalytic hypotheses were not premised on already known facts but were instead "the fruit of a new method of observation."[67] Half a century later, two analysts reached a similar conclusion that "the process of observation" was fundamental but largely taken for granted in psychoanalysis, Kohut's body of work on the topic proving the exception; regrettably, they noted, that work had promoted little discussion, with analysts merely assuming general agreement on to what the process referred.[68]

Kohut's argument was that the defining mark of the psychological was the centrality of introspection—defined as vicarious empathy—and empathy to its mode of observation (contrasting to this the physical, which could be known through the senses). Psychoanalysis was distinguished, he argued, by its "*scientific* use of introspection and empathy," in which Freud and his early collaborator Josef Breuer had, early on, excelled, and it could not be but a psychology employing an introspective-empathic stance.[69] The discipline had for too long entertained concepts rooted in biology—of the drives, for example—that were like so many foreign bodies in it. Kohut was well aware that introspection and empathy resonated with the non-Western and mystical, but he was determined to conscript both in the service of demonstrating that analytic depth-psychology posited a "new kind of objectivity, namely a scientific objectivity which includes the subjective."[70] Empathy was thus a scientific tool, in fact the only scientific tool available to the psychoanalyst. It was, he insisted, a "specific cognitive process."[71]

The radical implications of Kohut's initial 1959 brief for empathy were not immediately clear. In it, he relegated free association—Freud's fundamental rule—to an ancillary position in the analyst's armamentarium, among the "auxiliary instruments" to be mobilized in support of introspection and empathy.[72] Free association would be increasingly associated with the intellectual dimension of analysis, with a preference for thin insight and bloodless interpretation over robust engagement. Evenly-suspended attention was similarly demoted in the Kohutian analytic world, knocked from its pedestal to serve as mere handmaid to empathy, a technique that would focus the analyst's mind prior to empathy's supervention.[73] And, most daringly, Kohut took on the transference, arguing that the analyst did not function as a screen onto which the patient's internal structure was projected but a real presence and experienced as such. Psychoanalysis was like small particle physics, he would later suggest, with the analyst-as-observer part of the observational field. The discipline's objective truths only existed to the extent they accounted for the effects of the observational process. The analyst influenced the process "as an intrinsically significant human presence"—there were strong echoes of Ferenczi here.[74]

Critics would later suggest that Kohut was trading in ideal types, that Freud too had employed empathic understanding and, conversely, that even empathically attuned analysts perforce made extrospective observations—of behavior, in and outside of the consulting room.[75] As used in the post-Kohutian analytic literature, empathy signaled an immediacy of understanding between analyst and patient that was altogether fantastic, even undermining of the analytic process—understanding the meanings with which patients endowed events and narratives being among the aims of analysis. But to argue, as some critics have, that Freud's advocacy of free association derived from his doubts about the capacity of introspective self-observation to yield up the contents of the unconscious for analysis by analyst and patient alike is to engage in post-Kohutian anachronism. Freud was less troubled by the limitations of empathy than he was by the reflexive association of analysis with the practices of suggestion—telepathy and mediums, both of which he and Ferenczi discussed at length in their correspondence—that threatened to undermine psychoanalysis's hard-won scientific standing.[76] Kohut's critique of evenly-suspended attention recalls Ferenczi's charge that it amounts to "no attention at all."

Throughout his career, Kohut labored to keep empathy on the side of science. He asserted it was value-neutral—evident in the fact that the cunning among persons could mobilize it for hostile or destructive ends.[77] He stressed it was foremost a mode of data collecting, not to be called upon as a shortcut to understanding; the analyst-as-scientist properly ordered and scrutinized his assembled data just like any other scientist, calling on nonempathic capacities and skills.[78] He vacillated on whether empathy was an innate human capacity, close to intuition,[79] or a skill that could be developed only through assiduous training. Was it "God's gift bestowed only on an elect few?"[80] Was it—as one analyst suggested—"immaculate perception"?[81] Or was it a complex cognitive process?[82] What distinguished the skilled from the unskilled empathic observer? Critics argued that in Kohut neutral acceptance of empathy shaded into validating acceptance,[83] or, more strongly, that in Kohut's advocacy of empathy could be discerned an endorsement of gratification.[84] It is not surprising, in light of the history of empathy's psychoanalytic fortunes, that Kohut would invoke its normative dimension in the realm of the therapeutic, arguing that its "mere presence" had salutary effects.[85] Empathy was simply too protean a concept to resist this sort of conscription.

Kohut's invocation of empathy, as we've seen a fraught concept in the history of analysis, spawned decades of vigorous discussion and dissent, much of it focused around gratification and provision in the analytic setting—the same complex of issues mobilized as grounds for Ferenczi's exclusion from the analytic canon—and correspondingly little focused on the term's observational dimension, despite the fact this was just as subversive of classical analysis. Kohut defined the scope of the analytic field operationally, as comprising phenomena that were observable deploying empathy and introspection—"the observational tool of a new science." The method, he argued, defined "the contents and limits of the observed field," and errors and inaccuracies resulted when this connection between observational mode and theory was ignored. Consider his treatment of the drives, which were foundational to Freud's psychology. To be human, in a Freudian world, was to be roiled by lust and aggression; to be civilized was to constantly negotiate between satisfying these inner demands on the one hand and tamping them down on the other. Kohut argued that while Freud's "drive concept" was "derived from innumerable introspected experiences," a drive in itself was an abstraction that could not be observed, like other of Freud's concepts existing "in a no man's land between biology and psychology."[86] Later, his pen sharpened, the drive would become "a vague and insipid biological concept" that had had "significant deleterious consequences for psychoanalysis" and was, further, not properly part of it.[87] If analysts hewed more closely to what was knowable through the means of observation to hand, they would recognize that what they could observe, by means of vicarious introspection, was not the drive but the self, the person—loving, lusting, asserting.

Kohut explained that in 1959 he had naively assumed that every scientist proper shared with him the basic epistemological stance "that an objective reality is in principle unreachable and that we can only report on the results of specific operations." Classical analysis, from this perspective, was positivist in orientation, with the analyst functioning as arbiter of a reality that from the self psychological perspective did not in fact exist. Indeed, Kohut went so far as to propose that classical analysis and his self psychology stood in relation to each other much as did Newtonian physics and the physics of Max Planck.[88] Where the Freudian analyst was mechanistic and deterministic, seeking truth through intense scrutiny, Kohutians theorized themselves not as screens onto which patients projected their realities but as inextricably part of the observational field. In Kohut's construal, Freud's was a truth morality, and the patient of classical psychoanalysis correspondingly sought insight and mastery, whereas from his own perspective other values and aims mattered more.[89] The only analytic truth was that produced by, or between, analyst and patient, and it was necessarily limited by the former's access by means of empathy to the latter's interiority. Kohut's proclaimed interest was in "a new kind of objectivity" that would encompass the subjective,[90] and his explorations of this—of the point where observation and objectivity met—were as, if not more, antithetical to the foundations of classical analysis than his explorations of what we might see as the more sentimentally tinged dimensions of empathy, the empathy that lay observers might see as unqualified good. Psychoanalysts ought not be ashamed of empathy, Kohut held, for they, like all other empirical scientists, were observers of data that they sought to bring into causal relationship in pursuit of explanations. Other scientists might exclude them from the community of scholars, pridefully flaunting the superiority of their methodologies, but the analyst deploying "the irreplaceable instrument of observation" was as entitled as any other practitioner to the status of scientist.[91]

Whether or not Kohut's scientizing program for psychoanalysis, organized around empathy as a mode of observation akin to those deployed in other sciences, succeeded in finally securing the discipline's scientific status is an open, still debated question that is in any case beyond the historian's ken. What can be established is that psychoanalytic partisans of Kohut's self-psychological revolution saw his methodological and epistemic moves unmooring their discipline from its nineteenth-century positivist and mechanistic framework and moving it toward a twentieth-century constructivist orientation.[92] Talk of Heisenberg's uncertainty principle crept into the analytic literature, with one analyst going so far as to assert in a flourish of optimism a unity between microphysics and psychoanalysis on the question of observation, with both sciences equally attentive to "the interaction between observing instruments and the objects."[93] But the issue would not be so readily settled. However much Kohut wished to divest empathy of its therapeutic and moral resonances, these were too deeply interwoven into its history in the analytic realm. Kohut wrestled with this to the end of his life, in his last public address admitting—after militantly defending his career-long position that empathy was properly considered as the defining analytic mode of observation—that it referred also to a therapeutic action, that it was beneficial "in the broadest sense of the word." He added: "That seems to contradict everything I have said so far, and I wish I could just simply bypass it. But, since it is true, and I know it is true, and I've evidence for it being true, I must mention it."[94]

Notes

1. Heinz Kohut, "Introspection, Empathy, and Psychoanalysis—An Examination of the Relationship Between Mode of Observation and Theory," *Journal of the American Psychoanalytic Association* 7 (1959): 459–83.

2. Kohut, "Introspection, Empathy, and the Semi-Circle of Mental Health," *International Journal of Psychoanalysis* 63 (1982): 395–407.

3. Theodore Reik, *Listening with the Third Ear* [1935] (New York: Grove Press, 1948), quoted by Michael Franz Basch, "Empathic Understanding: A Review of the Concept and Some Theoretical Considerations," *Journal of the American Psychoanalytic Association* 31 (1983): 101. See also George W. Pigman, "Freud and the History of Empathy," *International Journal of Psychoanalysis* 76 (1995): 237–56.

4. Questions put by Ralph R. Greenson, "Empathy and Its Vicissitudes," *International Journal of Psychoanalysis* 41 (1960): 418–24; Basch, "Empathic Understanding," and "Kohut's Contribution," *Psychoanalytic Dialogues* 5 (1990): 367–73; Peter Shaughnessy, "Empathy and the Working Alliance: The Mistranslation of Freud's *Einfühlung*," *Psychoanalytic Psychology* 12 (1995): 221–31.

5. "*Enfant terrible*": Ferenczi, "Child-Analysis in the Analysis of Adults" [1931], in Ferenczi, *Final Contributions to the Problems and Methods of Psycho-Analysis*, ed. Michael Balint, trans. Eric Mosbacher (London: Hogarth Press, 1955), 127.

6. Ferenczi, 4 Aug. 1932, in *The Clinical Diary of Sándor Ferenczi*, ed. Judith Dupont, trans. Michael Balint and Nicola Zarday Jackson (Cambridge, Mass.: Harvard University Press, 1988), 184–87.

7. The notion of the "wise baby" was first adumbrated in Ferenczi, "The Dream of the 'Clever Baby'" [1923] in Ferenczi, *Further Contributions to the Theory and Technique of Psycho-Analysis* [1926], comp. John Rickman, trans. Jane Isabel Suttie (London: Hogarth Press, 1950), 349–50, and it appears in several of his subsequently published papers. Quotation from Ferenczi, 9 July 1932, *Clinical Diary*, 160.

8. Quotation from Freud, "Analysis Terminable and Interminable" [1937], in Freud, *The Standard Edition of the Complete Psychological Works of Sigmund Freud*, 24 vols., ed. James Strachey in collaboration with Anna Freud, assisted by Alix Strachey and Alan Tyson (London: Hogarth Press and the Institute of Psycho-Analysis, 1953–74), 23: 231; Freud to Ferenczi, 13 Dec. 1931, discussed below; all Freud-Ferenczi quotations in *The Correspondence of Sigmund Freud and Sándor Ferenczi. Volume 3, 1920–1933*, ed. Ernst Falzeder and Eva Brabant, trans. Peter T. Hoffer (Cambridge, Mass.: Harvard University Press, 2000).

9. Ferenczi, 7 Jan. 1932, *Clinical Diary*, 1.

10. Freud, "The Dynamics of Transference" [1912], *Standard Edition*, 12: 107.

11. In German, "*gleichschwebende Aufmerksamkeit*," often translated as "free-floating attention."

12. Freud, "Recommendations to Physicians Practising Psycho-Analysis" [1912], *Standard Edition*, 12: 111, 115–16.

13. Freud to Ferenczi, 4 Jan. 1928, *Correspondence of Freud and Ferenczi*, 331–33. Ferenczi incorporated portions of Freud's letter in his essay, "The Elasticity of Psycho-Analytic Technique" [1928], in Ferenczi, *Final Contributions*, 99. "Capacity for empathy" is Ferenczi's rendering of Freud's "tact."

14. Ferenczi, "Elasticity," 99, 95.

15. Freud to Ferenczi, 4 Jan. 1928, *Correspondence of Freud and Ferenczi*, 331–33.

16. Ferenczi to Freud, 15 Jan. 1928, ibid., 334–35 (brackets in English translation).

17. Ferenczi, "Elasticity," 87–101.

18. Harry Francis Mallgrave and Eleftherious Ikonomou, introduction to Robert Vischer et al., *Empathy, Form, and Space: Problems in German Aesthetics, 1873–1893* (Santa Monica: Getty Center for the Humanities, 1994), 1–85.

19. See M. J. Blechner, "Epistemology: Ways of Knowing in Psychoanalysis (Panel Presentation)—Differentiating Empathy from Therapeutic Action," *Contemporary Psychoanalysis* 24 (1988): 301–10.

20. Freud to Wilhelm Fliess, 31 Aug. 1928, in *The Complete Letters of Sigmund Freud to Wilhelm Fliess, 1887–1904,* trans. and ed. Jeffrey Moussaieff Masson (Cambridge, Mass.: Harvard University Press, 1985), 325; also quoted in M. Kanzer, "Freud, Theodor Lipps, and 'Scientific Psychology,'" *Psychoanalytic Quarterly* 50 (1981): 395.

21. Freud, "Group Psychology and the Analysis of the Ego" [1922 (1921)], *Standard Edition,* 18: 110.

22. Alix Strachey to James Strachey, 2 Jan. 1925, in *Bloomsbury Freud: The Letters of James and Alix Strachey, 1924–1925,* ed. Perry Meisel and Walter Kendrick (New York: Basic Books, 1985), 170–71; also quoted in Pigman, "History of Empathy," 244.

23. Freud, "On Beginning the Treatment (Further Recommendations on the Technique of Psycho-Analysis I)" [1913], *Standard Edition,* 12: 140.

24. Pigman, "History of Empathy."

25. Freud, "Recommendations to Physicians," 115, 118.

26. Freud to Jung, 6 Dec. 1906, in *The Freud/Jung Letters: The Correspondence between Sigmund Freud and C. J. Jung,* ed. William McGuire, trans. Ralph Mannheim and R. F. C. Hull (Princeton: Princeton University Press/Bollingen, 1974), 11–13.

27. Meeting of 30 Jan. 1907, in *Minutes of the Vienna Psychoanalytic Society. Volume 1: 1906–1908,* ed. Herman Nunberg and Ernst Federn, trans. M. Nunberg (New York: International Universities Press, 1962), 101.

28. For a treatment of the seductiveness constitutive of the analytic setting, see Glen O. Gabbard, "The Analyst's Contribution to the Erotic Transference," *Contemporary Psychoanalysis* 32 (1996): 249–73.

29. Josef Breuer and Sigmund Freud, *Studies on Hysteria* [1893–95], *Standard Edition,* 2: 302. Also cited by Gabbard, "The Early History of Boundary Violations in Psychoanalysis," *Journal of the American Psychoanalytic Association* 43 (1995): 1115–36.

30. Freud, "Observations on Transference-Love (Further Recommendations on the Technique of Psycho-Analysis III)" [1915 (1914)], *Standard Edition,* 12: 157–71.

31. Freud to Ferenczi, 11 Jan. 1930 and 11 Oct. 1920, *Correspondence of Freud and Ferenczi,* 380–81, 34–36.

32. Ferenczi, 1 May 1932, *Clinical Diary,* 92–95.

33. Freud to Ferenczi, 20 Jan. 1930, *Correspondence of Freud and Ferenczi,* 385–86.

34. Freud, *New Introductory Lectures on Psycho-Analysis* [1933 (1932)], *Standard Edition,* 22: 151.

35. Strachey, "Editor's Note," to Freud, "Analysis Terminable and Interminable" [1937], *Standard Edition,* 23: 211–15.

36. Freud, "Analysis Terminable and Interminable," 221–23.

37. Freud, "Sándor Ferenczi" [1933], *Standard Edition,* 22: 229; idem, *New Introductory Lectures,* 153; idem, "Sándor Ferenczi," 229.

38. Ferenczi, 1 May 1932, *Clinical Diary,* 92–95.

39. Ferenczi, "Confusion of Tongues between Adults and the Child" [1933], in Ferenczi, *Final Contributions,* 156–67.

40. Ibid., 161.

41. Freud, *The Diary of Sigmund Freud, 1929–1939: A Record of the Final Decade,* trans. Mi-

chael Molnar (New York: Scribner's, 1992), citing Freud to Anna Freud, 3 Sept. 1932, 131; also cited in Robert Kramer, "The 'Bad Mother' Freud Has Never Seen": Otto Rank and the Birth of Object-Relations Theory," *Journal of the American Academy of Psychoanalysis* 23 (1995): 312; and Arnold Wm. Rachman, "The Suppression and Censorship of Ferenczi's Confusion of Tongues Paper," *Psychoanalytic Inquiry* 17 (1997): 473.

42. Ferenczi's report of the meeting, given to Izette de Forest, author of *The Leaven of Love* (New York: Harper & Brothers, 1954), who in turn passed it on to Erich Fromm, who published it in his biography of Freud, *Sigmund Freud's Mission: An Analysis of His Personality and Influence* [1959] (New York: Harper & Row, 1972), 62–65.

43. Consider Freud to Max Eitingon, 29 Aug. 1932: "He must be prevented from reading his essay. . . . Either he will present another one, or none at all": in B. Sylwan, "An untoward event: Ou la guerre du trauma de Breuer à Freud de Jones à Ferenczi," *Cahiers Confrontation* 12 (1984): 108, cited in Rachman, "Suppression," 471.

44. Freud to Anna Freud, 3 Sept. 1932, in Freud, *Diary*, 131.

45. Freud to Max Eitingon, 2 Sept. 1932, *Correspondence of Freud and Ferenczi*, 442 n. 1.

46. Fromm, *Freud's Mission*, 65.

47. Freud to Ferenczi, 2 Oct. 1932, *Correspondence of Freud and Ferenczi*, 444–45.

48. Rachman, "Suppression," 474–75.

49. In Greenson, "Empathy and Its Vicissitudes," 418.

50. Ferenczi, "Elasticity," 87–89.

51. Freud to Ferenczi, 4 Jan. 1928, *Correspondence of Freud and Ferenczi*, 331–33.

52. Ferenczi, "Elasticity," 100.

53. Kohut, "The Future of Psychoanalysis," *Annual of Psychoanalysis* 3 (1975): 335–37.

54. Kohut, "Psychoanalysis in a Troubled World," *Annual of Psychoanalysis* 1 (1973): 14.

55. Ferenczi to Freud, 1 Sept. 1924, *Correspondence of Freud and Ferenczi*, 168–72.

56. Freud to Ferenczi, 13 Dec. 1931, ibid., 421–24.

57. Freud to Eitingon, 18 April 1932, in Ernest Jones, *The Life and Work of Sigmund Freud* (New York: Basic Books, 1957), 3: 171.

58. Not often noted was that Ferenczi did not kiss patients—rather, he allowed one, Clara Mabel Thompson (1893–1958; a well-known American analyst), to occasionally kiss him. See Ferenczi, 7 Jan. 1932, *Clinical Diary*, 1–4. See *Correspondence of Freud and Ferenczi*, 423 n. 2; and Ferenczi, *Clinical Diary*, 3 n. 3, for explications of the incident that led to the charge. Thompson had, in Ferenczi's words, "occasionally even kissed me," and had taken to claiming publicly that she was "allowed to kiss Papa Ferenczi, as often as I like": 7 Jan. 1932, ibid., 2.

59. Jones, *Sigmund Freud*, 3: 176–78.

60. Kohut, "Future of Psychoanalysis," 339–40.

61. Kohut to Kurt R. Eissler, 18 April 1974, in Heinz Kohut, *The Curve of Life: Correspondence of Heinz Kohut, 1923–1981*, ed. Geoffrey Cocks (Chicago: University of Chicago Press, 1994), 306.

62. Kohut, "Autonomy and Integration," *Bulletin of the American Psychoanalytic Association* 21 (1965): 854; idem, "The Psychoanalyst in the Community of Scholars," *Annual of Psychoanalysis* 3 (1975): 356.

63. Kohut to John E. Gedo, 26 Oct. 1966, in Kohut, *Curve*, 152–54.

64. See, for example, Kohut, *The Restoration of the Self* (Chicago: University of Chicago Press, 1977), 306 n. 14, where he quotes Ferenczi writing that "inner forces can only be perceived through introspection."

65. Quote from Kohut, "Semi-Circle of Mental Health," 396.

66. James H. Spencer and Leon Balter, "Psychoanalytic Observation," *Journal of the American Psychoanalytic Association* 38 (1990): 393.

67. Theodore Schroeder, "The Psycho-Analytic Method of Observation," *International Journal of Psychoanalysis* 6 (1925): 156. Schroeder's article is an exploration of empathy in the domain of the psychological and analytic; it has been cited only three times in the analytic journal literature.

68. Spencer and Balter, "Psychoanalytic Observation," 394.

69. Kohut, "Semi-Circle of Mental Health," 398. Leon Balter and James H. Spencer, "Observation and Theory in Psychoanalysis: The Self Psychology of Heinz Kohut," *Psychoanalytic Quarterly* 60 (1991): 364–65, argue that Kohut's definition of the psychological is limiting, excluding, for example, behaviorism, which employs "extrospection"—a term Kohut used—as an observational tool.

70. Kohut, "Semi-Circle of Mental Health," 399, 400, reflecting here on what he did not articulate clearly enough in his 1959 paper.

71. Kohut, "Community of Scholars," 356.

72. Kohut, "Introspection, Empathy, and Psychoanalysis," 464.

73. Kohut, "Forms and Transformations," 263.

74. Kohut, *How Does Analysis Cure?* ed. Arnold Goldberg (Chicago: University of Chicago Press, 1984), 37.

75. For example, Balter and Spencer, "Observation and Theory."

76. Indeed, Marjorie Brierley wrote in 1937 that in her mind, "empathy, true telepathy, is indispensable to sound analysis": "Affects in Theory and Practice," *International Journal of Psychoanalysis* 18 (1937): 267.

77. Kohut, "Semi-Circle of Mental Health," 396.

78. Kohut, *The Analysis of the Self: A Systematic Approach to the Psychoanalytic Treatment of Narcissistic Personality Disorders* (Madison, CT: International Universities Press, 1971), 300–1.

79. Philip S. Holzman, "Psychoanalysis: Is the Therapy Destroying the Science?" *Journal of the American Psychoanalytic Association* 33 (1985): 754–55.

80. Kohut, *How Does Analysis Cure?* 83. For empathy as innate, from the mother, see "Forms and Transformations," 262; as the product of training, "Community of Scholars," 352; as intuitive, *Analysis of the Self*, 302–3. See also Ferenczi, "Elasticity."

81. Donald P. Spence, "Perils and Pitfalls of Free Floating Attention," *Contemporary Psychoanalysis* 20 (1984): 38.

82. Basch, "Empathic Understanding," 111.

83. Balter and Spencer, "Observation and Theory," 372.

84. Theodore Shapiro, "Empathy: A Critical Reevaluation," *Psychoanalytic Inquiry* 1 (1981): 423–48.

85. Kohut, "Semi-Circle of Mental Health," 397.

86. Kohut, "Introspection, Empathy, and Psychoanalysis," 464, 465, 460, 479.

87. Kohut, "Semi-Circle of Mental Health," 401.

88. Kohut, *How Does Analysis Cure?* 36, 41–42.

89. Ibid., 54ff.

90. Kohut, "Semi-Circle of Mental Health," 400.

91. Kohut, "Community of Scholars," 353.

92. See, for example, Paul H. Ornstein, "Chapter 1. Introduction. Is Self Psychology on a Promising Trajectory?" *Progress in Self Psychology* 9 (1993): 4.

93. Maxwell S. Sucharov, "Chapter 11. Quantum Physics and Self Psychology: Toward a New Epistemology," *Progress in Self Psychology* 8 (1992): 202.

94. Kohut, "On Empathy (1981)," in Kohut, *The Search for the Self: Selected Writings of Heinz Kohut: 1950–1978*, ed. Paul H. Ornstein (Madison, CT: International Universities Press, 1991), 4: 530.

PART FOUR

Observing New Things: Objects

The essays in part 4 show us observers training their eyes on and devising ways of apprehending new phenomena. Some of these, like "society" and "the entire economy," were newly defined in the eighteenth and nineteenth centuries. Such abstractions could not be grasped in any possible visual field but could nonetheless be usefully plumbed, even if piecemeal, by expert observers. Others, like emotions and radiation, were ephemeral and invisible. Resourceful researchers devised ways to capture their traces and establish their reality. Observation in these essays—variously allied with measurement, recording, classification, and visualization—is a remarkably complex endeavor that we can here see broken down into its constituent operations.

The question of how best to observe the abstract entity known now as society animates Theodore M. Porter's essay on the conservative nineteenth-century French social investigator Frédéric Le Play. Le Play, a pioneer in establishing the genre of the social survey, was trained as an engineer in the postrevolutionary École Polytechnique, where he imbibed a reformist, problem-solving ethic and developed a penchant for statistics and empirical observation. This was quickly supplemented by a singular ability to "see" social relations—for example, in an 1840 study of a mining enterprise in the Harz Mountains, the webs of customary obligation and commercial exchange, of abstract share prices and concrete minerals that knitted together *patron* and worker into an organic whole. Le Play increasingly questioned the utility of statistical information, branding it as impersonal, collected by faceless officials and mobilized by the liberally disposed to address the poverty and perceived moral disorder of the poor in an age of unbridled capitalism. At the same time, he championed observations made in the context of hierarchical personal relations. A traveler extraordinaire, he sought out sites where tradition flourished and where rich and poor lived as members of an integrated community. From Siberia to Sweden to Spain, he carefully observed ordinary workers—whether it was the tradition-bound herder or the rootless modern nomad—and formulated a typology of progress that turned on a calculus of increased freedom on the one hand and moral danger on the other. He rejected a priori theory in favor of systematic observation of social realities, and modeled the social sciences—in their embrace of facts—on the natural sciences. Yet, in the social and political tumult of the 1860s and 1870s, Le Play recoiled. Even as this most cosmopolitan of observers continued to travel far and wide, to witness and record, in his writings he elevated ancient custom over the modern observational social science he'd been so important in defining.

Porter traces the arc of Le Play's career from engineer, that exemplary figure of modernity, through prolific, empirically minded social observer, to self-appointed venerable sage, seeking truth in the preliterate. He shows us a Le Play whose moral engagements overwhelmed his observational commitments, stressing how singular and paradoxically enduring these were. Along the way, Porter invites the reader to question the well-established opposition between dry statistics and personal observation, underscoring just how full of well-observed, and often chilling, specifics about factory conditions and life of the poor the parliamentary Blue Books were—even though they have come to epitomize the bloodlessness of the impartial social investigator confronted with the horrors of industrialism. Le Play admired the quality of the investigations the Blue Books transmitted to the public, and so too, Porter suggests, did Charles Dickens. Social investigator and novelist alike traded in vivid portrayals of experience. Indeed, in Porter's hands, the roots of social science are multiple and dispersed, to be located in Blue Book and novel, in social reportage and story, in on-the-ground observational practices as well as in the theoretical formulations that figure in standard accounts of disciplinary formation.

With Mary S. Morgan's essay we shift our lens from society to the economy, an entity that in its entirety economists first attempted to observe and measure in the 1930s and 1940s. Routinely conflated in twentieth-century economic thought, observing and measuring were at once distinct and interdependent operations: observing the economy—the complex relationships between parts and whole prompting analogies with mapping, doing jigsaw puzzles, and playing chess—was premised on first measuring its parts, which was itself premised on being able to see what it was that was being measured. It is this seeing that proves so unexpectedly complex here. Morgan's subjects devised a system of national income accounting (NIA) to guide researchers as they constructed pictures of nations' economies, instructing them on what was to be observed—incomes, products, expenditures—and how their observations were to be classified and recorded. NIA worked well enough for the Western economies for which it was developed. But, as the economist Phyllis Deane found out both in compiling statistics and in visiting Northern Rhodesian villages herself, in measuring only that economic activity that occurred outside the household, in the market, it proved completely inadequate to capture the vibrancy of African economies. The capaciously conceived households she observed scrambled economists' categories: was the household where one slept, ate, produced, or spent? In village economies, little money changed hands but economic activity was everywhere, with domestic relations between men and women commercialized, sites of negotiated exchange. Observation here demanded an eye alert to the anthropological as well as to the economic, imagination in addition to technical skill.

The epistemological and practical issues encountered by those who would observe human emotions, arguably among the most ephemeral of observables, are explored in Otniel E. Dror's essay here. The story he narrates is at once straightforward and surprising. On the one hand, we have a relatively familiar story of the development of apparatuses and techniques that displace the skilled human observer, minimizing the subjective element in order to yield new, patently objective, and quantifiable measures of what lay within in graphic form. Nineteenth-century observers of the emotions had relied for the most part upon their own emotional reactions to their subjects to gauge what the latter were feeling, mobilizing sympathy in the service of science. Sympathy, first cast as an innate capacity, then later as a form of disciplined experience, allowed one to experience what was roiling the other; imitation of the other's outward demeanor prompted a parallel mirroring of the other's interior. Observation of the other was in effect a form of self-observation. Newly developed late nineteenth-century technologies promised to routinize this transaction, hedged about as it was with associations of an excessive and by definition feminine sensitivity. Rather than plumbing their own subjective reactions, investigators could now deploy instruments such as the plethysmograph to measure visceral blood flows in the interior of their subjects' bodies and the cardiograph to make the heart's secrets visible. Subjectivity was sidelined, and the man of science could relegate the feminized work of feeling the other's pain to the machine. The dream, as one investigator put it, of "measuring directly the feelings of the human heart," appeared within reach.

On the other hand, Dror suggests that the machine did not so much displace sensitive, embodied observers as it internalized them, taking on the many registers in which they had practiced. Machines for observing emotion had to match the observational acuity of the most sensitive of persons; Ivan Pavlov, for example, argued against "very sensitive" vivisectionists that his machines could guarantee that his dogs felt no pain. The machine as exteriorized entity would ideally do the work of feeling that embodied observation demanded of investigators. Their reactions to subjects in pain would no longer disturb the observational field. Yet, as Dror shows, sympathy would not so easily be banished; in the laboratory setting, it turned out, emotions could overwhelm the technologies designed to capture them. Distressed, angry, fearful, and excited experimental subjects prompted reenactments of the premachine age observational mise-en-scène, with the observer's own fearful reactions once again forcefully inserted into the picture.

The technologies associated with photography figure prominently in Kelley Wilder's account of the Nobel Prize–winning scientist Henri Becquerel's attempts to represent in pictorial form the radioactive rays he discovered. Wilder follows Becquerel into the laboratory, watching him experiment with an array of different ways of capturing the elusive rays on film—mostly notably, he turns his photographic plates on their sides to produce precisely the images of the rays he wants. Photography here is not a passive but a volatile medium, demanding technical virtuosity and open to creative manipulation. Photographic images, likewise, are not straightforward representations of reality but ingeniously constructed exhibits. Images produced *by* X-rays were everywhere at the turn of the twentieth century; Becquerel, eager to distinguish his rays from the novel X-rays as well as to establish that they did indeed exist, photographed the rays themselves. In successfully doing so, he proved himself an exemplary observer of the invisible.

11

Reforming Vision: The Engineer Le Play Learns to Observe Society Sagely

THEODORE M. PORTER

In an 1840 inquiry into abuses of child labor, a committee of the British parliament voted four to two against requiring their expert witness to answer the question: "Have you seen the work of Mrs. Trollope, 'The Factory Boy,' and by your means as an Inspector, have you formed any opinion as to the correctness of her statements?"[1] The members of the committee clearly had formed opinions, for they treated her novel as commensurable with the eyewitness testimonies and statistical tables provided to them by factory inspectors. As free and responsible men, they did not limit themselves to evidence that would stand up in court, but recognized that the most crucial facts would often be concealed from them. Such assumptions regarding observation and social knowledge were typical of social science in its nineteenth-century form, which did not fear to infer from the results of local experience, often transmitted in stories.

Frédéric Le Play, whose career in social science began at just this time, shared their sense of the fluidity of social observation. He was, indeed, a particular admirer of British parliamentary inquiries, which he commended as more direct and more probing than statistics, yet, like statistics, resolutely empirical. Le Play's family budgets, the foundation of his monographic method of social science, embodied an intensely personal mode of observation, rooted in what he construed as traditional ties of seigneurs and *patrons* to the laboring poor. During the course of his career, he gave increasing preference to the inherited wisdom of the sage over methodical, cosmopolitan science as the ideal form of social observation. In this way, he unshackled truth from well-attested facts and upheld the epistemic validity of narrative.

In his maturity, then, Le Play claimed his rights as the modern founder of a scheme of observation at odds with the social doctrines and methods of liberal Europe. Yet to us it is clear that he could not stand outside the nineteenth century. His celebrated *méthode d'observation*, while steeped in archaism, drew from diverse observational practices of administration and reform. It began with travel, an indispensable practice in the engineering profession for which he was trained.[2] His investigations of new technologies of mining opened out into the examination of the skills and knowledge of craftsmen and the statistics of trade. The economy of mines included workers and their families as well as machinery, forests, and mineral deposits. He modeled his family budgets on statistical tables of trade, and he compared his interviews with miners and factory laborers to the work of parliamentary committees.

For a decade beginning about 1840, Le Play observed the economy of labor and markets through the lens of socialist politics. Later, he worked to refashion social science as profoundly conservative in its consecration of an imagined social order—which he worked tirelessly to establish as reality—anchored in paternalism, moral faith, and personal bonds of loyalty rather than in labor markets and bureaucratic rationality. The recovery of this kind of society demanded a new mode of observation, the study of cases as the basis for a tradition of inquiry, like those of law and medicine examined in this volume by Gianna Pomata and J. Andrew Mendelsohn. In his maturity, Le Play's "social science" looked outward in order to look backward, preferring oral traditions and personal engagement to detached inquiry.

Engineering Roots

Le Play was one of those early practitioners of social science—their name is legion—who folded his own path to intellectual enlightenment into the rationale for his science. The hero of his reminiscences was a young engineer, highly successful in his schooling but dissatisfied with the Saint-Simonian ideas that had seduced his fellow Polytechnicians, who benefited from the guidance of revered teachers in 1829 as he prepared to set off on his mining investigations. In the Harz Mountains in Germany, where they sent him, mining officials opened his eyes to a society organized on unfamiliar principles. His epiphany came in the form of a complex observation: that in the Harz mines the purpose of production was not wealth, but the maintenance of a God-fearing population. "Il n'y a rien à inventer," he declared much later; nothing new is required to fashion a social science that can relieve the terrible suffering of the present. We have only to observe, or, more precisely, to observe as of olde—not from a distance, but with the personal concern and charity of traditional elites. At their feet he had learned social science.[3]

By telling his story this way, Le Play accented a profound social conservatism as the polestar of his entire career. He also sharply distinguished his vaunted *méthode d'observation* from other modes of social science. He epitomized his career from the vantage point of its end as a series of voyages across eastern and northern Europe as well as industrialized Britain, in pursuit, we may say, of *temps perdus*, of all the old customs and unregulated relationships among unequals that stimulated men to observe God's commandments and assured them of their daily bread. A proper reverence for social observation, he held, undergirded the benevolent authority of fathers in the home and of *patrons* in the workplace. The best observers could be found in societies that had preserved the ancient virtues and had no need of reform. But France had sacrificed its traditions on the altar of revolution. Now, he wrote in 1864, the moment had arrived "to replace the conflicting theories that since 1789 have agitated it with opinions founded on methodical observation of social facts."[4]

Le Play spent his career in the Corps des Mines, the most elite of French engineering corps. As graduates of the École Polytechnique, he and his comrades were highly trained in mathematics, yet their ethos was more professional and administrative than scientific. At the School of Mines, where the cream of Polytechnique engineers received three years of advanced instruction, they spent just half of each year attending lectures. The remainder of their training was specifically practical, including two long journeys to observe and record the functioning of diverse mines and factories.

Le Play, a prodigy, completed his formal studies after just two years, then spent six months in 1829 touring the mines of Germany with his friend and comrade the Saint-Simonian Jean Reynaud. Their itinerary, covering four thousand miles, included a visit to the Harz Mountains. Le Play's diary from these travels is silent on labor conditions but is filled with technical descriptions and drawings of mines and machines. A few years later, he commented on the admirable efficiency with which the administration of the Harz mines managed its competition against Spanish mines, which knocked the bottom out of the lead market in the very year of his visit. The Harz mine officials had sent the Göttingen professor of mineralogy, J. F. L. Haussmann, to Spain to check out the competition—a tour Le Play reproduced a few years later. Haussmann concluded that current levels of production by the wasteful Spanish were unsustainable. The economies imposed by the Harz mining administration to preserve for workers the necessities of life during this unavoidable moment of hardship, a starstruck Le Play wrote in 1832, "could not be too much praised."[5] Evidently these wise and virtuous men had come a long ways since the years of French revolutionary occupation, referred to by a patriotic local historian as the *Matzhammelzeit*, the era of embezzlement.[6]

Le Play's first social investigations assumed an economic and statistical form. During the 1830s and 1840s, he applauded statistics for its undogmatic empiricism, and he collaborated with some of its best-known practitioners, such as the medical statistician Louis-René Villermé. By 1855, when he published the first edition of his study of European workers, his methods had evolved away from those of the census, which he now regarded as indiscriminately inclusive. Yet his family budgets were an outgrowth of his statistical work, and they even gained him the Montyon Prize for statistics in 1856 from the Académie des Sciences.[7]

At the Corps des Mines, statistics served as an administrative tool. The *Annales des mines* was concerned above all with the location and extraction of minerals, printing brief chemical excerpts along with much longer, original pieces on mining technology and on the geography and geology of regions rich in ores. Initially, the corps treated statistics as merely a form of bookkeeping, but in 1832, the neophyte Le Play introduced a more scientific commitment to statistics into the *Annales.* His study had been ordered by his superiors, one of whom, De Cheppe, issued a statistical proclamation to accompany it. In pursuing its mission to protect the interests at play in mining, De Cheppe explained, the administration of the corps could not rely exclusively on general principles, for many important choices hinge on statistics. "Theories and narrowly based systems can be deadly if no account is taken of the power of facts. . . . The language of figures has its own authority." Engineers supplied the Corps des Mines with numbers of the highest exactitude, gathered with admirable care, and these, he announced, deserved a place in the journal.[8]

That, we may infer, was the task assigned to Le Play, whose title spoke of "observations" on international commerce in minerals. For this young engineer, as for the postwar economists discussed in this volume by Mary S. Morgan, economic observation meant measurement. He laid out the quantities and values of production according to mineral substances (metals, salts, combustibles, construction materials), and subdivided them by categories of production such as goldsmith work and jewelry. French mineral exports, which were highly processed, incorporated much labor, the "first element" of wealth and the "first gauge" of prosperity. Le Play was not content with factual nuggets but looked for patterns to size up the competitive position of French mines and of the trades that depended on them. He summarized his results in a set of comprehensive import-export tables, for him the essential contribution of the memoir. Their bookkeeping form, a balance of income and expenditure, provided a template for his subsequent worker budgets.[9]

Le Play alluded in his memoir to the "admirable organization" of the Harz mines, which demonstrated the advantages of association (*l'état d'associations*) in the mineral industry. What exactly did he mean? Two decades later, in a note to the tables of his monograph on the Harz, he supplied the complicated details, giving some credit to the collective efforts of workers as well as to their bosses. The mineral rights in the Harz, he explained, belonged to companies that held shares as investments. They sold minerals to a foundry, operated for the profit of the sovereign, who also owned the forests. A commercial administration, organized by the state, managed the sale and purchase of materials. Factory profits were not distributed immediately to shareholders but were deposited into reserve funds, on which the workers could draw during natural disasters, wars, and revolutions. The *patrons* sold wheat to the workers at a fixed, government-subsidized price and provided free medical care as well as insurance against injury and sickness. The workers themselves organized pensions for the permanently disabled, funded by contributions from several sources. It was, Le Play argued, an exemplary system of foresight and protection against misfortune and market fluctuations. By the late 1840s, the Harz mining organization was gaining a reputation for wise social policy, and a delegate to the Frankfurt parliament of 1848 proclaimed that it had solved the worker question (*Arbeiterfrage*). Le Play spoke of a wise administration, looking after the needs of a laboring population that he characterized as "mediocre in energy and intelligence."[10]

In 1840, he published an essay on statistics for an encyclopedia edited by his friend Reynaud and the humanitarian socialist Pierre Leroux. He referred to statistical collection as one of the vital roles of government and an index of political enlightenment. To discuss the theory of government without statistics is like discoursing on combustion in ignorance of the composition of the atmosphere. It would be better to train statesmen in statistics rather than the literature of small towns in ancient Greece. Perhaps in a stable political order, administered by men of inherited wisdom and practical experience, statistics would be dispensable, but for France, which had overturned all the old structures in 1789, nothing could contribute more to good government.[11]

Le Play's growing faith in the wisdom of the ages was, as he understood it, fully compatible with the spirit of science, physical as well as social. In autobiographical moments, he liked to say that his method of social observation was merely the extension to a new field of natural-scientific observation. That resemblance, however, depended on a distinctive interpretation of science. In his metallurgical lectures and writings of the 1840s, as in the statistical essay of 1840, Le Play exalted intuition and skill over systematic learning. He was profoundly skeptical of abstract reason, the kind of thinking that, as he would later claim, had made a mess of French politics. The sympathetic, involved observer who puts his faith in experience is a far better guide than the abstract theorist. Even chemistry, he thought, suffered by being detached from real processes of mineral extraction. It has as much to learn from metallurgy as it can teach, because mines depend on a host of techniques unknown to theoretical science.

Le Play's unusual self-positioning with respect to scientific practice is nicely epitomized in a short paper of 1847 by Reynaud, who united Saint-Simonian technocracy with Swedenborgian vision. "De la métallurgie du fer par Swedenborg" begins by celebrating iron as the basis for economic prosperity and military might. Unfortunately, so long as workers shroud their methods in secrecy, the improvement of iron technologies must be abandoned to chance. But Swedenborg, a Swedish mining administrator, had held up a lamp in the darkness, and his 1734 treatise on iron could be ranked with the illustrations of the trades in Diderot's *Encyclopédie.* Swedenborg's contribution owed less to laboratory chemistry than to the immense knowledge of metalworkers, until then a variety of alchemy, requiring only to be systematized. The proper starting point of metallurgy, Reynaud concluded, is not the formal learning of chemists but the craft knowledge of workers. That insight was the starting point for "our excellent metallurgist M. Le Play," who had founded his brilliant school at the École des Mines on the principles of Swedenborg.[12]

Le Play, in turn, praised Reynaud's article in his book-length memoir on metallurgy in Wales for the *Annales des mines.* Swedenborg, he explained, had understood that the despised race of miners, working in obscurity, commanded knowledge as good as that of many sciences. Metallurgy involves distinct principles and modes of action, most of which are understood at the foundry and not in the laboratory. While he had undertaken to investigate and systematize this knowledge, he denied that it could be reduced to chemistry on a large scale. "As yet, no savant has put himself in a position to study the connection between these facts, typically so complex, and the elementary laws of the physical sciences." We cannot say whether the limits of rational science are a matter of principle or merely provisional, arising from a social divide between practical mining and theoretical science. Le Play spoke of the workers as coordinating thousands of phenomena with great precision and deploying "grandes lois naturelles" that science had barely glimpsed. Sometimes their methods, such as forcing hot air into furnaces, had proved themselves despite the unanimous skepticism of savants. "La pratique vaut mieux que la théorie" (Practice beats theory). These workers are a "repository of experience accumulated since the origin of civilization," able to direct "with exquisite tact the subtlest nuances in phenomena whose existence has not even been suspected by science." They were like nature itself, and science might learn more readily by studying their work than by examining the natural world directly.[13]

Location and Detachment

Even in his maturity, Le Play assigned reason an important role in the effort to apprehend nature, but it was a blunt instrument by comparison to skills and techniques, even those of humble workers. In the social domain, an overconfident rationalism presented dangers still greater than in chemistry and mineralogy. In the 1860s he spoke of Socrates as the founder of social science, articulating wisdom as old as civilization. Nevertheless, he insisted on the epochal significance of his method of observation, which he linked to modern economic change and social dislocation. That method reflected, in a distinctive way, the conundrums of the age. Social science as a nineteenth-century project was unmistakably anchored in practices of observation and recording, especially statistical ones.[14] Although closely engaged with concrete problems of administration and reform, it also aimed to detach knowledge from the limits of locality. Statistics meant surveying society from above in order to transcend the limits of direct inspection, capturing states, economies, and societies in a net of numbers.[15]

Around 1848, Le Play began to be skeptical of statistics, criticizing them as secondhand observations. If numbers derive their authority from the machinery of government, the statistician cannot be a proper observer. In the place of official statistics, Le Play now exalted official inquiries or *enquêtes* such as English parliamentary reports. These, in contrast to the work of faceless census takers, relied on "direct investigation." He did not mind that the parliamentary Blue Books were full of numbers, for his own budget studies were no less quantitative.[16]

Direct observation, in the form idealized by Le Play, renounced neutrality and detachment, preferring an encounter between persons whose lives were joined hierarchically. The best observer is a man with responsibility for the observed, as when a *patron* knows and looks out for his laborers or a landlord for his tenants. In his own investigations, Le Play was of necessity a cosmopolitan, coming in from outside, but he conspicuously allied himself with local elites, who in a way were party to the observation. He sometimes intervened in the lives of his informants, as in the case of a day laborer's wife in Vienna who, he noticed, always bought food for one meal at a time. Larger purchases, he advised the woman, could save the family up to 17 percent, and he offered to advance her the funds to begin buying for the future. (She at first seized the opportunity, but then repented, remarking that the family could deal with unavoidable hardship but that she could not hold back food she already possessed from her hungry children.)[17] Le Play celebrated the traditional economies of eastern and northern Europe for maintaining such personal ties even as they disintegrated in France. Yet in practice he did not turn away in despair from French industrial politics but campaigned on behalf of a specific managerial ideal, the supremacy of the benevolent *patron*.[18]

The independence of observer and observed is one typical criterion of objectivity. Nineteenth-century social science was rarely concerned with such objectivity, either in name or in substance. To be sure, increasing ease of travel as well as the relative anonymity of urban life offered new opportunities and incentives for impersonal observation, which might be defined in terms of the displacement of a peripatetic observer. The flâneur, who strolls about in order to experience an alien world, is, like the census taker, neutral and detached. Most nineteenth-century social observation, though, was of a different sort. Henry Mayhew, who wrote on the London poor in the mid-nineteenth century, studied people and places he knew well, and Friedrich Engels was guided into the streets and habitations of poor Irish laborers in Manchester by his lover Mary Burns, who knew them well. Urban statistical societies, as in Manchester, were made up mainly of local physicians and factory owners, working to clean up bad neighborhoods. In Paris, Alexandre Parent-Duchâtelet read archives, conducted interviews, and visited prisons and brothels in his study of prostitution. These were typical sources for social and medical reformers as they explored the strange urban world of disease, poverty, and vice. Rarely did numbers stand by themselves.

Le Play's own observing was almost global, extending from ragpickers in Paris to Muslim herders in Siberia. His goal was always to make these strange locales familiar by befriending elites and conversing with laborers. For social surveys, whether quantitative or not, the site of observation was most often the city, and in particular those spaces created by an industrializing economy in which working people were more or less sequestered from polite society. The explosion of statistical activity in France and Britain around 1830 was driven by anxieties about disease and moral disorder in these working-class populations. Their hovels and garrets and the streets where they gathered, though described by reformers and novelists, were barely accessible to respectable people, who experienced a frisson of danger when entering these alien spaces.[19] Statistics, with its depersonalizing tendencies, was a way of keeping the poor at a distance, and yet the enumerators went from door to door in their effort to penetrate the darkness.

Le Play, too, was moved by the modern condition of cities, but he sought remedy by escaping the modern metropolis to the towns and villages of distant lands. There he labored to recover a sense of the more integrated community, where *patrons* had close contact with working people and knew them individually. His *méthode d'observation* used family budgets to trace personal and economic relationships within villages or regions whose ancient customs had not yet been too much disrupted by modern life. Among the virtues of close observation was the respect it automatically conferred on traditional structures of society. Statisticians, by contrast, were typically liberal by disposition, and looked to promote progress by cleaning up houses and streets and by educating the poor to raise their moral and intellectual character.

Sites of Personal Knowledge

As a mode of observation, statistics meant classifying, counting, and averaging. It seemed, as Balzac complained in *Le curé au village*, to depict society as a heap of atomic individuals. According to a common assumption, statistics was most appropriate for persons deficient in individuality.

The presumed opposition of statistics to direct observation, which would be more personal and more humane, is engagingly caricatured in Charles Dickens's *Hard Times* (1854). Encouraged by his "deadly statistical recorder," the utilitarian schoolteacher Thomas Gradgrind performs social observations from the blue interior of an observatory that has no windows, only shelves packed with parliamentary reports. The novelist's voice condemns Gradgrind and his clan for their ignorance of the persons summed up in these documents. So, when his daughter Louisa escapes the pretentious estate of her husband Mr. Bounderby, a factory owner, to the little cabin of a worker, Stephen Blackpool, it comes as a revelation to her. "For the first time in her life, Louisa had come into one of the dwellings of one of the Coketown Hands; for the first time in her life, she was face to face with anything like individuality in connexion with them."

Dickens, who began his writing career as a parliamentary reporter, was familiar with official Blue Books. He must have understood that an observatory lined with committee reports admitted as much light as one filled with novels. At times, to be sure, these reports appear profoundly bureaucratic:

> The tabular forms have been filled up, and the queries answered by employers, in sufficient numbers and with sufficient completeness to permit of our stating, with a fair approximation to the truth, the number of children and young persons employed in the branches of manufacture in question, at all ages under 18, and their relative proportion to the adult workpeople; the usual hours of work; in what cases, and to what extent, over-hours or night work

prevail; and what amount of time is allowed for meals, and whether they are taken regularly or irregularly.[20]

At other times they used vivid language and emphasized direct experience in much the same fashion as did Dickens. In a way they were his rivals.

Frances Trollope's novel, *Life and Adventures of Michael Armstrong, the Factory Boy*, which provoked such consternation in the 1840 select committee, was understood by everyone as an exposé of actual factory conditions. Her avowed purpose was "to drag into the light of day, and place before the eyes of Englishmen, the hideous mass of injustice and suffering to which thousands of infant labourers are subjected, who toil in our monstrous spinning-mills. . . . The true but most painful picture has been drawn faithfully and conscientiously." Like *Hard Times*, *Michael Armstrong* is concerned less with laborers as such than with the enlightenment of middle-class people, in Trollope's case of ladies who, like the author herself, should be moved to escape their sheltered enclaves and experience with their own eyes, ears, and noses the suffering inflicted on the poor by their fathers and husbands, the managers and owners. Her most poignant passages cast the ladies in a role recalling parliamentary inquirers.

"What is the billy-roller, Sophy?" inquired Miss Brotherton, in an accent denoting considerable curiosity.

"It's a long stout stick, ma'am, that's used often and often to beat the little ones employed in the mills when their strength fails—when they fall asleep, or stand still for a minute."

"Do you mean, that the children work till they are so tired as to fall asleep standing?"

"Yes ma'am. Dozens of 'em every day in the year except Sundays, is strapped, and kicked, and banged by the billy-roller, because they falls asleep."[21]

Dickens notwithstanding, the Blue Books overflowed with personal experiences and hardships as well as impersonal statistics, and the committees did not disdain theater. Here is an item from the questioning of factory inspector Charles Trimmer in 1840:

2848. Do you think that little children are exposed to very great hazard, particularly female children, with their flowing garments, in going round from one part of the mill to another?—I think they are; there was a case occurred very recently at Stockport, where a girl was carried by her clothing round an upright shaft; her thighs were broken, her ankles dislocated, and she will be a cripple for life.

2849. What do you think would have been the expense of boxing off that upright shaft?—A few shillings.[22]

A committee report of 1863 added the voices of the children to those of officials. Although the document implies quotation, their words, reported by factory inspectors, are as stylized as those of children in Dickens. Typical is James Barnacle (no relative, surely, of Tite Barnacle of the circumlocution office in *Little Dorrit*), transmitted in staccato rhythms from the pottery works by Mr. F. D. Longe. "I am 12 years old. There are nine other boys here. They would be from 12 to 14. I have worked three years in the dipping house. My father is a sagger maker. I cannot read. I go to school on Sundays. We get 3s. or 4s. a week each."[23]

The parliamentary members of these committees knew that nothing could replace firsthand knowledge, because they understood the forms of deception and concealment to which factory inspectors—such as Leonard Hoerner (here)—were exposed.

> 428. Have there been any instances in which attempts have been made to conceal the children during your visits at the mill?—Yes, they have been concealed in wool bags.
> 429. And in what other places?—I have never detected it, but I was told that in a mill that I visited, believing children to be improperly employed there, after I had gone I was told, "If you had looked into the necessaries you would have found them full."[24]

This select committee, aware as it was of the limits of what could be learned officially, never challenged hearsay reports like this one, but welcomed the glimpse they offered behind the scenes. The really crucial observations, invisible to the eyes, depended on the stories of concerned neighbors.

Le Play, like Dickens, looked to patronal benevolence and paternalism for a solution to the social question. British parliamentary examiners were less credulous. Perhaps, one insinuated, the unreasonably modest fines assessed by courts for flagrant violations of child labor standards owe something to the financial interest of magistrates in the very mills where the violations occurred. And if mill owners appoint schoolmasters who teach on their premises, doesn't that make the schoolmaster [Gradgrind?] "the servant of the mill owner?" [Bounderby?][25] Le Play, to be sure, was thinking of what could be, or had been, and the yawning gap between ideal and reality was why *la réforme sociale* was so urgent. As in former times, he dreamed, *patrons* should live beside their employees or tenants and deal with them humanely. The desiccation of personal ties owing to Louis XIV's policy of drawing the nobles to Versailles, he argued, had been responsible for the angry attacks of peasants on their chateaux during the French Revolution.

Karl Marx, whose analysis of capitalism relied as much on empirical investigations in Blue Books as on theoretical traditions deriving from G. W. F. Hegel, Henri de Saint-Simon, and David Ricardo, was resolutely impersonal. His focus on the capitalist system and on categories of analysis such as "abstract labor power" contrasts with a study he cited repeatedly for empirical information by his collaborator Engels. *The Condition of the Working Class in England*, published in 1845, is more like a guidebook to the rapacity and the suffering characteristic of industrial manufacturing. Engels wrote in a personal voice, and so it is significant that he mixed anecdotes from Blue Books with what he had seen and heard with his own senses. Idle dilettantes, sentimental novelists, dutiful civil servants, and scientific socialists join hands in works like this, collaborating in a project of social observation that has supplied our most persistent images of the human impact of industrialization.

For Le Play, the official inquiry was the secret of good government in Britain. In contrast to Continental bureaucracy, which "combines the reality of power with an absence of responsibility," administration in Britain delegated authority to officials on the ground and embraced the activity of local elites. Official inquiries gave these men the opportunity to demonstrate their good and conscientious work to Parliament. They filled the role of *sages*, men of wisdom and experience, the eyes and ears of the nation.[26] No better observatory could be devised than the chambers wherein they communicated their observations directly to representatives of the supreme legislative authority in response to probing questions, and which were recorded in Blue Books and diffused by the press.

Their vivid reports were much valued by opponents of the capitalist order. Le Play himself condemned free markets, and he applauded government actions against the terrible abuses of workers, especially children, in unregulated steam-powered factories.[27] Marx gave fulsome praise to the officials who supplied his data, explaining to the German readers of *Capital* that its gruesome statistics (and occasional anecdotes) did not mean laboring conditions in England were worse than across the Channel. It was rather that the English alone uncovered the truth about factory life.[28] Parliamentary reports of 1863 and 1864 on the employment of children, heavily cited in *Capital*, are full of dark, satanic images, as of "a number of human beings pent up together, breathing over and over the same polluted atmosphere." At the Lucifer Match Company, the fumes cause necrosis of the jaw, as explained by John Pegge of the Royal College of Surgeons. "The sufferings of a patient in the earlier stages of the disease, and until it has run itself out, leaving the jaw quite dead and exposed, are intolerable. He will take almost any amount of narcotics with comparatively little effect." Factory inspector Scriven introduced a committee to young boys in the pottery trade who worked from 6 a.m. to 9 p.m., often in closely confined, windowless hothouses, with a cast-iron stove in the center, "heated to redness, increasing the temperature often to 130 degrees. I have burst two thermometers at that point."[29]

That bursting betokens the transformation of quantity into quality, as winged mercury unites furnace, instrument, and inspector.

The Ancient Wisdom

Le Play appealed to attractive images of what once was and might again be, an alternative to huddled masses in an indifferent world. Not until his last years did he merely revere the past, nor did he condemn industrialization as such. Until about 1850, mining techniques defined his profession, and for two decades after that he contributed to the planning and judging of international exhibitions, those showcases of new technology. Material progress was good in itself, even if it often had moral disorder as a by-product. In 1879 he wrote of the great promise of James Watt's steam engine and Richard Arkwright's spinning jenny, which could have benefited all classes of society were it not for one other terrible invention of the same era, Adam Smith's political economy, which dissolved customary obligations in a free, competitive market, depriving the poor of their economic security.[30] More and more, commercialism sapped the spirit of generosity of the English upper classes, which had been inculcated by liberal education and by travel abroad. Even so, he praised the English for maintaining loyalties across class lines that had decayed in France, and for preserving an alliance of religious faith with belief in science.[31]

The great French engineering corps were never disposed to favor free-market economics.[32] State engineers, including the young Le Play, embraced their own mission as expert planners who could protect a society against the hazards of unregulated exchange. During 1848, his correspondence with the liberal statistician George Richardson Porter provided occasion to express his low regard for English economic doctrines and his support of socialism and of working-class activism.[33] Dispirited by the excesses and then the failure of revolution, he began to look instead to owners and managers to create a better society. They should be modern, like engineers, in their exploitation of technology, but traditional in their acceptance of paternal responsibility for their workers. As an organizer of international exhibitions, Le Play backed prizes not only for technologically advanced production techniques but also for innovative arrangements to secure a decent life for workers.

Under the empire of Louis Napoleon, Le Play moved beyond his function as state engineer to become a confidant of the emperor and a member of the

Conseil d'état. The first edition of his monographs, appearing in 1855 in a folio volume from the Imprimerie Impériale, "par autorisation de l'Empereur," has the scent of an imperial blessing. This collection looks back to old times with a feeling of loss, yet is structured into a narrative of progress. The most primitive societies, such as he had observed in Siberia, depend on forced labor. Herdsmen and farmers have no free will but entrust all decisions to the *patron*, to whom they are passionately loyal. Le Play found them insurmountably opposed to change, even when it would benefit them. This system was not to be condemned out of hand, but its static character was a shortcoming. A better system would allow the development of individual liberty, as indeed was to be found at the next stage in social progress, typified by Sweden, which made employment relations permanent but voluntary. The spread of democracy, diffusing the authority of the superior classes, led ineluctably to further improvement, though also to heightened danger. Temporary labor contracts offer more freedom to workers while placing more responsibility on the father as head of family. Such a system, he declared, can raise moral character, as in Norway, certain Swiss cantons, and parts of Spain, and especially where a tradition of communal lands persists. But in England, Belgium, and France, where electoral politics and new industrial technologies have undermined the patronal role, rootless workers are degraded into "nomads of a new type" and become vulnerable to demagogues.[34]

And yet, with care, these flaws can be remedied. There is no solution to be had through individual discovery, "un seul jet d'eau du cerveau d'un penseur" (one jet of water in the brain of a thinker), but only through meticulous observation of social reality. His dismissal of the a priori theoretical insight was a rejection also of revolution, while systematized social observation implied a politics without polarization. Le Play's social science meant proceeding collectively and letting experience rather than dogma determine, for example, whether free exchange or state intervention leads to greater morality in an economic system. The path of observation would allow social science to advance through the same progressive phases as astronomy, physics, chemistry, and natural history. Since social facts are readily visible, requiring no great feats of instrumental precision, observation can support a public form of science.[35]

By 1864, Le Play was less optimistic about the progress of society. The social conflicts of the 1860s reminded him of the uncertainties of 1830, with turbulence everywhere and his friends declaring fealty to Saint-Simon. He remembered now that he had adopted then the strategy of René Descartes, doubting all but what he could personally demonstrate. Since real knowledge arises from observation, this Cartesian move had brought Le Play directly to his methodological goal. Immediately, however, he had realized (he now claimed) that his own eyes could not be the final authority. Like Descartes, he must hold his conclusions suspect until they had been verified by the most eminent men with the most universally recognized qualities. Indeed, by 1864 Le Play was coming to identify such authorities as uniquely privileged social observers.

The Sage as Observer

Thus, as Le Play distanced himself from the liberal traditions of personal freedom, direct observation assumed a new, diminished role. Social science became less a matter of empirical verification, and more of recognizing the timeless centrality of the human soul and its relations with God. The eternal truths of man and society would be revealed by looking within. In 1864, he identified Jean-Jacques Rousseau's theory of perfectibility and the modern faith in revolution as the two most deadly doctrines. Social problems were a consequence of original sin, which must be suppressed by fathers in each successive generation. The proper direction of reform was mainly to recover what had been lost.[36]

The second and most widely influential edition of *Les ouvriers européens*, published in six volumes from 1877 to 1879, aimed to reconcile a still more profound conservatism with the empirical methods he cherished. Horrified by the Paris Commune in 1871, he began to repeat unceasingly and with tedious prolixity his moral doctrine, that man needed nothing more in the world than to heed the Ten Commandments and to be assured of receiving his daily bread. The twin institutions of the patriarchal family and the patronal employer were perfectly suited to provide the basic human necessities, and with them "social peace."[37] Individuality as a positive force now disappeared from his social analysis. There are vast catalogs of plants and animals, he explained in 1879, and in chemistry an unlimited number of combinations can be produced in the laboratory, but moral science has witnessed nothing really new since the Decalogue. While the conditions of human life are, in their details, various, all families, and not only happy ones, participate in a shared human condition.

Le Play continued to sponsor systematic social observation, and the Société d'économie sociale that he had formed in 1856 moved toward the identification of monographs with a moderate politics of social reform. Emile Cheysson, his most prominent disciple, allied the monographic method with statistics and struggled to reconcile *patronage* with social engineering.[38] For Le Play, however, observation was becoming more and more an instinct rather than a set of techniques. Also, he found less and less variability to justify further studies. "In what concerns the two supreme laws of happiness and the two fundamental elements of the [social] constitution, scrupulous observers will be convinced by the study of a single monograph."[39]

Under the Third Republic, Le Play discovered in economically less advanced societies the key elements of a solution to the problems of France. Their remoteness from the modern economy and modern ideas was the very reason for their success. Paradoxically, the method of observation now required a French reformer to be cosmopolitan, to study economic life in remote locations rather than merely observing at home.[40] By contrast, men of the East and North had no need even to look outside themselves for understanding, because they had direct access to the wisdom of the ages. This ancient wisdom, for Le Play, was fundamental. It was possessed above all by *sages*, who persisted wherever a society had been insulated from the corruptions of the Occident. It is a blessing for them to understand no languages but their own. Although the monographic studies in his second edition became longer and more detailed as well as more numerous, his introductions to the volumes tended to drain the *méthode d'observation* of its purpose. They took a dim view of technological progress and of science. Le Play specifically condemned the new English "évolutionisme," calling it a preconceived idea and a threat to the observed truths of religion. Going beyond his longstanding view that wise men of the East understood by instinct all that social science could hope to discover in its more laborious fashion, he now claimed that real knowledge had to be passed along through an uninterrupted oral tradition. Codifying the ancient wisdom and writing it down as law was the beginning of the fall from grace.[41]

The West had fallen very far, and Le Play's introductions offered little hope that formal study, even of his own monographs, could bring redemption. Social science presents no possibility of a true experiment, as in natural science, hence no alternative but to reason from solutions that have worked in analogous circumstances. In practice, the empirical results available to a modern European cannot match the experience of listening to the sage as he interprets his own traditions, traditions that work.[42] These could not be passed along straightforwardly, like pencils or bits of information, but had to be assimilated, slowly. Wisdom was the key ingredient as well as the desired outcome of this form of observation, which presumed a reverence for tradition and a distrust of dogmatic assertions.

Still, Le Play proceeded quantitatively, as in 1833 and 1855.

Wisdom Congealed into Tables

Le Play's quantifying took the form not of counting persons but of assembling family budgets. When, as a mining engineer, he began this work, he conceived them as analogous to the chemical analysis of ores. In both cases there is a conservation principle, the principle of double-entry bookkeeping, that income equals outgo or expenditures. This requirement provided an automatic check to the figures. He wanted as complete an analysis as possible, and he insisted on assigning money values to items that were not traded in the marketplace, including labor, in-kind obligations between worker and *patron*, and some (but not most) unpaid services of the wife.

Le Play's budgets provided the basis for a microsociology in which the whole is made visible by a close analysis of a part.[43] He always sought the advice of the eminent in selecting a typical or representative family. An itemization of the family budget allowed him to form a picture of relationships spanning a whole community (fig. 11.1). Le Play put particular emphasis on items that were governed by custom rather than by direct market forces. These would include, for example, the right of a peasant to graze his cow or to haul firewood with the *patron's* cart. They demonstrated the persistence of traditional relationships, of personal bonds of loyalty and responsibility, in much of the world. Patronal benevolence was far more prevalent among Scandinavian factory workers or Harz miners than in urban France. His monograph on a *maître-blanchisseur* (master laundry worker) in the Paris suburbs, for example, showed no subventions of this kind at all.[44]

Equally revealing were the means for insuring the poor against unexpected hardship. Worker cooperatives, which gather contributions and make payments as need requires rather than accumulating funds, are adequate when probabilities are known, but where experience is inadequate to estimate the probabilities, as for example with loss of employment, such funds will inevitably run dry when times are bad. A cooperative could never enforce the level of abstinence and morality required to assure adequate resources in the worst of times, such as he had witnessed in the Harz in 1829. For this reason, the old system, in which the seigneur takes responsibility for cases of misfortune, was a better one, and Le Play took it as a model for relations between the *patron* and the worker. These studies consistently stressed the advantages of communities held together by bonds of affection between owners and workers, the poor and the rich. Bureaucratic solutions only interfere with the spirit of initiative, he thought, and any "corporate" response to poverty must lack the needed subtlety and sensitivity. A corporation, after all, can know nothing of the private world or *vie intime* of a needful family. It cannot substitute for

RECETTES.	MONTANT DES RECETTES.	
	VALEUR des objets reçus en nature.	RECETTES en argent.
REVENUS DES PROPRIÉTÉS.		
ART. 1er. REVENUS DES PROPRIÉTÉS IMMOBILIÈRES.		
Intérêt (4 p. 100) de la valeur de cette maison..........	61f 33	30f 67
Intérêt (4 p. 100) de la valeur de ce jardin..........	1 60	»
ART. 2. REVENUS DES VALEURS MOBILIÈRES.		
Intérêt (5 p. 100) de la valeur de ces outils..........	0 25	»
Intérêt (5 p. 100) de la valeur de ces objets..........	0 35	»
ART. 3. ALLOCATIONS DE CAISSES D'ASSURANCES MUTUELLES.		
(Ce droit ne donne actuellement aucun revenu.)..........	»	»
(*Idem.*)..........	»	»
Valeur de l'allocation supposée égale à la contribution annuelle de la famille..........4f 80. Cette recette, n'étant que la rentrée d'une somme égale versée à la caisse d'assurances, est omise ici, comme la dépense qui la balance (D. 5e Sen)..........	»	»
TOTAUX des revenus des propriétés..........	63 53	30 67
PRODUITS DES SUBVENTIONS.		
ART. 1er. PRODUITS DES PROPRIÉTÉS REÇUES EN USUFRUIT.		
(La famille ne jouit d'aucun produit de cette nature.)..........	»	»
ART. 2. PRODUITS DES DROITS D'USAGE.		
Bois (2,000 kil.) évalué sur pied à..........(5).	6f 21	»
ART. 3. OBJETS ET SERVICES ALLOUÉS.		
Remise (0f 065 par kil.) sur le prix marchand de 729 kil. de seigle achetés aux magasins domaniaux..........	47 38	»
Remise d'intérêt (1 p. 100) sur la somme de 2,300f 00 due à l'administration, pour l'acquisition de la maison..........(3).	15 34	7f 66
Remise (0f 30 par 100 kil.) sur le prix marchand de 1,900 kil. de bois achetés aux magasins domaniaux..........	5 70	»
Dépense faite par l'administration pour l'école des garçons : par famille d'ouvriers..........	9 00	»
Contribution de l'administration aux caisses qui allouent des secours médicaux et des subsides en argent aux malades : par famille d'ouvriers..........	4 20	6 75
Remise d'intérêt (1 p. 100) sur la somme de 40f 00 due à l'administration pour l'acquisition du jardin..........	0 40	»
Remise d'intérêt sur une somme de 2,340f due à l'administration pour l'acquisition de la maison et du jardin : voir ci-dessus..........	»	»
Dépense annuelle de l'administration pour secours médicaux et subsides en argent alloués aux malades : voir ci-dessus..........	»	»
Contribution annuelle de l'administration à la caisse qui alloue des pensions aux ouvriers vieux ou infirmes, aux veuves, aux orphelins jusqu'à l'âge de travail : par famille d'ouvriers..........	13 20	»
TOTAUX des produits des subventions..........	101 43	14 41

FIGURE 11.1. The upper right-hand quarter of Le Play's family budget for a family in the Harz Mountains, from the 1855 folio first edition of his *Ouvriers européens*. The tables, which include capital and income accounts, assign money values whenever possible. The list calls attention to services governed by custom rather than contract, demonstrating the persistence of a traditional economy.

a *patron* exercising personal charity for families attached to the same house and to the rigors of the same workshop.[45] Le Play's monographs, particularly in their revised form, do not present society statistically as a mass of individuals, but as families great and small, enmeshed in a web of mutual obligations. The truths that emerged from such study, he held, were deeper than those of the census.

Conclusion

Although the history of social science has usually been written with an emphasis on social theory, it was from the beginning a project of making and collecting observations. Much of this work was performed by government agencies and was bound up with systematic administrative interventions. Much of the rest—and this would include all of Le Play's later works of social observation—aimed to investigate and promote reform. Until very late in the nineteenth century, few proponents of social science were troubled that embracing the structures of social and institutional power might compromise its objectivity. They worried much more about the subordination of social science to theoretical prejudices, to radical doctrines of free-market liberalism or socialism.

The basic task of social science, as Le Play and many of his contemporaries conceived it, was to work out a form of observation appropriate to the new social and economic conditions of the nineteenth century. For him, this was also about identifying the kind of authorities who could accommodate and redirect the forces of social change. Through the 1860s and especially after 1871, he became disaffected with the modern world, and he argued with increasing conviction that the way forward was to recover what was valuable in the past. He proposed to study that past as something living, to be found not mainly in old books but in societies that had been insulated from the disruption and anonymity—that "modern nomadism"—of industrial society.

In this, the period of his greatest influence, he worked to reconstruct social observation in a way that would recognize and reinforce paternalistic authority. He put more and more emphasis on implicit, unarticulated forms of knowledge, the wisdom contained in parables passed down by tradition. He continued to proselytize on behalf of monographs, but what could they now contribute? We should look to social science not for original discoveries but for a glimpse of reflected light. Le Play now rewrote his own history to emphasize a wide gulf that separated his method of observation from other efforts in empirical social science, thereby refashioning himself as a revolutionary practitioner of social science—revolutionary in the antique sense of returning to ancient roots. Yet the trajectory of his life reveals him as very much a man of the nineteenth century. His paradoxical achievement was to reconstruct the tools of modern economic and statistical observation so as to reinvigorate an organic social order based on the implicit wisdom of sage elites who knew by instinct rather than through detached investigation and conveyed this wisdom in stories.

Notes

1. *Second Report from the Select Committee on the Act for the Regulation of Mills and Factories (13 April 1840)*, iv. The inspector was R. J. Saunders. On literature and social science, see Wolf Lepenies, *Between Literature and Science: The Rise of Sociology*, trans. R. J. Hollingdale (Cambridge: Cambridge University Press, 1988).

2. Harry Liebersohn, "Scientific Ethnography and Travel, 1750–1850," in Theodore M. Porter and Dorothy Ross, eds., *The Cambridge History of Science*, vol. 7, *Modern Social Sciences* (Cambridge: Cambridge University Press, 2003), 100–13.

3. Frédéric Le Play, *Les ouvriers européens (2e édition), tome premier, la méthode d'observation appliqué de 1829 à 1879, à l'étude des familles ouvrières* (Tours: Alfred Mame et fils, 1879), 110, 13–16; Fabien Cardoni, "Précis de la formation d'un ingénieur des mines, Frédéric Le Play de 1806 à 1830," in Antoine Savoye and Fabien Cardoni, coord., *Frédéric Le Play: Parcours, audience, heritage* (Paris: École des Mines, 2007), 12–41.

4. Frédéric Le Play, *La réforme sociale en France déduite de l'observation comparée des peuples européens*, 2 vols. (Paris: Henri Plon, 1864), vol. 1, avertissement, v.

5. Le Play, "Observations sur le mouvement commercial des principales substances minérales, entre la France et les puissances étrangères," *Annales des ponts et chaussées*, 2nd ser., 2 (1832): 501–45, 517–18; Michael Z. Brooke, *Le Play: Engineer and Social Scientist* (London: Longman, 1970), 7–9, 42–44; Françoise Arnault, *Frédéric Le Play: De la métallurgie à la science sociale* (Nancy: Presses Universitaires de Nancy, 1993), 21.

6. Friedrich Günther, *Der Harz in Geschichts-, Kultur- unad Landschaftsbildern* (Hannover: Verlag von Carl Meyer, 1888), 622. *Matzhammel*, Günther explains, is miner's slang—not restricted to the Harz—for *Veruntreuung* (embezzlement).

7. Guy Thuillier, "Economie et société dans la pensée de Le Play en 1844," *Revue administrative* 15 (1962): 481–86, 482.

8. De Cheppe, "Statistique," *Annales des mines*, 2nd ser., 2 (1832): 547–48. In *Reforme sociale*, 1: 41, Le play recalled that his early memoirs on commercial economy had been requested by his superiors.

9. Le Play, "Observations"; De Cheppe, "Statistique," *Annales des mines*, 2nd ser., 3 (1833): 693–702.

10. Le Play, *Les ouvriers européens. Études sur les travaux, la vie domestique et la condition morale des populations ouvrières de l'Europe, précédées d'un exposé de la méthode d'observation* (Paris: Imprimerie Impériale, 1855), 145, *Réforme sociale*, 1: 296. See also Christoph Bartels, *Vom Frühneuzeitlichen Montagnegewerbe zur Bergbauindustrie: Erzbergbau im Oberharz, 1635–1866* (Bochum: Deutsches Bergbau-Museum, 1992), 414–15.

11. Le Play, "Statistique," in Pierre Leroux and Jean Reynaud, eds., *L'Encyclopédie nouvelle*, 8 vols. (Paris, 1840), 8: 275–77.

12. Reynaud, "De la métallurgie du fer par Swedenborg," *Magasin pittoresque* (1847): 14–15, 26–27.

13. Le Play, "Des procédés métallurgiques employés dans le pays de Galles pour la fabrication du cuivre et recherches sur l'état actuel et l'avenir probable de la production et du commerce de ce métal," *Annales des mines* 3rd ser., 13 (1848): 3–224, 389–666, 11–23.

14. Porter and Ross, eds., *Cambridge History*, preface and introduction to part 1; Porter, "The Social Sciences," in *From Natural Philosophy to the Sciences: Writing the History of Nineteenth-Century Science*, ed. David Cahan (Chicago: University of Chicago Press, 2003), 254–290.

15. Eileen Janes Yeo, "Social Surveys in the Eighteenth and Nineteenth Centuries," in Porter and Ross, *Cambridge History*, 83–99.

16. Le Play, *Ouvriers européens* (1855 ed.), 3–4 ; "Frédéric Le Play: Anthologie et correspondance," ed. Stéphane Baciocchi and Jérôme David, special issue, *Les études sociales* 142–44, no. 2 (2005–6): 119–47.

17. Le Play, *Ouvriers européens* (1855 ed.), 128.

18. See Alain Desrosières, "L'ingénieur d'état ou le père de famille: Émile Cheysson et la statistique," in Desrosières, *Pour une sociologie historique de la quantification* (Paris: Presses de l'École des Mines, 2008), 257–87.

19. Andrew Mendelsohn, "The Microscopist of Modern Life," in *Science and the City*, ed. Sven Dierig, Jens Lachmund, and Andrew Mendelsohn, Osiris, 18 (Chicago: University of Chicago Press, 2003), 150–70.

20. Royal Commission on the Employment of Children in Trades and Manufactures not regulated by Law, *First Report* (1863), chair, Hugh Seymour Tremenheere, viii.

21. Frances Trollope, *Life and Adventures of Michael Armstrong, the Factory Boy*, 3 vols. (London: Henry Colburn, 1840), preface, chap. 14.

22. *Second Report* (note 1), 41.

23. Royal Commission on the Employment of Children *First Report*, (1863).

24. *First Report from the Select Committee on the Act for the Regulation of Mills and Factories* (1840), 27.

25. Ibid., 27, 77.

26. Le Play, *Réforme sociale*, 2: 237, 272. Chapter 64 is called "La réforme par l'enquête."

27. Le Play, *Réforme sociale*, 1: 310–11.

28. Karl Marx, *Capital* (1867), trans. Ben Fowkes (London: Penguin Classics, 1990), preface, 91.

29. Royal Commission on the Employment of Children, *First Report* (1863), xiv, xxviii, xlviii.

30. Le Play, *Ouvriers européens* (1879 ed.), 1: 123.

31. Alan Pitt, "Frédéric Le Play and the Family: Paternalism and Freedom in the French Debates of the 1870s," *French History* 12 (1998): 67–89; Le Play, *Réforme sociale*, 1: 35.

32. Theodore M. Porter, *Trust in Numbers: The Pursuit of Objectivity in Science and Public Life* (Princeton: Princeton University Press, 1995), chap. 6.

33. Brooke, *Le Play*, 14.

34. Le Play, *Ouvriers européens* (1855 ed.), 9–10, 292–94.

35. Ibid., 10–11.

36. Le Play, *Réforme sociale*, 1: 9, 11–12, 30–34. On Le Play's paternalism and French population politics, see Theodore Zeldin, *France, 1848–1945*, vol. 2, *Intellect, Taste, and Anxiety* (Oxford: Clarendon Press, 1977), 953–59.

37. Le Play, *La paix sociale après le désastre, selon la pratique des peuples prospères. Réponse du 1er Juin 1871*, 2nd ed. (Tours: Alfred Mame et fils, 1875).

38. Sanford Elwitt, *The Third Republic Defended: Bourgeois Reform in France, 1880–1914* (Baton Rouge: Louisiana State University Press, 1986), chaps.1–2; Porter, *Trust in Numbers*, 63–65.

39. Le Play, *Ouvriers européens* (1879 ed.), 1: 218.

40. Amanda Anderson, *The Powers of Distance: Cosmopolitanism and the Cultivation of Detachment* (Princeton: Princeton University Press, 2001).

41. See introductions to Le Play, *Les ouvriers européens*, 2nd ed., vol. 2, *Les ouvriers de l'Orient*, vol. 3, *Les ouvriers du Nord*, and vol. 4, *Les ouvriers d'Occident* (1877). On evolutionism, see Le Play, epilogue, *Les ouvriers européens* (1877 ed.), 3: 495–96.

42. Le Play, *Les ouvriers européens*, 2nd ed., vol. 6, *Les ouvriers de l'Occident* (1878), xlv.

43. On statistics and social policy in France, see Alain Desrosières, *La politique des grands nombres: histoire de la raison statistique* (Paris: La Découverte, 1993); François Ewald, *L'état providence* (Paris: Bernard Grasset, 1986), chap. 3.

44. Le Play, *Ouvriers européens* (1855 ed.), 268–69.

45. Le Play, *Réforme sociale*, 1: 390; 2: 25–28.

12

Seeking Parts, Looking for Wholes

MARY S. MORGAN

Prologue

For much of the twentieth century, observation and measurement have been deeply intertwined in the practices of economics. Ask an economist what "an observation" is and, until recently, the answer would have been "a statistical data point."[1] Ask an economist about problems of observation, and he will think about ones of measurement.[2] During the twentieth century, these two conflations came to seem so natural to economists, that the commentator has quite some difficulty separating them out again. The ability to do so relies on the fact that scientific observation involves not just processes of observing but ones of recording and reporting these observations. Alignment is required between the acts of observing and those of recording (and for both of course with other broader elements in the epistemology and ontology of a field). The necessity of such alignment may be one way to understand why, in empirical branches of economics, there is such a close colligation between *observation* and *measurement*: for economists, measuring creates or constitutes the alignment between *observing* and *reporting*.

The nature of the relation between act of observation and mode of reporting may, in turn, reinforce certain characteristics in the observer and on the observation process. In certain sites of observation, such as building history, observers share an ethos that it is the process of recording itself that opens the eyes to recognize particularities in the case observed. Such observers record their observations by carefully drawing every brick and stone and every last feature of them (such as the Berlin Wall) and then transcribing certain elements that indicate historical changes onto time/space charts.[3] Here, the initial act of recording is one in which the recording hand enforces the eyes to see things that they might otherwise neglect if they looked using other, more passive, recording media (for example, in this building history context, photographs). For twentieth-century economists observing the economy as a whole, the ethos is different, but equally particular: processes of recording must be ones that ensure not just numerical accuracy with respect to all the individual parts, but completeness with respect to capturing the whole. In contrast to the archaeological case, much of such recording is carried out independently and at long distance, as for example, when firms fill in forms about their activities for tax purposes that are then used by those in statistical bureaus to abstract particular entries as "observations" for economists. The link between the observing economist and the numerical recording is therefore reversed; rather than the economist's observation preceding recording, recording takes place in the statistical bureaus and precedes the economist's observation, and is often quite remote from it.[4]

Yet the disciplinary aims of numerical accuracy and completeness still work to link observer and a network of many separate recorders. Whereas the numerical accuracy largely depends on those making the initial recordings of the initial pieces, completeness largely depends on the skill of the economist in locating those many different recordings and fitting them together. Whereas earlier, economists had been active gatherers of data themselves, these statistical bureaus or economic observatories established in the later nineteenth and early twentieth centuries (some commercial and others public) largely took away from the economists the need to observe and record observations themselves.[5]

This essay focuses on a time when economists were adopting a new set of concepts for the economy as a whole, and on a place where economists wanted to use those concepts to measure an economy as a whole but found very few numbers already available and no fully functioning observatories. In consequence, they were forced to visit and look into those economies in order to see them firsthand. To observe "the economy" as a whole, they needed to measure its parts, but, as we shall find, in order to measure its parts, they had to find ways to see them. In looking for them, and from looking at them, they hoped to record their observations in a particular form, namely, in numbers that fitted their particular conceptual framework—though they did not always succeed. This episode provides the materials to probe the easy conflation of "observations" with "measurements" in twentieth-century economics, and to suggest what might be special about the nature of observation in the social sciences.

Observing a Whole Economy

The project of observing, or measuring, the economy as a whole arose in a new form in the later 1930s and 1940s with the development of national income accounting (NIA) systems of measurement associated in the United States with Simon Kuznets and in the United Kingdom with Richard Stone.[6] Their ambition was not just to measure the whole, but to do so in ways that would reveal hidden internal structures of the economy as suggested by the new concepts of macroeconomics. Their problem of observing the economy can be seen as similar to that of mapping a country: we cannot easily observe the country as a whole, for the problems of scale and size with respect to the human observer mean we can only see a little bit of it at a time, and we have difficulty even getting a perspective on that bit. Yet the analogy captures the idea that once we have made a map, we gain the sense that we can see not only the country as a whole, but its main surface features such as rivers and mountains as well—though of course, what we see is the record of the outcome of that process of observation. So, observing an economy is a bit like an exercise in cartography, where we are mapping economic rather than physical space and observing and recording the economic society at the same time.[7]

There is at least one very important difference: the small bits of the economic society that can be observed separately do not fit naturally next to each other as they might do in mapping. This makes it more like doing a jigsaw puzzle, but a very difficult one because we do not have the set of pieces, we do not have "the picture to go on" (that is, to guide choices about how to put them together), we do not even know the dimensions of the whole, let alone its shape, and there are no recognizable edge pieces to help us. The individual elements have to be defined, observed, recorded in numbers, labeled, and categorized before they can be fitted together in a way that makes sense This might seem a hopeless task, akin to the redactive and synthesizing task of making "a general observation" on the state of disease in the nation of eighteenth-century France that we find in J. Andrew Mendelsohn's essay in this volume. But twentieth-century economists held an advantage, for their jigsaw of economic numbers had to fit into an overall conceptual framework designed to ensure that the complete economy is covered. Only then could the economist "observe" a numerical picture of the economy as a whole and, more importantly, see something of its hidden internal relationships. Or, following our earlier analogy, it is as if, having mapped the surface of an economy by recording each little bit at a time and finding out how they can be fitted together, economists expect to reveal something of its internal and underlying, that is, its geological, structure. Somewhat more whimsically but no less daunting, in the early 1950s, the terrain was pictured as "a vast chessboard" on which could be laid out, one on each square, the people, resources, and economic claims of a nation's economy: "Our celestial economist [like Laplace's supreme mathematician with respect to the physical world] might, as time unfolded, picture to himself a succession of such giant chess boards."[8]

Neither the identity of the parts nor the aggregation of these economic accounts was self-evident. These original scholars, Kuznets and Stone, had devised the concepts and techniques of national income accounting (NIA) for their own "advanced" economies. And they doubted that one set of rules and definitions would enable them to measure all economies in the same way. Kuznets had argued for the particularity of each country in each historical time period and thus thought that comparison over space and time was not likely to be viable using NIA.[9] While economies might have certain features in common, categorizing the variations would be more important than observing the commonalities. And it was because of the differences they *perceived* in economic and social arrangements that economists involved in the initial development of NIA understood the project as a method for bringing the economy into observation. That is, while the accounting framework helped one to see an economy, it was not originally understood as a standardizing instrument that also enabled comparison between countries, because the concepts did not necessarily fit all economies in the same way and so their measurements would not be comparable.[10] Thus, following Stone's publication of his NIA approach in 1941 (with James Meade), the question immediately arose: were they "universally applicable," that is, could they be applied to "primitive" economies?[11]

To answer this question, "an experiment" began in the early 1940s to see whether the system could be applied to the British colonies, and the economist Phyllis Deane was hired to conduct this "test" by constructing national income accounts for two areas of Central Africa and Jamaica.[12] It is the African countries that concern us here. Following an interim report based on research carried out during wartime London (published 1948), she carried out field visits to Africa in 1946–1947 before her longer report (published 1953).[13] She then acted as a member of the advisory committee overseeing a similar project begun in 1950 to construct national income accounts for the Nigerian economy, which forms the second case study for this essay. This was a team project of two economists and a statistician: Alan Prest visited that economy twice, working with Ian Stewart, who spent a whole year in the field, and Godfrey Lardner of the Nigerian Secretariat in Nigeria.[14] Stone—now widely regarded as the father of national income accounting—was actively engaged on the advisory committee of both these colonial projects, along with his partner in this NIA work and at home, Feodora Stone.[15]

Looking Secondhand, Finding Numbers

The NIA system, as constructed by Stone and his collaborators, provided three things for the observing economist. First of all, it provided a framework for organizing a picture of the economy, one that not only brought the elements of the economy into range, but also created a form of perspective so that the economist could make sense of the whole, and did so in a way that ordered its elements and provided a means to understand their relationships. In fact, the accounts created three different ways of seeing the economy at once: the *income*, *output*, and *expenditure* perspectives, or as Deane promised in her first report of 1948, the NIA "provides a three-dimensional picture of the national economy."[16] These three alternative ways that economists visualize the economy provide theoretical or conceptual perspectives.

Second, national income accounting involved a new form of complete economic census that was supposed to measure the "aggregates" by counting separately and independently all the flows of incomes, products, and expenditures (in monetary terms) for each of these perspectival columns for the whole economy. We can see something of the task in table 1 from Deane's first report (fig. 12.1).[17] The aggregate *income* was recorded in column 1, made up from the income obtained by the different groups of actors in the economy: total wages earned by workers, profits gained by capital holders, rent gained by landlords, and so forth (the traditional economic categories). Column 2 recorded the aggregate *product* or *output* again within traditional

THE LOGIC OF THE FUNDAMENTAL TABLES

TABLE 1. The simplest case of the income-output-expenditure table

I	II	III
Net national income	Net national output	Net national expenditure
1. Rents	7. Net output of agriculture	14. Expenditure on goods and services for current consumption
2. Profits	8. Net output of mining	
3. Interest	9. Net output of manufacture	
4. Salaries	10. Net output of distribution	15. Net investment
5. Wages	11. Net output of transport	
	12. Net output of other services	
6. Total net national income	13. Total net national output	16. Total net national expenditure

FIGURE 12.1. From Phyllis Deane, *The Measurement of Colonial National Incomes: An Experiment*, National Institute of Economic and Social Research, Occasional Papers, 12 (Cambridge: Cambridge University Press, 1948), 9. Reprinted with the permission of Cambridge University Press.

categories, namely, agriculture, manufacturing, and services. Finally, it embodied the new Keynesian conceptual categories of *expenditure* in column 3: aggregate consumption, aggregate investment, etc. These columns—once constructed—could be manipulated to reveal the behavioral mechanisms thought to make the aggregates change over time, or they could be subdivided in other finer-grained ways to reveal different hidden structural features of the economy. As Deane wrote in 1953, the economic policymaker

> wants to be able to see each of the constituent items in the network of national economic activity not only as a separate feature of the national accounts, but also as a factor influencing and influenced by other activities. . . . A colony which has processed its economic data by producing a system of social accounts has marked out the chief outlines of its economy and made it possible to observe the structural content and changes therein as a connected picture, even though the uncertainty of some of the outlines may leave parts of the picture rather blurred.[18]

Since the framework was an accounting one, in principle, these three aggregates of *incomes*, *products*, and *expenditures* displayed in the columns should "balance"—that is, be equal to each other: a triple-entry economic bookkeeping.

Third, the accounting system came with a manual of instructions that contained the definitions of the bits or elements to be observed and recorded and told the economist how to treat them, that is, how to adjust them (if necessary), and how to categorize them into the various aggregate columns and boxes. So the manual is a field guide enabling the economist to recognize and classify the various incomes, outputs, and expenditures that they observe themselves or that have been recorded by others, while simultaneously providing a set of regulations for their treatment as measurements to ensure that the overall tables balance. As Meade and Stone suggested, their 1941 paper provided such a manual, but not an exhaustive one; there was still much work to be done by any economist applying the accounting:

> [A]n outline has been given of the main problems of definition which must be solved in order to ensure that the tables balance. There are, however, a thousand and one small problems of definition which arise in attempting to measure the individual items in the different tables.[19]

Since NIA had been invented to fit Western economies, the problems of application to those economies was to a considerable extent covered in the accounting framework and rule book.

For Deane, based in London during the war, looking at these African economies secondhand, it seemed at least worth trying (thus the terminology of "experiment") to construct national income numbers for them. She gathered together all the available reported data on her target economics from many different sources—however narrow in focus, however fragmentary, and however out-of-date.

Yet there were still a "thousand and one small problems of definition" to solve before observations could be recorded in the correct boxes within the NIA. Which activities counted as output? What was the difference between cultivating land and collecting wild food that should make one an economic activity (something was produced) and the other not? The convention she used was where a tangible good was produced, the activity counted: growing crops counted but collecting firewood did not! There were many problems about where to draw lines for recording purposes: around production for the market, or around all production that created goods? around a country or around its nationals? among others. This was desk work of meticulous accounting, of classifying and fitting together unruly numbers, not the speculative observation of economists' "armchair work" we see in Harro Maas's essay in this volume. And while this secondhand mode of looking from the desk was not an easy way to observe a whole economy, Deane nevertheless concluded that these new concepts of NIA based on "Western," "developed," or "advanced" economic experience might be adapted to illuminate the African economies.

The complete accounting system was designed to provide a discipline to ensure that each and every part of the economy was recorded and fitted into the framework somewhere, and not counted more than once. The demarcation criteria were critical for categorizing activities in order that at least some kinds of measurements could be recorded for each of the columns. In principle, the total measurement for the national economy should be the same from each of the three perspectives: they should all three end up "in balance." And, since each column was built up from "independent and distinct calculations" and the data were "differently derived and differently classified," the different columns would operate as a checking system to make sure complete coverage was gained.[20]

Unfortunately, this balancing check did not quite work out as expected. The two most obvious difficulties were, first, that Western concepts of NIA excluded goods and services produced and used only within the household, and, second, that economic activity consisted only of activities that lead to exchanges for money. Of course, in the context of Western economies, these two assumptions went along together, for typically goods and services produced and used within the household were not marked by monetary exchanges between household members.[21] Deane was quite aware that economies of Central Africa she was studying experienced considerable economic activity that would not be counted in NIA either because it was inside the household or was traded, bartered, or gifted without monetary exchange. Yet she also believed that these economies were not pure subsistence economies: they were economies in which most people's activities were mostly nonmonetized, but most people had some monetary income and some monetary transactions. In such economies, column 1, incomes, was not a good measure of economic well-being, which was better measured by column 2, outputs, provided that monetary values could be assigned to them. Thus, Deane essentially carried out double-, rather than triple-entry bookkeeping, collapsing—for practical rather than theoretical reasons—incomes and outputs (or production) into one column and having a second column of "consumption" (rather than "expenditures") because of the low level of monetized exchange.

Another independent set of checking came by using other "observers" to triangulate the evidence. So Deane's first attempt at constructing the national income accounts out of "secondhand" materials was sent out for comment to "a few informed observers," that is, observers in Britain and Central Africa who had firsthand experience of that region.[22] They were then revised and published in her 1948 preliminary report. Meanwhile, in 1946–1947, she had been out looking in the field: visiting copper mines, doing survey work in villages, and burrowing through the census office.

Looking Firsthand in the Field, Seeing Fog

In Deane's second report of 1953, after visiting Africa and observing some of her economies firsthand, we find a change in tone. She admitted that in such economies as those she was trying to measure, national income accounts were not such as to enable the investigator to see sharp lines and clear elements, but rather "a few large shapes in a thick fog." In her view, the problem lay not in

> the margin of error arising from inadequate statistical data that hinders most the application of national income estimates to practical policy purposes, it is the fog that surrounds the concepts themselves.[23]

Whereas in her first report the problem had seemed to be how to fit the secondhand recordings of observations made by others into the NIA concepts, here the difficulty was how to fit NIA concepts to the economy that she saw herself.

Deane argued that there is always a problem of the fit of economists' concepts to the activity they wish to observe and so measure, for such concepts tend to be "vague at the fringes." In a developed economy where most economic transactions are market transactions, the ones that don't quite fit the conceptual definitions are either excluded by definitional conventions (e.g., gifts) or because they are believed relatively small (e.g., barter). When those awkward nonmarket transactions at the edge of Western economic consciousness are the main activities in the African economy, the problem of observation using the definitions to guide the observer manifestly changes: "What is the fringe in one society, however, is not necessarily the fringe in another."[24] While this did not necessarily invalidate the overall NIA project, it did cause her huge difficulties in applying the conceptual apparatus within the framework for observing and recording the economic structure within her African economies.

The main problems in taking the NIA to Africa as a framework for observation and recording measurements continued to be the assumption that Deane had struggled with—namely, that nonmonetized exchanges and exchanges within the household were excluded. This lead to some startling paradoxes in the African context. For example, in a salient example that runs through this literature, a marriage payment in money might be included in the accounts of transactions because it was monetized, but the value of nearly all the staple food produced might not because it was not exchanged for money—precisely the opposite of what the "Western" national income investigator would want to count.[25] For such economists, the use of money to indicate an economic activity did not provide a valid account of the values of incomes, products, or expenditures, and where it was used, there were doubts about its role as a viable measuring stick of those activities.

Nevertheless, Deane still regarded this lack of conceptual fit as one of degree rather than intrinsic. She argued that the NIA concepts never fit exactly to any Western economy either, and that the problem might be considered equally to lie in the eyes of the beholder, the Western economist in Africa, for whom

> logical compromises . . . have to be made in practice. . . . based on an inadequate background of sociological data and are therefore more arbitrary than they would be for an investigator for whom the community's accepted ends of economic and social policy are part of his native background.[26]

Note the benefit here that Deane suggests comes from the close knowledge and engagement of a "native background," not a professionally distant and scientific knowledge, for in the social sciences, background knowledge is experiential knowledge that comes from living in a community rather than acquired by scientific means.[27] Close-up experience was necessary to make sensible observations and recordings, that is, to see through the considerable fog to make classification decisions.

In Deane's perception, the economies she studied were overwhelmingly "village economies," a type of economy that was semisubsistence, and for which the NIA concepts were not just ill designed but frankly "alien" yet must somehow be adapted to the task.[28] Her firsthand experience of the village economy came in visiting three areas in what was then Northern Rhodesia (now Zambia) a territory of 290,000 square miles, with a sample census count of 1.7 million people in 1950 (there had never been a full census), twenty-one different ecological/agricultural systems, and over fifty different tribes. In April–May 1947, with the help of a group of locals, anthropologists, and other expatriates, she undertook what she describes as her second "experiment" (following the first "experiment" of conducting secondhand observation in London), namely, to understand the "village economy" by observing it firsthand with the help of a survey questionnaire. The local anthropologists helped her in choosing "typical" villages, in designing the survey, and in introducing her to the villages. She clearly enjoyed working with the anthropologists and commented favorably on the

> enlarged viewpoint and the stimulus which can be gained from seeing one's data through the eyes of other observers while they are actively being collected and analysed, are advantages which an economist can give as well as receive.[29]

But, being a member of a different tribe from the anthropologists—for economists, recall, observation entailed measurement—Deane naturally sought to record numbers from her survey observations. As she said, while the qualitative data (whether people eat eggs or milk, or whether "little girls of more than six years of age regularly pound maize") provide "the flesh and the form," the quantitative material is "the skeleton" without which the qualitative data would be a "shapeless affair."[30]

The main difficulty in her survey work lay in defining, or perhaps finding, households, the base unit in the NIA within which (recall) exchanges were not counted. Here we find Deane grappling with the answers to her survey questions to turn her observations about the household into something typical that might be recorded in quantified form. For example, listen to this stream of observational statements about the village economy:

> Theoretically, a man made a hut, a garden and a granary for each wife and a garden and granary for himself. In practice, his newest wife probably shared a granary with him and worked on his field. She might even share a granary with another wife. More frequently, household equipment, such as a mor-

> tar and pounder, was shared, although there was usually a fairly definite understanding about the actual rights of ownership, Where a man earned a money income from his garden or as wages there was no recognized share for each household. He allocated it as he chose. Where a woman earned income through sale of produce or beer brewing the money was hers to be spent on her household. Sometimes two or more wives would combine households in preparing the day's meals. Hence, in practice the accounts of the six households in a polygamous group were usually so intermixed that for most purposes it was convenient to collect the data and present the accounts together.[31]

If the economic relations between husband and wives were complex, the economic lives of young bachelors and children were even more difficult to observe, let alone to describe. Recording the household's economic activity depended upon being able to categorize its observed behavior, but it proved almost impossible to pin down the household even as an observational unit:

> If one travels round the village on a person to person basis and asks each woman how many unmarried children she has she would include in her answer young children living with their grandmother, perhaps in another village. . . . If one travels on a hut to hut basis and asks each woman how many children she has sleeping in her hut the answer will include such children as those who habitually eat with their grandmother and others who are, say, the children of migrant brother.

How could she even define, let alone research, a household when her fieldwork observations told her:

> The principal difficulty in surveying was that the sleeping household, the eating household, the income household, the producing household, and the spending household all represented different combinations and permutations within one wide family group.

Whereas of course, for the economist,

> the ideal household for accounting purposes is the group of persons eating and sleeping under one roof and pooling their income.[32]

Observing the village economy made Deane deeply aware of the problems of applying NIA concepts to the African village economy. The household could not be well defined as a recording unit, and not even seen in one place at one time as an observable unit.

It was not just that the concepts did not fit the economy, but the incredible variety of economic experiences, both within villages and between those of different districts, made labeling, categorizing, and otherwise organizing her observations extremely difficult. Her observations just refused to fit easily into either her own background experience or her economic concepts. At the same time, the absence of the universal measuring stick of money made it difficult to quantify the things she did observe in her survey work into a coherent picture.

> It is doubtful whether it would ever be possible to define income so that it meant the same thing to the African villager and the European town dweller, or to the African in his subsistence habitat and the same African in temporary urban employment.[33]

Deane thus concluded that the problem of applying the NIA framework to the village economy, the semisubsistence economy, was not entirely a data problem but more of a conceptual problem, one that she had looked at in her London work but that she did not really see until her fieldwork made her more deeply aware that

> it is not clear what light, if any, is thrown on subsistence economies by a science which seems to regard the use of money and specialization of labour as axiomatic.[34]

Deane had underlined the limitations of theoretical concepts/categories of NIA and found that they don't enable you to see very far when you visit other kinds of economies for which they were not devised. She saw the problems as practical ones about how to draw lines between the things observed in order to place them into the already formed NIA categories, and since many of the village observations neither fitted those boxes nor could be recorded into quantified form, her survey work did not enable her to fill in much more of the jigsaw puzzle of the whole economy in the social accounts. Nor did the clash between conceptual and observable categories that Deane experienced from her direct observation of the economy lead her to a new set of conceptual definitions that would have enabled her to turn her observations into usable measurements. Indeed, her comment on her experience in the field seems to betoken a radical skepticism about economics.[35] Nevertheless, Deane's pioneering work formed something of a model for later scholars such as Prest, Stewart, and Lardner (PSL), and in turn, their development of the NIA equally proved a model for later workers in the field.[36]

Seeing with New Categories

Lardner, Prest, and Stewart, like Deane, gave an account of their attempts to record the Nigerian economy that fairly bristles with firsthand observational experience. And like Deane, they came to the conclusion that the overall concepts of "Western" (their label) national income accounting did not fit

Nigeria: the main field guide distinctions between categories could not be made in the field because the observer could not recognize and separate economic activities. The distinctions between production and use, between production and consumption, between production and distribution, between production in manufacturing and in agriculture, or between income, wages, and profits, could not be made in practical terms. Another problem they noticed lay in the lack of standardized and precise measures, in particular, the use of handy consumer items (such as the tin containers in which cigarettes were sold—the "cigarette cup") as measuring units. So neither "Western" concepts nor recognized standardized measurements were viable in Nigeria.

Where Deane, when looking from her desk in Britain, had created different categories within the perspective of incomes in her column 1, and her practical solution has been to collapse that column into the outputs column 2, PSL's move in the field was far more radical. Western economy guidelines were based on the market activity of the household, drawing a circle around it so as not to count the economic transactions within that circle. We have seen how problematic this was for Deane's observations and her recordings in the field, where most economic activity seemed to take place within the household, not beyond it, and how she had struggled in her fieldwork to pin down and record exactly what a household consisted of in order to value subsistence production. PSL's solution was to change one of the fundamental definitions in the field guide, namely, to move the boundary of what was to be counted inward and to draw a smaller circle:

> The contrast of our treatment with the "Western" one is that of drawing the ring round the individual as opposed to drawing it round the family.[37]

This tight circle enabled them to capture in their observations, and so to count, all economic transactions between individuals within the household group as economic activity for the NIA.

PSL's redefinition was both conceptually cleaner and more practical than Deane's solution, for it enabled them to make use of monetary transactions wherever they occurred. This included, infamously, making use of the bride payment as payment for all the services that a wife produced for her husband—not just gardening, farming, and cooking, but also childbearing and -rearing.[38] Their argument was not only consistent with the wider NIA requirement to count all economic activity, but also was based on their field observations, for the relationships between husband and wife were in many ways more commercialized than those in Western economies:

> [M]any cases have been known of wives suing husbands for debt; women's earnings from trade cannot be touched by husbands; food provided from

> women's own cultivation for the general use of the family is often on a loan basis and delicacies such as pastries are only provided for cash. . . . Altogether, it seems reasonable to argue that commercial transactions exist inside the family as well as outside it.[39]

It is paradoxical that it was the new concepts and measuring structures of the NIA that prompted economists such as PSL to recover in Africa the original Greek notion of economics as the activities of the household. Yet it is surely more deeply ironic—as Prest and Stewart noted—that the exchange relations they observed within the household in such "primitive" economies were far more monetized and so economically advanced or "rational" (according to the notions of twentieth-century economics) than those apparently rather primitive households, with a more gendered division of labor and power over money, of the "developed" or Western countries where these new concepts of NIA had been developed.

While finding bride prices was relatively easy, finding monetary valuations for other exchanges and outputs that would enable them to construct the NIA was often more difficult. Here they were more conservative and refused to rely on too much imagination to construct those missing numbers:

> Where goods and services are not marketed it is possible to go as far as asking what they might be worth if they were. To take the further step of inventing functional relationships such as demand and supply curves, or even appearing to invent such relationships, seems to us unwarranted in a country where consumption and production activities are inextricably mixed and where enterprises and households may often be synonymous. Further complications also result from the shadowy and ill-defined nature of many economic units. . . . Where complex economic transactions do not exist, little purpose is served by making them appear to do so.[40]

Whereas Deane had been unable to construct a separate production account from her London desk because so much of the village economy was not market-based or exchanged beyond the household, in the field PSL concentrated on the output column of the NIA and produced a figure of £597 million, which matched almost exactly their total expenditure column. In constructing the income column, however, they found a massive shortfall of 83 percent compared to those other totals. This was a direct outcome of their changing the definitions within the NIA, for their output account now included measurements for the intrahousehold production, which was of course substantial. And expanding the elements of economy brought under observation by including household production opened up another vast area of economic activity previously overlooked, namely, internal trading:

> There is no clear distinction between subsistence and trading activity. Rather, the two are inextricably bound together. . . . There is no tribe or group of villages which can be unequivocably labelled as subsistence producers only; much the most common situation is that of people and areas consuming part and selling part of their output either for internal use or for export.[41]

PSL's success in tracking the internal market, both in size and movements, brought another aspect of the economy into economic observation as can be seen in their commodity maps of internal trade (or "vulture's eye views," as they called them!), which even revealed a vibrant trade in some previously unreported commodities or unreported directions (e.g., fig. 12.2).[42] On this new basis, it was hardly surprising that their calculations of the average Nigerian per capita income was much higher than expected since they were counting substantial household production and substantial internal trade, which had been omitted in previous attempts to estimate the well-being of the country. So, by changing the definitions of the subunits, they had not only filled in a much larger section of the jigsaw of measurements, which they knew had previously been left missing, but had—in effect—expanded the area covered by the framework itself into areas they had not known about.

Observing this expanded area of the internal market had presented other challenges and involved a network of other economic observers within Nigeria. Like all Western economic investigators/observers on such NIA missions, Prest and Stewart were reliant not just on local economists (Lardner) but also on teams of local noneconomists, not in an any economic observatory, but people in the field doing other things, such as district officers and local state civil servants, agronomists, professional social scientists (especially anthropologists), and perhaps even more important, those active participants in the economy such as workers and managers in firms and traders at marketing boards. They relied on all these people as so many observation posts. While secondhand field reports such as the official statistics of ports of entry of goods (which captured the international exchanges) were the kind of record that would reach the Colonial Office in London, these local observers provided many other local reports that never traveled to London—surveys of canoe trips (note the well-rounded estimate of canoe traffic in fig. 12.2), of records of ferry passengers and freights, of bridge crossing points, of railway records, and so forth.[43] The records of observations from these different kinds of local observers and field points of observation were taken in, interrogated, and reassembled first in an evidence triangulation to make PSL's maps of trade, but then to find their places as pieces of the jigsaw puzzle of measurements for NIA.

Observations travel in packs, as Daniela Bleichmar has shown in her ac-

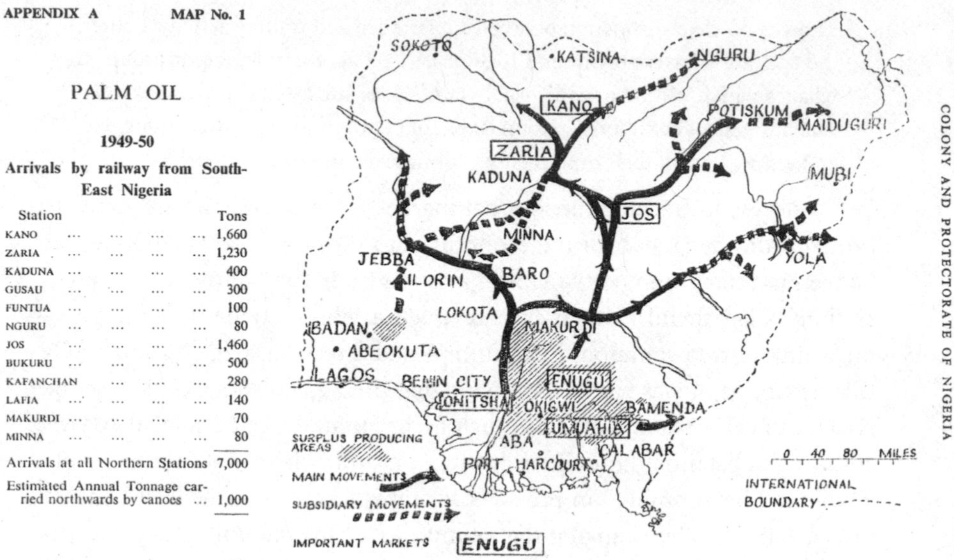

APPENDIX A MAP No. 1

PALM OIL

1949-50

Arrivals by railway from South-East Nigeria

Station	Tons
KANO	1,660
ZARIA	1,230
KADUNA	400
GUSAU	300
FUNTUA	100
NGURU	80
JOS	1,460
BUKURU	500
KAFANCHAN	280
LAFIA	140
MAKURDI	70
MINNA	80
Arrivals at all Northern Stations	7,000
Estimated Annual Tonnage carried northwards by canoes	1,000

98

COLONY AND PROTECTORATE OF NIGERIA

FIGURE 12.2. From Alan R. Prest and Ian G. Stewart, *The National Income of Nigeria, 1950–51*, Colonial Office. Colonial Research Studies, 11 (London: HMSO, 1953), 98.

count of "botanical travels."[44] Each observation made in the field comes with a group of related ones, for in economics, the pack is made of different bits, and assembly is more important than parsing or dissection. These related bits are not mere context or detail to be discarded, but the other pieces of a jigsaw puzzle of economic relations must be fitted together in a set of economic accounts to make sense of the economic life of the smallest retailer as much as of the largest mine. This fitting-together accounting problem is on a grand scale in the NIA. It was one that Deane was familiar with from her own desk work, and that Wolfgang Stolper, who followed in PSL's footsteps a decade later, commented on several times:

> I find that I am happiest when I can work with figures, push a slide rule or a calculating machine and make endless details fit into a grand pattern.[45]

This grand pattern was the creation of consistency in the NIA. This was not a simple macroeconomic adding-up problem—it entailed complicated accounting calculations using the perspectival frameworks from economic theory to create consistent observations of the whole from the many individual parts. Stolper thought this a matter that required not only an understanding of the conceptual space of NIA and an expertise with numbers, but also the faculty of imagination:

> I work, for a theorist, with imagination—not intuition, but just imagination. I have the ability of being able to extract a maximum of information from scanty data, but this requires the painstaking study of detail.[46]

These cases, covering the work of Deane, both at her desk and in the field, and PSL, form just two examples of a process in which a generation of young economists went into the field to observe and record the national income of almost all non-Western economies in the 1950s and 1960s. The purposes may have differed—they might have been part of colonial, postcolonial, development, or aid missions. But by making such economies "visible" as separate economies, the NIA concepts and measurements became important elements in twentieth-century discourses way beyond those of academic economics.[47]

Observations on Economic Observation

Scientific observation has been associated by commentators from the history and philosophy of science to sit closely with two different kinds of epistemic genres. As we see from other histories of observation in this volume, on the one hand the observer is situated in relation to theory and experiment, while on the other the observer is situated in relation to classification/categorization and dissection/assembly. Observing with national income accounting, with its theoretical perspectival structure and its jigsaw task of categorizing and assembly, places these economists within both these genres. Both the theoretical perspectival requirements for looking, and the categorization requirements of seeing, were needed to turn observations into quantitative records and assemble them together. These activities of observation, in turn, involved qualities of perception, imagination, and engagement in the observer.

For these early national income accountants, the accounting system was designed to provide the frame of reference within which all the observed elements could be fitted together to provide three different perspectival accounts of an economy and do so in a way that revealed aspects of the hidden internal structures and relations of that economy, even the size and shape of that economy, for there were no natural boundaries and "the economy" was an amorphous and largely invisible object. But economists who tried to observe this difficult object by staying at their desks and by using the secondhand reports as the basis for their accounts had to use not just their skill and economic intuition, but their economic imagination as well, to fill in the gaps they found in the records. In using such secondhand observations, economists found themselves quite strongly bound by the theory or concepts of the accounts—for remember it is the economist who must choose how to categorize the pieces of recorded data into the relevant boxes in the accounts. This created two pitfalls for the unwary economic imagination. One was the danger of imagining the behavior of the missing elements to make them fit exactly with the already given conceptual categories. The other was the danger of forcing the elements that have been observed to fit exactly to those existing conceptual categories because of the difficulty or inability to imagine new conceptual categories that would be more appropriate.

We can see that such dangers were recognized: both in PSL's refusal to imagine complex economic transactions where they did not exist, as well as in Deane's refusal to fill in the gaps that would fit secondhand reports to Western concepts. Yet both managed to steer an interesting middle path to create the alignment between theory and observation needed to make records of those observations. Stolper was willing to recognize the importance of imagination in order to fill in measurements for his missing observations, and, like Deane, was meticulous in working within the perspectival constraints provided by the national income concepts. Just as the Renaissance painters used the new linear perspective to constrain the representations offered by their imaginative recreations of history and myths to provide viewers with a sense of real-life observation, so economists used such tricks, or rules, of perspective given by national income accounting to line up their recorded observations of the real economic objects within the correct cells and columns.

Traveling to the site of observation turns the problems of imagination to those of perception. An economist who stays at home, like an artist of nature, avoids the cognitive dissonance involved in observing elements that don't fit your old experience but that you cannot quite fit into any new sense experience. Travel may broaden the mind, but the eyes may still take time to adjust. Europeans in Australia originally saw and painted landscapes with blue skies and brown earth containing trees with green leaves and brown trunks. Later generations came to see gray leaves and white trunks and burnt orange earth against purple skies. In a similar fashion, unfamiliarity gave way to new understanding for these economists who initially felt their task of observing the unfamiliar was quite akin to that of the anthropologists (who indeed helped them with their observations). They found the economic society they were studying so unfamiliar that they could not make full use of their experience of their own home economies to recognize and make sense of the economy of the countries they set out to observe and measure in terms that would fit the national income accounting framework.

Deane struggled to observe the Rhodesian village economy firsthand. These were struggles of perception: she looked from the viewpoint of her own life experience in Britain's economy and found it difficult, for example, to see the economic life of the African household in such a different form. Prest and Stewart took a more definitive step in seeing in Nigeria the household economy as a set of individuals who make economic exchanges. Lardner's experience here was surely invaluable for his economic perception was as a Nigerian. Since both Deane and Prest were close to the center of development of NIA, and since it was still a project in development, they felt able, indeed, found it a necessity, to adapt the accounting definitions and rules to their own ends of colonial social accounting in the field.

I have separated out these issues of imagination and perception and related them to conceptual knowledge and background experience for an important reason. It is a familiar point that all observation is theory-laden, that we make observations and see things against a background of our theories. Here we have two different kinds of background knowledge against which economists make observations. One is the abstract, conceptual knowledge of economic science, against which observations are made and to which they must be compared for matters of fit, which I have associated with the quality of the imagination, but an imagination suitably bounded by the requirements of conceptual perspective. Equally important for observing in the social sciences is life experience, which I associate with perception, for in social sciences everyone has general knowledge from the experience of living in their own economy and society.[48] This personal knowledge from observation and experience is also a valid kind of knowledge that informs an economist's observation of another economy. Deane, for example, referred to the importance of the eyes of the beholder: eyes that see familiar things are valuable just because they see them against a background of similar life experience and so can make sensible judgments about them. This can be contrasted with the value of eyes that notice things just because those things are unfamiliar, as these Western economists surely did during their African fieldwork. Deane commented on both these issues of perception: whereas the former eyes may have problems seeing new things, the latter have problems making sense of what they observe. Perception, molded by the observations of life experiences, is an interesting double-edged sword. Both kinds of preknowledge—scientific and experienced—play a role in how economists observe. Together they form the backgrounds from which economic observers look out, and the knowledge base within which they see things.

Distance also matters—in looking and in seeing. Scientific observation using the NIA in these cases has been portrayed as both a busy set of activities of *close engagement with the subject*, while as the same time, conversely, one requiring the economist to *take an objective, distant stance*. The activity of making economic observations from data, that is, from the secondhand reports of others, is an activity of sorting out the pieces, abstracting the relevant bits, categorizing them, and fitting elements together into their place in the pattern: it requires infinite patience and professional care. This all relies on a certain objectivity that comes from creating distance between the observer and the object. Objectivity here comes not from a professional stance of the social scientist but from the NIA, for it is that accounting framework that enforces the economist to place the seen objects into columns that will construct their three-dimensional perspectival account. But this focus on technologies of distance, while beguiling in many respects, seems to be equally balanced here by the virtues for the economist of close experience with the subject matter in ways that resonate with the histories of observation discussed in a number of other essays in this volume (especially in Elizabeth Lunbeck's essay on empathy, Otniel E. Dror's on observing emotions, and Theodore M. Porter's on Le Play's way of observing society).[49] It was this firsthand activity of observing that enabled the economists here to overcome the problems of fitting their economic concepts to economic life and to rethink those conceptual elements in order that the work of observing from their secondhand records might be more fruitful. While the accounting discipline created a consistency in overall perspective relevant for seeing the whole economy, here, just as in archaeological history where the recording hand enforces the eye to see, it was the firsthand, personal, looking in the field, that enabled economists to seek out, observe, and record those pieces that made up the whole.

Notes

1. Observation in economics requires multiple histories (as for other fields). Here it is understood as field- and desk work, using an accounting framework. Other histories suggest the following: a process of introspection, see John Neville Keynes, *The Scope and Method of Political Economy* (London: Macmillan, 1891); "armchair observation" (see Harro Maas, this volume); and as statistics, see Judy L. Klein and Mary S. Morgan, eds., *The Age of Economic Measurement*, annual supplement to *History of Political Economy*, vol. 33 (Durham: Duke University Press, 2001). The recent development of experimental economics may produce a new mode of observation in economics.

2. The classic example is Oskar Morgenstern, *On the Accuracy of Economic Observations* (Princeton: Princeton University Press, 1950).

3. I thank Johannes Cramer of the Technische Universität (TU), Berlin, for showing me hand-drawn recordings of the Berlin Wall, and Simona Valeriani for introducing me to these charts: "Harris Matrices," see Edward C. Harris: *Principles of Archaeological Stratigraphy*, 2nd ed. (London and New York: Academic Press, 1989).

4. Economists' experience may not be unusual: historians of science have long understood scientific observation as a process involving intermediating instruments, where observation

processes are carried out by technologies which also reverse the link between observing and recording compared to the practices of, say, the archaeologist.

5. Adolphe Quetelet had already linked modes of observing the planets and stars with those of man in the development of statistics, and we can think of the statistical bureaus of the twentieth century as sharing—for the social sciences—the same political overtones, mass projects of observation, and associated heavy requirements of calculation that have been taken as the hallmarks of nineteenth-century observatories: see David Aubin, Charlotte Bigg, and H. Otto Sibum, eds., *The Heavens on Earth: Observatory Techniques in the Nineteenth Century* (Durham: Duke University Press, 2010). The important point to note is that the form of measurement discussed in this chapter is not a statistical one, but an accounting one (even when the original bits of data come from statistical offices).

6. For Kuznets, see his "National Income," in *Encyclopaedia of the Social Sciences*, ed. Edwin R. A. Seligman, vol. 11 (New York: Macmillan, 1933), 205–44. For Stone, see James E. Meade and Richard Stone, "The Construction of Tables of National Income, Expenditure, Savings and Investment," *Economic Journal* 51 (1941): 216–33.

7. On earlier attempts to observe and record the economy as a whole, see, among a considerable literature: Sybilla Nikolow, "A. F. Crome's Measurements of the 'Strength of the State': Statistical Representations in Central Europe around 1800," (in Klein and Morgan, *Age of Economic Measurement*, 23–56) and the classic text, Paul Studenski, *The Income of Nations*, 2 vols. (New York: New York University Press, 1958).

8. Harold C. Edey and Alan T. Peacock, *National Income and Social Accounting* (London: Hutchinsons, 1954), 215 (bracketed phrase in the original), was one of the first practical manuals for economies where data were thin on the ground.

9. Kuznets, "National Income," 209.

10. Nevertheless, by the late 1940s, national income accounts had already become essential elements for various political actions: reconstruction aid (Marshall Plan aid, for example) required their calculation, for they not only provided a measure of general need but also identified specific needs.

11. The terminology quoted here and in the next paragraph comes from the opening pages of Phyllis Deane, *The Measurement of Colonial National Incomes: An Experiment*, National Institute of Economic and Social Research, Occasional Papers, 12 (Cambridge: Cambridge University Press, 1948).

12. Phyllis Deane studied economics during the late years of the Great Depression and went straight from her first degree into research, a move that would have been extraordinary for a female student but for the war years (see the interview with Phyllis Deane by Nicholas F. R. Crafts in *Reflections on the Cliometrics Revolution: Conversations with Economic Historians*, ed. John S. Lyons, Louis P. Cain, and Samuel H. Williamson [London: Routledge, 2008]).

13. See Deane, *Measurement*; and Phyllis Deane, *Colonial Social Accounting* (Cambridge: Cambridge University Press, 1953). The two areas of Central Africa she studied were then called Northern Rhodesia and Nyasaland.

14. I refer to the Prest, Stewart, and Lardner project in the text as PSL, though Lardner's name did not appear on the book. Alan R. Prest and Ian G. Stewart, *The National Income of Nigeria, 1950–51*, Colonial Office: Colonial Research Study, no. 11 (London: HMSO 1953).

15. Both these projects were overseen jointly by the Department of Applied Economics at the University of Cambridge (directed by Stone), the National Institute of Economic and Social Research (NIESR) in London, and the UK Government's Colonial Office. These intersections

go further: both Deane and Prest were members of the Cambridge University Department of Applied Economics, while Feodora Stone was also secretary of NIESR.

16. The terminology of "perspectives" draws on Bruno Latour's observation about the NIA in the context of his argument about immutable mobiles (see his "Visualization and Cognition: Thinking with Eyes and Hands," *Knowledge and Society* 6 [1986]: 1–40) but the sense is captured by Deane's own terminology here of a three-dimensional picture in *Measurement*, 8.

17. Deane, *Measurement*, 9.

18. Deane, *Colonial*, 3 and 8.

19. Meade and Stone, "Construction," 227.

20. Deane, *Measurement*, 8.

21. What this joint assumption might mean for NIA in developed economies was potentially substantial: Deane, *Measurement*, 19n reported figures for Sweden implying a 20–25 percent increase in national income if household activity was included. As we shall see, this was a much greater amount in Nigeria.

22. Both terms, Deane, *Measurement*, 4.

23. Deane, *Colonial*, 4.

24. Ibid., 116.

25. Ibid., 227.

26. Ibid., 128.

27. See Mary S. Morgan, "'Voice' and the Facts and Observations of Experience" (Working Papers on the Nature of Evidence: How Well Do 'Facts' Travel? 31/08 Department of Economic History, London School of Economics, http://www2.lse.ac.uk/economicHistory/pdf/FACTSPDF/HowWellDoFactsTravelWP.aspx), on the importance of this distinction for social science knowledge versus natural science.

28. Deane, *Colonial*, 130.

29. Ibid., 146.

30. Ibid., 119–20.

31. Ibid., 147–48.

32. Ibid., 148–49. Although she recorded some of the survey answers into quantitative form, she could not use them to improve the national income estimates.

33. Ibid., 227.

34. Ibid., 115–16.

35. Nevertheless, her radical skepticism did not stop her from later undertaking an even more heroic study to construct national accounts for the British economy going back to 1688!

36. These two projects prompted a number of similar studies in the 1950s and 1960s sponsored by the Colonial Office and the Colonial Economic Research Committee (part of an exit strategy from the colonies). Thus Peacock and Dosser, who constructed the NIA for Tanganyika (Alan T. Peacock and Douglas G. M. Dosser, *The National Income of Tanganyika, 1952–54*, Colonial Office: Colonial Research Study, 26 [London: HMSO 1958]), followed in paying tribute to Prest and Stewart. On PSL, see n. 14, above.

37. Prest and Stewart, *Nigeria*, 10.

38. Though the economic rationale seemed impeccable, it was of course, to Western economic eyes, somewhat shocking: they preferred to live with the paradox (attributed to Arthur Pigou) that measured national income should show a decline when a man married his housekeeper rather than to include exchange within the family.

39. Prest and Stewart, *Nigeria*, 10.

40. Ibid., 21.

41. Ibid., 9.

42. Ibid., 98–99.

43. Thus, for example, the insistence on fieldwork that Alan Peacock noted for the quality of NIA in colonial territories was highly dependent on accessing these local observers and their records (see Peacock and Dosser, *Tanganyika*, 3).

44. See the essay by Daniela Bleichmar in this volume, and at the meeting of the History of Scientific Observation group, Berlin, July 2007.

45. Clive S. Gray, ed., *Inside Independent Nigeria: Diaries of Wolfgang Stolper, 1960–62* (Aldershot: Ashgate, 2003), 141–42.Wolfgang Stolper, unpublished diary (available at the Duke Economists' Papers Project, Duke University Library), 121. Stolper worked in Nigeria in 1960–62, part of the time under Lardner: see Mary S. Morgan, "'On a Mission' with Mutable Mobiles," Working Papers on the Nature of Evidence: How Well Do 'Facts' Travel? 34/08, Department of Economic History, London School of Economics, 2008, http://www2.lse.ac.uk/economicHistory/pdf/FACTSPDF/HowWellDoFactsTravelWP.aspx.

46. Gray, *Inside Independent Nigeria*, 141–42.

47. See Daniel Speich, "Travelling with the GDP Through Early Development Economics' History," Working Paper on the Nature of Evidence: How Well Do 'Facts' Travel? 33/08, Department of Economic History, London School of Economics, 2008, http://www2.lse.ac.uk/economicHistory/pdf/FACTSPDF/HowWellDoFactsTravelWP.aspx; and Morgan, "On a Mission," for a more detailed sense of what this all amounts to.

48. See Morgan, "Voice."

49. On quantification as a technology of distance in the context of economic observation, see Theodore M. Porter, *Trust in Numbers: The Pursuit of Objectivity in Science and Public Life* (Princeton: Princeton University Press, 1995).

13

Seeing the Blush: Feeling Emotions

OTNIEL E. DROR

The nineteenth-century Scottish critic William Archer studied the production of blushes and sweat on the stage. In *Masks or Faces* (1888), the most important late nineteenth-century rebuttal of Diderot's *Paradoxe sur le comédien*, Archer addressed a series of questions to the leading actors and actresses of the British theater regarding their emotions: "Do you ever blush when representing bashfulness, modesty, or shame? Or turn pale in scenes of terror? Or have you observed these physical manifestations in other artists? Do you sweat 'in accordance' with 'the emotion experienced in the part'?" Archer's study of the players' passions was based on observations of the blush, pallor, and cold sweat as true indices of emotions since, unlike tears and the inflection of the voice, they were not "under the control of the will."[1]

Archer's logic resonated with that informing the contemporary investigations of Italian physiologist Ugolino Mosso, who observed his own fluctuating rectal temperature during variable emotions. Adopting the same line of reasoning as Archer, Mosso reported an increase of 0.7 degrees centigrade above his usual temperature on his return home from work in a state of aroused emotions.[2]

The renowned French psychophysiologist Alfred Binet, a contemporary of Archer and Mosso, also studied fleeting affective experiences, inducing emotions in subjects inside the experimental laboratory and observing the inflow and outflow of blood in the viscera and limbs.[3]

These diverse approaches to the study of emotions in three different national contexts testified to the emergence of a novel physiological framework for studying and observing the emotions in the laboratory, the theater, and the clinic. This new physiological framework redefined the nature of the observation of emotions while marginalizing a broad array of nineteenth-century approaches and techniques for observing the emotions, such as sympathy, embodied mimicry, and innate sensitivity.

In studying this major transformation in the observation of emotions, I distinguish between three different historical phases and two major paradigms of observation: observation in terms of "feeling" and observation in terms of "seeing." Observation in terms of feeling framed observation around the embodied-experiential reactions of observers. Observation in terms of seeing framed observation around the new technologies for gauging emotions. These technologies observed emotions inside the innards of the body—at the level of the internal milieu and the viscera.

Observation in terms of feeling was the dominant mode for observing experiential states prior to the advent of technologies for observing emotions. From the 1860s, observation of emotions was progressively framed around seeing through the mediation of technologies. During the 1920s, observation in terms of feeling made a comeback. The return of observation in terms of feeling and of the embodied-experiencing observer framed and engendered a category of "supremely extreme" emotions. Experimenters narrated the category of supremely extreme emotions in terms of their embodied-experiential reactions.

The shift from observation in terms of feeling to observation in terms of seeing during the last third of the nineteenth century reflected the convergence of the scientific study of the emotions and broader disciplinary and cultural transformations. These included the "split status of the face," the radical embodiment of emotions, the rise of the graphic method, and the modern "appropriation of subjectivity."[4] These broader developments framed and provided the context for the shift from feeling to seeing in observation.

The investigators of emotions who observed in terms of seeing during the last third of the nineteenth century critiqued a wide spectrum of existing traditions, modes, and systems for observing emotions. They argued that the viscera reflected a truer, deeper essence of emotion than methods that depended for their observations on the conscious awareness and self-reports of subjects or observers, or systems that depended on "superficial" facial expressions and gestures. They also suggested that observations of visceral emotions were easily converted into quantitative measurements, in comparison to the difficult process of quantifying verbal descriptions or facial expressions of emotions (the latter were quantified inside laboratories only during the twentieth century); that visceral emotions were independent of language and were hard to manipulate or skew deliberately; and that the involuntariness of visceral emotions presented a "natural" distinguished from the artificial/mannered/social expressions and gestures of emotions.

The emotions that were observed in terms of seeing through the mediation of technologies were no longer defined or identified in terms of subjective experience, overt gesture, facial expression, or verbal report. They were recognized in and through the viscera. These emotions were not unconscious or subconscious but were expressions of emerging *visceral-physiological* identities for emotions. Machines for observing emotions contributed to and consolidated this significant shift, whose implications extended into the twentieth century in the form of radically new conceptions of the emotions and in the development of the polygraph, biofeedback, the EEG, and various imaging technologies for observing the emotions. These developments also meant that in practice any emotion that was unobservable to the machine was no longer an emotion.

Despite the observers' representation of the new technologies in terms of a radical epistemic break in the very nature of observation, and despite the challenge to the practices, genres, categories, and meanings of observation in terms of feeling, the investigators who introduced the new technologies implicitly drew on and internalized existing models and modes of observation. In particular, investigators drew on the commonplace and visible facial blush (and the covert erection) and on the organ of the heart as signifiers of emotions. By observing what they defined in terms of hidden *visceral* blushes, and by tracing what they argued were imperceptibly minute perturbations in the organ of the heart, these investigators introduced a new model for and a different conceptualization of the very nature of observation, while drawing on familiar tropes of observing emotions. These novel types of observations of emotions, they argued, demanded expert knowledge and scientific instruments.

The twentieth-century return of observation in terms of "feeling" and the construal of a category of "supremely extreme" emotions reflected a second major shift in the conceptualization of emotions. From the 1910s, investigators redefined a variety of experiences in terms of the newly discovered hormone adrenaline and the category of "excitement." A myriad of nineteenth-century "feelings" and "emotions"—the pleasures of consumerism and of watching a nude body, the passion of a kiss, and the love between husband and wife—were all reconceptualized in terms of "emotional excitement" and its embodied essence of adrenaline.[5]

This shift to "excitement" and to adrenaline transformed the principle that had underpinned nineteenth-century technologies for observing emotions. If during the last third of the nineteenth century experimenters observed emotions by observing a variety of visceral blushes, then by the 1920s experimenters observed emotions by observing a variety of indices of adrenaline-sympathetic reactions. This shift in what investigators observed in observing emotions can be easily missed, since the graphic representations of emotions that were transcribed by emotion-gauging technologies project continuity. Nineteenth-century graphs of visceral blushes (that is, of nineteenth-century emotions) were often visually identical to twentieth-century graphs of adrenaline reactions (that is, of twentieth-century emotions). While nineteenth-century embodied-feeling observers of emotions sympathized—that is, viscerally blushed—in observing emotions, twentieth-century embodied-experiencing observers of emotions developed an adrenaline-sympathetic response in observing the adrenaline-sympathetic responses—that is, the emotions—of their subjects and animals.

The embodied-feeling paradigm of observation and the machine-mediated-seeing paradigm of observation were enacted through distinctive sets of practices, interventions, and manipulations. These multiple shop-floor enactments embodied distinct observational paradigms. These enactments argue against the traditional Bernardian opposition between observation and experiment in terms of "passive" observation versus the active manipulations and interventions of an experiment. This perspective on and framing of observation also suggests that in the realm of emotions the experiment can be conceived as a mode of observing. The experiment in emotion studies eliminated the haphazard and opportunistic nature of observations of emotions, which had dominated the earlier phase of this science. By creating techniques for inducing emotions at the experimenters' will, experimenters reduced the element of luck in observation.

I begin this essay with a brief and by no means exhaustive review of several premachine-age modes of observing emotions through feelings, paying particular attention to embodied-experiential observations. I then present the "blush" as a natural symbol and metaphor that underpinned and legitimated the new observational paradigm of seeing emotions. Then I examine the relationships between embodied observations (through feeling) and technology-mediated ones (through seeing). These distinctions can be read in terms of a nineteenth-century gendered distinction between seeing men of science versus feeling, *Sensitive* women. In the last section, I focus on the twentieth-century return of the embodied-experiencing "excited" observer and the emergence of the category of the "supremely extreme" emotion.

Since my major objective is to study the more salient features of different observational paradigms, I assume an analytical and conceptual perspective. In addition, I do not overly belabor the distinctions between emotions, feelings, sentiments, moods, and so forth. As Alexander Bain explained in *The*

Emotions and the Will (1859), "Emotion is the name here used to comprehend all that is understood by feelings, states of feeling, pleasures, pains, passions, sentiments, affections."[6]

Feeling Emotions

Prior to the advent of the new technologies, the *mise en observation* of emotions and experiential states encompassed a vast variety of different practices and techniques. Some investigators, like Herbert Spencer, assumed a linear relationship between the intensity of a felt emotion and its effects on the muscles of facial expression. These latter expressions were observable. Still others observed emotions by exclusion. Emotion was observed by examining all possible sources for the observed phenomenon, and when none were found, "emotion" had been observed—this was the practical logic of these types of observations.[7]

Many of the diverse practices for observing emotions worked through the body of the observer and his or her feelings and experiences. The experiencing-feeling self was the apparatus for observing emotions and other states of mind. These types of observations often depended on the observer's self-referential and reflexive adoption of a particular consciousness. Duchenne de Boulogne, the renowned investigator of emotions, for example, appealed to the observer's feelings in observing-*recognizing* facial expressions in his famous 1862 monograph *Mechanisme*; or as Duchenne had put it, how "do we feel" when looking at a particular facial expression.[8]

The dominant nineteenth-century mode of observing, knowing, and reading emotions was *sympathy*. As Bain explained in his chapter "Sympathy and Imitation":

> [I]t is to be reckoned a general tendency of our constitution, that when the outward signs of emotion are in any way prompted, the wave, passing into the interior, inflames all the circles concerned in the embodiment of the feeling, and gives birth more or less powerfully to the accompanying conscious state. The possibility of sympathizing fully with other minds depends upon this fact.[9]

Many embodied observations of experiential states worked by deliberately—not automatically and sympathetically—mimicking-embodying the observed Other. These types of observations shifted from a Bain-like vernacular approach in which observation was construed in terms of a natural (even reflex) and innate stream of experiences, to observation as a learned and active effort—that is, observation as disciplined embodied experience.

The observer—not necessarily yet a scientific expert—assumed the bodily demeanor of the subject, creating in himself the subjective experience of the Other and thereby feeling-observing the Other's state of mind. In *Sensation et mouvement* (1887), Féré introduced a model and principle of observing—or "reading"—other minds, which worked through this form of embodied epistemology. "If one can read the thought of one's interlocutor," Féré explained, "it is because in looking at him, one unconsciously assumes his expression, and the idea presents itself in consequence. . . . One has cited a diplomat who had the habit of imitating the expressions of people whom he wanted to figure out."[10] In *The Sublime and Beautiful*, Edmund Burke, as quoted by William James, had made a similar argument in speaking of the physiognomist Tommaso Campanella: "When he had a mind to penetrate into the inclinations of those he had to deal with, he composed his face, his gesture, and his whole body, as nearly as he could, into the exact similitude of the person he intended to examine; and then carefully observed what turn of mind he seemed to acquire by the change."

William James made a similar point in chapter 25 of *Principles of Psychology* (1890), "The Emotions." "Imitating in an exact manner someone else's gesture, face, and whole body," he argued, quoting from Dugald Stewart, "gives insight into the state of mind of the person thus imitated." Gustave Fechner had also presented this familiar nineteenth-century perspective in *Vorschule der Ästhetik*—"to go tripping and mincing after the fashion of women puts one, so to speak, in a feminine mood of mind."[11]

These varieties of embodied techniques for observing emotions challenged the subject-object divide and challenge some of our contemporary assumptions regarding the subject-object divide, gender, and Western knowledge. During this earlier phase it was the male-scientific observer who adopted and promulgated this form of embodied knowledge and the co-sharing of visceral or bodily feelings with the observed subject. The feminization of this form of knowledge and the construction of a clear opposition between a feminized-embodied mode of observing and a scientific machine-mediated mode of observing reflected the shift in the paradigm for scientifically observing emotions.

The distinction between a feminized and a masculine mode of observing emotions during this earlier period did not reside in embodiment per se, but in the distinction between three different phases in observation: looking, embodying, and self-observation. Premachine learned embodied techniques for observing emotions included as a first phase "looking at" the subject. Embodiment ensued on this looking, which was followed by self-observation (of one's own embodied state). Learned observation was thus enacted in terms of two distinct types of observation: observing the Other (with one's eyes) followed by self-observation. Men discriminated between these two phases of observation and between these two types of looking, and they implicitly mapped "looking" and "embodying" on the distinction between mind and body (the former observed the latter).

In contrast, contemporary feminized techniques for observing other minds sometimes worked by touching the body of the observed. "Touching" in order to observe emotions, rather than embodiment itself, presented a clearer demarcation and distinction between feminized and masculine modes of observing emotions.[12] These different distinctions are significant since the machines for measuring emotions would amalgamate and confuse these different modes of observing (and confuse the gendering of observation). The new technologies touched the body of the observed in mechanically embodying the emotions, in order to transform the felt emotion of the observed into a seen object of knowledge—the graph.

Seeing the Blush: Observing Emotion

> In the earlier books on Expression . . . the signs of emotion visible from without were the only ones taken account of. . . . The researches of Mosso with the plethysmograph have shown that not only the heart, but the entire circulatory system, forms a sort of sounding-board, which every change of our consciousness, however slight, may make reverberate.
>
> WILLIAM JAMES, *"What Is an Emotion?"* (1884)

During the eighteenth century, Marin Cureau de La Chambre, physician to Louis XIV, argued for the uselessness of a "window in front of the heart to see the thoughts & designs of men." Antoine-Joseph Pernety also disparaged the notion of a "Glass of Momus." A window into the heart, he argued, would show only a beating organ and nothing else. Johann Kaspar Lavater, the renowned eighteenth-century physiognomist, followed suit, dismissing "the dream of an artificial device for opening the profundities of the heart."[13]

During the nineteenth century, various commentators and investigators of the emotions, including Alexander Bain (1859), W. Stanley Jevons (1871/1888), Charles Darwin (1872), and Francis Ysidro Edgeworth (1881), rejected the possibility of technologies that could measure and record the emotions in exact quanta. Jevons "hesitate[d] to say," in the opening pages of *The Theory of Political Economy* (1871/1888), "that men will ever have the means of measuring directly the feelings of the human heart."[14] Jevons was clearly mistaken. His humility, moreover, contrasted with the hubris of the new science and its emerging technologies, which realized Edgeworth's fantasy of "a psychophysical machine, continually registering the height of pleasure experienced by an individual."[15]

In the important monograph *Fear* (1884), the first full-length study of emotions by a *laboratory-based* physiologist, Angelo Mosso presented the paradigmatic distinction between emotion-seeing machines for observing emotions and embodied-feeling observers of emotions. In speaking of his own difficulty in studying affective experiences, Mosso presented the quintessential challenge to the embodied observer of emotions—his sympathy with the observed: "Even men who have been hardened and accustomed to the sight of blood and of human misfortunes, are yet moved at the terrible picture of pain."[16]

In order to overcome or bypass the paralyzing effects of observing pain, Mosso made "use of instantaneous photographs in studying the expression of the face" during pain.[17] In Mosso's hands, photography was a technology that circumvented his—and humanity's—frailty of sympathy, which disturbed the observation of emotions and pains. Mosso's concern was not with subjectivity, distortion, overinterpretation, or system building, but with his personal incapacity to observe pain. Though the sight of pain was painful for Mosso, Mosso did not shed tears or blush in sympathy.

The machine-Mosso dyad allowed Mosso to parse between detached observation and personal-embodied feelings and to better manage the tensions and relationships between observing and feeling. It empowered and at the same time disempowered feelings (his experiences during the observation). On the one hand, the machine allowed Mosso to harbor and experience strong emotions of sympathy during the observation, since it continued to observe and record during Mosso's moment of sympathy. In this sense, it empowered Mosso's emotions, since it provided them—and Mosso—with a room of one's own inside Mosso's scientific body during the moment of observation.

At the same time, Mosso's emotional reactions to and during the observation of pain were no longer relevant to or the subject of Mosso's scientific observations on pain, which focused henceforth on the output of machines. Mosso could now observe pain—even though it pained him to study pain—by marginalizing his experience of pain and focusing on the machine-mediated pain. His very ability to make these types of distinctions in respect to the observation of emotions was partially indebted to the new machines that observed pain, to the different identities that were construed for pain, and to the premachine age distinction between body and mind or between sympathizing/embodying the observed-Other and observing one's own sympathizing body (with one's mind).

In construing the observation of experiential states in terms of an opposition between seeing and feeling (that is, feelings disturbed Mosso's seeing), and in privileging the seen over the felt, Mosso contributed to the decline of the observer's feelings in observation. The distinction between the observer's embodied-experienced reactions—"sympathy"—and the machine's embodied-mechanical reactions—"transcriptions"—signified the progressive shift from the observer's feelings as a technology for observing feelings to seeing (via the mediation of machines) as the observation of feelings.

Many of the early technologies for observing emotions framed the observation of emotions in terms of the natural symbol of the blush (or erection).[18] Investigators observed emotions by measuring visceral blood flows into and out of various organs and limbs. Emotions were observed in terms of "engorgements" and "vasomotor" changes—the dilation and constriction of blood vessels—in a variety of internal organs. The observation of emotions in terms of visceral blood flows was modeled on the familiar and natural facial blush and male erections. The transcription of these visceral blood flows into graphs and numbers represented the objective index of emotions and underpinned the new mode of observing emotions in (psycho)-physiology.

When the French psychophysiologist Alfred Binet, for example, depicted and represented the emotion of "fear" in graphic form during the 1890s, his plethysmograph amplified the minute swelling and shrinking of his subject's finger, which was due to the inflow and outflow of blood that occurred during the experience/experiencing of this emotion. These same exact blood flows during emotions created erections/flaccidity or blushes/pallor when they occurred in the spongelike tissues of the penis or the malleable tissues of the cheeks. Binet's machine simply amplified the minute erections/blushes of the finger. Fear, as it appeared in Binet's graphic representations, was thus a state of diminished blood flow, a contracted-flaccid-pallor state, visible in terms of a descending graph, which was interpreted and construed as displeasure (versus the ascending, erecting, blushing graph of the finger during pleasure) (fig. 13.1).

This model created the following logic and order: blush/pallor = inflow/outflow of blood = erection/flaccid = up/down = pleasant/unpleasant. It made perfect physiological, psychophysiological, psychophysical, and Helmholtzian sense and reflected basic structures of representation (e.g., up vs. down is pleasant vs. unpleasant is happy vs. sad). Maurice Schiff (1867), Mosso (1884), and Alfred Lehmann (1892) all observed inflows and outflows of blood in various organs and tissues in observing emotions.[19]

This general approach to the observation of emotions also extended to the observation of emotions in animals. It explains why the dog's or the rabbit's ears were favored sites for the physiological study of a variety of emotions,

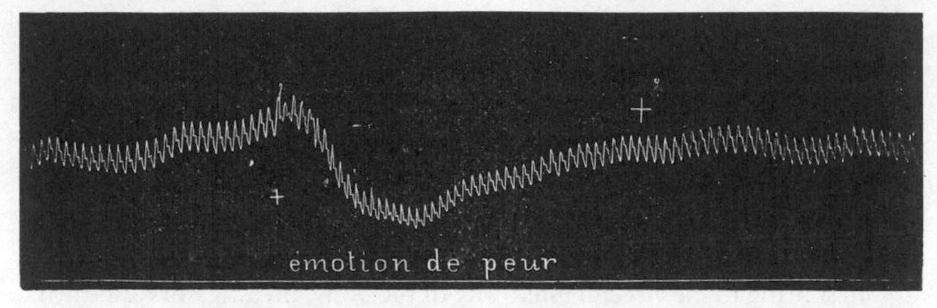

FIGURE 13.1. Emotion of fear. The effect of fear on the capillary trace: the word "serpent" was suddenly pronounced. A. Binet and J. Courtier, "Influence de la vie émotionnelle sur le coeur, la respiration et la circulation capillaire," *L'année psychologique* 3 (1896): 77.

from jealousy to anger, since their injection with blood was easily observable. These observable visceral blushes were framed by physiologists as a natural extension of Darwin's study of the exterior facial expressions of the emotions into the viscera. These Darwinian interpretations of the viscera were for the most part absent from Darwin's own study of *The Expression of the Emotions in Man and Animals* (1872).

The observation of visceral blushes and the construal of a blush-type reaction as signifying emotion in some essential and generic sense partly drew on the eighteenth-century tear. During the eighteenth century, tears functioned as a form of general signifier of affectivity, assuming different meanings depending on context and coded behaviors.[20] The blush presented a similar structure and comparable functions. It was a generic and quasi-universal natural sign that signified an affective response, although some investigators observed that it was present only in "civilized" white societies, and many noted that it was a particularly feminine type of response.

The overlap between the tear and the blush appeared throughout the nineteenth century. As Thomas Burgess explained in *The Physiology or Mechanism of Blushing* (1839): "Lachrymation or weeping, when produced by grief, is the result of an emotion more closely resembling that of blushing than any of the others to which man is subject."[21] Darwin and Maurice Schiff would make similar observations regarding the tears and the blush. "On pleure de joie et de douleur, comme l'on rougit" (One cries from joy and from pain, just like one reddens/blushes), Schiff explained in summarizing the identity between and analogous affective significances of the blush and tears.[22]

The shift to the observation of visceral blushes signified an important change from legibility to illegibility. If tears and visible blushes were "natural" indices of emotions, which could be observed by anyone, the new types of emotions—in the interiority of the body, on the visceral plane—were visible to and interpretable only by the physiologist and his new technologies.

Rather than presenting a language of expressions of the emotions that could serve as a universal mediator of mutual exchange-recognition of emotions, as Darwin's study of the universality of facial expressions suggested, these physiologists proposed an esoteric language of visceral blushes, which was universally expressed in the body but comprehensible only to a select few. What had been taken for granted, the natural ability to read emotions, was now represented as requiring expertise and technologies.

In shifting to the visceral interiority of the body, investigators appealed to Newton's color theory as a model for their own transition from "subjective" modes of observation, which relied on the embodied experiences of observers, to objective modes, which depended on instruments and empirical-physical observations. The new visceral emotions were presented as more reliable and authentic, and were positioned by these investigators in opposition to (a woman's) feigned tears, the superficiality and deceptiveness of facial expressions and gestures of emotions, and the poverty and distortions of language and verbal reports of emotions.

Observing Machines: Seeing Emotions

> In August 1879 Biffi showed in the Instituto Lombardo the heart of a youth, in the left wall of which, in the autopsy, he had found a needle sticking. . . . This instance shows how insensible the heart is, and yet, in the language of the poets and in the imagination of the people, it will always remain the centre of the passions and of feeling. . . . The heart is nothing but a force pump situated in the centre of the blood-vessels.
>
> MOSSO, *Fear* (1896)

In observing emotions through machines, investigators un-self-consciously transplanted to the machine and mechanized those features that they had rejected in the human-embodied observer. This implicit internalization of embodied skills for observing emotions into machines often went hand in hand with a reciprocal mechanization (a mechanistic modeling) of the multiple organic embodied sensitivities through which previous observers had felt-experienced (observed) emotions prior to the machine.[23]

As we have seen, the observation of experiential states before the age of machines often depended on the generation of experiences in the body of the observer. These embodied experiences constituted the observation of the experiences in the Other. As Bain had explained, the observer's embodied experiences of the Other's experiences—which he conceived as "sympathy"—depended on the observer's visceral-embodied mirroring of the observed. Sympathy was mediated by and worked through the creation of a mirrored visceral-embodied state in the observer. The new machine-mediated observations depended on a similar principle but presented a radical shift. Observation worked by transplanting the visceral activations of the observed subject not into the body of the observer, but into an exteriorized entity—the machine.

The machine literally projected the visceral activations of the observed subject outside of the body in terms of graphic tracings or numeric tables. Instead of observing through self-embodied experiences, observers looked at the embodied visceral physiology of experience through the mediation of machines and in terms of graphs and numbers. The mechanical effects of the blush (i.e., the shift of blood to the hand, finger, or brain) or of the beating heart literally pushed and moved the sensitive mechanical registering apparatus, which transcribed these visceral movements into graphs.

Claude Bernard, the French physiologist, for example, had applied Étienne Jules Marey's cardiograph in order to observe emotions (1865).[24] By translating the heart's "intimate functions" into omnipresent and permanent graphs on paper, Bernard literally "read in the human heart."[25] For Bernard, the (human) heart was both a mechanical pump and a sensitive organ that participated actively in the creation of emotions. Bernard observed the emotions by gauging the activity of the mechanical-sensitive heart with the aid of a mechanical-sensitive machine, the cardiograph, which he positioned directly above the heart on the chest wall. As Bernard explained, the accuracy and sensitivity of the machine increased the "less separated it is [from the heart] by the chest wall." This explained why it was "easier to read the heart of infants than to read that of adults, and why also it is naturally more difficult to read the heart of women than that of men."[26]

Bernard's technology for observing emotions was modeled on and embodied a mechanical rendering of the heart. It betrayed the secret impressions of the heart (that is, the emotions) by measuring the concealed impressions of the organ of the heart.[27] Bernard's technology was in certain respects an externalized duplicate or extension of the heart. Observation thus worked by transferring to machines and seeing, rather than by embodying and experiencing-feeling, the visceral perturbations of observed subjects. This basic underlying principle of observing in terms of "seeing" would continue to dominate the observation of emotions during the early twentieth century.

One major twentieth-century technique for observing emotions was the bioassay. In this procedure, a strip of visceral muscle (e.g., ileum), which had been removed from an animal, was exposed to the blood of an animal that had been emotionally excited. The emotional state of the excited animal was observed in the reactions of the in vitro exteriorized muscle (ileum) to the "excited" blood. These reactions were transcribed into a graph of emotions (fig. 13.2).

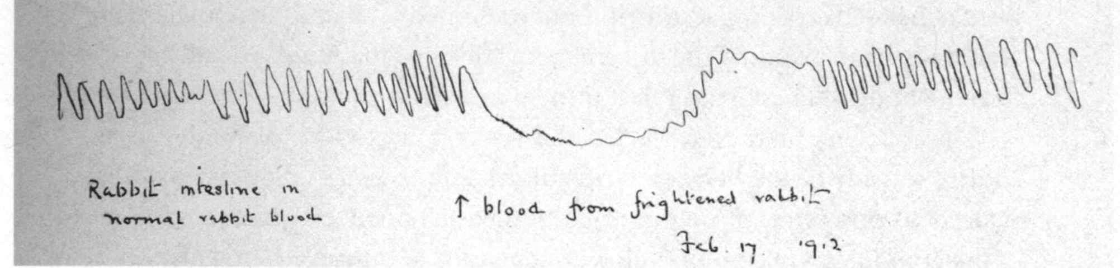

FIGURE 13.2. The effect on rabbit intestine of normal rabbit blood (*left*) and of blood from a frightened rabbit (*center*). G. W. Crile, "Studies in Exhaustion: III. Emotion," *Archives of Surgery* 4 (1922): 145.

Put differently, the *observed* activation of exteriorized viscera, rather than the *felt* activation of one's own viscera, was the observation of the emotion. Moreover, and significantly, the reactions of the externalized viscera (ileum) on its exposure to the blood from the emotionally excited animal mimicked and mirrored the visceral reactions of the excited animal during the moment that it had been emotionally excited.

This observational praxis, which worked by projecting—rather than feeling—the embodiments of the Other onto machines, also inhered in the representational logic of emotions. Nervous subjects created a nervous response in the machine—quivering dials or graphs. The serene subject, in full control of his or her emotions, created a serene response from the machine—a stable dial or unperturbed graph. Subjects, experimenters, or clinicians observed how the machine was in rapport with their own visceral experiences. The machine literally assumed, personified, and exhibited one's gut reactions. The familiar experience of one's own trembling body during fear, excitement, or apprehension was transferred onto the graph, which now literally trembled during these same experiences. During the twentieth century, some machines also embodied familiar *bodily gestures* of emotions in representing the behaviors of *visceral* organs. "Jerky" was now exhibited at the level of a patient's stomach; and an "odd convulsive blood pressure rise" indicated a kind of "emotional spasm," as William M. Martson explained.[28]

This explains why during the twentieth century various psychoanalytically oriented psychosomatic clinicians incorporated instruments for seeing emotions into their therapeutic interactions, despite the conspicuous opposition between the observational paradigm that undergirded these machines and psychoanalytic modes of observing.[29] Harold G. Wolff, Stewart Wolf, Harold Lasswell, Roy Grinker, and John C. Whitehorn fused emotion-detecting technologies for observing emotions and psychoanalytic precepts. These and other investigators modeled or likened the externalized graph of one's emotions and its contemplation by the subject to the psychoanalytical process. The graphic record of one's visceral reactions was analogized to a projection or transference of unconscious or repressed affects, which the subject could now literally see, acknowledge, and heal in the process. A technophysiological mode of observing emotions and a Freudian-inspired framework converged and created synergistic interactions rather than trivializing or even opposing each other.

This general framing of the observation of emotions also explained why some investigators argued that machines that directly measured visceral activations during emotions could observe an emotion before the individual experienced his or her own emotion, since the machine sometimes saw or gauged the visceral perturbations before the individual had time to experience his or her own emotion.[30]

Men's Seeing: Women's Feeling

> Johann Kaspar Lavater, the great eighteenth-century physiognomist, was endowed with a "feminine nature" and an "excessive sensibility."
>
> PAOLO MANTEGAZZA, *Physiognomy and Expression* (1914)

The new observers of emotions positioned their new technologies against a wide spectrum of female or feminized competitors. The transfer of embodied-feeling modes of observing to machines for seeing and their mechanization explained away a wide spectrum of suspect epistemologies, extrasensory phenomena, and the supposed mind-reading abilities of female or feminine *Sensitives.* The machines assumed these varieties of ways of observing in their very mechanisms. Instruments for measuring emotions literally manifested and projected the types of knowledge that had been construed as "mute," suspect, and dependent on "sensory refinement" or "bodily skill."[31] The new technologies, according to naturalistic detractors, observed other minds by picking up and exteriorizing (in graphic form) those very same "faint sensations" and bodily cues to which psychics and *Sensitives* were attuned.

C. P. Peirce, in the context of debates over the "personal equation," thought, as Ian Hacking has argued, "that his discovery that there is no minimum threshold 'has highly important practical bearings, since it gives new reason for believing that we gather what is passing in one another's mind in large measure from sensations so faint that we are not fully aware of them, and can give no account of how we reach our conclusions from such matters. The insight of females as well as certain 'telepathic' phenomena may be explained in this way.'"[32] William James, who was a prominent member and leader of the Society for Psychical Research, also referred in his lecture on "telepathy" to the recent "public exhibitors of 'mind-reading.'" "In most of these feats," James explained, "the agent is required to think intently of some act while he lays his hands on some part of the so-called mind-reader's person. . . . It is safe to assume that . . . the percipient is guided by . . . the agent's hands more or less unconsciously . . . so that muscle-reading, and not mind-reading, is the proper name for this phenomenon."[33]

The machines that measured emotions, that literally observed other minds, as Mosso or Hugo Münsterberg argued, worked by picking up and recording the same subtle physiological changes, to which, according to their scientific detractors and critics, "mind readers," *Sensitives*, psychics, and mediums were attuned.[34] As Mosso explained, to "see the curve of a single pulsation of hand or foot. . . . I can distinguish that of . . . one who is afraid and that of one who is tranquil."[35] His 1870s experiments on Michel Bertino portrayed a machine hooked up to Bertino's brain, whose output was a record of Bertino's changing emotions, which Mosso gauged by measuring inflows and outflows of blood in Bertino's skull.

The new technologies were also positioned in opposition to women antivivisectionists and "addled-headed men," who presented themselves as emotion-gauging machines in challenging the laboratory. These feminized sensitive observers identified pain, emotion, and suffering even where the great experimenters, despite their machines, had not detected pain or emotions. The men of the laboratory deployed their emotion-gauging machines to argue against emotion-sensitive antivivisectionists. They positioned the observations of the new machines against the observational acuity of sensitive emotion-gauging people.

Ivan Pavlov drew on the observations of machines in arguing against sensitive women-machines. "Very sensitive people," he explained, "have been apt [to be] upset" by some of his experiments:

> We have been able to demonstrate . . . that they [very sensitive people] were labouring under a false impression. Subjected to the very closest scrutiny, not even the tiniest and most subtle objective phenomenon usually exhibited by animals under the influence of strong injurious stimuli can be observed in these dogs. No appreciable changes in the pulse or in the respiration occur in these animals.[36]

For the experimenter, the observation of emotions through machines solved a problem. If *Sensitives* and a gamut of feminized competitors were naturally constituted for reading and intuiting the other, then the men who studied emotions had to become no less adept at observing the emotions and feelings of their interlocutors. Their new science of feelings and emotions demanded it. Their sensitivity, however, differed from their alter-types. Instead of becoming vessels for or embodying feelings (and "reading minds"), they *observed* feelings through technologies, which embodied the traits of sensitivity, passivity, and impressionability, to do women's work for them.

With his machine, the investigator of emotion was a man who was sensitive *to*—but not *of*—feelings: he was sensitive, but not emotional; he identified pain in the body, but remained in full control of his "medulla"; he established a caring rapport, but remained objective. The machine was thus situated culturally between the insensitive experimenter—who relied on the machine to observe emotions for him—and the female *Sensitive* or antivivisectionist, who was "over-sensitive" and gave false readings.[37]

The man of reason, science, and medicine could thus retain the capacities of the man of feelings, could engage in empathic knowledge production; could be sensitive—even more than a *Sensitive*, since his machines reacted to and visualized the minutest feelings or pains. Instead of observing by embodying, sympathizing with, or feeling the Other, experimenters observed the output of machines. The graphic outputs of machines, moreover, did not evoke feelings in the scientific observer. I have found no allusions in this nineteenth-century science that watching the movements of the graphic machines evoked feelings. This was in stark contrast to the literature that studied facial expressions, which often explicitly depended on the emotive reactions (feelings) of the observer as an essential element in identifying/interpreting the represented emotion.

Observation was thus framed by these men in terms of the control over the observer's emotions and the elimination of the experiencing-embodied observer. Observation was thus implicitly construed in terms that were directed not at nature but at the observer. Observers implemented and operationalized "scientific observation" in terms of self-management of and self-control over their own bodies and subjectivities. This logic of observing harked back to the very early history and the origins of the notion, even the prenotion, of observing—that is, "observance." The shift from feeling to seeing signified a shift from self-observation to self-observance.

The Observers' Excited Feelings

> When we witness any deep emotion, our sympathy is so strongly excited, that close observation is forgotten or rendered almost impossible; of which fact I have had many curious proofs.
>
> CHARLES DARWIN, *The Expression of the Emotions* (1872)

The transference of the principle of sympathy into machines for seeing emotions and its embodiment in terms of mechanical contrivances blurred the distinctions between seeing machines and feeling observers. This blurring of boundaries and the challenges to the distinctions between feeling and seeing appeared in conjunction with the development of machines for observing emotions during the late nineteenth century. During the 1920s, these challenges became more apparent and consolidated around the experimental category of the "supremely extreme" emotion.[38] Machines for seeing emotions were sometimes described as overreacting in embodying—that is, mechanically sympathizing with—the observed emotion. Instruments were described in terms of the pull of the emotion, the vulnerability of technologies, and their victimization by emotions, and in terms of the machine's overt embodied (mechanical) presence, rather than in terms of its transparency, resistance, resilience, and subject-object divide. These failures of machines to observe in terms of "seeing" enacted a category of vehement or intense emotions.

During the nineteenth-century phase of this science, for example, extreme trembling during emotions abolished any possibility of using the psychogalvanometer or the pneumograph, as Alfred Binet discovered during the late 1890s.[39] Alfred Lehmann also encountered an emotion that overwhelmed his technologies—as the registering stylograph "hit the edge of the cylinder from time to time" due to the overwhelming displeasure experienced by the subject—O.[40] Mosso had already noted in the 1880s that "when strong emotions such as fear are concerned, one must have recourse to other methods of writing the pulse, as the animal is very uneasy and tries to escape. . . . this is a question as yet little considered in physiology."[41]

During the twentieth century, the "supremely extreme" emotion was represented and observed in terms of the observed annihilation of the apparatus for observing. Instead of being observed through and in terms of the machine's scientific output (e.g., as graphs or numeric tables), the emotion was observed through the machine's destruction. As Carney Landis reported from his laboratory during the mid-1920s, a really angry subject attacked the technology, tore the apparatus from his arm, and threw it at the experimenter.[42] These particular moments, which signified the laboratory's failure to maintain the subject-object divide, were progressively defined and construed in terms of a category of the intense/extreme.

The experimenter's embodied experiences also reappeared. The disembodied observer of machine-mediated-seen emotions emerged in terms of his excitements during the observation of intense/extreme emotions. These visibilities of twentieth-century observers harked back to premachine-age modes of observing. During these observations, the experimenter narrated his own emotions and how he had been recruited, often against his will, into the observation itself. These descriptions of experiencing-excited observers of emotions appeared in various early twentieth-century reports.[43] They contrasted with the late nineteenth-century image of the detached, transparent, and disembodied observer who observed emotions through the machine.

There was thus a recurring trope or mode of writing about the self during the observation of "supremely extreme" emotions inside the laboratory and clinic. This mode of writing the self into the observation of an extreme worked by positioning the observer inside the observation—in spite of himself. The visibility of the twentieth-century observer (of extremes) was reminiscent of the nineteenth-century sympathizing observer, since the observer's reaction was portrayed in its involuntariness, as a reflexlike response to the observation, rather than in terms of agency and will.

Unlike the premachine-age nineteenth-century embodied-experiencing observer of experiential states who mimicked and/or sympathized with the observed, twentieth-century experiencing observers described the observer's feelings in terms of "emotional excitement" and a fight-or-flight adrenaline response. The "supremely extreme" threatened the observer and/or threatened to overrun the order/control of the laboratory. This type of breach in the subject-object divide was distinct from the dissolution of the subject-object divide that had typified various embodied modes of observing emotions during the premachine era. The nineteenth-century embodied-"feeling" observer of emotions sympathized—that is, viscerally blushed—in observing emotions, while the twentieth-century embodied-"excited" observer of emotions developed, like his subjects, an adrenaline-sympathetic response in observing the adrenaline-sympathetic responses—that is, the emotions—of his subjects and/or animals.

The attempt to seal this breach in the subject-object divide explains why twentieth-century experimenters who worked with—what they defined in terms of—"supremely extreme" emotions often observed through techniques that observed the emotion after the fact. We can thus identify the emergence of a split in the category of observing in terms of seeing. One category often depended on online graphic machines that were directly connected to the subject's body; the second category included various practices for collecting specimens (urine, blood, and tissue samples) from the body of the subject/animal. These latter specimens were examined outside of the body itself long after the emotion had subsided. These types of after-the-fact observations, in contradistinction to those enabled by online machines, were often deployed by those same experimenters who reported reactive embodied emotions while *seeing* the unmediated emotion.

These latter observers first reduced the emotion to the behavior of an exteriorized and isolated muscle tissue—the "bioassay"—which was then represented as a graph. The violent rage was observed in the controlled graph. An important series of conversions, whose significance went far beyond any particular protocol or even the culture of laboratories, took place in and through these particular technologies for observing (extreme) emotions.

There was a dramatic affective difference between observing an animal that almost ate you alive, as John F. Fulton described one laboratory cat, versus observing the bioassay or the transcription of the cat's "rage."[44] As their correspondence shows, experimenters were not afraid of the graph, as they clearly were of real enraged animals or subjects. While subjects lost self-control and, at times, challenged the observer, the inscriptions were always well behaved.

It was only when these same experimenters reported in their often unpublished texts what "really" happened outside of the machine and inside the space of the laboratory itself that we become aware of the emotions that were evoked in the experimenters themselves during the observation of extreme emotions. Through their private narratives experimenters bypassed the graphic technologies and "seeing" and reintroduced feeling into the observation. The self-referential reports of observers' excited feelings and the reports of overwhelmed technologies construed the category of the "supremely extreme."[45]

Conclusion

Many of our contemporary approaches to observing, measuring, detecting, and recording emotions emerged during the late nineteenth century with the shift to machines that observed visceral emotions. Emotion was redefined in terms of that which could be observed objectively, as a measured—and thus visible—physiological entity, rather than that which was experienced and felt by the subject or the observer of emotions. Lloyd Ziegler, writing in the 1920s, expressed well this new ontology in explaining that "one of the most striking facts is the failure of some patients to recognize any emotion though readings clearly indicated a rise of metabolism"—testifying to the presence of an emotion.[46] Or as John C. Whitehorn, director of the Phipps Psychiatric Clinic at Johns Hopkins University, explained, "some kind of a distinction need[ed] to be maintained . . . between the outer emotional talk and behavior and the inner emotional reaction"—as observed via Whitehorn's cardiotachometer.[47]

The return of the embodied observer during the twentieth century reflected the shift in the dominant underlying paradigms of emotions from a nineteenth-century blush paradigm (which depended on the sensitive body and was observed in terms of inflows and outflows of blood) to a twentieth-century adrenaline paradigm (which was observed in terms of adrenaline-sympathetic reactions).[48] The significance of this shift from blush to adrenaline for observation can be presented from the perspective of the changing status of the nude or denuded woman in experiments on emotions. In nineteenth-century scientific studies of the emotions, the blushing disrobed woman figured visibly as a subject of observations—be she a nude model, an exposed slave, or a disrobing patient. During the twentieth century, the nude in emotion studies underwent an important shift. Instead of observing the blush in disrobed women, experimenters often observed the emotional—that is, adrenaline—reactions of male subjects who gazed at pictures or drawings of female nudes. This shift from the blushing and observed female nude to the adrenaline-activated male observer of nudes turned the limelight in a different direction. It significantly altered the geometry of vision, and it directly implicated the male experimenter who was also privy to these nudes. These twentieth-century experiments with the nude focused not on male blushes or erections in response to the female nude (with rare exceptions), but on measurable adrenaline reactions.

Notes

I thank my colleagues and co-contributors to this volume who have commented, inspired, and provided wholehearted friendship during our collective work on "The Histories of Observation." I also acknowledge the wonderful support of the Scholion Interdisciplinary Research Center in Jewish Studies at the Hebrew University of Jerusalem.

1. William Archer, *Masks or Faces?* [1888] (New York: Hill and Wang, 1957), 166–68.

2. Ugolino Mosso, "Influence du système nerveux sur la température animale: Recherches," *Archives italiennes de biologie* 7 (1886): 306–40.

3. A. Binet and J. Courtier, "Circulation capillaire de la main dans ses rapports avec la respiration et les actes psychiques," *L'année psychologique* 2 (1895): 87–167.

4. Jonathan Crary, *Suspensions of Perception: Attention, Spectacle, and Modern Culture* (Cambridge, Mass.: MIT Press, 1999), 97–98; idem, *Techniques of the Observer: On Vision and Modernity in the Nineteenth Century* (Cambridge, Mass.: MIT Press, 1990), 148; Stephen Kern, *The Culture of Love: Victorians to Moderns* (Cambridge, Mass.: Harvard University Press, 1992); and Lorraine Daston and Peter Galison, *Objectivity* (New York: Zone Books, 2007).

5. Otniel E. Dror, "Afterword: A Reflection on Feelings and the History of Science," *Isis* 100 (2009): 848–51.

6. Alexander Bain, *The Emotions and the Will* [1859], ed. Daniel N. Robinson (Washington, D.C.: University Publications of America, 1977), 3.

7. See Otniel E. Dror, "Das Gefühl in der Maschine," in *Begriffsgeschichte der Naturwissenschaften: Die historische Dimension naturwissenschaftlicher Konzepte*, ed. Ernst Müller and Falko Schmieder (Berlin and New York: Verlag Walter de Gruyter, 2008), 275–86.

8. See G. B. Duchenne de Boulogne, *The Mechanism of Human Facial Expression* [1862], ed. and trans. R. Andrew Cuthbertson (Cambridge: Cambridge University Press, 1990), 54.

9. Bain, *Emotions*, 215–16. On sympathy, see also Alison Winter, *Mesmerized: Powers of Mind in Victorian Britain* (Chicago: University of Chicago Press, 1998), chap. 12.

10. Ch. Féré, *Sensation et mouvement: études expérimentales de psycho-méchanique* (Paris: Félix Alcan, 1887), 15 (all translations from the French are my own).

11. For these citations, see William James, "The Emotions," in idem, *Principles of Psychology* (New York: Henry Holt, 1890), chap. 25, reprinted in Carl Georg Lange and William James, *The Emotions* (Baltimore: Williams and Wilkins, 1922), 114–15.

12. I note that even touching was not an absolute criterion, since male physicians touched their patients in order to gauge the pulse and the emotions.

13. Barbara Maria Stafford, *Body Criticism: Imaging the Unseen in Enlightenment Art and Medicine* (Cambridge, Mass.: MIT Press, 1991), esp. 84–129, citations on pages 85, 87, 91, 96.

14. W. Stanley Jevons, *The Theory of Political Economy* [1871] (London: Macmillan, 1888), 11. On Jevons, see Harro Maas, "Disciplining Boundaries: Lionel Robbins, Max Weber, and the Borderlands of Economics, History, and Psychology," *Journal of the History of Economic Thought* 31 (2009): 500–17.

15. Cited in Margaret Schabas, "Victorian Economics and the Science of the Mind," in *Victorian Science in Context*, ed. Bernard Lightman (Chicago: University of Chicago Press, 1997), 72–93, on 82.

16. Angelo Mosso, *Fear*, 5th ed., trans. E. Lough and Frederich Kiesow (London and New York: Longmans & Green, 1896), 201.

17. Ibid.

18. For the notion of "natural symbol," see Mary Douglas, *Natural Symbols: Explorations in Cosmology* (London: Barrie & Jenkins, 1973).

19. M. Maurice Schiff, *Leçons sur la physiologie de la digestion* (Florence and Turin: Hermann Loescher, 1867); and Alfred Lehmann, *Die Hauptgesetze des menschlichen Gefühlslebens* (Leipzig: O. R. Reisland, 1892).

20. Anne Vincent-Buffault, *The History of Tears: Sensibility and Sentimentality in France*, trans. Teresa Bridgeman (New York: St. Martin's Press, 1991).

21. Thomas Burgess, *The Physiology or Mechanism of Blushing* (London: John Churchill, 1839), 181.

22. Schiff, *Leçons*, 1: 270; and Charles Darwin, *The Expression of the Emotions in Man and Animals* (New York and London: D. Appleton and Co., 1872), 168.

23. On the sensitivity and individuality of machines, see Graeme J. N. Gooday, "Instrumentation and Interpretation: Managing and Representing the Working Environments of Victorian Experimental Science," in Bernard Lightman, ed., *Victorian Science in Context* (Chicago: University of Chicago Press, 1997), 409–37, on 409. On the "individuality" of machines, see Jimena Canales, "Exit the Frog, Enter the Human: Physiology and Experimental Psychology in Nineteenth-century Astronomy," *British Journal of the History of Science* 34 (2001): 173–97, esp. 194–96. See also Henning Schmidgen, "Time and Noise: The Stable Surroundings of Reaction Experiments, 1860–1890," *Studies in History and Philosophy of the Biological and Biomedical Sciences* 34 (2003): 237–75.

24. Claude Bernard, "Sur la physiologie du coeur et ses rapports avec le cerveau," in Bernard, *Leçons sur les propriétés des tissus vivants*, coll., ed., and arranged by M. Émile Alglave (Paris: Germer Baillière, 1866), 421–71.

25. Ibid., 437.

26. Ibid., 439–40. Thus, even after the shift to machines, women remained less readable and therefore less reliable.

27. Anon., "How the Body Betrays the Mind," *Literary Digest* 48 (1914): 153–55.

28. William Moulton Marston, *The Lie Detector Test* (New York: Richard R. Smith, 1938), 131. See also Thomas L. Stedman, ed., *Twentieth Century Practice: An International Encyclopedia of Modern Medical Science*, vol. 4 (New York: William Wood, 1895), 579; and David Brunswick, "The Effects of Emotional Stimuli on the Gastro-Intestinal Tone: II. Results and Conclusions," *Journal of Comparative Psychology* 4 (1924): 225–87, on 232–33.

29. On psychoanalysis and its emphasis on talking and expressing the affect in words, see Kern, *Culture of Love*, 135–36.

30. Fred W. Eastman, "The Physics of the Emotions," *Harper's Magazine* 128 (1914): 297–303.

31. See Lorraine Daston, "Scientific Objectivity with and without Words," in *Little Tools of Knowledge: Historical Essays on Academic and Bureaucratic Practices*, ed. Peter Becker and William Clarke (Ann Arbor: University of Michigan Press, 2001), 259–84, on 264.

32. Ian Hacking, *The Taming of Chance* (Cambridge: Cambridge University Press, 1990), 203–5.

33. William James, "Telepathy" [1895], in James, *Essays in Psychical Research* (Cambridge, Mass.: Harvard University Press, 1986), 119–26, on 119–20.

34. On the *Sensitive* and her qualities, see R. Laurence Moore, *In Search of White Crows: Spiritualism, Parapsychology, and American Culture* (New York: Oxford University Press, 1977).

35. Mosso, *Fear*, 99.

36. I. P. Pavlov, *Conditioned Reflexes: An Investigation of the Physiological Activity of the Cerebral Cortex*, trans. G. V. Anrep (Oxford: Oxford University Press, 1927), lecture 2, 30.

37. For the important, but poorly articulated, cultural tensions between "the two poles of vulgar boyish cruelty and effeminate sentimentality" during this period, see Philip J. Pauly, *Biologists and the Promise of American Life* (Princeton: Princeton University Press, 2001), 187. On these tensions, see also Martin S. Pernick, "The Calculus of Suffering in Nineteenth-Century Surgery," in *Sickness and Health in America: Readings in the History of Medicine and Public Health*, ed. Judith Walzer Leavitt and Ronald L. Numbers (Madison: University of Wisconsin Press, 1985), 98–112.

38. For the notion of "supremely extreme," see Norman E. Freeman, "Decrease in Blood Volume after Prolonged Hyper-activity of the Sympathetic Nervous System," *American Journal of Physiology* 103 (1932): 185–202.

39. Binet and Courtier, "Circulation."

40. Lehmann, *Hauptgesetze*, 98.

41. Mosso, *Fear*, 114.

42. Carney Landis, "Studies of Emotional Reactions: II. General Behavior and Facial Expression," *Journal of Comparative Psychology* 4 (1924): 447–509, on 486.

43. For some of these descriptions and avowals, see Emanuel M. Bogdanove, "'Gullible's Travails': or How I Eventually Discovered the 'Implantation Paradox,'" in *Pioneers in Neuroendocrinology II*, ed. Joseph Meites, Bernard T. Donovan, and Samuel M. McCann (New York: Plenum Press, 1978), 53–74; J. F. Fulton and F. D. Ingraham, "Emotional Disturbances Following Experimental Lesions of the Base of the Brain (Pre-chiasmal)," *Proceedings of the Physiological Society* (April 27, 1929): xxvii–xviii, *Journal of Physiology* 67 (1929); T. R. Elliott, "The Control of the Suprarenal Glands by the Splanchnic Nerves," *Journal of Physiology* 44 (1912): 374–409; Fulton to Cannon, 9 Feb. 1929, folder 399, box 28, series I, Manuscript Group 1236, John Farquhar Fulton Papers, Yale University; Cannon to Kast, 6 Sept. 1932, folder 1289, box 94, Walter

Bradford Cannon Papers (H MS c40), Harvard Medical Library in the Francis A. Countway Library of Medicine, Boston, Mass.; and John C. Whitehorn, unpublished autobiography, p. 24, folder 1, box 1, John C. Whitehorn Papers, Acc. 73–40, American Psychiatric Association.

44. Fulton to Cannon, Feb 9, 1929.

45. These developments, I suggest, comprise one aspect of the genealogy of the return of the body and the denigration of representation in recent literature. See, e.g., Stephen Lying, ed., *Edgework: The Sociology of Risk-Taking* (New York and London: Routledge, 2005); and Hans Ulrich Gumbrecht, *Production of Presence: What Meaning Cannot Convey* (Stanford: Stanford University Press, 2004).

46. Lloyd H. Ziegler and B. S. Levine, "The Influence of Emotional Reactions on Basal Metabolism," *American Journal of Medical Sciences* 169 (1925): 68–76, on 73.

47. John C. Whitehorn, "Emotional Responsiveness in Clinical Interviews," *American Journal of Psychiatry* 94 (1937): 311–15, on 314–15. This central motif was expressed by countless observers.

48. These themes are fully explored in my book manuscript, *Blush, Adrenaline, Excitement: Modernity and the Study of Emotions, 1860–1940* (currently under revision for the University of Chicago Press).

14

Visualizing Radiation: The Photographs of Henri Becquerel

KELLEY WILDER

What are the two photographs in figures 14.1 and 14.2 for? Are they observations, or are they objects of observation? Do they illustrate observational practice or explain it? Are they experiments, or are they evidence of experiments?

In photography's usual slippery fashion, they functioned in all of these capacities at different times, and often several of them at once. They serve for us as a document of seminal investigations on radioactivity, but numerous captions could describe them in other ways. They are photographs (or more correctly, photolithographs from photographs) made by French physicist Henri Becquerel using first a uranium salt and second a crumb of radium. They are also two of a series of sixty photographs Becquerel used to illustrate his 1903 *Recherches sur une propriété nouvelle de la matière* [Investigations of a New Property of Matter].[1] Figure 14.1 is recognizable to historians of physics as the first visual image of the radioactivity emitted from a uranium salt. For Becquerel, it became the visual evidence of his decisive moment of "discovery," even though at first he exhibited it with less lofty aims, as an image of rays analogous with Wilhelm Conrad Röntgen's newly announced X-rays.[2] Figure 14.2, on the other hand, is a relatively unknown image that marks a turning point in Becquerel's understanding of the mysterious emissions. It is not only more complex than the first image, but it was made with a specific brief in mind—to make visible the deviation of radioactive rays by a magnetic field. These photographs are only two of hundreds of what Becquerel called "observations" (photographic, electric, phosphorescent, magnetic, and fluorescent) in a series of experiments on radiant bodies. The images are what he called "mis en évidence," or "making (radioactivity) perceptible," in this case visualizing it through photography. This chapter is about how each

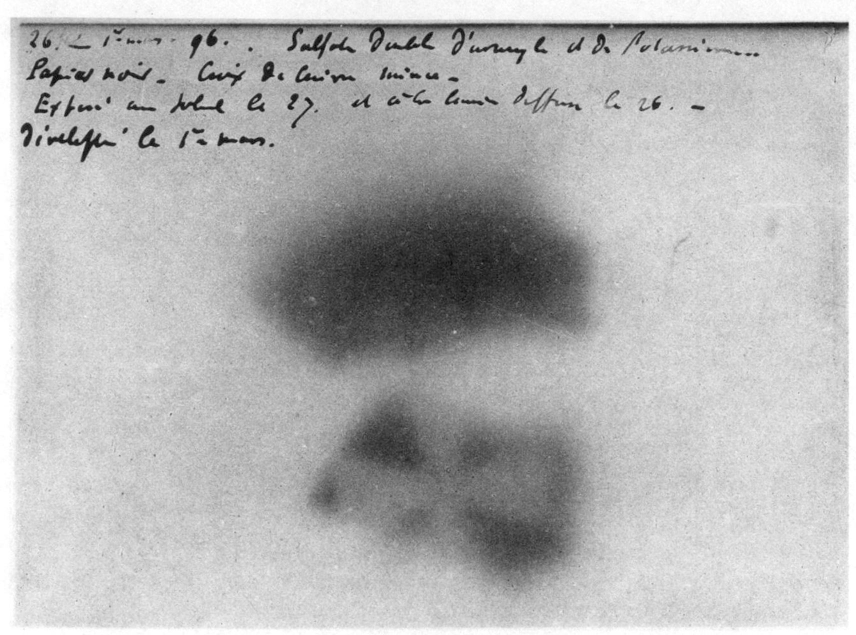

FIGURE 14.1. Photograph of the radioactivity emitted from a uranium salt. Photolithographic copy (made by inter-positive) of a glass plate negative, February 1896. From Antoine Henri Becquerel, *Recherches sur une propriété nouvelle de la matière: activité radiante spontanée ou radioactivité de la matière*, Mémoires de l'Académie des Sciences de l'Institute de France (Paris: Firmin-Didot, 1903), plate 1, fig. 1.

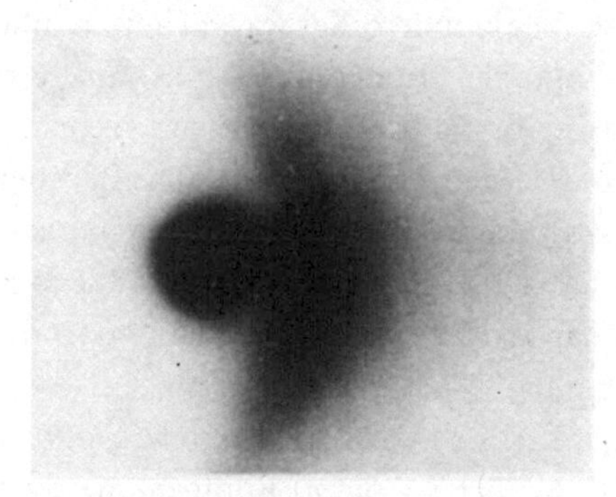

FIGURE 14.2. Photograph of radioactivity being deviated by a magnetic field. Photolithographic copy (made by inter-positive) of a glass plate negative, circa December 1899. From Antoine Henri Becquerel, *Recherches sur une propriété nouvelle de la matière: activité radiante spontanée ou radioactivité de la matière*, Mémoires de l'Académie des Sciences de l'Institute de France (Paris: Firmin-Didot, 1903), plate 6, fig. 29.

of these photographs came to represent an important stage of Becquerel's photographic method, and how that method exhibits typical, although not entirely successful, photographic observation.

Antoine-Henri Becquerel, known as Henri, was the third member of a scientific dynasty that dominated French science for a century and a half.

Henri and his grandfather, Antoine-César, his father Alexander-Edmond, and his son, Jean were all members of the Académie des Sciences, and all held the chair for physics at the Museum of Natural History in Paris. Between them they contributed to such diverse fields as electrochemistry, optics, photography, agriculture, radioactivity studies, and quantum theory. Henri's father, Edmond, also helped to found the Société française de photographie.[3] Edmond and Henri Becquerel were a dominant force in adapting photography to scientific studies during its first and very formative fifty years.

Edmond Becquerel concentrated his research on spectra of the three categories believed to exist at the time: luminescent, chemical, and thermal. Henri Becquerel inherited his post and title from his father, and perhaps more importantly, took over his father's laboratory, including his entire photographic apparatus and several ongoing projects on phosphorescence. Just as it was important to Victor Henri's microcinematographic habits that he was working in Etienne-Jules-Marey's former physiology lab in the Collège de France, this inheritance exerted a great influence over Henri Becquerel's working methods.[4] In addition to the laboratory materials, Henri also inherited his father's conviction that he could visually deduce the qualitative differences between the luminous, chemical, and thermal spectra. Thus visualizing was an important tool for Becquerel's method of thinking about radiant matter long before he exposed photographic plates to radioactive materials.[5] It was quite likely the reason why photography was such an attractive tool, not only for its efficiency in doing the work of visualizing, but for its robustness in deploying a visual method of working.

Visualizing with photography is not some kind of mechanical observation akin to the mechanical objectivity so ably defined by Daston and Galison.[6] Since cameras, lenses, emulsions, and other photographic equipment register images that lie so far outside the scope of the human senses, it is also never *merely* a matter of making visible the previously invisible.[7] To make a thing visible to the unaided human eye out of objects too small, too large, too fast, too slow, and too far outside the spectrum of human vision is only the precondition for observation, and dozens of small decisions go into the process of giving these ephemeral phenomena form. All of these decisions impinge on the final shape of the visual output that then becomes an object of study, or of contention, or even of historical import. The photographs, as visualizations, are a formal or one could say pictorial manifestation of what existed in Becquerel's mind as well as being material objects (glass, lithographic plates, and silver gelatin prints) that react to radiation of all sorts in very different ways.[8] Visualizing is thus a combination of Jean Perrin's "mis en observation," the setting up of the conditions under which things can be observed, with the learned practice of imaginative photographic manipulation.[9] Unlike Perrin, who borrowed his techniques from other established scientific disciplines, Becquerel used novel and frequently contested photographic techniques, many of which he and his father developed themselves. An examination of Becquerel's techniques of visualization in the context of scientific observation provides a rare opportunity to investigate the tension between photographic visualization, observation, and experiment, especially the dichotomies alleged by many late nineteenth- and twentieth-century scientists and philosophers of science: passive observation versus active experimentation, and subjective versus objective forms of representation.

For Becquerel, photography was not just about generating visual material from a new and invisible source or radiation, or about recording what he saw in permanent or semipermanent form. It was about how he gave radioactivity a materiality where it had none, and how he chose to describe its physical characteristics in pictorial rather than numerical form. He went beyond using photography as an instrument merely to detect the presence or absence of radioactive emissions. He used photographs as a tool for thinking—a method for understanding the physical nature of the rays by giving them visual form. That is, he not only visualized objects *with* the rays, as one does with X-rays, he made visualizations *of* the rays, and this is the single greatest difference between figure 14.1 and figure 14.2.

Photographic methods employed by scientists are multiple and complex. There was never *a* photographic method, but many individual methods, each particular to a time, a scientist, or laboratory, and most importantly, to a phase in the history of photography.[10] The incessant renegotiation of the worth of photographic methods requires that the question, what is *this* photographic method (good) for? be answered afresh in each new context. From notorious cases like the Venus transits, to lesser known examples like Becquerel's photographic method for investigating radioactivity, the authority photography appealed to ranged to an astonishing degree: from mimetic representation on the one extreme to numeric on the other.[11] Photography's flexibility and its stubborn resistance to fixed definition render it a difficult tool partly because of enduring claims to its simplicity and transparence.[12] Although Becquerel was a sophisticated photographer, it is difficult to pin down his understanding of the relationship between his photographic method and his observational practice. In the *Recherches* and in many articles published in the *Annals of the French Academy of Sciences*, he referred sometimes to the ability of photography or the photographic plate to "observe" radiation, sometimes to the photographs as recorded "observations," and throughout he referred to his photographic method as a type of experimental method. He also often wrote about having "observed" the photographs. Perhaps this "confusion" simply arose out of the indeterminate nature of photography in the late nineteenth century. It was, after all, a bridging medium, neither wholly art nor wholly science, but a complex mixture of the two, employed at a time when various forces were pressing the two disciplines apart.[13]

Scientists who used photography in the nineteenth century, even the latter decades when the first industrial standards for emulsions began appearing, were using a volatile and often unpredictable medium that more than occasionally left them open to charges of nonreplicable experiments. Before the late 1880s, no reliable method existed for determining the sensitivity of photographic plates, even ones manufactured by a reliable company. The scope of photography's capabilities was also quite simply unknown at the time, and each passing week in the 1890s brought yet another view of its seemingly limitless compass. Although Becquerel was obviously testing the "rays" that are known to us as radioactivity, he was also testing the photographic plate. In the course of his work, he placed crystals directly in contact with the plates, wrapped the plates, sometimes in paper and sometimes in aluminum, put them in various containers to protect them from daylight, and allowed them to expose for varying length of time (some less than an hour and some for several days); he used multiple photographic plates in a single experiment, observing then the effect on first one plate then another, and finally, he added direction to his images. His interest in the action of different rays on different emulsions led him to try albuminized photographic plates and other gelatin-based emulsions, even daguerreotypes, but he always returned to his *Lumière Bleue*, a gelatin bromide dry plate manufactured by the Lumière firm in Lyon, and the very height of photographic technology at the time.[14] Like many scientists, Becquerel was well aware of the foibles of different emulsions, and of his ability to control the outcome of his photography by controlling the emulsion; either by using different ones, or by heating or cooling the emulsion, or by different sorts of developing. Becquerel's most successful experiment on photography was not, however, made on emulsions, but on the photographic setup. As his understanding of the nature of radioactive emissions grew, he adapted his photographic setup, turning it on its side, in order to better visualize the heterogeneous nature of the radiation he was studying.

In order to understand the significance of directionality in Becquerel's photographic method it is important to go back not only to the origins of his method, but also to the origins of his father's method. They are both, after all, just part of one long tradition of using the chemical action of radiation (light or radioactivity in this case) to study the composition of it.

Edmond Becquerel had added to the usual list of studiable spectra (luminous, thermal, chemical) the phosphorographic spectrum. He believed not only that there were different types of spectra, but also that the differences between them were qualitative, and that these qualitative differences could and should be made visible. Edmond and Henri Becquerel invented and used many photographic methods, most of which were based on this belief and on the habits of studying light in their laboratory.[15] It was Edmond Becquerel who made his photographic methods as much about the study of the photographic effect as it was about the study of any radiant body. Photochemistry, or emulsion science as it was later known, was, for this Becquerel, a separate but equal part of studies on light.[16] Henri Becquerel grew up in a laboratory where photography wasn't monolithic. It was as variable as any instrument with many buttons and knobs, and it required constant adjustment. In photography, the Becquerels had found the perfect vehicle for their experiments in light and physical chemistry. Edmond Becquerel devoted nearly a third of volume 2 of *La lumière* to photochemistry and the photochemical effect, employing his nearly thirty years of research on photographic emulsions.[17]

Because photography was an equally interesting field for experiment, Henri Becquerel could conceive of adjusting it, even to the extent of turning it to face a different direction or exposing two plates at once to see if they exhibited different markings. He saw the results of these adjustments or experiments as photochemical observations of a radiant substance. He introduced radioactivity to the plates by different methods and controlled for extraneous radiation (like visible light or heat). He then settled on a method he felt produced reliable visualizations of the rays, that is, visualizations that matched up his concept of radiation with the physical form of the rays. Becquerel's observations take two forms: observations about the reaction of the emulsion to radioactive substances and observations based on the photographically visualized rays. Stringing this activity out on paper as I have done here gives a skewed impression, though, of the proximity in which Becquerel's observing and experimenting took place, which in actuality met at the interface of the photographic plate.

Seeing the Invisible in Two Different Ways

Becquerel's discovery of the unknown and invisible rays was a matter of both accident and design.[18] He was in attendance at the Academy of Science on 20 January 1896, when Henri Poincaré announced Wilhelm Conrad Röntgen's phenomenal discovery of X-rays. Like many scientists in the audience, Becquerel immediately set out to reinvestigate the qualities of luminescent and phosphorescent substances following Poincaré's suggestion that luminescence and the X-rays had something to do with one another.[19] Given Röntgen's discovery of a new invisible and penetrating ray, any number of other bodies might have been emitting similar and as yet undetected penetrating rays. Becquerel's expertise and experience with luminescent and phosphorescent bodies suggested his first series of experiments, which contained many of the basic components of his photographic method.

> A Lumière gelatin-bromide photographic plate is wrapped with two sheets of very thick black paper, so that the plate will not be affected by exposure to the sun for a day. A sheet of phosphorescent material is placed on the exterior of the paper wrapping, and it is then exposed to the sun for several hours. When the photographic plate is later developed, the silhouette of the phosphorescent substance can be seen, appearing in black on the plate. If a piece of money or a stencil is interposed between the phosphorescent substance and the paper, the image of these objects appears on the plate.[20]

This, however, only describes the images made prior to figure 14.1, when Becquerel was convinced that exciting a substance with sunlight to emit a phosphorescent glow was the key to generating penetrating rays similar to Röntgen's.[21] Becquerel then tried to replicate his own results by using the same crystals and the same type of plates with the same precautionary wrapping. He also meant to expose them under the same conditions, but as so often happens in these situations, "the sun only appeared intermittently." The serendipity of cloudy Parisian skies and a sun-dependent experiment produced images that alerted him to the presence of photographically active, paper-penetrating rays that worked in the absence of sunlight, and made the image shown in figure 14.1.

> . . . I developed the photographic plates on the 1st of March, expecting to find very weak images. The silhouettes appeared on the contrary very strong. I immediately concluded that the action had continued in the darkness and arranged the following experiment:
>
> At the bottom of an opaque cardboard box I placed a photographic plate. Then, on the emulsion side, I rested a strip of uranium salt, a convex strip that only touched the gelatin bromide at a few points. Beside it and on the same plate I placed another strip of the same salt, separated from the gelatin bromide surface by a thin sheet of glass; this entire operation taking place in a darkened room, the box was closed, then enclosed in another box, then placed in a drawer.[22]

Contrary to his expectations, this experiment produced what he claimed was the first "spontaneous" photograph, that is, one generated without any known cause.[23] "Spontaneous" was an adjective that had been applied to photography ever since its publication in 1839, because photographs were drawn by "nature's pencil" and not the human hand. Becquerel's use of "spontaneous" is an attempt to lend credibility and novelty to these images—a crucial exercise, if he was to succeed in proving the novelty of his discovery. These rays were only interesting as an area of study if they were not Rontgen's rays, and Becquerel was able to elucidate the difference by redefining the accepted notion of "spontaneous" as it had been applied to photography, excluding photographs made by X-rays. Röntgen added the energy of the Crookes tube to make photographs, while Becquerel added nothing to his uranium salt, but it made photographs too. Becquerel, in trying to distinguish his rays from Röntgen's, emphasized their difference at the level of photographic production.[24] His images were not made by "exciting" any response by either light or the cathode ray tube. They appeared in the dark, merely in the presence of a substance considered inert, and thus "spontaneous."

Becquerel's new photographs might have been made by unknown causes, but the photographic setup was an old standard. The radioactivity, acting more or less like the sun, was used to "illuminate" an object, in this case a sheet of copper in the shape of a cross. Where the cross blocked some radioactivity from reaching the plate, the emulsion was less affected and a shadow formed, showing the outline of the cross. William Henry Fox Talbot had used the same setup in 1834 for making photographs of botanical specimens without a camera. It is a top-down model where the sun (or other illumination) is directly above, and preferably absolutely perpendicular to, an object resting directly on the chemically sensitive photographic plate or paper.[25]

Figure 14.2 is decidedly different, partly because it was made with an entirely different substance, and partly because it was part of a new line of enquiry. Polonium and radium, new substances discovered by Marie and Pierre Curie in 1898, reinvigorated studies of these rays that were analogous to X-rays but somehow different. In the autumn of 1899, Friedrich Oscar Giesel, a chemist in Braunschweig and one of the few scientists who took up the study of these new Becquerel rays, as they were known, succeeded not only in observing a curious S-shaped deviation caused by sending the rays of polonium and radium through a magnetic field; he also photographed it. Becquerel, eager to pursue this work further, noted that although Giesel had made important chemical observations about the rays, he had nonetheless neglected to investigate their physical properties completely. It was this last, Becquerel wrote, that would show the evidence of some sort of reflection or refraction, and by that, of deviation. Becquerel was convinced that there was more to these rays than he had discovered with the uranium crystal.[26] With this goal in mind, he pursued a set of experiments with fluorescent material intended to depict the rays themselves, or more accurately, to depict the direction of their deviation. The photographs he then made of the same experiment were meant to "explain the previous observations":

> Effects observed in a plane parallel to the field.
>
> Photography, in the sharpness of detail it records, gives quite superior results to fluorescence for the study of [electrical] fields. I carried out in particular the following experiments:
>
> 1. A photographic plate, wrapped in black paper, is placed horizontally, parallel to the field, between two poles placed at a distance of 45 millimeters. After exciting the electro-magnet, the active material is deposited on the plate, equidistant from the poles. After some minutes the material is removed and the plate is developed. A very strong impression is noticeable, that has not been produced uniformly around the source, but is completely to the right of the field (to the left for an observer looking at the + pole). Apart from the black spot that marks the position of the radiant source, the maximum impression is distributed on a narrow strip connecting the poles. The maximum spread corresponds to the direction normal to the field passing by the radiant source which, in this case, is very much in the middle of the field. The maximum intensity of the impression also occurs in this direction. On both sides of this maximum the curve is inflected and connects to the poles at a nearly normal direction to the polar surfaces, on the same side as compared to the center of these surfaces. Other attempts showed that the spread and the curvature decreased as the magnetic field intensified. . . .
>
> This experiment explains the previous observations made with fluorescent screens and, in particular, the absence of phosphorescence in an intense field on a cylindrical screen whose radius is greater than the average diameter of the inflected trajectories of the most active rays.[27]

Reading the account of this experiment, it is easy to see why figure 14.2 differs from figure 14.1 in its "look." It is a direct result of their different subject matter. Although both figures 14.1 and 14.2 show photography detecting radioactivity, only figure 14.2 shows photography also depicting its trajectory. The first photograph is intended to delineate the edges of an object, *visualizing with* the rays. The second takes the rays themselves as the pictorial subject, *visualizing* the rays. It could only have been achieved by upsetting the normal top-down method of making a photograph without a lens. Here the source of exposure lies oblique to the plate. The rays do not meet the photographic plate perpendicularly, but skim along it, nearly parallel. This shift in Becquerel's perspective on the rays says a great deal about the sort of qualities he expected to discover and the sorts of qualities he valued in rendering them visible. The difference between visualizing with and visualizing of need not be about an increase in the manipulation, although that is often a necessary step. It is instead about the type of observation that he wanted to generate. In figure 14.1, Becquerel's photographic plate observed the presence or absence and the penetrability of radioactive rays. In figure 14.2, the observation consists of bringing together what is physically there, the material part that consists of rays, with the vision of the rays' directionality that existed in his mind alone. In this capacity his photographic method excelled, knitting together what existed in his mind and in the room seamlessly into one concrete image that consists of not only an observation, but an explanation.

The Photographic and Electroscopic Methods

When he accepted his joint Nobel Prize in 1903 with Pierre and Marie Curie, awarded for the discovery of radioactivity in uranium salts, Henri Becquerel described his photographic observations as a "photographic method." Photographic methods existed in many sciences pre-Becquerel and proliferated post-Becquerel but this one is distinguished by his assertion that it was "primarily a qualitative one" that worked in tandem with his "electrical method," a method that provided him with "numerical data."[28] This was not an attempt to divide experiment from observation. For Becquerel, both methods allowed observation and measurement. The question is what he meant by qualitative, and how it might describe the special relationship of visualization, experimentation, and observation in his photographic work.

In the 30 March report in which he announced the discovery of spontaneous photographs, Becquerel described his electric and photographic methods for the first time, distinguishing the results obtained by the two. He also clearly stated that the rays of uranium salts penetrated more substances, particularly metals, than X-rays, setting up X-ray images as the foil for Becquerel-ray images.[29] Before this publication, Becquerel had limited his conclusions to ascertaining that radiation was emitted from the uranium crystal, and that the radiation was not unlike the radiation studied by Röntgen. He confirmed Charles Henry's and G. H. Niewenglowski's findings, all of which seemed to sustain Poincaré's original hypothesis. Becquerel's discovery that the rays could be emitted even without energy entering the system, "spontaneously" as he called it, was the first indication that his rays differed from the Röntgen rays emitted from a Crookes tube and were therefore genuinely novel. Like many things that are defined, at least at first, by what they are *not*, it was important for Becquerel to establish that his rays were *not* the same as Röntgen's rays. For these purposes, he did not confine himself just to photography. By March he was certain that something about the uranium rays was different, noting that the "invisible radiation emitted by the salts of uranium have the property of discharging electrified bodies."[30] Following this research meant all but giving up his photographic method, and he would not return to it for many months. It seemed that the electric or electroscopic method (he used the terms interchangeably) would be more fruitful for investigating the properties of the rays emitted from uranium.

Other scientists, like Charles Henry, William Crookes, Niewenglowski, S. P. Thompson, Julius Elster, Ernst Rutherford, and Hans and Friedrich Geitel working together, who employed photography to detect Röntgen rays and, as they soon came to be known, Becquerel rays as well, also used photographic plates in various capacities in their experiments. To date, Becquerel's photographs are the only known remnants of these experiments, conserved as they were in a series of expensive and technically demanding photolithographs. Many, like Thompson, found photography efficient but time consuming and too inaccurate and pursued electrical methods instead.[31]

Although the electrical method seemed to support Becquerel's general hypothesis that the rays he had discovered differed fundamentally from Röntgen's rays, the data occasionally contradicted his photographic results. He made no secret of the hope that the electrical method would give him more exact insights into the nature of these new rays, and he concentrated his attention on perfecting the instrument in the years 1896–1899.

> *b. Electroscopic Method.*—The discharge of a body electrified by rays that have penetrated various screens leads to the same conclusion. I have already shown that the quartz absorbs less radiation from uranium salts than from the rays of the Crookes tube.[32]

Becquerel developed many electroscopes himself, and the Curies, in whose laboratory he often worked, invented several more. The most basic of these consisted of a conductor for the electricity and two leaves of gold, which, when receiving an electric current, would separate and gradually, and as the charge diminished, fall back together. The electroscope proved a tricky instrument, and Becquerel filled his notebooks with design sketches, modifying it almost daily in 1896. His greatest problem was to achieve the perfect translation of radioactivity to electric charge, to be sure that he was measuring the power of the rays and not just the conductivity of his own electroscope. He was able to compare, just as he did with his photographic method, the relative intensities of the rays as they passed through different substances. In addition, he was able to understand and quantify the speed at which the two pieces of gold leaf fell back to their normal closed positions, what would now be roughly equivalent to the rate of decay.

Becquerel admitted that his first measurements with the electroscope were inadequate, but they were no more problematic than his photography. In spite of the problems, he was able to formulate some differences between the new rays and ultraviolet light, due to the different reactions to positive and negative charges.[33] With an electroscope, Becquerel could answer the question of whether rays penetrated a substance or were present at all. But this was no more than he could do with photography. Where the electrical method did surpass photography was that it could answer questions about the exact numerical relationship between the impenetrability of one substance in comparison to another.

In 1896, Becquerel tested photographic and electrical methods side by side. But by 1897, he was experimenting almost exclusively with the electroscope. In 1897–1898 he established the speeds of discharge, the distance at which a body could be discharged, and the role of air in the discharging action.[34] He concluded that all uranium salts emitted radiation of the same type, although of apparently different strengths, and that this was an atomic property of uranium. He was able to establish the relative strengths of rays emitted from various uranium compounds, and he also posited that spheres of charged uranium, acting under the influence of their own radiation, ionized gases. This trait he counted among the fundamental properties of radioactivity. Despite these positive results, he was still not aware of a very basic characteristic of the rays, that they were in fact complex. In the autumn of 1898, acknowledging that both Röntgen and Becquerel had noticed discrepancies in the homogenous state of radioactive rays, Rutherford published his findings on the emission of alpha and beta types of radiation from uranium. Rutherford employed, as Becquerel did, both electrical and photographic methods. He also made an almost offhand remark that beta rays were primarily responsible for photographic action at a distance from the source.[35]

Becquerel's wholehearted return to the photographic method from his electrical experiments has several plausible explanations. He was aware that the numbers he extrapolated from using the electroscope sometimes said less about the radiation than about the conductivity within his apparatus.[36] Although the photographic method gave Becquerel an important point of direct contrast between the Röntgen rays and his new radiation, there was no apparent need to use the method further unless a purely visual method of investigation was going to lead to further discoveries about alpha and beta rays while simultaneously explaining their uniqueness in comparison to X-rays. The success of X-ray imaging in capturing the hearts and minds of the public and scientists may have played a part in Becquerel's search for a visual equivalent to his electroscopic experiments. In 1896 and 1897, only a

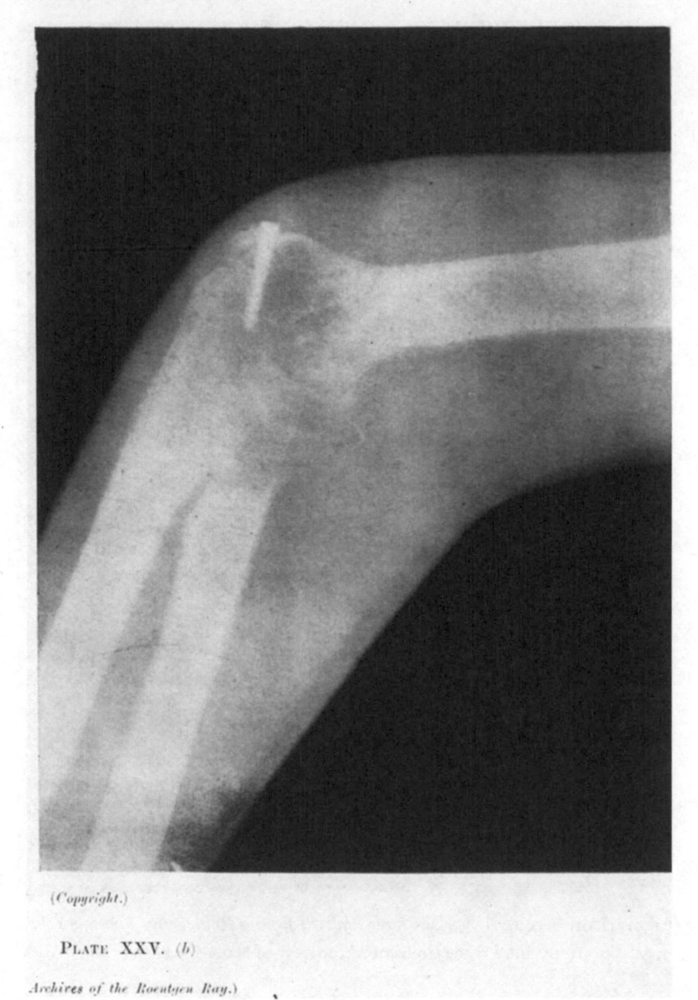

FIGURE 14.3. Thomas Moore, *Fractured Olecranon [After Surgery]*, in *Archives of the Roentgen Ray* 2, n. 1 (1897): plate 25b, photomechanical reproduction.

few physicists believed that Becquerel's rays were truly distinct from X-rays. This is in contrast to the literally thousands of papers published on X-rays and the thousands of images circulated to show the characteristic visual traits of X-ray images.[37] X-rays have the advantage of being somewhat pictorial, even if the objects in them appear quite abstract and unorthodox. It is against the backdrop of this flourishing visual culture of the X-ray that we should view images like figure 14.2. Nothing could differ more from Röntgen's photographs than those Becquerel produced from 1899 to 1903.

One of the differences is the subject matter. Becquerel photographed the rays themselves, whereas X-ray photographers photographed objects with

FIGURE 14.4. Henri Becquerel, *Various Rays Emitted from a Radioactive Substance through a Slitted Screen*, 1903, gelatin silver print. Private Collection, courtesy of Hans P. Kraus, Jr., New York.

the rays (fig. 14.3). There was no real-world equivalent to Becquerel's photographs, and nothing like them had existed before. Even electrical discharges on photographic material, often called Lichtenberg figures, had a visual culture stretching back to the 1770s.[38] Visualizations of radioactivity were by their very nature highly constructed in a way that no scientific photograph had so far been, because the radioactivity needed to be constrained and formed into a shape in order for it to appear at all on the photographic plate. This shape was constrained not only by Becquerel's imagination, but by the nature of photographic and radioactive materials. It carried a great deal of weight, serving not only as an object of observation but as an explanation of foregoing observations. It was mean for both internal use and external dissemination. Becquerel's photography made the qualities of radioactivity perceptible (*mis en evidence*) by creating images like the one in figure 14.4, a recreation of Paul Villard's splitting of the gamma and beta radiation by a magnetic field. Although he could have made cameraless photographs after the example of several different accepted methods, he developed a wholly new visual "look" that recreated and explained his electroscopic findings about the rays.

We have so far discussed two possible visual iconographies that Becquerel used, the one, with the rays set perpendicular to the plate, and the other, with the rays running along the plate. There is a third, and well-known, photographic iconography, and this was also produced without a camera. It was the format of images made in spectral analysis, the vertical strips that resemble a supermarket barcode. Becquerel had employed this in his invention of *phosphoraphie*, and it provided information, visual and otherwise, as to the nature of the radiation. In spite of his experience with the format, Becquerel did not pursue this line of research or the familiar format it offered.[39]

His innovation instead called for a singular piece of equipment that would enable him to send out rays nearly parallel to a photographic plate and to corral them in such a way as to create the appearance of solid bodies out of ephemeral radiation. This was a lead holder that contained his radiant source, allowing the radiation to escape only out of small slits, often positioned to run the length of the long side of the plate. The resulting photographs exhibit many of the formerly "invisible" qualities: the alpha and beta radiation; absorption of the alpha radiation; nondeviability of the gamma radiation; deviability of beta radiation, and the nonpolarization of radiation. In this case the photographic plate, while receiving radiation pointing directly at it, was as blind to the tripartite quality of radioactivity as Becquerel. It was only when Becquerel, hearing of the proof of the heterogeneity of the rays, constructed a method for making the radioactivity travel along the plate that the qualities of the complex rays could be registered. In other words, these images differ from the earlier radioactivity images because Becquerel gave them a specific material form, one that communicated his concept of radiation.

Recall figure 14.2 and its description: the photograph "explains the previous observations made with fluorescent screens." In other words, it was not Becquerel's first observation of the phenomenon, but rather a re-observation. On 7 December, he had staged a similar experiment using phosphorescence to ascertain the deviation of the radiation emitted from radium. Having done this, he proceeded to photograph the observations he *knew* he could recreate. Like an experiment, this observation was, and needed to be, replicable. The photography here is not equal to "observing" but is both a precondition for and an explanatory part of observing. It is also the key to understanding the difference between the qualitative photographic method and the quantitative electrical method. It may also be the reason Becquerel went to such great lengths to ensure a clear differentiation between the two.

Conclusion

It comes as no surprise that Becquerel used photography so extensively. The discovery of X-rays was the result of serendipitous exposure of photographic plates to an active Crookes tube, and Becquerel was one of the innovators of scientific photography. Becquerel, however, not only visualized something new, he visualized it in a whole new way, and he stuck to his method even when it might have been clear to him and to others that radioactivity, in stark contrast to X-rays, would not generate an extensive iconography.[40] Perhaps no one could have imagined that a hundred years later radioactivity would be primarily associated with the auditory click of a Geiger counter or the mushroom cloud. Surely no one could have imagined radioactivity as Becquerel did.

Henri Becquerel, steeped in the photochemical and photographic traditions, followed his convictions and his training. He made photographs within the rubric of an experimental method, a photographic method, that produced intentionally polysemic images. In some instances, Becquerel experimented on the photographic plates, sometimes he observed with them. He also explained and repeated observations with photographs recording them for purposes of dissemination. Little tension appeared to exist for him in investing his photographs with these meanings. Familiar as he was with photographic practice, Becquerel knew that a certain amount of work was necessary to make photographs appear at all.[41] This knowledge hardly invalidated his use of photography; instead, it gave him the power to control it and to inject a certain amount of imaginative practice. The old adage, begun perhaps by William Henry Fox Talbot in 1839, that photography causes things to "register themselves," becomes in this instance not a case of *whether* it registers itself, but *how* it registers itself. Becquerel's experimental setup determined the form radioactivity would take in each image. If visualizing is a necessary precondition for observation in the case of radioactivity, then it follows that in this case the precondition for making visual observations of radioactivity lay in the experiments conducted on the photographic plates and setup. The photographic plate became an interface for experiment, illustration and observation.

From the first, Becquerel needed to imagine that photography told *a* truth but not necessarily *the* truth. From 1899, he ran experiments using both electric and photographic methods singly and in combination. "The activity was either observed by photography, or measured by the electroscope . . . ," he wrote in the summary of these experiments, opposing measurement and quantitative description with visual observation and qualitative description.[42] Both are driven by experiments, but to very different ends. If Becquerel wasn't measuring with his photographic method, what exactly was he doing? It was certainly much more than looking at real events, and also more than just making them visible to the human eye. He formed photographic visualizations of radioactivity to make an argument that both explained his observations and set out clearly the differences between his rays and X-rays. His observations were half-practicing photography, and half-creation of knowledge to think and argue with. He depended on these images to present the evidence of his findings. Becquerel's visual ingenuity lies at the heart of his discovery, and it is his changed vision of radioactivity that first revived and then formed the core of his photographic method.

Mid-nineteenth-century texts on photography beginning with Talbot's *Pencil of Nature* and continuing on through the century claimed that the medium embodied much of what was desirable in the ideal observer: patience, tirelessness, fidelity, speed, accuracy, and a lack of bias. Scientific texts that advocated photography took up this rhetoric and wielded it in defense of their decision to use such a complicated and often deceptive medium.[43] Becquerel's finely manipulated photographic method reminds us that photographic plates, quite in contrast to the rhetoric describing them, require a great deal of intervention in order to produce images. In the case of radioactivity, its presence in a room will cause a silver-bromide emulsion (either on film or on glass) to darken. But in order to achieve more than a foggy mass, in order to achieve a photograph that could provide useful qualitative observations about radiation, Becquerel needed to use his photographic method to experimentally test a theory. He needed to point the rays in a particular direction, and contain parts of them to control their contact with the plate.

Becquerel's attempts to observe this new substance required a creative and innovative plan of visualization, one that he solved by employing a photographic method. With this, he was able to bring together his theories on radioactivity and his knowledge of the three different types of rays, with the material effect of radioactivity on photography, forming striking images that immediately set his own research apart from work on X-rays. Figures 14.1 and 14.2 worked for Becquerel as observations. Figures 14.2 and 14.4, however, show Becquerel's visualizing of radioactivity using the full strength of his imaginative photographic method to exhibit the physical qualities of radioactivity.

Notes

1. Antoine Henri Becquerel, *Recherches sur une propriété nouvelle de la matière: activité radiante spontanée ou radioactivité de la matière*, Mémoires de l'Académie des Sciences de l'Institute de France (Paris: Firmin-Didot, 1903).

2. For a full account of Becquerel's experiments, see Nahum S. Kipnis, "The Window of Opportunity: Logic and Chance in Becquerel's Discovery of Radioactivity," *Physics in Perspective* 2 (2000): 63–99.

3. Founded in 1854, the Photographic Society of France (SFP) is one of the oldest and most influential in the world and remains active to this day.

4. See the essay by Charlotte Bigg in this volume.

5. Throughout this chapter, I have referred to the photographic glass plate negatives as simply "plates" as they are known to photohistorians.

6. This argument was first formed in Lorraine Daston and Peter Galison, "The Image of Objectivity," *Representations*, 40 (1992): 81–128, and fleshed out in Daston and Galison, *Objectivity* (New York: Zone, 2007).

7. Joel Snyder, "Visualization and Visibility," in *Picturing Science, Producing Art*, ed. Caroline A. Jones and Peter Galison (New York: Routledge: 1998), 380.

8. Bruno Latour, "How to Be Iconophilic in Art, Science and Religion?" in ibid., 425.

9. See the essays by Charlotte Bigg and Lorraine Daston in this volume.

10. See Jennifer Tucker, *Nature Exposed: Photography as Eyewitness in Victorian Science* (Baltimore: Johns Hopkins University Press, 2005), chap. 5.

11. Microfilm, for instance, functions as an authoritative copy of a book, because it reproduces the text sufficiently for us to accept it as a mimetic copy. It looks so much the same as a matter of fact, that we hardly conceive of microfilm as photography at all. Less common are photographs that rely on numeric measurements for their authority, but they do exist. Photometric photographs, Raman spectroscopy, and parts of the astrographic catalogs are authoritative photographic methods that rely on the ability of the photograph to be accurately measured. This is not to imply that photographs can be wholly numeric, but that the 'important' information is numeric and not pictorial. See Kelley Wilder, *Photography and Science* (London: Reaktion Books, 2008).

12. Simplicity dominated the photographic sales pitch from early in the 1840s (when it was far from simple) to the mid-twentieth century. Transparency, as described by Joel Snyder, "Nineteenth-Century Photography and the Rhetoric of Substitution," in *Sculpture and Photography: Envisioning the Third Dimension*, ed. Geraldine Johnson (Cambridge: Cambridge University Press, 1999), 21–34, has dominated theoretical discussions in one way or another since the 1970s.

13. Lorraine Daston, "Fear and Loathing of the Imagination in Science," *Dædalus* 127, no. 1 (1998) 84–85.

14. Fonds Becquerel, notebook titled "1896 II Uranium," "Bibliothèque Centrale du Muséum national d'Histoire naturelle.

15. Soraya Boudia rightly argues that habit and familiarity with a certain way of doing things

explain at least some of his experimenting and observing practice. Soraya Boudia, *Marie Curie et son laboratoire: Science et industrie de la radioactivité en France* (Paris: Éditions des Archives Contemporaines, 2001), 43–45.

16. Edmond Becquerel, *La lumière, ses causes et ses effets* (Paris: Firmin Didot, 1867–68).

17. Ibid., 50–121. Edmond Becquerel, as Jean-Baptiste Biot's assistant, was one of the first scientists to experiment on both W. H. F. Talbot's and L. J. M. Daguerre's discoveries in 1839.

18. For a discussion of chance in Becquerel's work, see Nahum S. Kipnis, "The Window of Opportunity: Logic and Chance in Becquerel's Discovery of Radioactivity," *Physics in Perspective* 2 (2000): 63–99.

19. Henri Poincaré, "Les rayons cathodiques et les rayons X," *Revue générale des sciences pures et appliquées* 7 (1896): 52–59.

20. Antoine Henri Becquerel, "Sur les radiations émises par phosphorescence," *Comptes rendus* 122 (1896): 420–21.

21. The images that would have appeared given these descriptions would have matched figure 14.1. In this particular case, it was not the form of the final photograph that held the crucial information but instead the process of making that photograph.

22. Becquerel, "Sur les radiations invisibles émises par les corps phosphorescents," *Comptes rendus* 122 (1896): 501–3.

23. Becquerel, *Recherches*, 3.

24. This spontaneous action with regard to the ionization of air was said by Lord Kelvin to be perhaps the most wondrous part of Becquerel's discovery. It doesn't seem to have been recognized at the time as such, however, partly because, as Kelvin notes in a letter of 1903, of the discouragement by the academies of any individual presenting anything that smacked even vaguely of perpetual motion. Kelvin to Becquerel, 4 Dec. 1903, manuscript letter, Institut de France, Fonds Becquerel.

25. For more on this, see Kelley Wilder, "Photography and the Art of Science," *Visual Studies* 24 (2009): 163–68.

26. Becquerel, *Recherches* (1903), 127–28.

27. "M. Giesel, à qui l'on doit de très importantes observations surtout au point de vue chimique dans le domaine de la radioactivité, ne paraît pas s'être attaché ultérieurement à une étude physique plus complete du phénomène important qu'il venait de mettre en evidence." Becquerel, *Recherches*, 132–33.

28. Antoine Henri Becquerel, *Sur une propriété nouvelle de la matière, la radioactivité* (Stockholm: P. A. Norstedt and Sons, 1905). English translation in *Nobel Lectures: Including Presentation Speeches and Laureates' Biographies; 1901–1921 Physics* (Amsterdam: Elsevier, 1967).

29. Antoine Henri Becquerel, "Sur les propriétés différentes des radiations invisibles émises par les sels d'uranium, et du rayonnement de la paroi anticathodique d'un tube de crookes," *Comptes rendus* 122 (1896): 762–67.

30. Becquerel, "Sur les propriétés différentes," 762–67.

31. Ernst Rutherford, "Uranium Radiation and the Electrical Conduction Produced by It" *Philosophical Magazine* 47 (1899): 3.

32. Becquerel, "20 March 1896," 763.

33. Becquerel, *Recherches*, 18.

34. Antoine Henri Becquerel, "Recherches sur les rayons uraniques," *Comptes rendus* 124 (1897): 438–44.

35. Rutherford, "Uranium Radiation," 13.

36. Becquerel, *Recherches,* 29.

37. For an overview of the visual and material culture of the X-ray, see Monika Dommann, *Durchleuchtete Körper: Die materielle Kultur der Radiographie 1896 bis 1930* (Chronos: Zurich, 2003).

38. Kelley Wilder, "Fotografie als wissenschaftliche Beobachtung: Arthur von Hippel und Fred Merrills Lichtenbergsche Figuren," in *Einfach komplex—Bildbäume und Baumbilder in der Wissenschaft,* ed. Andreas Janser and Marius Kwint (Zurich: Museum für Gestaltung, 2005), 64–65.

39. Klaus Hentschel, *Mapping the Spectrum: Techniques of Visual Representation in Research and Writing* (Oxford: Oxford University Press, 2002), chap. 6.

40. Some of this iconography is covered in Charlotte Bigg and Jochen Hennig, eds., *Atombilder: Ikonographie des Atoms in Wissenschaft und Öffentlichkeit de 20 Jahrhunderts* (Göttingen: Wallstein-Verlag, 2009).

41. Recent work by Robin Kelsey, Jennifer Tucker, Joel Snyder, Christoph Hoffmann and Peter Geimer give examples of just how much.

42. Becquerel, *Recherches,* 288.

43. Alex Pang, "Technology, Aesthetics, and the Development of Astrophotography at the Lick Observatory," in *Inscribing Science: Scientific Texts and the Materiality of Communication,* ed. Timothy Lenoir and Hans Ulrich Gumbrecht (Stanford: Stanford University Press, 1998), 222.

PART FIVE

Observing Together: Communities

Observation is the oldest and most pervasive form of collective empiricism: the collection, transmission, and distillation of the experience of many inquirers into weather proverbs, astronomical tables, medical regimens, botanical descriptions, socioeconomic statistics, and a myriad other findings. No one person and no one lifetime would suffice to discover the subtle correlations and cycles that regulate nature and society. It is not enough that many eyes and hands be enlisted in the task of sustained observation; their efforts must be coordinated. Sustained observation creates communities, real and virtual. In antiquity, observational communities stretched over time: the Babylonian astronomers who scanned the heavens for centuries; the sailors, shepherds, and farmers who watched the weather over generations. But starting in the early modern period, observers also made concerted attempts to organize themselves into communities distributed over space. The essays in this section explore how these synchronic communities worked to recruit, discipline, motivate, and coordinate observers—with significant consequences for the kind of observations produced.

Daniela Bleichmar examines the point of intersection between scientific observation and empire in her essay on the efforts of eighteenth-century Spanish botanists to canvas the plant riches of the New World. In the commercial context of the period, bioprospecting for spices, drugs, and other vegetable commodities was both "big business and big science"; the Spanish crown poured resources into expeditions, botanical gardens, and the production of lavish illustrated floras. The sheer scale of the enterprise, both in terms of the geographic distances and number of new plant species involved, posed formidable challenges of organization: how to calibrate the eyes and hands of so many far-flung travelers, artists, and botanists? In the imperial metropolis of Madrid, Casimiro Gómez Ortega, the director of the Royal Botanical Garden, recruited correspondents and trained the travelers who would man the expeditions and gathered in the specimens, descriptions, and, above all, images they sent back in an attempt, not always successful, to control the field from the center. In colonial Bogotá, the priest and naturalist José Celestino Mutis schooled and supervised a small army of native artists in the production of some 6,700 finished folio plant illustrations. Globe-spanning observer communities were intrinsically fragile and, in the case of eighteenth-century botany, largely held together by traffic in images.

J. Andrew Mendelsohn's essay on the medical-meteorological inquiry launched by the Paris-based Royal Society of Medicine in 1776 confronts the problems of coordinating observers by close examination of what the center actually did with the reports from the field sent in by physicians from all over France. The innocuous word "redaction" in fact stands for a series of paper-and-ink procedures that turned the sprawling, motley, uneven reports of the individual observers into observations that were at once succinct and general. By extracting, summarizing, and correlating, the Paris redactors crystallized "general observations" out of the welter of particular observations, of varying quality, that poured in from the provinces. Although the provincial physicians were guided to some extent by the questionnaires devised by the Royal Society, most of the work of calibration took place after, not before, observation. Both questionnaires and redaction did more than impose a standardizing template on observation; they also dissolved diseases into elements that could be compared along dimensions that cut across geographic regions and disease categories. This community achieved collective observation neither by training nor close supervision nor social cohesiveness; rather, procedures of sifting and editing conducted at the center solidified its results.

In contrast, the community of British observers of seaweed and algae described in Anne Secord's essay enjoyed no official government support, reported to no central authority, and were linked by bonds of shared enthusiasm and familial and friendly connections. This was a community that could overlook differences of class and confession to embrace an ex-shoemaker on the basis of his remarkable talents as an observer of marine algae. But it was also a community steeped in the climate of vigilance and anxiety created by the Napoleonic wars: to be a seaside observer was to watch for enemy ships as well as new species of seaweed; the "lower orders" of the vegetable kingdom were repeatedly intertwined by metaphor and analogy with the lower orders of society. Just as the threat of French invasion from without and subversion by radicals from within militarized turn-of-the-nineteenth-century British society, so even botanical taxa were likened to an army. The conduit through which these analogies between nature and society flowed were the practices of observation themselves, which drew upon the cultural habits of watchfulness and classification routinely applied to people and plants alike. These observational practices also cemented the affective bonds of observers to their objects and to one another: the complexity of seaweed varieties and the difficulty of collecting and preserving them placed a premium on painstaking, patient, discerning observation that only other devotees could assess and appreciate.

15

The Geography of Observation: Distance and Visibility in Eighteenth-Century Botanical Travel

DANIELA BLEICHMAR

The Naturalist as Observer

An unsigned portrait painted at the turn of the nineteenth century in the city of Bogotá, now the capital of Colombia, depicts a naturalist in the act of conducting a botanical observation (fig. 15.1, plate 3). The man, José Celestino Mutis (1732–1808), was arguably the foremost naturalist working in the Spanish Americas at the time.[1] The painting invites the viewer into Mutis's study, allowing us to witness him at work. Mutis is shown sitting at a table wearing his priestly robes, deeply engaged in the pursuit of his craft. His focused gaze fixes on the viewer with weary patience, as if we had just burst into his study of muted grays and browns and interrupted his silent labor. He has lifted his head, but his body remains hunched over in concentration, eager to resume the examination of the flower he holds up toward him. A branch of the same plant lays ready to be pressed between sheets of paper—once dried, it will become a herbarium specimen. Books scattered around the table serve as sources of corroboration for describing and classifying the plant. The books outline the task at hand: if the plant that Mutis examines has already appeared in a publication, he will determine whether it has been assessed correctly or whether the entry needs emendation. Mutis, a long-term resident of the region, knew well that previous travelers had often passed through the land so quickly that they might have made mistakes in their descriptions of a plant given the briefness of their encounter, or might have missed it altogether. Any discrepancy between published materials and the specimen that Mutis examined would provide a chance to contribute to the literature with a correction. Even better for Mutis would be if the plant did not appear in any

FIGURE 15.1. Salvador Rizo? *José Celestino Mutis*, circa 1800, oil on canvas, 48.8 × 36.2 inches (124 × 92 cm). Real Academia Nacional de Medicina, Madrid.

of the existing sources on South American flora, giving him the opportunity to become the discoverer of a new species if he published its description.

The magnifying glass that Mutis holds in his right hand calls our attention to his powers of observation. It hints at the specialized accoutrement that the naturalist owns and deploys in his work, but also suggests that this simple technology can merely augment what is the truly magnificent machinery at work: the naturalist's eyes. The magnifying glass, posed between Mutis's eyes and his object of study, serves as a symbol of the acute observational capacities that characterize him as a botanist. This is not simple looking but rather expert, disciplined, methodical observing. The naturalist's job, the portrait claims, is to observe. And this expert looking is connected to other activities: collecting (as evidenced by the plant, books, and herbarium), comparing and classifying (the books), and writing and drawing (the pen in Mutis's right hand and the sheet of paper before him on the desk).[2]

The portrait not only addresses the process and fruits of observation but also hints at some of its goals and rewards. Like the magnifying lens, the flower that Mutis so attentively considers functions as an attribute celebrating him. It is carefully presented to the viewer for consideration, painted in a bright red that vividly stands out against the muted colors that dominate the portrait. This particular plant has a starring role in the portrait because it is a specimen of *Mutisia*, a new American genus discovered by Mutis—that is, identified by him as not previously described by a naturalist in publication. Mutis found the plant in South America and sent a pressed specimen, as well as a couple of drawings showing details of the floral structure, to Carl Linnaeus. Although Mutis himself did not publish the plant's description, Carl Linnaeus the Younger named it in his honor when he first published it in the *Supplementum plantarum* (1782).[3] Thus, the portrait celebrates Mutis's talents as a botanical discoverer and relates them to his capacities as an observer.[4]

As a source for writing the history of scientific observation, this portrait is as interesting for what it shows as for what it does not show—and it is on the latter that this essay will focus. The painting characterizes observation as an individual, solitary act of concentration, a regime of attentiveness that requires withdrawal from worldly distractions and that relates to an older meaning of the word used today almost exclusively in the context of religious observance. But while the painting suggests that Mutis worked alone and in isolation, the flower intimates his links to a populous world that extended far beyond this darkened room. Mutis participated in an international network that connected naturalists, artists, collectors, physicians, and imperial and colonial administrators throughout the globe. This collective process affected the temporality and geography of observation. Natural history observation did not occur in a single session or location, but rather over extended periods of time, sequentially, and in various settings. It implied a series of comparisons and conversations, as naturalists attempted to see something that had not been seen before, to correct what someone else had seen, and to describe so that others could see what they had. Comparison was a multimedia affair, since naturalists contrasted live plants, dried specimens, printed and manuscript images, and textual reports, a complex process in which the various media often were in contentious discord with one another. The painting itself is the result of multiple acts of observation in multiple places and of the observation of diverse objects: Mutis's repeated looking at specimens of the plant in South America, his comparison of these specimens with published botanical descriptions (textual and pictorial), and Linnaeus's examination of the drawings and pressed specimens that he received from Mutis—in addition to the artist's own examination of his subject. The portrait is also the product of multiple trajectories going back and forth across the Atlantic, which include Mutis's initial voyage from his native Spain to South America; his shipment of letters, pressed specimens, and drawings to Linnaeus in Uppsala (as well as others sent to Madrid); the news of the honorific naming sent from Sweden back to Mutis in America; and, finally, the portrait's own voyage across the Atlantic.

This essay analyzes those aspects of late-Enlightenment natural history observation that the portrait does not immediately suggest. A first section focuses on collective empiricism, examining the operation of long-distance networks of observation in the Spanish empire in the last decades of the eighteenth century.[5] For naturalists interested in information from faraway places, the question was, how to observe at a distance? At the core of the notion of observation lay an individualistic rhetoric of *autopsia*—the process of having experienced or witnessed oneself, with one's own eyes. However, observation often was a collective endeavor, one that drew on the firsthand analysis of secondary materials in various media or on the firsthand observations of others. If distance presented a challenge to naturalists located far away from the materials they wanted to study, it could also represent an opportunity for those naturalists who, like Mutis, managed to travel. A second section of the essay investigates the ways in which observations themselves traveled. Long-distance collective empiricism depended on the circulation of objects, words, and images. Images, I argue, were the privileged media for embodying and transporting both the object of an observation and the observation itself. Images—used together with other media—allowed naturalists to bridge the distance between geographic locations across the world and between the field and the cabinet. The essay concludes with a section that reflects on the relation between distance, visibility, and invisibility in the form of various types of erasures produced by long-distance collective empiricism. While group observation could make inaccessible objects visible, it often also led to conflicts regarding authorship and authority.

Wandering Eyes: Imperial Observational Networks and Scientific Travel

As with other naturalists at the time, Mutis's hard-earned visual expertise was the result of lifelong training. Born in 1732 in the southern Spanish city of Cádiz into a family of booksellers, Mutis studied surgery and medicine, both of them fields that relied on the diagnostic eye. Between 1757 and 1760 he worked in Madrid as an instructor of anatomy, again an occupation concerned with observation. During those years he also attended botanical lessons at the Royal Botanical Garden of Migas Calientes, established in the outskirts of Madrid in 1755. Mutis's time at the garden proved pivotal both to his observational abilities and to his career. During the second half of the century, the garden's directors and instructors—José Quer, Miguel Barnades, Antonio Palau, and Casimiro Gómez Ortega—carried out a major overhaul of Spanish botany. Their goals were to improve the garden's collections and reputation and to train a new generation of Spanish botanists that would collaborate on this effort. To this end, they instituted a formal plan of lessons to train young men as botanists. Most of their students were physicians, surgeons, and pharmacists, as interested in the identification and classification of plants as in their practical uses. The garden's instructors also published translations and Spanish editions of major botanical works of the time, as well as their own botanical manuals. These publications often focused on the process of developing observational skills.

After assuming the garden's chief post in 1772, Gómez Ortega began an even more ambitious program.[6] One of his goals was to increase the popularity and prestige of botany. In 1774, he oversaw the garden's move from Migas Calientes to a central location in the Paseo del Prado, in a new elegant setting in the very heart of fashionable Madrid and near other recently formed institutions like the Royal Natural History Cabinet (est. 1771) and the San Fernando Fine Arts Academy (est. 1752).[7] He also strengthened the garden's botanical lessons, using the textbook he had published.

Gómez Ortega also sought to increase the garden's collections. He and Antonio Palau maintained an active correspondence with naturalists throughout Spain and Europe, exchanging seeds with individuals and with major botanical gardens. They enlisted contributors across the peninsula and the empire, sending them requests for samples and information and rewarding their collaboration with encouraging letters and the title of honorary or contributing member—much as André Thouin did at the Paris Jardin du Roi or Joseph Banks at Kew.[8] Gómez Ortega also publicized the garden's holdings in order to facilitate exchange with other institutions and to specify to contributors which plants were no longer desired. In 1796 he published an inventory that ran to thirty-four tightly spaced pages, totaling about 3,100 plant species; by 1803, the updated inventory had grown to forty pages.[9] In the intervening years, Gómez Ortega published not only descriptions of a hundred new rare plants but also a list of *desiderata*, naming the many different seeds that the garden would like to receive from correspondents.[10]

Under Gómez Ortega, the Royal Botanical Garden embarked on a global mission to investigate the floras of Spain's vast overseas territories in the Americas and the Philippines. The goals of this project were both taxonomic and economic: to publish the description of new plant genera and species and also to identify and successfully commercialize profitable natural commodities such as cinnamon, tea, pepper, or cinchona.[11] Gómez Ortega was particularly interested in useful or valuable foreign plants that could be successfully imported and cultivated in Madrid or other regions of Spain—an attitude shared by other naturalists of the time, among them Carl Linnaeus, who dreamed of acclimatizing the pineapple to Sweden, and Joseph Banks, who desired, more realistically, to cultivate plants in regions with similar climates.[12] "Examining her true interests," Gómez Ortega explained in a lecture in 1770, Spain "prefers to the laborious American gold and silver mines other fruits and natural products that are easier to acquire and no less useful in increasing prosperity and wealth."[13] Botanists and ministers alike hoped that a better-known and efficiently administered empire would furnish rich revenues by allowing Spain to compete with trade monopolies maintained by other nations. The Dutch, for instance, controlled the pepper, cinnamon, and nutmeg trades, while the French did the same with coffee and the British with tea. This climate of international economic and political competition created opportunities for naturalists to sell their services to interested patrons. Botanical expertise became a highly valuable form of knowledge: in the eighteenth century, botany was big business and big science.

The garden's network of contributors extended throughout the Spanish empire, drawing on its massive administrative apparatus and on a longstanding tradition of sending and receiving questionnaires, reports, and collections.[14] In 1779, Gómez Ortega published a set of guidelines to recruit correspondents and instruct them on what materials and information to send, and how to package them so that they would survive the long voyage to Madrid.[15] Each governor, viceroy, or *intendente* of Puerto Rico, Santo Domingo, Havana, Louisiana, Yucatan, New Spain, Santa Fe, Peru, and Caracas received six copies of this *Instrucción*, with orders to pass them on to those individuals better suited to carry them out in the territory under his authority. The *Instrucción* was reprinted and redistributed in 1787. Two years later, new royal orders required colonial administrators at every level to produce reports on potentially useful or valuable natural productions in their regions and to send back to Spain both information and samples. Colonial administrators quickly put these royal orders into effect, and responses flowed from points throughout the Spanish empire in the form of live and preserved specimens, drawings, and textual reports containing observations.[16]

Once in Madrid, these materials were assessed by naturalists, physicians, and pharmacists belonging to an extensive institutional network that included not only the Botanical Garden but also the Natural History Cabinet, the Royal Pharmacy, and numerous royal and naval hospitals. However, at times these shipments or reports proved puzzling to their readers in Madrid. For instance, a report sent from Puerto Rico in March 1790 claimed that "nutmeg" grew in the region of Coamo, noting that "its fruit is similar to the European oak's acorn even in its bitter taste, [and] it is used to treat colic pains and muscle spasms."[17] How were Gómez Ortega and his contacts at the Ministry of Indies to interpret this report? The fruit's taste and therapeutic effects were far from conclusive for identifying the tree as nutmeg. The report's author obviously lacked any botanical training, since he provided only vernacular names for the trees he described and failed to include any Latin names or relevant taxonomical data. While Gómez Ortega would have liked nothing better than to hit upon nutmeg in the Spanish empire, this report alone did not provide enough trustworthy information to decide on a course of action. He needed confirmation from an expert observer.

Collaborating closely with José de Galvez, minister of Indies between 1776 and 1787 and a leading reformer under King Charles III, Gómez Ortega helped coordinate seven natural history expeditions to multiple points of the empire.[18] These expeditions employed over fifteen naturalists and about four times as many artists, who compiled observations, shipped collections of seeds and plant specimens (dead or alive), and produced images over several decades, from the late 1770s until the beginning of the independence wars in Latin America in the 1810s. Their geographical coverage extended throughout most of the Americas, from Patagonia to California and up along the coast of the Pacific Northwest all the way to Alaska, as well as in the Philippines and Australia. The expeditions worked closely with members of the imperial administrative apparatus and also capitalized on the availability of interested individuals. From town to town in the Americas governors, treasury officials, physicians, surgeons, pharmacists, clergymen, and many others collaborated with the expeditions.

Gómez Ortega staffed these expeditions with men who had received botanical instruction at the Madrid garden who were eager to sign up for the voyages. Over the course of the eighteenth century natural history became a global project, and European naturalists hungered for observations and specimens from distant parts of the world. Travel provided a young botanist with opportunities to make his name, especially if he traveled to less well-known regions. Carl Linnaeus contributed to making travel normative when he described his experience of traveling to Lapland as central to his scientific development and sensibility, and recommended travel to any good naturalist. He also scattered around the world young disciples—his "apostles"—sending them to collect botanical information from Africa, Asia, and the Americas. Most of them did not return from their travels, succumbing to feverish tropical deaths.[19] He nevertheless provided a model for men like Gómez Ortega, who emulated his example and hoped for better results.

While travel could be perilous and was often unpleasant—as voyagers constantly remind us in their writings—observation and travel were closely linked. The very point of scientific travel was to observe. Forms of the word "observation" appeared frequently in the titles and subtitles of natural history publications. The term was particularly prominent in travel narratives, in which it provided the motive for voyaging, the activity to be conducted during the journey, and the result to be brought back. As explained in the orders given to pharmacist Fusée Aublet as he set off to explore the natural history of the French American colony of Guiana, "The whole purpose of your mission must be to see everything and to examine everything."[20] Mutis and other voyagers to the Americas agreed that the most significant problem with European knowledge of American flora was that the majority of available textual descriptions and images were inaccurate or incomplete and could not be trusted. Naturalists in Europe desperately needed more exact and complete descriptions, figures, and specimens, and for these they depended on travelers—and not just any traveler, but those who could provide accurate observations and descriptions as well as appropriate collections of specimens and images.

Even better than travelers, European naturalists agreed, would be long-term residents in foreign regions whose observations could be trusted. A traveler moving from one region to another would not have the opportunity to observe specimens that were out of season, might fail to observe a particular item thoroughly or to fully understand what he had witnessed due to his superficial knowledge of the region or his incapacity to identify reliable informants, or might reach inaccurate conclusions. There was a temporality to observation. Moving too quickly through a territory did not allow for sustained observation over a longer period, led to mistakes, and exposed the traveler—and any author who used his material—to criticism. In order to produce a satisfactory observation, a naturalist needed to look again and again over time. This had as much to do with the life cycles of the species observed as with a botanist's developing expertise over time. The Spanish imperial observational network, with a corps of trained observers and a vast imperial network that could facilitate their travels and provide local expertise, was uniquely positioned to fulfill the global ambitions of European natural history.

Moving Pictures: Transporting Observations

Mutis was exactly the rare type of correspondent that European naturalists desired: a trained and attentive observer, a long-term resident of the Americas, and an enthusiastic correspondent.[21] Mutis traveled to the Spanish Americas in 1760 in the capacity of personal physician to the incoming viceroy of the New Kingdom of Granada (a territory that corresponds to present-day Venezuela, Colombia, Panama, Ecuador, and part of Peru), and never returned to Europe, remaining there until his death in 1808 at the age of seventy-six. For twenty years after arriving in New Granada, Mutis had a varied career. He worked as a physician and as professor of mathematics, astronomy, and natural philosophy at the university in Bogotá (Colegio Mayor de Nuestra Señora del Rosario), where he reputedly was the first person to teach Copernicus and Newton in the Spanish Americas. After the expulsion of the Jesuits from all Spanish territories in 1767, he actively participated in the kingdom's educational reform. He also served as a mine administrator for almost ten years in the late 1760s and 1770s in two separate locations. All of these activities gave him ample opportunity to conduct observations of the flora, fauna, and geology of New Granada.

For Mutis, as for other traveling naturalists, observation was not just something he did but part of who he was, a *habitus* and a way of life, a regime of constant attention and investigation that extended to all spheres. He began a travel journal from the moment his voyage started, noting natural history observations as well as reflections and anecdotes from his daily life.[22] He composed long lists noting all the plants he saw, keeping track of their taxonomic characters and comparing them to published textual and pictorial descriptions. Over the years, he kept meticulous observation journals in which he recorded two sustained observation projects, one about ants and one about the "sleeps and vigils" of flowers, which entailed carefully tracking the opening and closing of multiple flowers and correlating their state to the time of day and weather on an hourly basis over a space of many weeks.[23] Both of these observation projects involved the kind of obsessive attentiveness and ingenious *mise-en-scène*—or rather, *mise en observation*—that blurs the line between scientific observation and experiment (see chap. 6).

Mutis relied not only on his own observations but also on conversations with a wide range of people from various social and ethnic groups, whom he questioned about their knowledge of local flora and fauna and their medicinal uses. Mutis recorded their responses in his journals, though he seems to have valued them mostly as an opportunity to scornfully rail against the stupidity of popular knowledge.[24] A hot-tempered, somewhat cantankerous man who complained of "the bitterness produced by dealing with people," Mutis far preferred the solitary study of nature and the type of disembodied conversations he maintained with a vast network of correspondents throughout the Americas and Europe as well as with the books in his impressive library.[25] He proved an important contributor to his European correspondents, sending Linnaeus about 250 herbarium specimens between 1767 and 1778, as well as two sizable collections of images; he also sent rich collections to Gómez Ortega.[26]

In 1783, Mutis received authorization to direct the Royal Botanical Expedition to the New Kingdom of Granada. Mutis headed the expedition for twenty-five years, until his death in 1808; the project continued after that until 1817, when all materials from the expedition were sent to Madrid because of the independence war. Mutis promised his patrons that the expedition would promptly yield useful and valuable information in the form of natural commodities, which Spain could use to break the trade monopolies held by European competitors. To this end, he assembled a team composed of *herbolarios*, or plant collectors, as well as artists and botanical contributors, and together they diligently attempted to locate American varieties of cinnamon, tea, pepper, and nutmeg, as well as new types of the valuable antimalarial cinchona (the source of quinine and a prized Spanish monopoly). He spent years studying different types of cinchona and became embroiled in a heated priority dispute regarding a new variety of the plant. He investigated American varieties of cinnamon and pepper. He also monitored European periodicals to keep track of British trade in tea and located a South American plant he tirelessly and unsuccessfully promoted as a potential substitute, the so-called Bogotá tea.

But Mutis also devoted enormous efforts to another end, one with less obvious economic or utilitarian application: the production of visual representations of American plants. Over the years, the expedition employed somewhere between forty and fifty artists, thirty of them working simultaneously at one point—an enormous size for an artistic workshop of any kind, and unheard of for one dedicated to scientific illustrations. While Mutis and a handful of botanical collaborators penned only about 500 plant descriptions, the much larger artistic team created a staggering total of almost 6,700 finished folio illustrations of plants and over 700 detailed floral anatomies.

In today's world of image databases and laser printers, it can be hard to grasp the dedicated labor it took to craft one of these paintings, let alone thousands. A single image involved a close collaboration among plant collectors, botanists, and entire teams of artists who specialized in the various steps it took to achieve a finished illustration. This process took several days. Each image embodies not only a plant but also multiple observations, decisions, negotiations, and types of expertise. Mutis did not work alone, as his portrait suggests, but rather supervised a large operation. He hired artists from Bogotá, Madrid, and Quito; he obsessed about how to train them and control their work, imposing a strict work schedule based on a nine-hour day, six days a week, for forty-eight weeks out of the year. Some of these artists worked in the expedition for the entirety of their adult lives. Mutis had strong ideas about both botanical and artistic aspects of the images, and got into monumental fights with those painters whose work ethic or results did not satisfy him. He recruited Spanish artists from the Madrid Fine Arts Academy of San Fernando, only to later find that they actually had strong opinions about art and were not as malleable as he had hoped. His solution to the problem of finding a large number of docile painters was to establish a free drawing school where young prepubescent boys could be trained as botanical draftsmen.[27] Given the existence of this extensive visual archive, the enormous efforts to which Mutis went to employ, train, and supervise his painters, and the frequent discussion of the production and uses of natural history illustrations in his journals and correspondence, it is clear that images were of central importance to the expedition's exploration of American nature.[28]

Moreover, Mutis was not alone in his visual voracity. Every single one of the Spanish natural history expeditions, without exception, employed artists and created an enormous number of illustrations. Indeed, the bulk of their work took the form of illustrations, yielding many more images than collections or textual descriptions, manuscript or published. As a group, these expeditions alone created about 13,000 drawings. The expeditions suggest that images held a starring role in eighteenth-century botany, especially long-distance botany. It is telling, for instance, that although the naturalists from the various Spanish expeditions only rarely corresponded with one another, the very few letters that they did exchange invariably mentioned their painters. In late 1788, Vicente Cervantes, a naturalist in the New Spain expedition, replied to a letter in which Mutis had asked about the talents of two Mexican artists who had recently joined the team. Cervantes answered the query not only with words but also by sending images that would demonstrate the draftsmen's skills.[29] Similarly, the lone letter that naturalist Luis Née, member of the Malaspina circumnavigation expedition, sent to Mutis addressed the incorporation of a botanical draftsman to the team. Writing from Ecuador in 1790, Née expressed his satisfaction with the artist and described him as good, patient, possessing the "foundations" of botany, and skilled at defining all the parts of a plant, especially the fruit and flower. "The drawings I have taken care to direct to this day," Née enthused, "include nothing beyond what is necessary for any systematist to know the [plant's] class and order. Adding a methodical description seems sufficient to know the plant that is being presented."[30]

It is worth noting that Née writes of "directing" the painter's work, indicating that, while artists were crucial members of an expedition, they were ultimately subservient to the authority of naturalists. The draftsman acted as the expedition's hand, hired to produce the images that naturalists desired; the naturalist served as its eye, selecting the object to be depicted, indicating which traits to focus on and which to disregard, and imposing the particular vision with which to approach and represent nature. The botanists' control over the artists' production and use of time also extended to regulating their bodies by mandating where and when they should travel, as well as to controlling their productivity by allocating work supplies. The desired images were botanical objects, not artistic creations, hence the insistence on depicting those characters considered important by the botanists and on avoiding a decorative painterly style—a key concern of natural history illustration at the time. Botanical considerations mandated the content, style, even the size of the image. In this and other expeditions, naturalists daily supervised and directed artists, evaluating whether a drawing qualified as finished and satisfactory or whether it needed correction. Figure 15.2 and plate 4, for example, show naturalist Luis Née's correction to José Guío's watercolor drawing of a plant, written directly across the image: "The fruit should be green: the painter made an error."[31] Without botanists to guide them, artists were useless, since "making a perfect drawing does not have to do with representing the visible parts of a plant but rather with knowing their location, direction, size, and shape."[32] Naturalists considered themselves the true authors of the drawings, with artists as their needed but subordinate amanuenses.[33]

Naturalists placed such a value on visual material because of the work that images did in the process of conducting, embodying, and transporting observations. Natural history observations were recorded and made portable in words and images. When naturalists set off on their voyages, they took with them printed images.[34] Illustrated books provided a visual and verbal

FIGURE 15.2. José Guío (Malaspina expedition), *Rubus radicans* Cav., watercolor, 1790, 11.8 × 19.3 inches (30 × 49 cm). A manuscript annotation from botanist Luis Née indicates, "The fruit should be green, the painter is mistaken" ("El fruto ha de ser verde, se equivoca el pintor"). Real Jardín Botánico, Madrid, ARJB VI/40.

vocabulary that was shared by naturalists throughout and beyond Europe. They presented standards against which naturalists could gauge the value of their own work, as well as models for them to emulate or react against. Thus, books helped to define and arbitrate a community of competent and relevant practitioners.[35] Printed images also defined the traveling naturalist's job by demarcating what they should accomplish in their voyages, namely, to describe any local productions not included within the European printed inventory of global nature, to rectify any discrepancies, and to resolve incomplete or erroneous descriptions. Books provided naturalists with the illustrations they needed to approach nature, with parameters for producing new images, and with a medium for presenting their own contributions to natural history. The traveling naturalist's way of seeing involved a constant triangulation among image, text, and specimen, using books to interpret what they saw in a field and producing their own texts and images to respond to what they read as well as to contribute new information. Naturalists also used images as working tools, to record or work out what exactly they were observing. When working out a taxonomical point for themselves, naturalists produced notes that tend to include diagrams to clarify the problem at hand as well as its solution. When writing letters, especially to correspondents who had never been in the region they were describing, naturalists appended drawings to get their point across. In the eighteenth century, images provided an entry point into the exploration of nature, functioned as a key instrument for producing knowledge, and constituted the foremost result of natural investigations.

The most important work that images did for eighteenth-century natural history was to abstract information, visually incarnate expert observations, and mobilize across distances plants that remained in key ways unseen and unknown, even three centuries after Europeans first encountered New World nature. For eighteenth-century naturalists, the manufacture and use of images was the central practice through which nature, particularly foreign and exotic nature, was investigated, explained, and possessed. Natural history was fundamentally a visual discipline, based on the observation and representation of specimens that sometimes were "out there" in the field, other times "in here" in collections, and yet other times in the hybrid domesticated space of illustrations, in which exterior and interior, the field and the collection, were collapsed into a single paper nature that was always and perfectly available for virtual exploration. The natural history illustration, with its flower always in bloom, its fruit permanently ripe, its animal suspended in clarity and permanence, was at once the instrument, the technique, and the result of natural history as a field of study.

Observing at a Distance: Geography, Authorship, and Erasures

Thinking about these trajectories from the field to the cabinet, from South America to Europe, highlights the choice on the part of the artist who painted Mutis's portrait (fig. 15.1) to depict the naturalist inside a completely enclosed study, without even a window through which we might see the American landscape. Although Mutis made his career by voyaging to South America, he is depicted as an armchair or cabinet naturalist. The production of scientific facts is thus constructed as a process privileging the intellectual and physical tasks of observing and classifying over the manual labor of procuring the specimens themselves, and the indoor cabinet over the outdoor field. As Dorinda Outram has argued, eighteenth-century natural exploration was characterized by a tension between two tendencies: on the one hand, an impulse to move, to know by traversing and experiencing, thus to know in a personal and embodied way; and on the other an impulse to stay put and have knowledge come to one and join the corpus of what is known, to examine multiple specimens and compare.[36] Natural history illustrations served to bridge the gap between these two impulses, erasing the distance between the cabinet and the field, Europe and the larger world.

Juxtaposing Mutis's portrait to another painting produced by an artist from his workshop further elucidates the transatlantic circulation of people, specimens, and images (fig. 15.3, plate 5). This portrait depicts the renowned Spanish botanist Antonio José Cavanilles (1745–1808), director of the Royal Botanical Garden of Madrid between 1801 and 1804 and a long-time correspondent and supporter of Mutis. The painting shows Cavanilles in profile,

FIGURE 15.3. Salvador Rizo? *Antonio José Cavanilles*, circa. 1800, oil on canvas, 33.9 × 26 inches (86 × 66 cm). Museo Nacional, Bogotá, Colombia. Collección del Museo Nacional de Colombia, Bogotá, Colombia. Photograph for Museo Nacional de Colombia by Juan Camilo Segura.

from the waist up, sitting before a worktable wearing his priestly habit. With his left hand, Cavanilles points to a botanical illustration that he studies with unblinking attention. The image is clearly identifiable as one of the works produced by the Mutis expedition. Gazing attentively at the image, the botanist observes the various parts of the plant and immediately transforms his visual analysis into a textual taxonomic description, which he writes with a quill pen in a notebook that lies open on the table. Eye and hand work in coordination, image produces text. Set against a dark background and the lustrous velvety black of Cavanilles's garments, the light-colored pages pop brilliantly. The botanical illustration is as much a protagonist of this painting as the man rapt in its study. It serves to connect the naturalist in Spain and the artist in America, erasing the distance between them. Although the portrait is unsigned and undated, the artist cleverly inscribed himself into the painting. The name visible at top of the image, *Rizoa*, points to the identity of both a South American plant and a South American artist, Salvador Rizo (1762–1816), who was the expedition's lead artist and Mutis's second-in-command, and in that capacity directed the artistic workshop that produced illustrations exactly like the one Cavanilles is examining. The painting is in fact most likely Rizo's way of thanking Cavanilles for naming this American genus after him.[37]

Rizo's portrait of Cavanilles celebrates both artist and naturalist through a botanical identity, much like Mutis's portrait honored him through the *Mutisia*. The replacement of a botanical specimen with an image is significant because it demonstrates how Mutis and Rizo expected naturalists in Europe to use the expedition's illustrations. Mutis, based in South America, could observe multiple fresh specimens over the years and work with an artist to create an image that presented a composite result of all those observations. This would be impossible to achieve with a single dried specimen, which would inevitably include the accidental particularities of that single plant—a leaf broken or damaged by an insect, a flower that had began to wilt by the time the plant was collected, or a specimen collected when the flowers had not fully bloomed. Herbarium specimens could rot, get eaten by insects, or suffer damage and lose portions of the plant. The picture, by comparison, incarnates not only the American plant but also the multiple specimens and observations that allowed naturalist and artist to produce an idealized version of it.[38] This composite specimen makes travel unnecessary: the painting allows Cavanilles to sit at his desk in Europe and observe South American flora "firsthand," using this rendition to classify and name it. Manuscript evidence demonstrates that Mutis had precisely this use in mind for his images. In a letter, he articulated the potential of images to allow long-distance knowing by seeing. "No plant," Mutis explained, "from the loftiest tree to the humblest weed, will remain hidden to the investigation of true botanists if represented after nature for the instruction of those who, unable to travel throughout the world, without seeing plants in their native soil will be able to know them through their detailed explanation and living image."[39] The illustration, then, not only stands in for the object it represents but also supplants the very act of travel, erasing both geography and distance. This erasure is underscored by the depiction of both naturalists indoors, in similarly furnished spaces that make it impossible to distinguish between Mutis's on-the-ground experience of American nature and Cavanilles's long-distance experience from Madrid. The painting itself is the result of a traveling image: since neither the American artist nor the Spanish botanist crossed the Atlantic, Rizo's portrait of

Cavanilles must have been based on a published engraving that transported Cavanilles's likeness to Bogotá, in the same way the expedition's paintings conveyed South American flora to Europe.

The two paintings erase not only place and space but also, through their focus on the lone botanist, the collective nature of long-distance observation. Mutis's large team of botanists and artists has disappeared, as have Cavanilles's correspondents. It might have taken a village to compose an observation, but ultimately it was one man who got credit for it. This was especially so in taxonomic matters, since the "discoverer" of a genus or species had the privilege of naming it and having his own name attached to it in turn: botanical Latin names are followed by an abbreviated form of the name of the person who first described that type in publication, so that the name "*Rizoa* Cav." honors both Rizo and Cavanilles, "*Mutisia* Linn. f." honors both Mutis and Linnaeus the Younger. There was an inherent tension in collective empiricism between the need for extensive networks and the practice of crediting a sole individual as the author of the observation itself, a tension that traveling naturalists keenly felt.

The link between authorship and priority in publication pitted traveling naturalists in competition against each other. The naturalists in the various expeditions often traveled on overlapping routes, which placed them at risk of duplicating one another's efforts. Traveling naturalists also had to contend with the possibility of getting trumped in publication while they voyaged. To prevent this, they had to remain as up-to-date as possible with the state of botanical publication—something that could prove hard to do while traveling, making them dependent on their European correspondents for updates—and to communicate their findings back to Europe as rapidly as possible. In March 1784, shortly after the official start of the New Granada botanical expedition, Mutis wrote to Gómez Ortega asking for news of the Royal Botanical Expedition to Chile and Peru (1777–88), inquiring about naturalists Hipólito Ruiz and José Pavón's "new genera and species, and whether some of mine are showing up there."[40] Mutis explained that he realized it might be difficult to adjudicate new discoveries when both expeditions concurred in their findings and that, while naming rights undoubtedly belonged to the first discoverer, priority might be hard to establish. His anxiety concerned not only future findings, but also the possibility of having his work of the past two decades, from his arrival in New Granada in 1761 to the official start of the expedition in 1783, invalidated by Ruiz and Pavón's observations, which would receive priority if published first, regardless of whether they duplicated his own previously unpublished work. Fearful of this possibility, Mutis offered to remit a catalog with the historical record of his findings, scrupulously noting the date for each observation based on his personal journal; for future discoveries, correspondence would serve this role.

The tensions were not only among travelers. Although naturalists in the cabinet and the field depended on one another—Gómez Ortega and Cavanilles, in Madrid, needed to receive materials from their correspondents, while naturalists scattered throughout the empire needed the men in Madrid in order to publish their findings—authorship could prove contentious even among such close collaborators. Naturalists in Madrid had to show appropriate gratitude, lest their correspondents abandon them. All of the hundred new and rare plants described in Gómez Ortega's *Novarum aut rariorum plantarum* (1797–1800) grew in the Madrid Royal Botanical Garden thanks to the correspondents who had sent seeds or live plants. Without them, there would have been no book, and Gómez Ortega acknowledged his debt by noting in the entry for each plant the name of the person who had sent it to Madrid. The natural history expeditions were responsible for the vast majority of the new plants in the publication, and they receive a prominent mention in the first page of the preface to the work. Cavanilles acknowledged his correspondents in the same manner in his publications, and both botanists used naming as a strategy for rewarding contributions.

The race to publication, not surprisingly, created rivalries and alliances, with Gómez Ortega favoring Ruiz and Pavón and in competition with Cavanilles and Mutis. Heated disagreements arose between Ruiz and Cavanilles, who attacked one another viciously in print.[41] The dispute concerned not only the priority of some of the discoveries but also the validity of descriptions of new genera and species and their classifications within existing or new categories, based on the accuracy or error of each botanist's observations and the materials they had used to conduct them. Ruiz accused Cavanilles of stealing his and Pavón's work by publishing new types that they had labored to identify and collect in South America, taking for himself credit that was rightfully theirs. He claimed that they had shipped seeds and pressed specimens to Madrid so that, upon their return, they themselves could consult these materials as they prepared their publication, and it was not for someone else to "profit from our work, publishing it as his own and with little exactitude."[42] He criticized Cavanilles's observations, noting that he had reached erroneous conclusions based on the examination of faulty herbarium specimens or of plants grown from foreign seeds in the Madrid garden, noting that they responded to the foreign soil and showed marked differences from their regular appearance in their home regions, whereas Ruiz had been able to examine live plants.[43] In any case, Ruiz reminded the reader that he and Pavón had already explained in publication that, although many of the genera they had identified in America had already been published by other authors when the travelers returned to Europe, most of these descriptions "were copied from imperfect descriptions, which also lacked plates, or were taken from plants born from seeds spread through European gardens and thus degenerated, or from mere skeletons [herbarium specimens]."[44] Faulty materials had yielded inaccurate observations, which only the experienced traveler could correct.

Cavanilles replied forcefully. He defended the use of dried herbaria as substitutes for live plants, pointing to the practical advantages they presented over travel. "As active as the botanist may be," Cavanilles explained, "and even if he consumes his life with travel, he will never be able to see more than a small portion of vegetables, compared to the countless number that exist, and in the end, to reach a deeper knowledge, he will resort to the *hortus sicus* or herbarium, which the princes of science acknowledged as useful and necessary."[45] Or did Ruiz presume to disagree with Linnaeus, de Jussieu, Lamarck, and Smith?[46] All naturalists, Cavanilles acknowledged, relied on travelers, since it was the travelers who collected and pressed specimens to form herbaria. Thus, if descriptions based on herbarium specimens had problems, the blame lay with the travelers who had formed the collections and not with the naturalists who had used them in good faith.[47] Cavanilles's defense of his work method, especially of the validity of his authorial role based on collections rather than a voyage, turned into a disquisition on the fraught relationship between travel and empiricism.[48] A voyager, he claimed, faced with dozens of new plants each day, having to examine and describe them on the go, and on top of that direct the work of his artists, was not in the best position to observe and reach conclusions. He recognized the work that Ruiz and Pavón had accomplished as voyagers, but criticized the conclusions of their observations, claiming they invalidated their role as authors. Authorship, he argued, resulted from the correct taxonomic determination of a plant, and nothing else:

> It is a manifest mistake to think that someone who publishes the plants that others gathered but did not examine is appropriating their work, because he leaves them the portion of glory that they deserve for traveling and pressing plants, and takes for himself only the glory resulting from [the plants'] examination and [his] scientific works. The author is not he who picks up plants and seeds and ships them without due examination: the true author of a plant is only he who makes it known to the public . . . . It is not the same thing to be a traveler and to be a Botanist, nor to see plants and to be a competent Judge to determine their fructification, genus, and species.[49]

According to Cavanilles, paradoxically, travelers could not be the best observers of the regions they visited. Only by retreating into the controlled space of the cabinet—as in his and Mutis's portraits—could a botanist observe properly. If distance was an obstacle to overcome, it was also necessary.

Images played an important part in both bridging and creating distance. In their function as mediators between the field and the cabinet, images not only embodied and transported observations but also carried out multiple erasures—of place, of distance, of time, of human actors. Like the indoor setting in Mutis's and Cavanilles's portraits, the blank white background characteristic of the natural history illustration makes it impossible to know whether a plant grew in South America or Europe; indeed, it makes that geographical information irrelevant to its study. The natural history illustration depicts a decontextualized, isolated specimen upon the white background of the page, a background that both frames and erases. Given the impressive powers of the naturalist's eyes to identify and classify, it is remarkable just how much these trained eyes chose not to see and not to show. The naturalist's gaze was extraordinarily selective not only about what it noticed but also about what it disregarded. For that reason, the visual culture of natural history presents a great paradox: the very point of observing, drawing, and publishing images was to place before European eyes little-known natural productions from distant lands. Specimens were collected, described, drawn, and published precisely because they originated in the Americas, Asia, or Africa. However, this information was not included in their depictions. More than mere representations, images acted as visual avatars replacing perishable or untransportable objects that would otherwise remain unseen and unknown outside of their local setting. Images defined nature as a series of transportable objects whose identity and importance was divorced from the environment where they grew or the culture of its inhabitants. Pictures were used to reject the local as contingent, subjective, and translatable, favoring instead the dislocated global as objective, truthful, and permanent. Efforts to make global nature visible always involved making parts of it invisible.

Notes

1. Marcelo Frías Núñez, *Tras El Dorado vegetal: José Celestino Mutis y la Real Expedición Botánica del Nuevo Reino de Granada (1783–1808)* (Seville: Diputación Provincial de Sevilla, 1994); A. Federico Gredilla, *Biografía de José Celestino Mutis* [1911] (Bogotá: Plaza & Janés, 1982); and María Pilar de San Pío Aladrén, ed., *Mutis and the Royal Botanical Expedition of the Nuevo Reyno de Granada*, 2 vols. (Barcelona: Lunwerg Editores, 1992).

2. Daniela Bleichmar, "Training the Naturalist's Eye in the Eighteenth Century: Perfect Global Visions and Local Blind Spots," in *Skilled Visions. Between Apprenticeship and Standards,*

ed. Cristina Grasseni (New York: Berghahn Books, 2007), 166–90; Brian Ogilvie, *The Science of Describing* (Chicago: University of Chicago Press, 2006).

3. On honorific naming, see Londa Schiebinger, *Plants and Empire: Colonial Bioprospecting in the Atlantic World* (Cambridge, Mass.: Harvard University Press, 2004), 194–225; and Mauricio Nieto Olarte, *Remedios para el imperio: historia natural y la apropiación del Nuevo Mundo* (Bogotá: Instituto Colombiano de Antropología e Historia, 2000), chap. 2.

4. On scientific portraiture, see Ludmilla Jordanova, *Defining Features: Scientific and Medical Portraits* (London: Reaktion, 2000).

5. On collective empiricism, see Lorraine Daston and Peter Galison, *Objectivity* (New York: Zone Books, 2007), 19–23.

6. Carmen Añón Feliú, *Real Jardín Botánico de Madrid, sus orígenes, 1755–1781* (Madrid: Real Jardín Botánico, 1987); Francisco Javier Puerto Sarmiento, *Ciencia de cámara. Casimiro Gómez Ortega (1741–1818), el científico cortesano* (Madrid: CSIC, 1992).

7. Andrew Schulz, "Spaces of Enlightenment: Art, Science, and Empire in Eighteenth-Century Spain," in *Spain in the Age of Exploration, 1492–1819*, ed. Chiyo Ishikawa (Omaha: University of Nebraska Press, 2004), 189–227.

8. Richard Drayton, *Nature's Government: Science, Imperial Britain, and the 'Improvement' of the World* (New Haven: Yale University Press, 2000); Emma Spary, *Utopia's Garden: French Natural History from Old Regime to Revolution* (Chicago: University of Chicago Press, 2000), esp. 49–98.

9. Casimiro Gómez Ortega, *Elenchus plantarum horti regii botanici matritensis* (Madrid, 1796); Antonio José Cavanilles, *Elenchus plantarum horti regii botanici matritensis* (Madrid, 1803).

10. Casimiro Gómez Ortega, *Novarum aut rariorum plantarum Horti Reg. Botan. Matrit.* (Madrid, 1797–1800) and *Index seminum plantarum quae in R. Matr. H. desiderantur* (Madrid, 1800).

11. Daniela Bleichmar, "Atlantic Competitions: Botanical Trajectories in the Eighteenth-Century Spanish Empire," in *Science and Empire in the Atlantic World*, ed. Nicholas Dew and James Delbourgo (New York: Routledge, 2008), 225–52; Nieto Olarte, *Remedios para el imperio*; Francisco Javier Puerto Sarmiento, *La ilusión quebrada. Botánica, sanidad y política científica en la España Ilustrada* (Madrid: CSIC, 1988). More generally, see Schiebinger, *Plants and Empire*; and Londa Schiebinger and Claudia Swan, eds., *Colonial Botany: Science, Commerce, and Politics* (Philadelphia: University of Pennsylvania Press, 2005).

12. Drayton, *Nature's Government*; John Gascoigne, *Science in the Service of Empire: Joseph Banks, the British State and the Uses of Science in the Age of Revolution* (Cambridge: Cambridge University Press, 1998); Richard Grove, *Green Imperialism: Colonial Expansion, Tropical Island Edens, and the Origins of Environmentalism, 1600–1860* (New York: Cambridge University Press, 1995); and Lisbet Koerner, *Linnaeus: Nature and Nation* (Cambridge, Mass.: Harvard University Press, 1999).

13. Casimiro Gómez Ortega, "Oración gratulatoria al tomar posesión de su plaza de académico supernumerario [Real Academia de la Historia]" [1770], in Puerto Sarmiento, *Ciencia de cámara*, 54.

14. Antonio Barrera-Osorio, *Experiencing Nature: The Spanish American Empire and the Early Scientific Revolution* (Austin: University of Texas Press, 2006); Daniela Bleichmar et al., eds., *Science in the Spanish and Portuguese Empires, 1500–1800* (Palo Alto: Stanford University Press, 2008), 271–89; Paula De Vos, "Research, Development, and Empire: State Support of Science in the Later Spanish Empire," *Colonial Latin American Review* 15, no. 1 (June 2006):

55–79, and idem, "Natural History and the Pursuit of Empire in Eighteenth-Century Spain," *Eighteenth-Century Studies* 40, no. 2 (2007): 209–39.

15. Casimiro Gómez Ortega, *Instrucción sobre el modo más seguro y económico de transportar plantas vivas por mar y tierra a los países más distantes* (Madrid, 1779, reprinted 1787).

16. Bleichmar, "Atlantic Competitions."

17. Tiburcio Rodríguez, "Relacion indibidual de los Arboles utiles, de Fabricas, Tintas, y Medicinales, que se enquentran en el Partido de Coamo (Puerto Rico)," 1 March 1790, Archivo del Real Jardín Botánico de Madrid (hereafter ARJBM), I/5/2/6.

18. See, among many others, *La expedición Malaspina, 1789–1794*, 9 vols. (Madrid: Lunwerg Editores, 1987–96); Daniela Bleichmar, *Visible Empire: Colonial Botany and Visual Culture in the Hispanic Enlightenment* (Chicago: University of Chicago Press, forthcoming); Iris H. W. Engstrand, *Spanish Scientists in the New World. The Eighteenth-Century Expeditions* (Seattle: University of Washington Press, 1981); Antonio González Bueno, ed., *La Expedición botánica al Virreinato del Perú (1777–1788)* (Madrid: Lunwerg, 1988); Nieto Olarte, *Remedios para el imperio*; B. Sánchez, Miguel Ángel Puig-Samper, and J. de la Sota, eds., *La Real Expedición Botánica a Nueva España, 1787–1803* (Madrid: Real Jardín Botánico, 1987); María Pilar de San Pío Aladrén, ed., *El águila y el nopal. La expedición de Sessé y Mociño a Nueva España (1787–1803)* (Madrid: Lunwerg Editores, 2000); Juan Pimentel, *La física de la monarquía. Ciencia y política en el pensamiento colonial de Alejandro Malaspina (1754–1810)* (Madrid: Doce Calles, 1998); Marie Louise Pratt, *Imperial Eyes: Travel Writing and Transculturation* (New York: Routledge, 1992); Arthur Robert Steele, *Flowers for the King: The Expedition of Ruiz and Pavon and the Flora of Peru* (Durham: Duke University Press, 1964); and the references in n. 1.

19. Koerner, *Linnaeus*, 95–139; and Staffan Müller-Wille, *Botanik und weltweiter Handel: Zur Begründung eines natürlichen Systems der Pflanzen durch Carl von Linné (1707–78)* (Berlin: Verlag für Wissenschaft und Bildung, 1999).

20. Jean Baptiste Christophe Fusée Aublet, *Histoire des plantes de la Guiane françoise*, 4 vols. (London, 1775), 1: xiii. See also chap. 3, this volume.

21. José Celestino Mutis, *Archivo epistolar del sabio naturalista Don José C. Mutis*, ed. Guillermo Hernández de Alba, 4 vols. (2nd ed.; Bogotá: Instituto Colombiano de Cultura Hispánica, 1983).

22. José Celestino Mutis, *Viaje a Santa Fe*, ed. Marcelo Frías Nuñez (Madrid: Historia 16, 1991). See also chap. 5, this volume.

23. José Celestino Mutis, *Diario de observaciones de José Celestino Mutis (1760–1790)*, transcription and prologue by Guillermo Hernández de Alba, 2 vols., 2nd ed. (Bogotá: Instituto Colombiano de Cultura Hispánica, 1983), and idem, *Escritos científicos de José Celestino Mutis*, ed. Guillermo Hernández de Alba, 2 vols., 2nd ed. (Bogotá: Instituto Colombiano de Cultura Hispánica, 1983).

24. On the incorporation and erasure of contributions from local populations to natural history in the Americas at the time, see Susan Scott Parrish, *American Curiosity: Cultures of Natural History in the Colonial British Atlantic World* (Chapel Hill: University of North Carolina Press, 2006); and Neil Franklin Safier, *Measuring the New World: Enlightenment Science and South America* (Chicago: University of Chicago Press, 2008).

25. Quoted in Gredilla, *Biografía*, 272.

26. All of the specimens that Mutis sent to Linnaeus, as well as some of the images, are available online through the Linnean Society of London (http://www.linnean.org/).

27. On the complex relationships between botanists and artists at the time, see Daston and Galison, *Objectivity*, chap. 2, esp. 84–98.

28. Daniela Bleichmar, "Painting as Exploration: Visualizing Nature in Eighteenth-Century Colonial Science," *Colonial Latin American Review* 15, no. 1 (June 2006): 81–104.

29. Vicente Cervantes to José Celestino Mutis, Mexico, 27 Dec. 1788, ARJBM, III/1/1/83, ff. 6r–6v. Reproduced in *Archivo Epistolar*, 3: 219–23.

30. Luis Née to José Celestino Mutis, Guayaquil ca. 22 Oct. 1790. ARJBM, III, 1/1/230, folio 2r. Reproduced in *Archivo Epistolar*, 4: 74–76.

31. Sometimes, artists talked back: see Daston and Galison, *Objectivity*, 93–94.

32. Jorge Escobedo, Visitador Superintendente General del Perú, 29 June 1788, quoted in Hipólito Ruiz, *Relación histórica del viage, que hizo a los reinos del Perú y Chile el botánico D. Hipólito Ruiz en el año 1777 hasta el de 1788, en cuya época regresó a Madrid*, ed. Jaime Jaramillo Arango, 2 vols. (Madrid: Real Academia de Ciencias Exactas Físicas y Naturales, 2nd. ed. 1952), 469–70.

33. Daston and Galison, *Objectivity*, chap. 2, esp. 84–98; and Kärin Nickelsen, *Draughtsmen, Botanists, and Nature: The Construction of Eighteenth-Century Botanical Illustrations*, Archimedes Series, 15 (Dordrecht: Springer, 2006).

34. Daniela Bleichmar, "Exploration in Print: Books and Botanical Travel from Spain to the Americas in the Late Eighteenth Century," *Huntington Library Quarterly* 70, no. 1 (March 2007): 129–51.

35. For an earlier instance, see Paula Findlen, "The Formation of a Scientific Community: Natural History in Sixteenth-Century Italy," in *Natural Particulars: Nature and the Disciplines in Renaissance Europe*, ed. Anthony Grafton and Nancy Siraisi (Cambridge, Mass.: MIT Press, 1999), 369–400.

36. Dorinda Outram, "New Spaces in Natural History," in *Cultures of Natural History*, ed. Nicholas Jardine, James A. Secord, and Emma C. Spary (Cambridge: Cambridge University Press, 1996), 249–65. See also chap. 8, this volume.

37. Antonio José Cavanilles, "Descripción de los géneros Aeginetia, Rizoa y Castelia," *Anales de Ciencias Naturales* 3 (1801): 132–33.

38. On scientific illustrations and the choice between depicting types versus specimens, see Lorraine Daston and Peter Galison, "The Image of Objectivity," *Representations* 40 (1992): 81–128.

39. José Celestino Mutis to Juan José de Villaluenga, president of Quito *Audiencia*, 10 July 1786, ARJBM III/2/2/196 and 197; reproduced in *Archivo Epistolar*, 1: 316.

40. José Celestino Mutis to Casimiro Gómez Ortega, 31 March 1784, ARJBM, III/1/2/32.

41. Antonio José Cavanilles, *Colección de papeles sobre controversias botánicas* (Madrid, 1796); and Hipólito Ruiz, *Respuesta para desengaño del público a la impugnación que ha divulgado prematuramente el presbítero Don Josef Antonio Cavanilles* (Madrid, 1796).

42. Ruiz, *Respuesta*, 66.

43. Ibid., 10, 11, 14, 16, 25, 27, 35, 37, 39, 45, 65.

44. Ibid., 65.

45. Cavanilles, *Colección de papeles sobre controversias botánicas* (Madrid, 1796), 7.

46. Ibid., 11.

47. Ibid., 7–8.

48. Ibid., 10–13.

49. Ibid., 13.

16

The World on a Page: Making a General Observation in the Eighteenth Century

J. ANDREW MENDELSOHN

What the gathering of powerful savants in the King's library in the Louvre heard read to them aloud by the physician Claude-Antoine Caille sometime after four o'clock in the afternoon of 19 March 1782 hardly seems to our ears a scientific observation or indeed anything meriting such a forum:

> The dominant [atmospheric] constitution of this [past] year presents a long-lasting drought, & comparing it with that of 1780, both present an excess in their temperature: the first in humidity, the second in dryness. . . . The humoral constitution, which should be regarded as the material cause of the intermittent and remittent fevers of 1780 & 1781, was bilious & it took on an atrabilious character in the autumn of 1781.[1]

This was a scientific observation, of a kind called in its time a "general observation," as opposed to a "particular" one. It is as succinct in size as it is general in scope: it pertains to all disease and climate over the whole of France. This chapter is about how such general observations were made.

That they were made is a point I wish to advance about early modern science. The eighteenth century is known for putting the world on many pages—the thirty-five-quarto volumes of Georges-Louis Leclerc, comte de Buffon's *Histoire naturelle*, the *Encyclopédie*, huge medical *historia* and nosologies of thousands of disease species. These shared in the common project of *historia* and Baconian "gigantesque inventories."[2] Observing, collecting, and ordering were its widely pursued activities. But vast collections sustained another mode of science whose dual watchwords were *precise* and *general*. It appropriated and developed methods for achieving these seemingly contradictory goals, methods of generalizing rather than classifying, fusing rather than distinguishing and naming, all the while remaining "exact." Generalization, as we shall see, required certain forms of collectivity: a community of observers was not enough; its observations had to be made commensurable with one another and synthesized. And this required certain techniques. The instruments and practices of general observation were not of glass and metal and spirits of mercury, nor were they perceptual habits and cognitive practices of attention and memory.[3] They were of paper and ink. This chapter will illustrate the three most important ones, devoting a section to each: the extract, the précis, and the table. Together, as we shall see in a fifth section, these enabled a prestatistical form of averaging, in words rather than numbers.

But first, what were general observations, why did they matter, and what made them difficult to achieve?

Midway between universal laws of nature and the particulars of things and cases, times and places, a general observation could concern a kind of natural object or a phenomenon extended in time and space over many and varied particulars. The one read aloud in the Louvre that March afternoon in 1782 summed up the raining, shining, fevering, vomiting confusion of weather and illness over the largest national territory in Europe. This combination of compression (what they called "the most precise" or "the most succinct") and panoramic scope of application (what they called "the most general") rendered it of the greatest scientific and practical interest to its audience.

Its audience was also its author. They were the leadership of the newly founded Royal Society of Medicine, which the government had made responsible for developing knowledge and policy concerning human and animal disease. France, ever more awash in the debt that would help bring on the Revolution, faced a series of economically devastating epizootics and epidemics. Crucial to effective intervention, according to revived Hippocratic doctrine, was knowledge of climate in relation to disease. Hence in 1775 the *philosophe*-economist-statesman Anne-Robert-Jacques Turgot, as controller-general of finance, had ordered a national medical, topographical, and meteorological inquiry via questionnaire to be distributed by each intendant to physicians in his province. The following year this survey was made permanent as a new society to which, during its seventeen years of existence, over five hundred provincial physicians and surgeons were elected. In return for the honor, they were to submit monthly observations.[4]

Three years and 150 weekly Louvre meetings after Caille's observational report of 1782, he was ready to raise it to a still higher level of generality:

> The bilious constitution originates in the extraordinary dryness of 1778, & the excessive heat of 1779. It lasted from the latter year until the very long and very

> cold winter of 1784, which makes a period of five years. All of the diseases that prevailed during this time were more or less involved in this constitution.[5]

"Constitution," a Hippocratic term revived especially by the "English Hippocrates" Thomas Sydenham, essentially meant dominant pathological type. Depending on which of the four bodily fluids or "humors" was observed to predominate in illnesses, the humoral or morbid constitution could be inflammatory (blood), catarrhal (phlegm), bilious (yellow bile) as here, or, as apparently in the autumn of 1781, atrabilious (black bile). Such an observation retained features of the particular: bound to time and place, it was not yet of the form "bilious constitutions generally . . ." or "constitutions generally. . . ." But that is clearly where the society was headed; in fact, it offered in 1783 a prize of 600 livres for the best answer to the question whether the atrabilious constitution "has an existence distinct" from the others and "what is its influence in the production of epidemic diseases."[6] All this mattered because, as observation showed and Caille emphasized, successful treatment required knowing the constitution. And tracing its dependence on climate promised predictive power.

But that was easier said than done.

Province of Poitou, village of Saint-Maurice-le-Girard, 7 November 1778, high noon, Dr. Jean-Gabriel Gallot stands at his barometer. It reads 27 inches, 9 lignes, down from 9.5 lignes that morning.[7] This is one of about 1.3 million individual acts of meteorological observation that were fused into the general observations read in the Louvre or published by the society during the rise and reign of the bilious constitution—which ought to have been synthetic challenge enough.[8]

But so is this: "The 7th [of November] W[esterly wind]—until 10th with rain from time to time."[9] This was the full extent and exactitude of the meteorological observations made the same day (and the next three) by another correspondent of the society, Dr. Souquet in Boulogne-sur-Mer. Such was the lack of uniformity entailed by having a research staff of devoted but untrained and scattered amateurs.

Gallot was among the society's most regular and prolific provincial observers. This he could achieve only "on horseback, often at a gallop, always at a trot, covering sometimes more than 10 leagues in a day." Galloping was Gallot's only way to get home every day in time to make his noon meteorological observations and also to observe so many patients, an activity prolonged after dark when at home by candlelight he wrote up their case histories in his journal.[10] Quarterly, he drew together his "nosological observations" and submitted them to the society; others did so monthly; some submitted annual reports as well or instead. Behind the five-year general observation read in the Louvre in 1785 lay over 2,000 such medical reports and thousands of case histories, each made of dozens of bedside observations over days or weeks or months.[11]

Staggering scale of collection in itself was not unprecedented. Medical case histories, or *observationes*, had since the sixteenth century been written by the hundreds—literally published in *centuriae*.[12] But the Paris Royal Society of Medicine was different. It did not publish thousands of *observationes*. It aimed to fuse them. And just as observers like Gallot and Souquet differed in their meteorological observing and recording habits, so too did their medical case histories and reports range widely in form and content. Journal editors and the commentator-compilers of *observationes* could put up with such lack of uniformity. But an enterprise of synthesis could not.

The problem, then, was dual: how to make many voices one; but also the more fundamental problem of how to get beyond particulars to the general.

When in the 1960s *Annales*-school historians stumbled on this mass of data "slumbering" in the archives, they understandably assumed it had not been processed in its own time. The disease observations seemed impenetrable thickets of medical prose. The meteorological observations, on the other hand, could be fed into a computer and were.[13] Yet this had in effect already been done. The society *did* process the data, meteorological and medical. And in order to do so, it had standardized, as far as possible, the production of that data. But all this happened on eighteenth-century terms, not twentieth-century ones.

Extract

If the culture and practice of producing exchangeable and collectable scientific observations were epistolary,[14] and that of their coordinated production and inventory was administrative,[15] that of their evaluation, analysis, and generalizing synthesis was, I argue, editorial. The society exemplified all three. Originally calling itself "Society and Correspondence of Medicine," its monthly reporting forms were headed "Correspondence" (fig. 16.1), its operations were run from a "bureau" filled with files and registers, and its regulations included a whole section on *Rédaction.* This was a pale word for what contemporaries described metaphorically as distillation or digestion of particulars into general knowledge, yet it is a word that takes us to how this was done.[16] In extant manuscripts of this process, the initial perceptual act all but disappears into a palimpsest of hands and layers of redaction, all organized at the society's bureau into a working archive run by the anatomist Félix

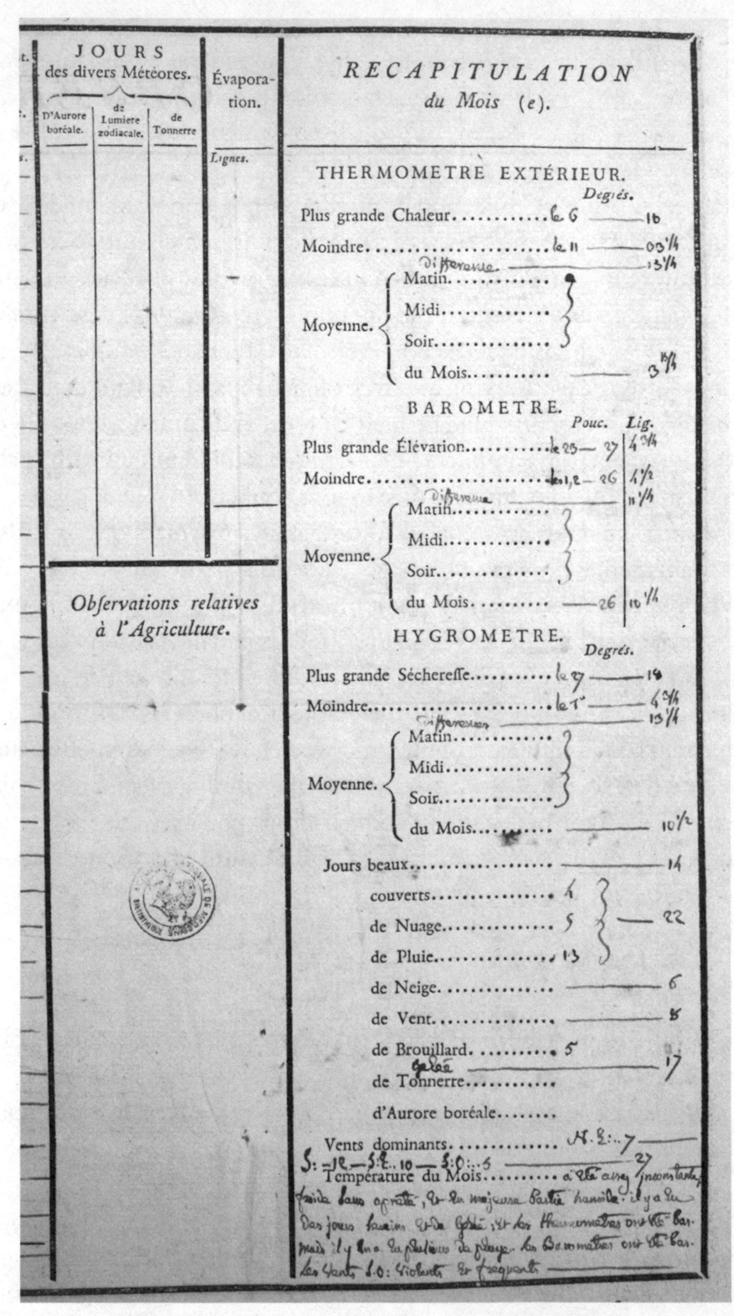

JOURS des divers Météores.

D'Aurore boréale. | de Lumiere zodiacale. | de Tonnerre

Évaporation.

Lignes.

Obſervations relatives à l'Agriculture.

RECAPITULATION du Mois (e).

THERMOMETRE EXTÉRIEUR.

Degrés.

Plus grande Chaleur............le 6 — 10

Moindre.........................le 11 — 03 1/4

Différence — 13 1/4

Moyenne. Matin............. Midi............. Soir............. du Mois.......... — 3 1/4

BAROMETRE.

Pouc. *Lig.*

Plus grande Élévation..........le 23 — 27 | 4 3/4

Moindre........................le 1, 2 — 26 | 4 1/2

Différence — 11 1/4

Moyenne. Matin............. Midi............. Soir............. du Mois.......... — 26 | 10 1/4

HYGROMETRE.

Degrés.

Plus grande Sécherеſſe..........le 27 — 18

Moindre.........................le 1er — 4 3/4

Différence — 13 1/4

Moyenne. Matin............. Midi............. Soir............. du Mois.......... — 10 1/2

Jours beaux.................. — 14

couverts.............. 4

de Nuage.............. 5 — 22

de Pluie.............. 13

de Neige.............. — 6

de Vent............... — 8

de Brouillard.......... 5

de Tonnerre (Gelée)........... — 17

d'Aurore boréale........

Vents dominants............ N.E.: 7 — 27

S: 12 — S.E. 10 — S.O. 5

Température du Mois.......... a été assez inconstante, froide sans âpreté, et la majeure partie humide. il y a eu des jours sereins et de gelée, et les thermomètres ont été bas. mais il y a eu peu de pluie. les baromètres ont été bas. les vents S.O. violents et fréquents.

FIGURE 16.1. Example of the Royal Society of Medicine monthly reporting form as completed by one correspondent for January 1785. Detail shows the far-right column, headed "Recapitulation of the Month," a fill-in-the-blank procedure for making a general observation. From "Corréspondence de la Société Royale de Médecine. Observations Météorologiques faites à St Diez, Province de Lorraine, Par M. Félix Poma, Mois de janvier 1785." Fonds Société Royale de Médecine, Bibliothèque de l'Académie Nationale de Médecine, SRM 156 dr. 3. By permission of the Académie Nationale de Médecine, Paris.

Vicq d'Azyr—whose positions as perpetual secretary and "first corresponding physician" also made him "principal redactor"—assisted by his clerk who labeled each observational report {Constitution} or {Epidemics} or {Practical Medicine} and so on.[17]

Synthetic redaction's basic and versatile unit of production was the "extract" (*extrait*). This was adopted from previous, more common uses: extracting information from government or church registers; excerpting from books into one's private notebook or commonplace book[18]; abstracting books for journals. The term extract had migrated from manufacturing technology to paper technology: "That which is drawn out by means of chemistry. *An extract of pearls, an extract of rhubarb*. . . . That which one extracts from a book, from a register. *Extract from the registers of Parlement, baptismal extract*. . . . The abridgement, the summary of a trial, of a book, &c."[19] These origins were visible in the society's collective publications subtitled "extracted from the correspondence" or "registers" and in its collective work (*travaux*), which was to make and read aloud at meetings an *Extrait raisonnée* of the best of ancient and modern medical writing *and* of the observation reports submitted by the provincial correspondents.[20]

One would-be correspondent was Dr. Prêvot de la Caussade, former army physician, residing in Montauban. He submitted an annual report of climate and disease for 1776. The clerk in the society's bureau has labeled it "Constitution" in big letters and "no. 47" (possibly no. 47 for that year's constitution).[21] Paris medical graduate and society member Charles-Jacques Saillant was assigned its redaction. Prêvot's report runs to ten pages; Saillant's "extrait" to four. So there was something to extract: "here's what it reduces to."[22] Here we will see extracting *not* producing general observation—and how and why an observer's effort at synthesis, replete with *observationes*, could fail.

Prêvot opens with general observations: the year 1776 in Montauban saw generally mild weather; illness, even the autumn epidemic fever, was also mild. This summary is followed by four detailed cases of the fever. So far so good. The trouble is that the cases do not illustrate the fever: it was mild, yet Prêvot relates three *fatal* cases and one severe one. Clearly these marked his practice and experience as a doctor ("I should not hide [the fact] that I myself lost a case to the fevers"). Prêvot lacks the redactive techniques, which we shall meet below, for selecting an exemplary "particular" observation to illustrate the "general" one. He did generalize about the year 1776. But on paper he does not know how to identify and include the descriptive features that would make generalization useful instead of vague.

The second half of the report details three more cases. These are unrelated to the dominant fever, but they *are* related to one another. A synthesis? No, a theme: Prêvot uses these cases to explore the problem of classifying disease. This was not the point, Saillant knows. So he dutifully salvages the case histories (the very currency of Hippocratic medicine and not to be thrown away), extracting them in condensed form from Prêvot's annual report, and deletes all else, dissolving the ties that bound them together around the effort at better treatment and prognosis through better disease denomination.

Thus did two modes of science—the one classifying cases, the other aiming to generalize them—sharply diverge even as they intersected in the extractable observational unit of the case history. And thus did extracting, as practiced here by Saillant, not necessarily yield synthesis and generalization. We turn now to examples in which it came increasingly close to doing so.

Besides the activity around its bureau and Louvre meetings, the society had another center of redaction, which dealt with numbers as well as words. This was the Observatory at Montmorency near Paris, run by Father Louis Cotte, priest of the Oratory and pioneer of modern meteorology. He "redacted" the weather observations submitted in the printed monthly reporting tables (or regrettably in other form) which the society distributed to its correspondents (fig. 16.1). Though himself no physician, he also redacted the disease observations recorded in the blank box headed "Observations nosologiques" that was the verso of the printed monthly form. Both sides of these forms, by the thousands, and many other observational reports bear his signature mark: "Extr."

What did he extract and what did he do with the extracts? Like the hapless Prêvot, redactor Cotte too aimed at that synthesis, the annual report. Cotte's published "Results for the Year 1776" is a list of diseases by place, a simple table. (Adjacent appeared a table of weather observations by place, more on which below.) This annual disease table was based on exactly the same sort of table that he produced for each month of the year. Hence each annual disease entry is an extract of the entries in each of 12 monthly compilations of extracts from all correspondents from that place—in short, a compilation of extracts from compilations of extracts. The general Parisian nosological year of 1776 according to Cotte consisted of "smallpox, catarrhal affections, pneumonia, diarrhoea, dysentery," five items. In extracting these from the Paris entries in his monthly tables, he left behind sixteen items, including measles and "benign" smallpox, but also more complex designations such as dysenteric diarrhoea with putrid fever and cases from which he seems to have extracted "catarrhal affections," such as catarrh accompanied by fever and bloody expectoration and degenerating into pseudopneumonia, into rheumatisms, or into sciatic gout. From 1779, Cotte added a new rubric at the end of the annual table: a short list of the diseases prevailing *in France as a whole* that year. This was obviously the result of a third round of extracting (from the dominant diseases in each place) and compilation.[23]

Like redactors, observers could use the same procedure—on their own observations. Dr. Gallot, whom we left galloping between barometer and bedside, submitted a "Mémoire on the Diseases Prevailing in 1776."[24] It looks nothing like Cotte's "Results of the Year 1776: Diseases." The one is a text, the other a table. Yet they are in many ways identical products of an identical procedure. Gallot's annual report was extracted from his monthly reports that were, in turn, extracted from the cases recorded nightly in his journal. The medical year is not as condensed as in Cotte's table. More author than redactor, Gallot extracts less severely: he keeps the diseases in monthly sequence and retains features like relapse and resistance to treatment. But the main difference is simply that Cotte extracted the "prevailing diseases" from sixteen geographical locations (and later many more), Gallot from one.

Besides illustrating the role of extracting in moving observation from the particular toward the general, there are two points to be made here. Gallot the graduate of Montpellier medical school and Cotte the priest of the Oratory and pioneer of meteorological research and its quantification had totally different backgrounds of higher education, scientific discipline, and working life. In their participation in the society's inquiry, the one cultivated the old art of medical observational writing; the other, the newer art of the table. Yet they used the same techniques of synthesis and with similar outcomes. This suggests that the redactive paper and ink practice of extracting, not medicine or meteorology or quantification or even table making, was most decisive for the history of general observation. A bookish, clerkish, and editorial technique designed and normally used to manage existing knowledge was making new knowledge. The point is not only that the same technique was used in a new way. Rather, managing the old could be continuous with making the new, an outcome one might expect from a practice perfected and applied to natural philosophy by Renaissance humanists.[25]

The second point is that the redactive practice of general observation—iterative extracting—was so far failing to constitute a general observation. What the leading *médecin-observateurs* of the late eighteenth century called the crucial "nuances"[26] of disease were getting lost (Cotte) or stuck (Gallot) in the process instead of being generalized. And the outcome was the same whether the observer was the unsatisfactory Prêvot or the master extractor Cotte or the quick-learning Gallot. Prêvot too had been able to list the main diseases of 1776 in his area: intermittent fever, apoplexy, paralysis, epilepsy, smallpox, measles, pleurisy, cholera morbus. And such a list was a dead end. The proof is that after all Cotte's extracting, the society was unable to include disease, only weather, in the annual prose *Résultats généraux* based on his tables.[27]

So why were the redactive practices of general observation not yielding general observations of the kind worth reading aloud in the Louvre, observations that revealed general disease patterns, preserved enough detail to be useful, enhanced predictive power, and helped guide treatment? The answer, as we are about to see, has to do with what was being extracted.

Précis

A general observation that did all those things was called a précis. The term referred to a published genre and to a list–like part of a publication or manuscript. Caille called the five-year bilious constitution of France he read in the Louvre a *précis historique*; this ten-page précis included a one-page, numbered, four-point *précis* of the chief disease features and their meteorological and topographical circumstances. Clues to how it was made can be found in one working précis in manuscript:

> . . . 3. The precursor symptoms were a general exhaustion, a malaise . . . 4. The shivers were quite marked in some patients. 5. The symptoms succeeded one another rapidly and promptly became serious. 6. There was vomiting in many cases. 7. The head especially was affected and the delirium more or less long; the chest appeared less often attacked. . . .[28]

The twenty-five numbered observation statements from which this quotation is taken formed a second part of Dr. Gallot's report on the year 1776. The first part, as we saw, was essentially a list of diseases; the second, a list of symptoms and features that *cut across* disease categories. This was a major difference: the first list was a dead-end collection; the second was on its way to being a synthetic general observation. But whence the difference? How and why did Gallot write two such different reports in one?

The first he did of his own accord; the second answered a questionnaire. Vicq d'Azyr, aware of the enormity of "collection and redaction" he would soon face, had drafted a questionnaire and published it within a sixty-page instruction booklet (*Mémoire instructif*) distributed to physicians throughout the country. The loosely formulated six queries of Controller-General Turgot, who looks a mild taskmaster by comparison, could be answered in many ways and at many lengths.[29] Vicq d'Azyr's thirty-seven specific questions could not. They could do much more than elicit information. They could ideally standardize observation, made and written, where even a standardized form, with spaces for answering a few broad questions like Turgot's, did not.[30] But this would entail more work and less autonomy for the observer. Among those willing and able was Dr. Gallot. To the question whether the skin was observed to be dry and hot, Gallot answered: "8. The skin was dry among those not experiencing critical sweats." This, then, turned observation into yes/no questions. Other questions rendered observation multiple choice. Questionnaire: "Was the prevailing fever of a petechial nature [red spots on the skin] or catarrhal or inflammatory?" Gallot: "18. One can say in general that the fevers tended more to the catarrhal and putrid than the inflammatory." Or almost multiple choice.[31]

The key point is that answering such questions did much more than standardize observation. It disintegrated individual illnesses of uncertain relationship into some thirty comparable and recombinable elements: features such as symptom; anatomical locations affected; sequence of events; presence or absence of crises; and "character," which could be degree of severity, or degree of simplicity or complication, or humoral type (bilious, inflammatory, catarrhal, atrabilious) or other kind of pathological process (putrid, verminous). Gallot found, he wrote, that his responses linked malignant fevers, intermittent and particularly double tertian fevers, and serious catarrhal fevers. His answers thus formed, he realized, "a précis of the most essential of what I have noticed" of the most serious diseases of the year.[32] Thus did Gallot, simply by being conscientious in following instructions, move from making a *diseases list* for 1776, in which most particulars were lost (and a few remained in tension with generalization), to making a *disease précis* for 1776, which was both far more general *and* at the same time retained the particular.

The final point to be made here is that this difference resulted from the same method, extracting, differently applied. A précis could not be more different from the jumbled collection of excerpted disease names in a list—or excerpts in a commonplace book or abstracts in a journal. Extracting, in its usual use, tended to yield many excerpts from many books. Here, extracting yielded in effect the précis of many books—that is, of many cases. Whence the difference? Our survey of annual reports for 1776 reveals three kinds of extracting. First, one could extract units: what Saillant extracted from Prêvot's report were cases, the *observationes* themselves but, like abstracts of books, condensed. Second, one could extract names of categories into which the units had been sorted: what Cotte extracted from cases and reports (and Prêvot and Gallot from their own cases to make such reports) were names of disease categories to which the cases belonged. Third, one could extract elements of which the units were made: what the thirty-seven-point questionnaire, in Gallot's hands, extracted from cases were their symptoms and features. Only the last of these applications of extracting could yield a synthetic general observation. For a fusion of *observationes* could only be a fusion of elements of which they were made. Questionnaire or no questionnaire, making a précis can be reconstructed as the result of reading each case history through the same extracting *stencil*—an explicit or tacit uniform set of specific yes/no and multiple-choice questions—and combining the results from many cases. (This was only likely to produce results if the act of observation itself had been preshaped by a questionnaire or by its unwritten equivalent in shared standards of detail.) The innocuous-looking précis broke down the *observatio*, the fundamental, exchangeable unit of knowledge. That would explain why historical change in observational medicine was slow: the epistemic genre of the case history stood in the way of, even as the inexhaustible observing and writing activity it generated made possible, synthetic generalization.

Using the paper and ink techniques of extract and questionnaire, Gallot had unintentionally produced, as an afterthought to the first part of his 1776 report, a *précis* of general features of epidemic disease in his canton. Yet this précis remained more list than general observation and it was twenty-five items long, not very precise. Gallot still lacked three things. To find out what they were, we turn to a more experienced general observer, from whom the very idea and design of the society's detailed questionnaire had come.

> *March* [1779]
> The *catarrhal ailments became more common* than in the previous month.
> There were many [*plusieurs*] phlegmonous [*fluxions*] of the eyes.
> Many sore throats.
> Many cases of pseudo pleurisy, *more inflammatory than putrid.*
> Many cases of pneumonia of the *same character* [as the pleurisy].
> A few [*quelques*] ephemeral fevers.
> A few tertian fevers and verminous fevers. . . .
> A few gouty rheumatisms. . . .
> ——————The morbid constitution [*constitution maladive*] was catarrhal inflammatory.[33]

At first glance, this March section of an annual report for 1779 seems just the sort of disease list submitted by the society's mediocre correspondents. But note the italicized phrases. Cutting across two dominant disease categories is a common "character": inflammatory. This character, and the catarrhal character indicated in the first line, would have been extractable from case histories via questionnaire-like redaction, the intermediate précis of which are not represented in the report.

This report was submitted to the society by the inventor of the long and detailed disease questionnaire. This inventor was, tellingly, a committee. This will further the argument that generalization in science depended on certain forms of collectivity.

The Academy of Sciences, Arts, and Letters of Dijon was among the most active of the provincial academies in eighteenth-century France. Hugues Maret was its perpetual secretary from 1764 until his death in 1786. This made him central to the pursuit of science by coordinated, centralized collective and science by committee. A physician, Maret sat on the academy's medical committee. During an epizootic in Bourgogne, the committee designed a thirty-five-item questionnaire to guide observation of animal disease; it was distributed by the government to physicians throughout the province. So innovative was this that Vicq d'Azyr printed the questionnaire in his *Mémoire instructif* and modeled it on the society's own, thirty-seven-item questionnaire for the observation of *human* disease, the one used by Gallot.

In their annual reports, Maret and the committee took two steps that Gallot, in 1776, did not yet know to take: they privileged one extract ("character") over any other shared feature, whereas Gallot had listed them all. And, second, Maret's group used these combined dominant characters to name a collective yet single whole, the "morbid constitution." No speculative entity, the constitution was a general observation. No "theory," it was constituted by iterative paper and ink practice. (This is how it could be central to a medical culture and enterprise that was bent on expunging "theory" and "system" from medicine and science.) The morbid constitution, the summation under the line at the end of the list, is nothing more than a final extract.

Summation under a line at the end of a list: this format evokes numerical calculation. One dimension is missing from this chapter's reconstruction of generalization in the age of observation as redaction: quantity.

Quantifying without Counting, Averaging without Numbers

The route to a general observation required knowledge of more and less, common and rare. Society correspondents' reports, though usually devoid of numbers, are replete with quantifying language. In Gallot's 1776 précis alone, we find two ranges of quantifiers: the adjectives *a few, many, almost all*; and the adverbs *often, less often, rarely*.[34] For March 1779, Maret and his committee have listed diseases of "many" cases and those of "a few." The morbid constitution is derived from the character of the diseases of many cases. Number was not absent because medical observation was purely qualitative: we have just seen it was not. Nor was number absent because physician-observers failed to be exact: quantifying without counting was *better* than with. Counts of cases would have been *too particular* and thus *less* conducive to generalization than a set of verbal quantifiers like those used by Maret and Gallot, each of which in effect encompassed a range of counts.

Quantifying alone did not yield the general out of the particular. Some form of averaging had to take place. This was because quantifying cases *within a category* could yield, say, "many cases of pneumonia," but not the nature ("character") of the category, the fact that those cases were "more inflammatory than putrid." This required averaging cases rather than counting them. Just as averaging with numbers requires repeated observations, so too did averaging without numbers. It is no coincidence that the society's medicometeorological observer who was most experienced in and devoted to observing the morbid constitution wrote that in medicine one must have observed "the same phenomenon a hundred times" to know it.[35] Such an observer found novelty in repetition, nuanced difference in apparent sameness, yet differences that, through extract and précis, numberless quantification and averaging, yielded higher orders of similarity: the changing character or "general character" of a disease or of different diseases related through their "involvement" (*participation*) in what thereby took empirical shape as the prevailing morbid constitution.

Such an observer was Jean-François Durande, who authored the case histories included in the annual reports submitted by Maret and the Dijon Academy's medical committee: "In the two cases of pneumonia that I have chosen from among many, one sees inflammation dominate at the beginning of the illness, then the catarrhal humor bringing on the most violent misfortunes [*accidens*] in affecting the chest, and finally the obstruction of the digestive tract necessitating [the use of] evacuants. A few other pneumonia cases [*fluxions de poitrine*] were more inflammatory, others more bilious."[36] Described here in words is essentially a bell-curve distribution of three characteristics observed in an unspecified number of cases of pneumonia. All cases displayed all three humoral characters to some extent and at different times. A few were more inflammatory (dominated humorally by the blood), and a few were more bilious (bile). Most, however, were predominantly and most seriously catarrhal (phlegm). Of these average cases, two were chosen to be published as *observationes* exemplary of average pneumonia in Dijon and environs in 1777 and, in turn, of the predominantly catarrhal constitution of 1777.

General observation selected from what *historia* collected; the one for comprehensiveness (all that had happened, all occurring variants), the other for exemplarity. Whereas compilers of *observationes* amassed as many as possible per disease category, the society's generalizing observers reported as few as a single case per category. Each such *observatio* was thus at once a particular observation (case history) and in effect a general observation, giving the "idea" of a disease or its character during a certain period.[37] Thus selecting an exemplary case fulfilled the same function as making a précis and depended on the same prestatistical averaging activity.

But whence this averaging without numbers, this procedure so important to generalization? The eighteenth century witnessed the rise of a scientific activity in which averaging in both numbers and words was being pursued widely and regularly enough to be able to shape medical (and other) observers' mental habits. This activity was meteorology.

Figure 16.1 shows the society's printed form for monthly reporting: meteorological observations in the table on the front, nosological ones in the blank box on the back. The meteorological table is apparently a template for making daily observations. It wasn't. It was a template for making a general observation.

Daily observations were supposed to be recorded in a logbook (*registre*), a "model" page of which was published in the *Mémoire instructif*: nothing more than six lines drawn down the page to make seven columns, headed day, hour, barometer, exterior thermometer, interior thermometer, winds, state of the sky.[38] From this log, which was typical of early modern weather recording,[39] the observer would copy daily observations onto the reporting form. On the form, daily observations were entered by row (one row per day); general observations were obtained by column, each of which stood ready at month's end to be summed up in some way. But how? And then what? The answers to these questions were provided in the last column. Headed "Recapitulation of the Month," its printed headings and structure guide the observer through a fill-in-the-blank procedure. The procedure mixes numerical averaging of measurements (thermometer, barometer, hygrometer), summation of qualitative descriptors (how many days clear, overcast, cloudy, foggy, and so on, hence the importance of having a standard range of these on the form), and an estimate of dominant winds, either by *coup d'oeil* at the winds column on the form or, as in the example in figure 16.1, by totaling the number of days for each wind direction. Altogether this yields at the foot of the recapitulation column the "Temperature of the Month," into which also figure the total quantity of rain and evaporation, which the horizontal lines at the foot of the relevant columns (to the left of the recapitulation column) invite the correspondent to sum. "Temperature," or *température générale* as it was also called, was not a numerical average, but a synthetic general observation in words, at once qualitative and quantitative, using binary combinations of the Aristotelian four qualities—hot, cold, moist, dry—or sometimes another quality such as mild or variable and additional summary comments. Though

Année 1777.

Constitution de Janvier.

L'Athmosphère	Maladie observée.
La pesanteur de l'air a très peu varié dans le cours de ce mois; elle a toujours été moyenne. dans les douze premiers jours plus considérable pendant les huit suivants l'air étant alors bien pesant, moindre du 21. au 25. le baromètre indiquant que l'air étoit léger et un peu au dessus de la moyenne pendant le reste du mois.	Les maladies qui ont regné dans ce mois ont toutes été Catharrales. il y a eu des rhumes, beaucoup de fluxions de differentes espèces, beaucoup de fausses pleuresie dans les premiers jours et quelques unes sur la fin. la plupart de ces fausses pleuresies étoient compliquées de fievres vermineuses. cette espèce de fievre a été plus commune sur la fin du mois et n'étoit alors que rarement accompagnée de points de coté.
Il y a eu quelques balancements rapides du mercure à differentes époques mais aucun de bien considérable.	un mal de gorge gangreneux a regné epidemiquement dans quelques villages de la plaine dont le terrein étant très bas est souvent très humide (1).
L'elevation la plus grande a été de 27 P. 6 L 3/6 et la moindre de 27 P. " 1/6 dont la difference est seulement de 6 L 1/6.	(1) Comme j'ai envoyé à la Société royale la description de cette maladie, que d'ailleurs elle ne differoit point de celles de cette espèce qu'on a deja observées, je crois ne devoir pas en donner ici l'histoire.
La température a été en général froide pendant tout le cours du mois, mais elle a été très froide	

FIGURE 16.2. Table of correlation, or divided page juxtaposing morbid and meteorological constitutions, from an annual report to the Royal Society of Medicine. From "Histoire météoro-nozologique pour l'année 1777," by Hugues Maret with "observations cliniques" by Jean-François Durande, submitted by the Comité Médical de l'Académie de Dijon. Fonds Société Royale de Médecine, Bibliothèque de l'Académie Nationale de Médecine, SRM 166 dr. 5, no. 1. By permission of the Académie Nationale de Médecine, Paris.

one of meteorology's great early quantifiers, even Cotte designed his published, number-filled synthetic monthly and annual meteorological tables to yield, in the final column, the "Temperature" in the form "hot and dry," "cold and humid," "mild and humid," and so on. The society's "correspondence" form, in short, was a paper and ink computer program for a human computer (until the twentieth century, the word referred to a person doing calculations) to make a general observation.

But was there a real parallel between meteorological averaging and a nosological version of it? That parallel is enacted on paper in Maret and the Dijon committee's annual reports. Figure 16.2 shows a typical page, like the one from which we saw the March 1779 entry on disease and morbid constitution. Headed "Constitution of the Atmosphere" and "Diseases Observed," two columns juxtapose parallel-running results of an averaging procedure, each yielding what was called a "constitution."

Writing and numerical calculation were not separate modes of science, qualitative and quantitative.[40] Interwoven and not considered separately by the society and its observers, they belonged to the larger method and object of general observation.

Tables of All Sorts

The process has now been reconstructed to the point at which it yielded not one but *two* general observations, one of the atmosphere, the other of disease. Each on its own was useful: by the 1780s, Cotte could see trends in his annual results and was forecasting long-term weather;[41] meanwhile, the society began using its hard-won knowledge of morbid constitutions to guide therapeutics throughout France.[42]

Yet climate was, after all, thought to be determining disease, and a general observation encompassing and relating both was supposed to be possible. This chapter began with one such. Recall that Caille's five-year bilious constitution of France ran from the dry and hot atmospheric constitution of 1778–79 to the cold of 1784. General disease characteristics under its sway (and therefore crucial treatment indications) tracked extreme swings in temperature (the general quality, not the thermometer reading) in the winter of 1782 and thereafter a trend toward humidity and then back to dryness.[43]

To make this dual kind of general observation required a last item of paper and ink technique: the divided page, as in the meteorological and nosological columns of the Maret group's annual report (fig. 16.2). The divided page was an instrument of correlation. Instead of correlation, they used words like "rapprochement" and "join," which referred to a paper rather than mental operation.[44] A variant with the same effect was the printing and binding of Cotte's tables of disease "results" interleaved with his foldout tables of meteorological "results." The divided page, or facing pages, brought together and juxtaposed observations of totally different phenomena.

Divided page better describes this instrument than *table*, which is too generic. Tables came in various sorts, even two or more in one. All fulfilled at least the contemporary dictionary definition, which specified above all a pedagogical origin and function: "A sheet, a board on which doctrinal, historical or other matters are digested and methodically reduced and shortened so that one can see them more easily and all in one view. . . . *He teaches grammar, philosophy by tables. He puts all the sciences, all the arts into tables.*"[45] When evaluating and preparing to synthesize reports by multiple observers and/or over multiple years, society redactors would extract keyword phrases and array them on a page by author or place and year, sometimes adding further columns. They called this a *table raisonnée.*[46] The society's monthly "meteorological observations" form (fig. 16.1) combines this extractive, reductive, "all in one view" function of the *table raisonnée* with that of a fill-in-the-blank procedural table for redacting a general observation. It is thus two sorts of table in one. Finally, figure 16.3 shows a unique extant manuscript tabular instrument constructed by Cotte in his earliest work on the society's inquiry. It is three sorts of table in one. Like the *table raisonnée*, it arrays on a page extracts from five (mainly three) observers' reports: the "A" columns are extracted from Cotte's own observations at Montmorency; the "B" columns from Gallot's at Saint-Maurice-le-Girard, and so on. Secondly, like the society's monthly reporting form (designed and printed only later), it organizes the redaction of a general observation: from the maxima, minima, and averages in the A, B, and C geographical location columns to the general "temperatures" in those places. Lastly, the lower right portion of the table is a divided-page table of correlation like that of Maret and his committee: a rapprochement of temperature (as redacted on the rest of the page) and diseases (extracted from the corresponding nosological reports).[47] This is the prototype paper and ink instrument Cotte used to make his published, interleaved disease and weather "results" tables.

In the case of medicine and meteorology, the instrument of correlation was equally an instrument of separation. For many physicians already habitually wrote climate-disease causal relationships into their observations. "This month's cold & humid and inconstant weather has doubtless occasioned the colds and the engorgements of the chest that prevailed."[48] Disentangling and juxtaposing the meteorological and the nosological on a divided page rendered causal correlation a research question instead of a questionable platitude.[49]

Resultats des observations Météorologiques faites à Montmorency, à St Maurice le Girard en Bas Poitou, à Nantes, à Villefranche en Beaujollois, et à Molsheim en Alsace Depuis le Mois de Janvier Jusqu'au Mois d'Octobre 1776.

Mois 1776.	Jours de la pl. gr. chal. A	B	Jours du plus grand froid. A	B	D	Plus grande chal.r A	B	C	Plus grand froid. A	B	C	D	E	Chal.r moy.e A	B
						Degr.	Degr.	Degr.	Degr.	Degr.	Degr.	Degr.	Degr.	Degr.	Degr.
Janvier..	5.	6.	28.	20-21.	31.	8 3/8	8.	. . .	-15 1/4	-10	. . .	-10.	-8 1/4 le 1er fevr.	-3,3.	-0,7.
février..	26.	12-16.	1.	1.	1.	10 1/4	10.	10.	-13.	-4.	-15.	-3 3/4		4,8.	6,1.
Mars...	24.	25.	3.	12.	2-3.	16.	17.	15.	-1.	-1.	3.	2.	. . .	6,7.	6,8.
Avril...	25.	25.	9.	6.	10.	20.	17.	16.	-0 1/2	1.	3.	3.	. . .	8,5.	8,8.
Mai...	30.	31.	9.	2-9.	. . .	22 3/4	20.	20.	1.	1.	3.	. . .	. . .	9,5.	9,4.
Juin...	21.	27.	7.	5.	. . .	24.	21 1/2	25.	5 1/4	5.	9.	. . .	. . .	13,7.	13,1.
Juillet...	5.	4.	1.	29.	. . .	27 1/2	24.	24.	9 1/8	9.	13.	. . .	. . .	16,0.	15,8.
aoust...	5.	2.	26.	7-9.	. . .	26.	29.	26.	8 3/4	8.	10.	. . .	. . .	15,0.	15,5.
Septembre.	14.	13.	20.	22.	. . .	19 1/2	18.	19.	3 3/8	3 3/4	5.	. . .	. . .	11,5.	10,5.
Octobre..	1.	9-23.	31.	7-14.	. . .	17 5/8	17.	. . .	-0.	3.	. . .	. . .	. . .	9,3.	10,0.

Mois 1776.	Jours de la plus gr. élév. du merc. A	B	D	Jours de la moind. élévat. du merc. A	B	D	Plus gr. élévat. du merc. A	B	C	D	Moindre élév. du merc. A	B	C	D	Élév. moy. du merc. A	B	D
							po. lig.	po. lig.	po. lig.	po. lig.	po. lig.	po. lig.	po. lig.	po. lig.	po. lig.	po. lig.	po. lig.
Janvier..	3.	3.	3.	13.	12.	13.	28.0 1/2	28.1.	. . .	28.3.	26.4.	27.3.	. . .	27.3.	27.6,9	27.8,7	27.9,9
février..	15.	15.	15.	11.	11.	11.	28.0 1/2	28.1.	27.11.	28.4.	26.11 1/2	27.0 1/2	26.10.	27.3.	27.6,9	27.9,5	27.11,0
Mars...	31.	31.	30.	9.	9.	9.	28.4 1/4	28.3.	28.1.	28.7.	27.5 1/4	27.5 3/4	27.4.	27.8.	27.10,5	27.11,2	28.2,0
Avril...	1.	1.	1.	8.	17.	10.	28.3 3/4	28.3.	28.1.	28.7.	27.5 1/2	27.8 1/2	27.4.	27.9.	27.11,3	28.0,1	28.2,9
Mai...	13.	13.	. . .	5.	8.	. . .	28.3 3/4	28.2.	27.10.	. . .	27.3 1/2	27.8.	27.5.	. . .	27.10,11	27.11,4	
Juin...	19.	19.	. . .	12.	12.	. . .	28.2 1/4	28.0 1/2	27.9.	. . .	27.7.	27.9.	27.6.	. . .	27.10,6	27.10,7	
Juillet...	29.	29.	. . .	6.	6.	. . .	28.2 1/4	28.0 1/2	27.11.	. . .	27.6 1/4	27.8.	27.7.	. . .	27.11,1	27.10,7	
aoust...	1.	25.	. . .	31.	31.	. . .	28.1 3/4	27.11 1/2	27.11.	. . .	27.7.	27.7.	27.7	. . .	27.10,9	27.9,9	
Septembre.	21.	21.	. . .	26.	26.	. . .	28.2 1/4	28.0.	27.10.	. . .	27.4 3/4	27.5.	27.5.	. . .	27.10,0	27.9,0	
Octobre..	9.	9.	. . .	6.	25.	. . .	28.2 3/4	28.1.	. . .	. . .	27.8.	27.8.	. . .	. . .	27.11,6	27.10,4	

Mois	Vents Dominants. A	B	C	Nombre des Jours de Pluye. A	B	C	D	Températures. Montmorency.	St Maurice-le-Girard.	Villefranche.	Maladies. Montmorency.	St Maurice-le-Girard.	Villefranche.
Janvier..	N.E.	N. et N.E.	. . . .	10.	9.	. . .	12.	Très froid et sec.	Froid et sec.	. . . .	Rhumes, pleurésies, petites véroles.	Grippe.	
février..	S.O.	N.O. et S.O.	S.	23.	24.	14.	25.	Froid, très humide.	Froid, et humide.	Froid et humide.	Maux de gorge, fièvres putr. pet. véroles.	Idem.	Point de Maladies.
Mars...	N. et N.E.	N.E. et N.O.	N.	9.	7.	4.	5.	Froid, assez sec.	Doux et sec.	Froid et sec.	Quelques rhumes, plus de pet. véroles.	Grippe diminue.	Idem.
avril...	N. et N.E.	N.	N.	8.	3.	5.	5.	Froid et sec.	Chaud et sec.	Idem.	Pet. véroles, fluxions de poitr. glandes, et dans les environs fièvres pourprées.	Encore quelques grippes.	Idem.
Mai...	N.E. et N.	N. et N.O.	N.	9.	10.	4.	. . .	Idem.	Froid, très sec.	Idem.	Fluxions, érésipèles, pet. véroles.	Affection aux glandes cervicales avec fièvre.	Idem.
Juin...	S.O. et N.E.	S. et S.O.	S.-O. et N.	16.	17.	15.	. . .	Chaud et humide.	Doux et humide.	Chaud et humide.	Point de Maladies.	. . . . . .	Idem.
Juillet...	N.-S.O. et O.	S.	N.	12.	7.	8.	. . .	Chaud et sec.	Chaud et sec.	Chaud et sec.	Quelques petites véroles.	. . . . . .	Quelques petites véroles.
aoust...	S.O.	S. et S.O.	N.-S. et O.	16.	10.	11.	. . .	Chaud et humide.	Idem.	Chaud et humide.	Point de Maladies.		Pet. vér. fièvres putrides vermin. pleurés. fièvres intermittentes petites maladies avec des dysenteries.
Septembre.	S.O. et N.E.	N.O. et S.O.	O. et N.	14.	9.	16.	. . .	Froid et humide.	Froid et sec.	Froid et humide.	Pet. vér. dans les environs.	.	
Octobre..	N.E.	S.E. et N.E.	. . . .	6.	5.	. . .	. . .	Chaud, très sec.	Doux et sec.	. . . .	Pet. vér. fièvres tierces et doubles-tierces, maladies de poitrine dans les environs.	Quelq. fièvres intermittentes, douleurs rhumat. des reins, quelques Cholera morbus.	

FIGURE 16.3. Three-in-one working table made by Louis Cotte for redacting meteorological and disease observations made in several locations into correlated general observations. Louis Cotte, “Résultats des observations météorologiques faites à Montmorency, à St Maurice le Girard en Bas Poitou, à Nantes . . .” for January–October 1776. Fonds Société Royale de Médecine, Bibliothèque de l’Académie Nationale de Médecine, SRM 189 dr. 5, no. 9. By permission of the Académie Nationale de Médecine, Paris.

Dividing the page seems too trivial to be called a technique. Yet it was crucial, and only the best observers used it—and then rarely. Why? Because it only made sense using general observations, and we have seen how much redactive work this entailed. To record, say, cloudiness of sky and cloudiness of urine in adjacent columns would have indeed been to correlate weather and disease, but absurdly so. Observers did look from their barometers to their patients, but soon learned not to expect correlation at that level: "We did not observe that the frequent swings of the mercury in the barometer, nor its considerable descent after the 20th [of November 1776], influenced the nature and events of disease."[50] The most instructive falling short of this kind (because it occurred at the higher levels of generalization, achieved through grinding pen and paper labor) is Cotte's own system of synthetic tables. General observations of climate in terms of the Aristotelian four qualities could conceivably be correlated, as physicians tried to do, with general observations of disease *character* in terms of the Galenic four humors, but not promisingly, as Cotte tried to do, with the prevailing disease *species* and *genera* (smallpox, measles, intermittent fevers, and so on).

The page in figure 16.4 looks very different from Cotte's tables and the Maret committee's columns. It is a text divided by year headings and accompanied by notes in the left margin. It is in fact a page from an ambitious table of general observation and correlation, Dr. Gallot's "Tableau of Prevailing Diseases from 1776 to 1786, with That of the Meteorological Results during the Same 10 Years."[51] Each nosological year has been extracted and condensed from his many-paged annual reports into a mere paragraph, devoted mostly to constitution and character rather than disease species and genera. The adjacent notes in the left margin are in fact each year's "Temperature" (dry and cold, or dry at the start and humid at the end, etc.) plus the maxima, minima, and mean thermometer and barometer readings and the dominant winds. This is the same divided-page instrument as in the Maret committee's annual reports, which, however, ran to twenty or more pages rather than the half-page paragraph into which Gallot has condensed each year.

It does not matter that the meaning of annual and seasonal climate-disease correlations remained elusive (and to some extent still does), nor that the interpretations ventured by the society belonged to an Aristotelian and Galenic framework of four qualities and four humors, which, though at last underpinned by millions of measurements using modern instruments and by thousands of case histories, medicine would increasingly leave behind in the nineteenth century. What matters is that the self-discipline of the strictly divided page undid both individual habit and collective tradition such that doctrine became investigative question; and that exact paper and ink

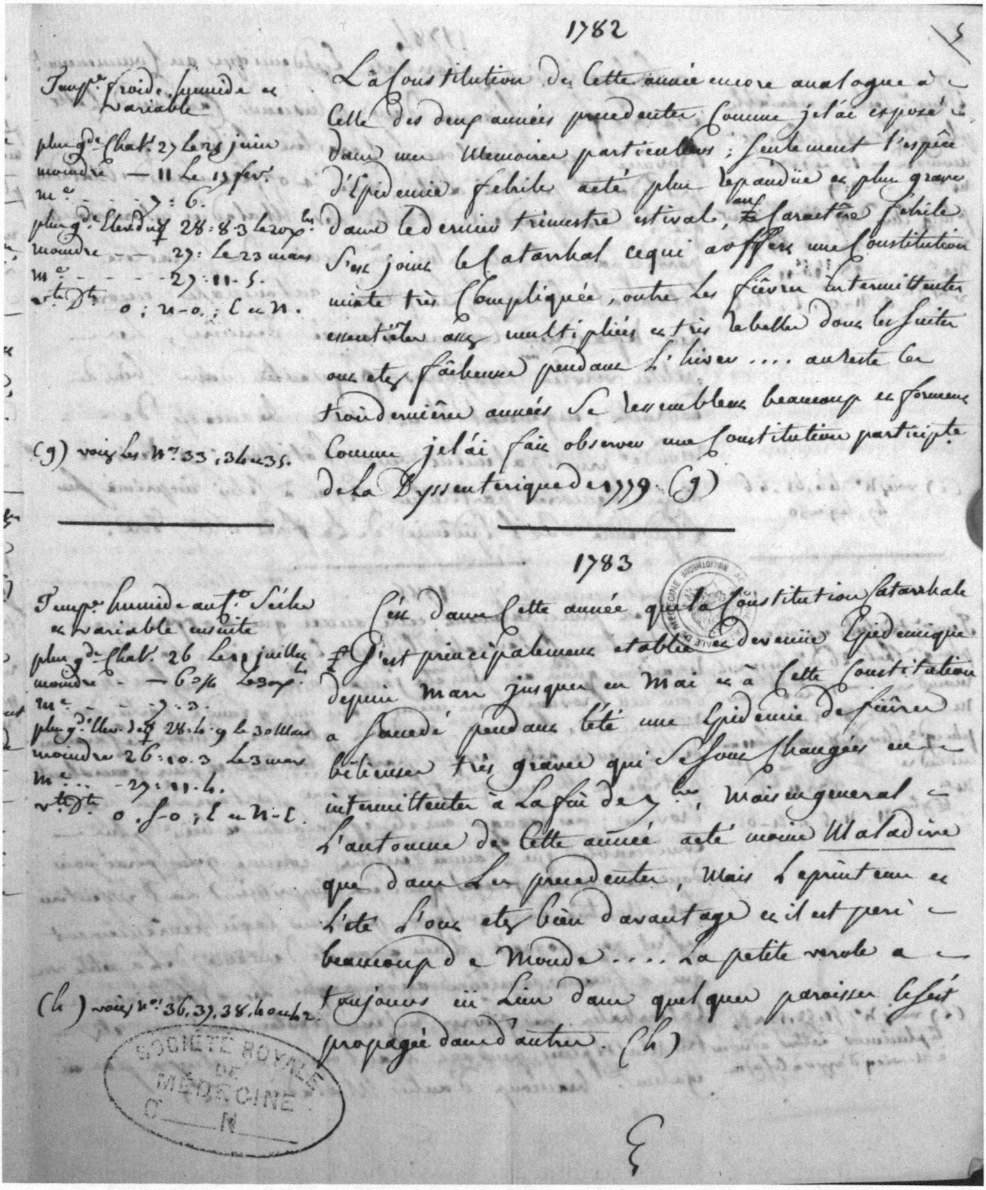
1782

Tempe. froide, humide et variable
plus gde chalr. 27 le 25 juin
moindre — 11 le 13 fev.
me. — 7: 6.
plus gde elevation 28: 8: 3 le 20
moindre — 27: le 23 mars
me. — 27: 11: 5.
vts. Dts. o; n-o.; l et n.

La Constitution de cette année encore analogue à celle des deux années precedentes comme je l'ai exposé dans un Memoire particulier; seulement l'espèce d'Epidemie febrile a été plus repandue et plus grave dans le dernier trimestre estival, au caractère febrile s'est joint le catarrhal ce qui a offert une constitution mixte très compliquée, outre les fièvres intermittentes essentielles assez multipliées et très rebelles dont les suites ont été fâcheuses pendant l'hiver .... au reste les trois dernières années se ressemblent beaucoup et forment comme je l'ai fait observer une constitution participante de la Dyssenterique de 1779. (g)

(g) voiez les no. 33, 34 et 35.

1783

Tempe. humide au commt. sèche et variable ensuite
plus gde chalr. 26 le 2 juillet
moindre — 6 3/4 le 30
me. — 7. 3.
plus gde elevation 28: 4: 9 le 30 mars
moindre 26: 10: 3 le 3 mars
me. — 27: 11: 4.
vts. Dts. o. s-o; l et n-l.

C'est dans cette année que la Constitution catarrhale s'est principalement établie devenue Epidemique depuis mars jusqu'en mai et à cette constitution a succédé pendant l'été une Epidemie de fièvres bilieuses très graves qui se sont changées en intermittentes à la fin de 7bre, mais en general l'automne de cette année a été moins Maladive que dans les precedentes, mais le printems et l'été l'ont été bien davantage et il est peri beaucoup de Monde ... La petite verole a toujours eu lieu dans quelques paroisses et s'est propagée dans d'autres. (h)

(h) voiez no. 36, 37, 38, 40 et 42.

FIGURE 16.4. Table made by Royal Society of Medicine correspondent Jean-Gabriel Gallot for condensing and correlating ten years of meteorological and disease observations, showing the page for 1782 and 1783, one of seven total pages. From Jean-Gabriel Gallot, "Tableau des Maladies Regnantes depuis et y Compris 1776 jusque en 1786, avec Celui des Resultats Meteorologiques Pendant les mêmes 10 années. Extrait des observations adressées à la Société Roiâle de Medecine et faites à St Maurice le Girard, Bas Poitou." Fonds Société Royale de Médecine, Bibliothèque de l'Académie Nationale de Médecine, SRM 189 dr. 5, no. 60. By permission of the Académie Nationale de Médecine, Paris.

techniques and habits of general observation arose and could be applied to medicine's abiding core problems of knowledge, which would remain challenges of producing, out of cases, general observations and correlations of symptoms, internal pathological events, and environmental influences.

Conclusion

The workings of observation as process usually remain mysterious and hidden between act and result. Here we have been able to watch those workings. Making a general observation, in practice, was redactive, the use and development of existing paper and ink instruments and techniques, above all, those of the extract, the précis, and the table. Two of the instruments described here did not come directly from redactive practice. The detailed, yes/no, multiple-choice questionnaires printed in the society's *Mémoire instructif* were not redactive but preredactive. They built synthetic redactive potential—the possibility of generalization—into the act of observation and/or its recording. Not surprisingly, they were designed not in the context of daily observation or that of administration (Turgot had issued only six broad queries), but by those directly facing the task of synthetic redaction, such as Maret and the Dijon Academy's medical committee, Vicq d'Azyr and the society. Another of the instruments described here, the society's printed table for correspondents' monthly "Observations météorologiques," was not an existing form of redaction, but redaction *on a form*.

Because these paper and ink techniques and forms of observational knowledge and rearrangement are primitive compared to, say, a computer database, they may seem negligible. Yet for something negligible, observers certainly went to much trouble redacting the publishable prose they preferred to write into forms that would rarely if ever be public. In fact, Cotte reported that he was *unable* to "get results" unless he "reduced to tables" observations that correspondents had "written out in full."[52] Not having redactive technique in the eighteenth century would have been like not having graphical representation in the nineteenth and twentieth centuries. A Gallot would have had to stand helpless before the accumulated mass of his ten years of case histories and meteorological measurements.

So valuable it was, in the age of redactive general observation, to get everything onto one page that Gallot concluded his 1776–1786 tableau with a one-paragraph "resumé" of the whole decade: the tableau showed the dominance of a constitution, catarrhal and bilious. We have come full circle to the bilious constitution read aloud by Caille in 1785, but with the minor difference that we are still in Poitou (actually in Saint-Maurice-le-Girard in Bas Poitou) and not in the Louvre. To get from Poitou to the whole of France, Caille and the other redactors will need to "extract from the registers" of the society material from more than eighty memoirs on epidemic pneumonia alone as well as from Cotte's annual tables. Yet a good deal of their work would already have been done, by the observers as we have witnessed it, and we now know too what the rest of that work would have looked like.

Read year by year, précis by précis, Gallot's "tableau" shows him writing ever less in terms of diseases and ever more in terms of the general character and constitution of disease. The tableau thus recapitulates the ten-year-long coming into being of a general observer. How this happened is suggested in the subtitle: "Extracted from the Observations Addressed to the Royal Society of Medicine and Made at St Maurice le Girard, Bas Poitou." The society used the same formula—"extracted from the correspondence"—in subtitling publications based on the collective inquiry, and these usually appeared as the work of as many as eight "commissioners" rather than an author.[53] Though not a collective, Gallot had in effect become one. Ten years increasingly spent redacting and not only authoring case histories had trained him. Author had become redactor, extractor from his own observation archive. This was how an individual observer could best make a general observation: act like a committee, a bureau, an editorial staff.[54]

These, the organizational forms for operating the inquiry, were a kind of collective, the kind it took to make a general observation, the kind that made such an observation thinkable. This differed from the kind of collective that made *observationes* and collected them: the medical and scientific republic of letters, a collegial network of authors writing, exchanging, collecting, ordering, and commenting on case histories. By contrast, the society's observers, though volunteers submitting their observations as "correspondence" by post, were caught in a centralized web rather than being points of intersection in an epistolary network. The more they gave up the voice of the author for the more voiceless work of the extractor, the more they gave up the virtuoso observer's autonomy to the templates of detailed questionnaire and tabular form, the more they and their observations could succeed in becoming part of general observation.

Notes

I thank all the contributors to this volume for encouragement and discussion of versions of this paper and the staff of the Bibliothèque de l'Académie Nationale de Médecine, Paris, for facilitating my research.

1. Claude-Antoine Caille, "Mémoire sur les fièvres rémittentes & intermittentes qui ont régné en 1780 & 1781," *Histoires et mémoires de la Société royale de médecine* 8, *Mémoires* (1790): 24–37, on 25–26. Hereafter *HSRM* for *Histoires* and *MSRM* for *Mémoires.*

2. Gianna Pomata and Nancy G. Siraisi, eds., *Historia: Empiricism and Erudition in Early Modern Europe* (Cambridge, Mass.: MIT Press, 2005); Daniel Roche, *Le siècle des lumières en province: académies et académiciens provinciaux, 1680–1789*, 2 vols. (Paris: Mouton, 1978), 1: 382.

3. For a remarkable example of a community of and at attention, see chap. 17 in this volume; and Lorraine Daston, "On Scientific Observation," *Isis* 99 (2008): 97–110.

4. See generally Caroline Hannaway, "Medicine, Public Welfare and the State in Eighteenth Century France: The Société Royale de Médicine of Paris, 1776–1793" (Ph.D. diss., Johns Hopkins University, 1974); idem, "The Société Royale de Médicine and Epidemics in the Ancien Regime," *Bulletin of the History of Medicine* 46 (1972): 257–73; Jean-Paul Desaive et al., *Médecins, climat et épidémies à la fin du XVIIIe siècle* (Paris: Mouton, 1972); Roche, *Siècle des lumières*, 1: 283–85, 316; 2: table 47, map 38. For the wider context, see Ludmilla Jordanova, "Earth Science and Environmental Medicine: The Synthesis of the Late Enlightenment," in *Images of the Earth: Essays in the History of the Environmental Sciences*, ed. L. J. Jordanova and R. S. Porter (Chalfont St. Giles: British Society for the History of Science, 1979), 119–46; Theodore S. Feldman, "Late Enlightenment Meteorology," in *The Quantifying Spirit in the Eighteenth Century*, ed. Tore Frängsmyr, J. L. Heilbron, Robin E. Rider (Berkeley: University of California Press, 1990), 143–77.

5. Caille, "Mémoire sur les fausses fluxions de poitrine bilieuses, & principalement sur celles qui on régné dans plusieurs cantons de la France en 1782, 1783 & 1784: Extrait de la Correspondance de la Société royale de Médecine," *MSRM* 5 (1787): 37–47, on 41.

6. *Prix distribués & annoncés par la SRM, dans sa séance publique, tenue au Louvre le Mardi 26 Août 1783* (Paris: L'Imprimerie de Ph.-D. Pierres, 1783), 6.

7. Gallot, "Observations météorologiques," Nov. 1778: Fonds Société Royale de Médecine (SRM), Bibliothèque de l'Académie Nationale de Médecine, Paris: SRM 189 dr. 5, no. 162.

8. Basis of calculation: in 1778–84, over forty of the correspondents in France were submitting all or most of the seventeen daily weather observations prescribed by Cotte; *HSRM* 5 (1787), foldout table "Résultats de l'Année 1783" after page 246.

9. Souquet, "Observations sur les Maladies . . . Température de l'Air," Oct.–Dec. 1778: SRM 137 dr. 27, no. 9.

10. Louis Merle, *La vie et les oeuvres du Dr. Jean-Gabriel Gallot, 1744–1794* (Poitiers: Société des antiquaires de l'Ouest, 1962), quotation from a contemporary on 9.

11. Basis of calculation: over forty correspondents provided reports, usually monthly, extracted by Cotte; *HSRM* 5 (1787), foldout table "Maladies Dominantes" after 246; correspondents' reports included between zero and six case histories; many also reported on particular epidemics.

12. See chap. 2 in this volume.

13. Desaive et al., *Médecins.*

14. See chaps. 2 and 3 in this volume. For the eighteenth century: Andrea Rusnock, "Correspondence Networks and the Royal Society, 1700–1750," *British Journal for the History of Science* 32 (1999): 155–69. For medicine: "Medical Correspondence in Early Modern Europe," special issue of *Gesnerus* 61, nos. 3–4 (2004).

15. In this volume, see chap. 3 for the seventeenth and eighteenth centuries and chap. 11 for the nineteenth. Classic: Roche, *Siècle des lumières*, vol. 1, esp. 382–83. Recent: Bernhard

Siegert and Joseph Vogl, eds., *Europa: Kultur der Sekretäre* (Zurich and Berlin: Diaphanes, 2003).

16. [Félix Vicq d'Azyr], *Mémoire instructif sur l'établissement fait par le Roi d'une commission ou société & correspondance de médecine* (n.p.: n.d., [1776]), Réglement, section VII, 55–56; "digest," 45. On the society's bureau, see Hannaway, "Société royale de médicine," 103.

17. *Mémoire instructif*, Réglement, 56.

18. Ann Blair, "Humanist Methods in Natural Philosophy: The Commonplace Book," *Journal of the History of Ideas* 53 (1992): 541–51; Helmut Zedelmaier, *Bibliotheca Universalis und Bibliotheca Selecta: Das Problem der Ordnung des gelehrten Wissens in der frühen Neuzeit* (Cologne, Weimar, and Vienna: Böhlau Verlag, 1992), chap. 2.

19. *Dictionnaire de l'Académie Française* (1694), s.v. "extrait."

20. *Mémoire instructif*, Réglements, section 2, "Extraits," 53.

21. Prêvot de la Caussade, "Observations for the year 1776," SRM 172 dr. 11, no. 5 (a).

22. Saillant, "Extrait du mémoire de M. Prevot," SRM 172 dr. 11.

23. Louis Cotte, "Maladies" for Sept.-Dec. 1776 and "Résultats de l'Année 1776," *HSRM* 1 (1779): 174–82; "Suite des Résultats de l'Année 1779: Maladies Dominantes," *HSRM* 3 (1782), foldout table after 178. Cotte had just published *Traité de météorologie* (Paris: L'Imprimerie Royale, 1774) based largely on extracting all meteorological observations published by the Paris Academy of Sciences since 1700 and those in its "registers"; he also extracted from his parish registers (see v, xii–xiii).

24. Gallot, "Mémoire sur les maladies regnantes en 1776," 2–3, SRM 189 dr. 5, no. 7.

25. Blair, "Commonplace Book"; Ann Blair, "Reading Strategies for Coping with Information Overload, ca. 1550–1700," *Journal of the History of Ideas* 64 (2003): 11–28.

26. Louis Lepecq de la Cloture, *Observations sur les maladies épidémiques* (Paris: Vincent, 1776), xxxvi.

27. *HSRM* 1 (1779): 183–84, and all volumes through *HSRM* 8 (1790): 85–86.

28. Gallot, "Maladies regnantes en 1776," 4.

29. Responses in SRM 133, 135, 142, 144, 149, 160, 170, 172, 183.

30. In some provinces, clerks wrote the six questions spaced out on a blank form to be filled in by physicians: e.g., Languedoc: SRM 172 dr. 13, no. 4. On the failure of questionnaires to standardize inquiry, see Marie-Nöelle Bourguet, *Déchiffrer la France: la statistique départementale à l'époque napoléonienne* (Paris: Éditions des Archives Contemporaines, 1989). On earlier questionnaires, see chaps. 3 and 15 in this volume.

31. *Mémoire instructif*, 28–29, "collection and redaction," 44; Gallot, "Maladies regnantes en 1776," 4.

32. Gallot, "Maladies regnantes en 1776," 4.

33. Hugues Maret, Jean-François Durande, and Comité médical de l'Académie de Dijon, "Année 1779": SRM 166 dr. 5 no. 2. Emphasis added.

34. Gallot, "Maladies regnantes en 1776," 4–5.

35. Lepecq, *Observations*, cxxxii.

36. "Observations de M. Durande" in Maret, Durande, and Comité médical de l'Académie de Dijon, "Histoire Meteoro-nosologique pour l'annee 1777," 15–20, on 18, SRM 166 dr. 5, no. 1.

37. Gallot, "Observations Nosologiques," Jan.–March 1779, SRM 189, dr. 5, no. 29.

38. *Mémoire instructif*, final page.

39. Andrea Rusnock, "Hippocrates, Bacon, and Medical Meteorology at the Royal Society,

1700–1750," in *Reinventing Hippocrates*, ed. David Cantor (Aldershot: Ashgate, 2002), 136–53, on 140–45.

40. Cf. Frängsmyr, Heilbron, and Rider, *Quantifying Spirit.*

41. See note 27.

42. For example in *Refléxions lues dans la séance tenue au Louvre, par la Société royale de médecine, le 18 Septembre 1781; sur la nature de la constitution de cette année, & le traitement des maladies qu'elle a occasionées* (Paris: Imprimerie de Ph.-D. Pierres, 1781).

43. Caille, "Fausses fluxions de poitrine bilieuses," 41.

44. "Rapprocher" from Society redactor Jean-Jacques Paulet, "Correspondance de Picardie. Observations faites à Boulogne sur mer par M. Souquet," 1v, SRM 172 dr. 2; "j'y joindrai" from Gallot, "Tableau des Maladies Regnantes depuis et y Compris 1776 jusque en 1786, avec Celui des Resultats Meteorologiques Pendant les mêmes 10 années," 1, SRM 189 dr. 5, no. 60.

45. *Dictionnaire de l'Académie Française* (1694), s.v. "table"; in the fourth ed. (1762), "tout d'une veuë" becomes "d'un même coup d'oeil" (at a glance).

46. Many examples can be found in SRM 172.

47. The accompanying text calls this a "Table of comparison" with a "coup d'oeil" function; Cotte, "Observations météorologiques 1776. Rapport du P. Cotte," SRM 189 dr. 5, no. 9.

48. Barrère, "Observations nosologiques," March 1783, SRM 166 dr. 1, no. 3.

49. For example: Maret et al., "Année 1777," 24r.

50. Maret et al., "Année 1777," 1v.

51. Gallot, "Tableau."

52. Cotte, "Rapport du P. Cotte."

53. For example: *Refléxions* (1781); its eight *commissaires* named in *HSRM* 3 (1782): 174–75.

54. So well had Gallot grown into this role that under the Revolution, elected to the National Assembly, he became secretary of its health committee; Merle, *Gallot*, chaps. 3–4; Dora B. Weiner, *The Citizen-Patient in Revolutionary and Imperial Paris* (Baltimore: Johns Hopkins University Press, 1993), 93–95.

17

Coming to Attention: A Commonwealth of Observers during the Napoleonic Wars

ANNE SECORD

> I am quite of opinion that the cultivation of any Science should form as it were one commonwealth, in which a similarity of taste and pursuit should be sufficient to authorize each member to ask every reasonable and practicable assistance from every other; and in such a business as botany, where so much depends upon observation, and where the observations still wanted are so infinitely numerous, it is particularly desirable. This idea has led me to wish a coalescence between the marine botanists of the Eastern and Western coasts . . .
>
> —WILLIAM WITHERING TO DAWSON TURNER, 27 January 1798

> Some have aptly enough compared
> A CLASS . . . . to an Army;
> An ORDER . . to a Regiment;
> A GENUS . . . to a Company;
> And a SPECIES to a Soldier.
>
> —WILLIAM WITHERING, *An Arrangement of British Plants*[1]

Every One Like a Soldier

William Withering's use of military units to explain the divisions by which the vegetable kingdom was classified was particularly apt, even in a work that aimed to attract female as well as male readers, in the edition of his botanical compilation published when Britain was at war with France. It was not that being at war was new; since the 1740s the British had been almost continuously engaged in military action. What was different about the revolutionary and Napoleonic Wars of 1793–1815 was the way they were experienced within Britain. According to an Edinburgh observer, by 1795, "when every citizen was a soldier, and every thing military the rage, it was the fashion for female relatives of the noblemen and gentlemen, who bore commissions in the regulars, fencibles, and volunteers, to assume the uniforms of the respective corps to which their fathers, husbands, and brothers belonged."[2] Britain became an "armed nation," with numerous volunteer militias, ready to act in case of an enemy attack.

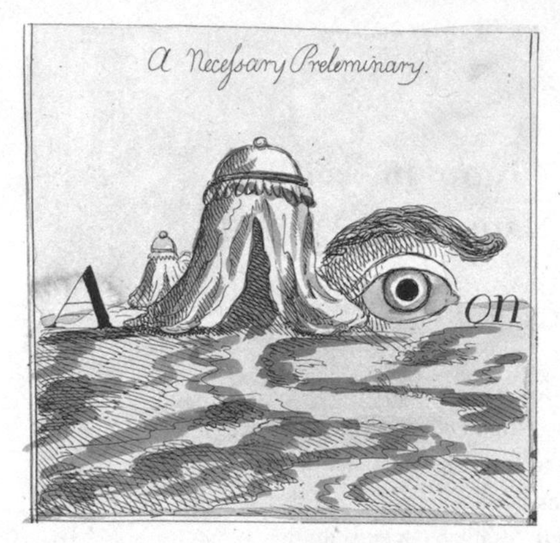

Let ENGLISHMEN
keep a watchful Eye upon
FRENCH SPIES,
who are employed to pull down or deface
all Loyal and Patriotic Papers.

Nichols and Son, Printers, Red Lion Paffage, Fleet Street.

FIGURE 17.1. *Left*, one of a set of six "Puzzles for Volunteers!!" (1803). In representing "attention," this puzzle calls for attention in the very act of observing the visual pun. Curzon Collection, b.12 (2), Bodleian Library, University of Oxford. Similar "Puzzles for Patriots!!" also underlined the need to observe attentively. Both sets of puzzles are reproduced in Franklin and Philp, *Napoleon and the Invasion of Britain*, 96–7. *Right*, broadside [1803]. John Johnson Collection, French Wars and Revolutions, folder 4 (66), Bodleian Library, University of Oxford.

Fear of a French invasion resulted in the most potent impact of war in Britain, namely, the regimes of watchfulness it engendered. The need for defensive lookouts capable of signalling the approach of the enemy reached its apogee with the building of martello towers along the coast from 1803, but four years earlier the War Office had secretly commissioned the astronomer William Herschel to build a spy telescope to be mounted on a castle wall in Kent in the hope that it would provide the earliest possible warning of a French invasion fleet or troop-carrying Montgolfier balloons.[3] The need for vigilance was fueled not only by fears of invasion but also by social unrest within Britain. Controversially including the poor, volunteer corps were under constant surveillance in case the aim of replacing seditious tendencies with loyalism by arming the populace did not overcome lower-class radicals' predilection for French revolutionary principles but encouraged them instead to turn on those who had armed them.[4] Within these corps, attention was "A Necessary Preliminary," while broadsides warned of the dangers of not being attentive (fig. 17.1).

Military spectacle informed more than just women's fashion; the order it conveyed affected factory, prison, and school discipline, while the importance attached to vigilance was reflected in the thought of even those most removed from the immediate effects of war. In this sense, war—even when experienced at a distance—persists in the "mode of daily living and a habit of mind."[5] Thus a Quaker pacifist could advocate a system of education in which "every one, like a soldier, must be upon the alert—& like soldiers, all at the same moment—thus the attention is kept in constant exercise and no idlers can live amongst them."[6] Fostering this form of attention required not only watching others but disciplining oneself in self-vigilance. During the war, such practices of watchfulness were deemed critical as much for safeguarding against infiltration by spies and invasion by the French, as for heeding evangelicals' calls to watch the state of one's soul as well as those of others. With respect to historicizing scientific observation, these habits of vigilance—usually seen as paradigmatic elements of regimes of enforcement and disempowerment—were as much ingrained through natural history, as naturalists, especially those exploring nature at the edges, adopted and self-acted a system of extraordinary surveillance.

The success of practices for standardizing scientific observation from the 1830s makes it easy to overlook the ways in which nature and society were explicitly intertwined in observational practices during the war. This way of seeing crossed social and political boundaries. The Northumberland wood engraver Thomas Bewick made clear his antiwar sentiments in his 1804 work on water birds, in which, as well as fine engravings of the creatures being described, he included vignettes showing weary, disabled soldiers who, rather than receiving a hero's welcome home, were forced to search for work.[7] This view contrasted with the seductive techniques employed by recruiting parties to drum up volunteers. Across the country poor men were "taken out" of themselves and led to doom by the lure of brightly colored uniforms and martial music,[8] a process Bewick mocked in his description of the goose. When geese are taken to market in droves as large as nine thousand, "it is," Bewick claimed, "a most curious spectacle to view these hissing, cackling, gabbling, but peaceful armies, with grave deportment, waddling along (like other armies) to certain destruction." The goose might be the "most vigilant of all sentinels," but Bewick brings home the deceptive glamour of war in the vignette that follows this description—"The Churchyard Cavalry"—in which children playing at soldiers are shown riding gravestones (fig. 17.2).[9]

Fears of being deceived by appearances informed the ways of seeing in this period, when there was rising concern about how to deal with those who lacked connections of any sort—vagrants in particular and those who inhabited liminal places in general. The investigations of naturalists at this time reveal that their processes of observation are part of a set of wider cultural habits of comparing, ranking, and determining boundaries, which was similarly applied to specimens, people, spaces, and nations. Just as historians have begun to argue that empire in this period is always about edges and encounters, and that attention should shift from imperial centers to the places where

FIGURE 17.2. "The Churchyard Cavalry." Enlarged (original wood engraving, 3.25 × 1.75 inches). From Thomas Bewick, *Water Birds* (Newcastle: Edward Walker, 1804), 304.

collections are made or new forms of knowledge encountered (particularly during wars with France),[10] so the processes of natural history observation are best revealed in confrontations with the unfamiliar or unexpected.

Rather than looking at voyages of exploration and the observation of new places, however, I focus on observers who, unable to leave Britain during the war, only went as far as the coast—notably early seaweed investigators. For these naturalists, the shore marked not only the boundary between land and the unknown immense space of the ocean, but also the limits of standard botanical techniques and Linnaean classification. Here, observers were confronted with ambiguity. Seaweeds bore no resemblance to terrestrial plants, went through a puzzling sequence of changes during their life cycle, and required rapid examination as they changed dramatically once removed from the sea. Beyond their variability and the difficulties of observing them in situ, seaweeds also confounded the kingdoms of nature, several appearing remarkably close to those animal forms whose vegetable-like nature led to their being labeled zoophytes. Moreover, these plants that frequently appeared "disguised" required, for their full examination, the use of magnifiers, which also possessed the potential to "deceive." Most disconcertingly, in a science of observation like botany, the knowledge of seaweeds had to be built up from partial glimpses—sometimes of only battered, incomplete, and hard-to-recognize specimens found among the "rejectamenta" of the sea.

Given the extent of Britain's coast and the propensity of the same seaweed to vary in different locations, establishing a community of observers was essential. Marine botanists had to develop ways to ensure that their observations resulted in the stabilization of the objects of their study before any attempt at classification could be made. This required the development of acute powers of attentiveness and the continual appraisal of other observers. Little wonder that Withering, Britain's foremost botanical compiler, was eager for a "coalescence between the marine botanists of the Eastern and Western coasts." Without the emergence of such a community, no one observer, no matter how adept in discerning the scientific objects of marine botany, could describe seaweeds in ways that allowed them to be detected by other observers. Publications not only gave credit for discoveries of new species but also assessed and monitored observational skill itself. Nowhere was this more important than when observing seaweeds, with their immense capacity to deceive the viewer. Scrutiny of the natural and social as an ensemble informs the ways of seeing in this period, and the analogies made and comparisons drawn reveal the purposes of such processes of observation. In presenting a portrait of a scientific community at work, dealing with the constraints imposed by war, I aim to show that observation is historicized, as are the emotions it arouses, in the intertwining of learned visual techniques and the visual habits that characterize a particular society at a particular time.

The Order of Observers

When, in 1802, Matthew Flinders, commander of the British expedition to circumnavigate Terra Australis, first looked upon a coral reef, he saw "wheat sheaves, mushrooms, stags horns, cabbage leaves, and a variety of other forms, glowing under water with vivid tints of every shade betwixt green, purple, brown, and white; equalling in beauty and excelling in grandeur the most favourite *parterre* of the curious florist." In seeing this "new creation, as it was to us" as "imitative of the old," Flinders made visual sense of these novel sights by noting resemblances and drawing analogies with the things he knew.[11] Although publication of Flinders's account was greatly delayed owing to his six-and-a-half-years' detention on the French island of Mauritius after being forced to make landfall there, even when he wrote these words in his diary scientific investigators had already learned not to see in this way at all. "Why should it be thought impossible, that the submarine plants, like the animals of that element, should have powers and properties new, original, and peculiar to themselves?" early inquirers asked. Marine botanists were warned of the necessity of laying aside "all comparisons and ideas of analogy taken from plants growing on land," for seaweeds presented a baffling series of transformations in their modes of existence for which it was "not easy to account on philosophical principles."[12] Moreover, their "occult situation and exposure to unseen currents baffle the most acute researches!"[13]

Seaweeds were thus ambiguous objects not only because as cryptogams their mode of reproduction was "hidden" but also because of the medium in which they existed. Any detailed examination had to be undertaken rapidly before the plants could dry out. Once out of the sea, marine plants could change dramatically in color and form, some dried specimens curling up or splitting beyond recognition. Many lost their green or brown hues and turned black on herbarium sheets and, if not dried quickly enough, could end up as putrid smelly masses. In contrast to botanists' usual practice of studying dried specimens in the herbarium, the primary place for the observation of seaweeds was therefore the shore.

Even if able to get to the coast, marine botanists required acute visual skills. Seaweeds, once stranded on the beach, were bleached by the sun and, even when fresh, frequently "disguised." When the discovery of the very rare *Fucus asparagoides* was announced to the botanical world, the plant was described as a gelatinous lump on the shore but of "exquisite beauty" when floating in water. The first discoverer of this "highly elegant" seaweed "among the rejectamenta of the sea on the Yarmouth beach" was Lilly Wigg, a shoemaker turned schoolmaster, whose accuracy of observation and strenuous autodidactism had resulted in his becoming "so eminently skilful in detecting, as well as in preserving, specimens of marine *algae*."[14] Thoroughly aware that it was all too easy to pass over shapeless gelatinous masses on the beach little suspecting the beauty that lay within, expert algologists did not make the same mistake with observers themselves. Thus Wigg became highly prized and admired as a collector, despite his "puritanic brown locks . . . which so much belie his name" and which gave him an "uncouth appearance."[15]

The beauty of seaweeds in water, captured in the illustrations that were so essential to their study, can distract *us* from the activity of collectors rummaging through the rejectamenta on the beach, and the difficulties this involved during the twenty-two years that Britain was almost continuously at war with France. During this time, those seaweeds known as marine wracks, which grow mainly in the intertidal zone, were associated less with the aesthetic sensibilities of the middle or upper classes than with the poor, and the significance of seaweed to the British agrarian economy as fertilizer, fodder, and food was reflected in the laws and local customs pertaining to "coastal commons." Moreover, the production of kelp (the alkaline ash of burnt seaweed), of strategic importance in the manufacture of glass, increased dramatically when the war interrupted supplies from the Mediterranean.[16] Kelp manufacture, with its burning pits dug in the shore and its noxious fumes, made such coasts unpleasant, while the threat of invasion by the French made fearful even those beaches that would later become the pleasure grounds of the Victorian seaside tripper.

Beyond this, British botanists often (and in stark contrast to the visibility of marine plants in the clear waters off Australia or in the Red Sea) had to look for "Sea-weeds withering on the Mud," as the Suffolk botanist and poet George Crabbe graphically put it.[17] This was no imaginative leap on Crabbe's part, but reflected his interest in seaweeds and his connections with other marine botanists.[18] Crabbe's poetry, which surveys manners and morals, including those of the poor, cannot be separated from his botany. The vision engendered by this entanglement is most acute in his 1810 poem *The Borough*, where Crabbe's interest in the lower orders of nature, such as seaweeds, comes together with one of his most vivid and shocking portraits of the poor, that of the amoral fisherman Peter Grimes, who murders his child apprentices. The novelty of Crabbe's portrayal was that he did not analyze Grimes's character, but conveyed the situation through precisely observed details as Grimes becomes almost as one with the hot slimy mudflats and entangled weeds when eventually shunned by his own community.[19] Crabbe's point was not to excite blame or pity for so despised a creature as Grimes but, through his closely observed portrait, to expose the inadequacy of the systems for dealing with the poor, namely, those which allowed pauper boys in workhouses to be sold for a pittance.

In dealing with both the poor and botany, Crabbe decried "systems," holding that "love of method" too often "serves in lack of sense."[20] He deplored reductionist taxonomies that resulted in parish paupers having to wear labels and thus becoming an undistinguished mass, seen as nothing more than a drain on the local rates as food shortages due to bad harvests and military blockades resulted in a precipitous rise in the number of the destitute and a massive rise in crime. In trying to comprehend the disorder produced by war, Crabbe pointed to the need to observe well; in the case of the poor this meant focusing on individuals to see the real causes of crime.[21] This, in turn, as with the kind of botany that interested him, meant going to those places avoided by those who shunned both the lower orders of society and of nature. But for Crabbe, the discoveries made in such places outweighed all objections: "give me a wild, wide Fen, in a foggy Day," he enthused, "with quaking Boggy Ground and trembling Hillocks in a putrid Soil: Shut in by the Closeness of the Atmosphere, all about is like a new Creation & every Botanist an Adam who explores and names the Creatures he meets with."[22]

Bewick, too, was wary of systems in natural history, seeing them as "skeletons injudiciously put together."[23] Far more observation was needed before classifications could be constructed, but, he warned, gaining an "acquaintance" with water birds meant traveling through the "faithless quagmire, amidst oozing rills, and stagnant pools," while only the imagination could dwell on the travels of the great sea birds over the oceans.[24] Similar difficulties confronted the seaweed observer, and Wigg was first made known to the botanical community for discoveries among "rejectamenta" on the beach and on "muddy salt marshes."[25] His finds were communicated by Thomas Jenkinson Woodward, lord lieutenant of Suffolk, who worked for four years with the Reverend Samuel Goodenough, treasurer of the Linnean Society of London, to show the inadequacies of what Linnaeus took to be the fructification of seaweeds and upon which he had based his classification of these plants. In calling upon the society "not to impute these corrections of our great master Linnaeus to any sinister views," their justification was short: "We see errors, we state them." By pointing out that the detection of these errors derived from more acute observation, they appealed to nature as arbiter. "We are confident of nothing," they concluded, "but that we have stated what we have actually seen."[26] But such a strategy, although it might assuage anxiety about abandoning Linnaeus's system of classification, was risky when dealing with plants as deceptive as seaweeds.

Finding so acute an observer as Wigg was possibly Woodward's greatest discovery in marine botany, one that he acknowledged when encouraging the young banker Dawson Turner of Yarmouth to turn his attention to a sustained study of seaweeds. After civil defense plans were instituted for evacuating coastal districts and enlisting the assistance of adult males in the event of a French invasion, Woodward found himself busy with the "troublesome supplementary militia" when he would rather have been examining Turner's "delicate & minute marine productions." The best he could do was outline the difficulties of seaweed studies, encourage Turner to contact the main marine botanist in the west of England, and express his regret that a "better situation" could not be found for Wigg.[27] The need to maintain a community that would never have come together but for a shared interest in seaweeds shows how the base was turned into gold by observational skill, just as botanizing in marginal places offered, as Crabbe had made clear, the pleasures and credit for uncovering a "new creation," albeit one that disgusted "the florists & the Ladies."[28]

When Woodward's duties with the Diss Volunteers, his distance from the sea, and deteriorating eyesight eventually led him to "leave the field to younger soldiers," Turner not only followed his advice about joining the seaweed community but also ensured Wigg's continued participation in it by giving him a job in his bank as a clerk.[29] This act was amply repaid by Wigg's being able to supply fresh specimens when Turner embarked on his monographic study of seaweeds, but it also reflected Turner's prior indebtedness to Wigg. Few had been as fortunate as the young Turner in coming under Wigg's practical tutelage, and when, years later, as Britain's foremost algologist, Turner fully described a rare species he had named *Fucus Wiggii*, he took care to explain that it was on behalf of "the marine botany of England" that he honored "my friend and original instructor in this department of science, Mr. Lilly Wigg."[30] The debt to Wigg, Turner made clear, was both personal and communal.

If Turner's interests had been sparked by his living on the east coast, for others, the war itself provided opportunities for observing seaweeds. Robert Brown collected marine algae when serving with the fencibles in Ireland, while Major Thomas Velley, "who by the accidents of War" was stationed on the south coast of England, "very scientifically employed the leisure, which his military Profession afforded him, in prosecuting his researches in this neglected tribe of Plants."[31] For botanists with the right connections, diplomats and naval officers could be recruited to collect from those far-flung places of strategic importance to Britain. The watchfulness of a nation at war, in some cases at least, provided an observational context that contributed to the study of marine plants.

This culture of defensive vigilance could also be a hindrance if strangers were viewed with suspicion: English botanists in Scotland were mistaken for French spies,[32] while working-class radicals believed that Britain under the governance of Goodenough's former pupil, Henry Addington (Lord Sidmouth), suffered one of the most repressive regimes of surveillance, with spies employed by the Home Office to watch for seditious activity.[33] For Goodenough, the distemper of the times was manifest in the lower orders of society. In response to his servant neglecting to take a letter to the post, Goodenough declared that the "base think themselves in every respect equal to the honourable," and he could only hope that with peace "we shall have more subordinate times."[34] Such views show the delicacy with which Turner had to act with respect to class boundaries in order to ensure that Wigg, a Dissenter and republican, was part of the community of marine botanists.[35]

A Watchful Eye

Class boundaries and the close examination of all individuals had emerged as the desiderata for seaweed investigators toward the close of the eighteenth century. If one of the aims of expert observation is to discern and stabilize scientific objects for a community of researchers,[36] the intricacies of determining the features that would allow marine algae to be arranged in families also involved deciding what defined the living beings you were looking at. In 1797, the Cornish gentleman John Stackhouse determined that seaweeds were plants, in the view of the British at least, in a remarkable series of experiments in which he claimed to have identified, isolated, and germinated their seeds. That Stackhouse's experiments on the propagation of algae from seed had by 1841 never been attempted by anyone else[37] perhaps shows that they were as dependent upon his circumstances as on his observational skill—his opportunities for experimentation resembling the extraordinary resources detailed in Mary Terrall's account of René-Antoine Ferchault de Réaumur in this volume. Having inherited the estate of Pendarves in Cornwall and married an heiress, Stackhouse commenced building Acton Castle on the Cornish coast in 1775. Not only did the rocks in nearby Stackhouse Cove abound in seaweeds, but the castle itself—referred to by Stackhouse as his "Marine Box"[38]—was purpose-built for their detailed examination. In order to observe marine plants in as natural a state as possible, Stackhouse had baths let into the floor of one of the lower rooms in which he could keep his freshly collected specimens. While the living forms of seaweeds were thus preserved indoors, in order to preserve his own health from increasing attacks of rheumatism and gout, Stackhouse had a large bath cut for himself in the rocks of Stackhouse Cove, situated so as to fill with the incoming tide.[39] In 1804, however, a victim to his gout, Stackhouse resorted to taking the waters internally, and his move to Bath put an end to his experimental work.

Stackhouse had been "spurred" to take "a farther peep into the *arcana* of marine plants" in order to explode the "most unphilosophic" view of Major Thomas Velley, who, based upon his own microscopic examination of what he took to be "minute grains or seeds" and Gmelin and Gaertner's views that seaweeds were unisexual or asexual, had proposed that seaweed reproduction was effected by "a self-sufficient power."[40] As for what Velley described as "the occult Principle wch. produces the Fructification," Stackhouse conceded that a complete knowledge of the process was unlikely "as in addition to other difficulties attending the use of high magnifiers, it is more than probable in marine subjects that the parts might collapse on being exposed to the air, or even to the strong light necessary for making the observations."[41] Not disturbing the "medium" in order to observe a phenomenon if not its mechanism, echoes the pleas of Spiritualists with respect to their experimental setup, as described by Michael D. Gordin in this volume. Goodenough and Woodward had also stated that "from the impracticability of examining these plants while actually growing in their native element, it is probable that the manner in which the impregnation is performed may ever remain among the arcana of nature."[42] Nonetheless, armed with "a compound Microscope with 6 different magnifying powers," Stackhouse began "a course of Observations on the Fructification of every species which has offered itself in fruit."[43]

The challenges were immense. Despite the superior magnifying powers of his compound microscope, Stackhouse was well aware that optical instruments show both how you can see better and how you can be deceived. Indeed, his own concern was with the higher powers of magnification. Using his "weakest power at first, to guard against optical deception," he would only then apply his "highest powers" to the same object, he explained to his readers, while privately he admitted that he was careful to have "always one of the Pillar Microscopes with a single lens at hand to guard against Deception." On the other hand, he appreciated the drawbacks of too little magnification, believing that possession of a microscope of greater powers might have led Velley to agree with his observations.[44] Moreover, when Stackhouse's continued observations led to a "curious" revelation, namely, that many species of marine algae exhibited more than one mode of fructification, this fact was confirmed only when his astronomical friend Edward Pigott, who was also interested in seaweeds, "contrived with a part of his Telescope a Microscope of high powers."[45]

It was, Stackhouse reported, Joseph Banks's doubts that "the pear-shaped bodies described and figured by me, as they appear in the compound microscope, might not be real seeds," that inspired his experiments. Since the "extreme minuteness" of the bodies in question made it impossible "to dissect their component parts with sufficient accuracy, in order to insure conviction," Stackhouse instead found the means to imitate nature—to carry out a mimetic experiment in order to observe a natural phenomenon.[46] "I resolved to procure, if possible, the spontaneous discharge of the seeds in sea water," he explained, "in order to submit them to a more accurate examination. I likewise conceived the idea that I might close my experiment by sowing the seeds on sea pebbles." It took a week before the plants he had carefully detached from rocks and then placed indoors in wide-mouthed glass jars had fully discharged their seeds. During this time, Stackhouse had changed the sea water in the jars every twelve hours with the help of a siphon, noted that the seeds were suspended in a heavy syrupy mucus that made them sink, and, after isolating some of the seeds, witnessed the "explosion or bursting of one of these seeds . . . which agitated the water considerably under the microscope." Confident that this demonstrated *"that marine plants scatter their seeds in their native element without violence, when ripe, and without awaiting the decay of the frond,*" and that the mucus ensured that they were carried down to rocks on which they could grow, Stackhouse moved on to the next stage of his experiment. He drained off the greater part of the water from the jars and poured the remainder, containing the seeds, onto sea pebbles and fragments of rock taken from the beach. These he let dry in order that the seeds might "affix themselves." "I then fastened strings to them," Stackhouse explained, "and alternately sunk them in sea water in a wide-mouthed stone jar, and left them exposed to the air, in order to imitate as nearly as possible their peculiar situation between high and low water-mark, and when the weather was rainy I took care to expose them to it." Within a few days it was clear to the naked eye that a thin membrane had developed, which eventually resulted in shoots.[47]

Stackhouse's continuing researches into the fructification of seaweeds were dependent not only on the possession of a powerful microscope but also on being able to find seaweeds in a state of fructification. To this end, he was greatly helped by the publication of a "Calendar of marine Plants" by Turner, which showed that many species did not fructify until midwinter, when many botanists had left the coast. This, Stackhouse believed, was why the fructification of many species had gone unnoticed.[48] The impetus behind Turner's publication was to educate collectors, many of whom assumed that whatever they collected on a brief trip to the coast was in a state of perfection and who thus proliferated errors by misinterpreting accidental swellings on the plant as the fruit.[49] Even with this information, it was still necessary to "wait year after year" for those specimens that rarely fruited.[50] Turner's list "admitted nothing that has not been the result of my own actual observation . . . along the Norfolk shore," observations, he acknowledged, that were often made in company with his "worthy friend Mr. Wigg."[51]

Despite Stackhouse's initial confidence that the "era is not distant when the Families of these marine plants will be properly arranged, and when clear distinct *essential Characters* will be prefixed to each,"[52] it soon became apparent that prior to any such order being discernible it was necessary to fully observe every plant that had ever been labeled a seaweed as well as the new ones being discovered. In such a pursuit the help of other observers was essential, and botanists emphasized the need to recognize and judge visual acuity separately from theoretical concerns. Thus, although Stackhouse continued to consider Velley "unphilosophic," he clearly shared Withering's view that Velley possessed "habits of accurate and nice discrimination."[53] And it was this clear observational skill, regardless of the conclusions drawn from the observations, that had led Stackhouse, when expressing the wish that similar microscopic investigations into the fructification of seaweeds "may be set on foot in diffᵗ. places," to regret that Velley, who "has seen with his glasses much of what establishes my generic character," had "not Leisure to attend to this."[54] By 1797 Velley recognized that he could continue to publish only "if the distracted state of affairs should hereafter assume a more peaceable aspect."[55] In the event, the war thwarted each attempt Velley made to produce a second fascicule of his work before his untimely death in a carriage accident in 1806.

The war also affected Stackhouse's continued attempts to arrange the marine algae. During a brief interval of peace in 1802 Stackhouse had taken the opportunity to travel to Paris, where his examination of seaweed collections revealed that French "marine Botany is as yet in its Infancy."[56] When war commenced again the following year, Stackhouse complained that "the Din of arms . . . discomposes even those who are not military"; it also put an end to the brief correspondence he had enjoyed with the French seaweed expert Jean Vincent Félix Lamouroux.[57] Stackhouse deplored the resumption of hostilities, but at least he was able to do so from the safety of his own home. His friend Pigott, who had also traveled to France in 1802, was not so lucky. Arrested and detained at Fontainebleau, it was not until 1806 that appeals for his return to England were successful.[58] But even in England the effects of the war could put a stop to scientific pursuits: "my Note Book has been enlarged by my Summers Investigations," Stackhouse confessed to Turner, "but I have no time or inclination to revise or digest my thoughts. That must be left for more tranquil times."[59]

By 1806, Stackhouse's ardor for the task of arrangement had been rekindled by the work of Daniel Mohr, professor of botany at Keil who, like him, wished to amend Linnaeus's system.[60] Confident that his microscopic distinctions provided the means to establish an arrangement based on natural characters and affinities between marine plants, Stackhouse proposed that Linnaeus's genus *Fucus* be replaced with sixty-seven new genera, of which only six had appeared in his *Nereis Britannica.* The remainder were described in his "Tentamen marino-cryptogamicum," which Stackhouse sent to the Société Impériale des Naturalistes de Moscou in 1807 "in return for an unsolicited Honor imposed on me by making me a Member." Although it was published in 1809, copies did not circulate until 1815, owing to Napoleon's Russian campaign and the burning of Moscow in 1812.[61] The delay prompted Stackhouse to produce a second edition of his *Nereis Britannica,* which was published in 1816, but this was eclipsed by Lamouroux's arrangement of sea plants.[62] Unaware of Stackhouse's "Tentamen," Lamouroux referred only to the genera described in the 1801 edition of *Nereis Britannica.* Despite this, it is unlikely that Stackhouse's arrangement would have been accepted by British botanists. From the start, objections were raised against a system in which the genera were "founded on microscopic parts" rather than on some more obvious feature, and Goodenough rejoiced when he first heard that Turner was to tackle the genus *Fucus*, since he neither liked "Col^{l}. Velley's nor M^{r}.——— (the Cornish Gentleman's) ideas upon the Subject."[63] Moreover, although Turner would later describe Lamouroux's arrangement as "ingenious" and "comprehensive,"[64] he did not adopt it, and considered Lamouroux, like the French in general, to be "rapid in conception & manner, inclined to build theories upon weak foundations, & in his opinions somewhat positive."[65]

By this point Turner's work was languishing. He had delighted the botanical community with his intention to describe and figure all the specimens of the genus *Fucus* that he could obtain, and by promising "to throw as much light as lies in my power upon the division of the submersed Algae into new genera."[66] In making even a preliminary attempt at arrangement Goodenough and Woodward had stated that they were "thoroughly conscious of our imbecility and ignorance."[67] But Turner threw down the gauntlet to observers and strained observation to the utmost by declaring that all the submerged algae should be "thrown into a general mass, paying no respect to the genera as they now exist, all of which comprise plants of the most anomalous nature."[68] From the outset, Turner's monograph, published in parts from 1808, became a record of his doubts and difficulties in classifying seaweeds. Nonetheless, the work was widely praised and high hopes were raised. In an article on "Fuci" written in 1815, the *Edinburgh Encyclopaedia* discussed the work of "that most scrupulously exact naturalist Mr Turner of Yarmouth." Drawing attention to the beauty of the drawings, the number of distinguished botanists and travelers who had contributed specimens, and Turner's "ample and luminous" descriptions in both Latin and English, the *Encyclopaedia* declared that in "no botanical production was there ever greater attention paid to minute accuracy." And there, in print, is forever held out the great promise of Turner's careful researches:

> Every classification of fuci must, in the present state of our knowledge of them, be to a certain extent artificial; but from this author, as near an approach to a natural arrangement as possible, may confidently be expected.[69]

This confidence was, in part, bolstered by the coming of peace. A fellow algologist told Turner of his delight on hearing that "this wondrous & glorious turn which has taken place in foreign affairs, has already enabled you to resume your Continental Correspondence, & thereby served to revive your botanical ardor," while Robert Brown longed to see Turner's "concluding fasciculus, wch. will, of course, contain your ideas on the subdivision of this overgrown genus."[70] John Galpine's *Synoptical Compend of British Botany* only partially adopted Lamouroux's arrangement "because a better arrangement . . . is soon expected from the able hand of Mr. Dawson Turner."[71] Turner, however, remained cautious. When the *Fuci* was eventually finished in 1819, it was "painful" for him to acknowledge that it was incomplete, owing to the great mass of species continually being discovered. Moreover, since the specimens he required from all the world's oceans "can only be casually procured through the kindness of friends," no length of time, in his opinion, would have enabled him to finish the work as he had wished.[72] Until all the species could be brought under one view, Turner's grand aim of "reducing the Marine Algae in general under natural families, in a well organized system" was, he declared, impossible.[73]

Seas of Thought

Turner may have failed to reduce the families of seaweeds to a "well organized system," but the same could not be said of his own family and those he included within it. During the unsettled period of the wars with France, pressure for a renewed commitment to religious and moral principle was often expressed through attitudes to nature and the family. Losing oneself in the wonder of nature was thought to counteract selfish individualism, while maintaining the home as a place of tranquility, comfort, and regularity encouraged contemplation. In the Turner home, moreover, nature and family came together in a direct way, for this was where Turner worked on his botanical researches and kept his natural history collections. Despite the strict regularity of the household—Turner reading a chapter from the Bible after breakfast, his family always busy—it was not a home in which "boisterous spirits" were constrained.[74] Nonetheless, the schoolboy Charles Lyell, seeing Mary Turner and her daughters etching at 6:30 in the morning, considered their industry far greater than the effort he expended at schoolwork.[75] Facilitating and organizing the work of others was as important to Turner as his own efforts in striving to achieve botanical order.

The strenuous effort to spend time well, together with Turner's struggle to regard the death of his seven-year-old son in a fire in 1806 "in the true light which I believe that my duty to my Maker requires," reveal his leanings towards evangelicalism.[76] While seriousness and domesticity may have been underlined by the "ostentatious uxoriousness" of those in government, it was also a time of unpredictable and heightened emotion in politics, with numerous cases of suicide and insanity among members of Parliament as well as the madness of the monarch.[77] The interpretation of afflictions such as personal bereavement or the insanity of a king as examples of the way in which "God smiteth those he loveth," and thus as tests of faith, was characteristic of the evangelical revival of the early nineteenth century. Evangelicals, who interpreted Napoleon's victories as a sign of the sinfulness of the British nation, considered that a thoroughgoing reform of society was required, a spiritual quest that led to constant self-vigilance as well as close scrutiny of others.[78]

Even for those not actively touched by evangelicalism, the watchfulness it engendered, together with the vigilance encouraged during the war, made it a powerful way of seeing during this period. Monitoring the state of one's soul was encouraged, particularly through the contemplation of nature, as demonstrated by the reprinting of earlier evangelical texts such as those of the Reverend James Hervey. His "contemplations on the sea" (fig. 17.3) reveal the "supreme uncontrollable power" of God through the observation that "a low bank of despicable sand" can not only sustain but "curb the rage of furious assaulting seas." When, at low tide, the observer discovers this sand to be smooth and firm, he hopes that such "will be the case with this soul of mine, and the temptations that beset me." While the shore in the tidal zone represented the firm sand of faith, contemplation of the full extent of sand and the immensity of the unknown ocean served as a reminder of the need for humility and modesty even in those who have "extraordinary insight into the mysteries of nature"[79]

The shore, by marking the divide between the known and the unknown, in both spiritual and physical terms, could also inspire the pleasurable terror denoted by the sublime. While touring the southern coasts of England, Turner forgot his disappointment in being unable to collect seaweeds at one spot by becoming

> wholly absorbed in the contemplation of the majestic objects with which we were surrounded before us lay the sea smooth as glass . . . while on every other side we were surrounded with huge dark unformed masses of rock sufficient to strike an awful terror into the most superficial observer . . .

Contemplating cliffs on the Cornish coast where, with a "tremendous roar," the water rushed into the "frightfully hollowed" out rocks, Turner "was insensibly led to reflect & moralize upon the equipoise every where presented by nature in all her works where no situation is so low as to be destitute of advantages nor so high as to be exempt from inconvenience." Since Turner loved "to reason on great things from small," he concluded his journal entry by quoting from the poet and philosopher James Beattie, whose writings contained his imaginative response to nature and a defense of religion. All this reconciled Turner to the flat sandy beaches of "my native Norfolk where

CONTEMPLATIONS ON THE SEA.
"Hitherto thou shalt go, but no farther"
vide Let. 9.
London, Published by Thomas Kelly, Paternoster row, June 12-1813.

FIGURE 17.3. "Contemplations on the Sea," from James Hervey, *Theron and Aspasio* (London: Thomas Kelly, 1813), 2: 1.

travelling is attended with far more expedition less trouble & less danger."[80] For others, however, the Norfolk beaches seemed immensely dangerous places, being likely spots for a French invasion of Britain. "In the imminent Danger that threatens y^{r}. Coast I bestow many an anxious thought on you," Stackhouse informed Turner, telling him that "I write principally to let you know how much we are interested for you & to beg, if an actual landing of French takes place that you will immediately send me a Line."[81]

Differences between Cornwall and Norfolk were also written into the descriptions of seaweeds, with the same plants from the west invariably larger than those in the east. Observational comparisons were used that must have been familiar enough to convey immediate meaning while also allowing calibration. Thus variations of size in one species were conveyed by plants from Cornwall having branches "exceeding the thickness of a crow's quill," while those from Yarmouth "scarcely equal that of the sparrow." Other measures of thickness included the quills of the swan, goose, blackbird, and wren, as well as human hair, packthread, hog's bristles, and a "common walking stick." The sizes of tubercles in different species were conveyed by references to "turnip-seed," "poppy-seed," and "mustard-seed" down to those smaller than "the smallest pin's head."[82] Such descriptions served not only to reassure observers of the relative visibility of hard-to-see features but also to render the liminal and ambiguous more familiar.

Botanists hoped that an order of nature, comfortingly arranged in natural families, would emerge from the close scrutiny of individual seaweeds, but the reality was often different. The lower orders of plants, notorious boundary crossers, would sometimes present a form so obscure as to confound observers. Edmund Burke had deemed obscurity sublime precisely because it was a frustration of the power of vision, and thus induced pain by making viewers strain to see that which cannot be comprehended.[83] The pain of trying to see ambiguous sea plants was expressed even in the last volume of Turner's work, when the remarkably lichen-like alga *Fucus Pygmaeus* presented a difficulty "still more humiliating to our pride, and placing in a stronger light the nothingness of all our progress in science, by shewing how Nature sometimes defies our attempts to determine the genus or even the order of plants to which individuals belong."[84] That the confrontation of this obscurity and consequent "nothingness" could tip over into the grandeur and terror characteristic of the sublime was because the inexplicability of phenomena that arrest the attention served to reveal the immensity and wisdom of God and nature in comparison to the puny efforts of an individual human life.

Turner's taxonomic concerns, and his realization that he would not succeed in dividing the "submersed Algae into new genera" as he had set out to do, were quelled both by his belief that it was better to wait until the *correct* order of nature could be established, since this alone would reflect God's purpose in the world, and by knowing that he had, as he stated in his farewell to his readers,

> laid before them a set of figures, upon the accuracy of which they may rely; and which, as representations of things that are, will, through every change of human opinions, retain an undiminished value; while they may serve, in the hands of some more able, and more fortunate successor, as the ground-work of that which he had hoped to have accomplished himself.[85]

During the unsettled period of war, especially, the attraction of seeing clearly enough to capture what was constant and valuable, particularly in those things that baffled the understanding, is obvious. Turner was inspired by and held forth to others the fantasy of achieving the total visibility (or overview) that would permit the construction of order. Fears that there might be no order underlying the chaos were mitigated in part by the experience of the pelagic sublime, which privileged obscurity and mystery.

Conclusion

Analogies between the natural and the social, this essay has suggested, deserve particular attention during the revolutionary and Napoleonic Wars, for they reveal that scrutiny of other social groups and overseeing one's conscience were part of the same regime as natural history observation and classification work. So, what should we make of the illustrations of individual seaweeds in Turner's monograph? They form a series that does not represent classificatory order but rather, like Hervey's contemplations, Crabbe's portraits of the poor, or Bewick's birds, aimed to make the reader look closer and observe more clearly. Hard-to-see or ambiguous individuals were to be scrutinized in order to place them more clearly in the collective, and their presentation in a sequence was designed to enhance this process. The images of nature by Turner, Bewick, and others (regardless of differences in status and politics) were meant to last for all time, but were thought to do so precisely because they embodied specific social and natural taxonomies. They were as much about continuity, tradition, and permanence as progress. Turner's confidence that his figures would survive "every change of human opinions" derived, of course, from a very time-specific union of personal and communal regimes of watchfulness and self-surveillance. The work of early marine botanists shows that concerns about the moral ugliness exhibited by natural beings that did not readily fall into families and the fears of being deceived by ambiguous appearances could be assuaged by a focus on visual acuity.

The investigations of marine botanists were characterized not only by the honing of visual skills in order to describe seaweeds but also by their constant self-surveillance and the continual appraisal of other observers, both past and present. Thus, while Réaumur's ideas were dismissed as erroneous, he still attracted adjectives such as "immortal" on the basis of his observational powers, which "being all practical, are truly valuable."[86] This judgment of individuals in the context of a community of observers revealed that "a quick, accurate & discerning eye" was not enough.[87] To guard against deceptive appearances (and thus against false sympathy for the poor, the disguises of spies, and the misclassification of ambiguous natural objects) required strenuous effort. Observation combines performance and product; fact, deriving from feat, as Gillian Beer reminds us, implies the thing done as much as the thing categorized.[88] "True Science," declared Stackhouse in 1802, "consists of accumulating Facts," and while he dismissed the opinions of an author he conceded that his book "considered as a Collection of Facts is extremely valuable."[89]

Establishing the facts of marine botany was slow and laborious in the extreme, the processes of which resulted in affective bonds to both objects and observers. It is thus not surprising that it is among his descriptions of seaweeds that Turner reported and mourned the deaths of two of his best young observers.[90] This portrait of a community at work cannot be fully captured by the notion of a network. Rather, in seeing how attentive and accurate observation could survive time and changing interpretations, readers became potential participants in a community bound together not so much by specific views on seaweeds as by their understanding and valuing of a culture of observation in which a detached clarity of vision existed in a context of emotional and social engagement.

Notes

A research grant from the British Society for the History of Science enabled me to visit archives in London and Norwich; I am very grateful to the Society for this generous support. Permission to quote from manuscripts has been granted by the Master and Fellows, Trinity College, Cambridge; the Linnean Society of London; the British Library; Norfolk Museums and Archaeology Service; and Christopher Barker. For further information, I am indebted to Mark Philp, Ian Caldwell, and John Edmondson.

1. Withering to Turner, 27 Jan. 1798, Dawson Turner Correspondence, Trinity College, Cambridge (hereafter TCC); W. Withering, *An Arrangement of British Plants*, 3rd ed., 4 vols. (Birmingham, 1796), 1: 6.

2. Hugh Paton, ed., *A Series of Original Portraits and Caricature Etchings, by the late John Kay*, 2 vols. (Edinburgh: Hugh Paton, 1838), 2: 230.

3. Richard Holmes, *The Age of Wonder* (London: Harper Press, 2008), 200. For illustrations of modes of invading Britain as fantasized by the French and feared by the British, see Alistair Horne, *Napoleon: Master of Europe, 1805–1807* (New York: William Morrow, 1979), 64–65; and Alexandra Franklin and Mark Philp, *Napoleon and the Invasion of Britain* (Oxford: Bodleian Library, 2003).

4. J. E. Cookson, *The British Armed Nation, 1793–1815* (Oxford: Clarendon Press, 1997); Mark Philp, ed., *The French Revolution and British Popular Politics* (Cambridge: Cambridge University Press, 1991).

5. Scott Hughes Myerly, *British Military Spectacle* (Cambridge, Mass.: Harvard University Press, 1996), 154–56; Mary A. Favret, *War at a Distance* (Princeton: Princeton University Press, 2010), 14.

6. J. Gurney to Turner, 2 July 1809, TCC.

7. Thomas Bewick, *History of British Birds*, vol. 2, *Water Birds* (Newcastle: Edward Walker, 1804), v.

8. Linda Colley, *Britons: Forging the Nation, 1707–1837* (London: Yale University Press, 1992), 307–8.

9. Bewick, *Water Birds*, 301–4.

10. Maya Jasanoff, *Edge of Empire* (New York: Alfred A. Knopf, 2005). See also Greg Dening, *Islands and Beaches* (Honolulu: University of Hawaii Press, 1980).

11. Matthew Flinders, *A Voyage to Terra Australis*, 2 vols. (London: G. & W. Nicol, 1814), 2: 88, 367.

12. Samuel Goodenough and Thomas Jenkinson Woodward, "Observations on the British Fuci, with particular Descriptions of each Species," *Transactions of the Linnean Society* 3 (1797): 235; Dawson Turner, *A Synopsis of the British Fuci* (London: J. White, 1802), 1: xxii.

13. Goodenough and Woodward, "Observations," 168.

14. Turner, *Synopsis*, 2: 365; James Sowerby and J. E. Smith, *English Botany* (London: J. Sowerby, 1790–1814), 12: 847.

15. J. E. Smith to Woodward, 20 July 1793, and Woodward to Smith, 23 July 1793, Linnean Society of London (hereafter LSL).

16. Jeanette M. Neeson, "Coastal Commons: Custom and the Use of Seaweed in the British Isles, 1700–1900," in *Ricchezza del Mare, Ricchezza dal Mare Secc. XIII–XVIII. Atti della 'Trentasettesima Settimana di Studi' 11–15 aprile 2005, a cura di Simonetta Cavaciocchi*, Istituto Internazionale di Storia Economica 'F. Datini,' Prato ([Florence]: Le Monnier, [2006]), 343–67. For kelp manufacture, see David Brewster, ed., *The Edinburgh Encyclopaedia* (Edinburgh: William Blackwood, 1830), 10: 17–18; and Alain Corbin, *The Lure of the Sea* (Cambridge: Polity Press, 1994), 203–6.

17. George Crabbe, *The Borough: A Poem in Twenty-Four Letters* (London: J. Hatchard, 1810), 5.

18. Crabbe, a "more than adequate botanist" according to D. E. Allen, *The Naturalist in Britain* (London: Allen Lane, 1975), 127, had been curate to the Reverend Richard Turner, uncle of seaweed expert Dawson Turner. Crabbe contributed to Dawson Turner and Lewis Weston Dillwyn, *The Botanist's Guide* (London: Phillips and Fardon, 1805), 2: 536–76, but burnt his own intended botanical work before publication (*The Life of George Crabbe by His Son* [London: Oxford University Press, 1932], 128, 158–59).

19. Crabbe, *The Borough*, 301–14.

20. George Crabbe, "The Learned Boy," lines 309–27, in *George Crabbe, Tales, 1812 and Other Selected Poems*, ed. Howard Mills (Cambridge: Cambridge University Press, 1967), 363.

21. René Huchon, *George Crabbe and His Times* (London: J. Murray, 1907), 64. For pauper badges, see also Sandra Sherman, *Imagining Poverty* (Columbus: Ohio State University Press, 2001), 237.

22. Thomas C. Faulkner, ed., *Selected Letters and Journals of George Crabbe* (Oxford: Clarendon Press, 1985), 51–52.

23. Bewick, *British Birds*, vol. 1, *Land Birds* (Newcastle: Sol. Hodgson, 1797), iii.

24. Bewick, *Water Birds*, vii, xii.

25. T. J. Woodward, "Descriptions of two new British Fuci," *Transactions of the Linnean Society* 2 (1794): 29; Sowerby and Smith, *English Botany*, 3: 205.

26. Goodenough and Woodward, "Observations," 93, 235.

27. Woodward to Turner, 24 Feb. 1797, TCC.

28. Woodward to J. E. Smith, 29 March 1793, LSL.

29. Woodward to Turner, 21 Feb. 1799, TCC; Hampden G. Glasspoole, "A Memoir of Mr. Lilly Wigg," *Transactions of the Norfolk and Norwich Naturalists' Society* 2 (1874–79): 269–74.

30. Dawson Turner, *Fuci, sive plantarum Fucorum generi a botanicis adscriptarum icones descriptiones et historia. Fuci; or, Colored Figures and Descriptions of the Plants Referred by Botanists to the Genus Fucus*, 4 vols. (London: John and Arthur Arch, 1808–19), 2: 84.

31. D. J. Mabberley, *Jupiter Botanicus: Robert Brown of the British Museum* (London: British Museum [Natural History], 1985), 31, 34; John Stackhouse, *Nereis Britannica; Containing all the Species of Fuci, Natives of the British Coasts* (Bath: S. Hazard, [1795]–1801), v.

32. Joseph Hooker, *A Sketch of the Life and Labours of Sir William Jackson Hooker* (Oxford: Clarendon Press, 1903), xiv.

33. *Oxford Dictionary of National Biography*, s.v. "Henry Addington"; Iain Bain, ed., *A Memoir of Thomas Bewick Written by Himself* (London: Oxford University Press, 1975), 137.

34. Goodenough to Turner, 23 March 1802, TCC.

35. Glasspoole, "Memoir," 273.

36. Lorraine Daston, "On Scientific Observation," *Isis* 99 (2008): 98.

37. W. H. Harvey, *A Manual of the British Algæ* (London: John van Voorst, 1841), xxiii.

38. Stackhouse to Turner, 8 June 1799, TCC.

39. Ian Caldwell, "John Stackhouse (1742–1819), and the Linnean Society," *Linnean* 24 (2008): 39, 43.

40. Stackhouse to J. E. Smith, 16 Sept. 1795, LSL; Stackhouse, *Nereis Britannica*, v; Thomas Velley, *Coloured Figures of Marine Plants, Found on the Southern Coast of England . . . to which is prefixed An Inquiry into the Mode of Propagation Peculiar to Sea Plants* (Bath: S. Hazard, 1795), 5, 6.

41. Stackhouse to J. E. Smith, 28 May 1795, LSL; Stackhouse, *Nereis Britannica*, vi.

42. Goodenough and Woodward, "Observations," 115.

43. Stackhouse to J. E. Smith, 16 Sept. 1795, LSL.

44. Stackhouse, *Nereis Britannica*, ix; Stackhouse to J. E. Smith, 3 Dec. 1795, LSL.

45. Ibid., xxvi.

46. Ibid., x. For the notion of mimetic experimentation, see Peter Galison and Alexi Assmus, "Artificial Clouds, Real Particles," in *The Uses of Experiment*, ed. David Gooding, Trevor Pinch, and Simon Schaffer (Cambridge: Cambridge University Press, 1989), 225–74.

47. Stackhouse, *Nereis Britannica*, x–xi.

48. Ibid., xxvi.

49. Dawson Turner, "Calendarium Plantarum marinarum," *Transactions of the Linnean Society of London* 5 (1800): 127.

50. Goodenough and Woodward, "Observations," 231.

51. Turner, "Calendarium," 128, 127.

52. Stackhouse, *Nereis Britannica*, vii.

53. Stackhouse to Turner, 1 July 1797, and Withering to Turner, 7 Oct. 1797, TCC.

54. Stackhouse to J. E. Smith, 3 Dec. 1795, LSL.

55. Velley to Turner, 14 June 1797, TCC.

56. Stackhouse to Turner, 18 Sept. 1802, TCC.

57. Stackhouse to Turner, 7 April 1803 and 6 Jan. 1806, TCC.

58. Stackhouse to Turner, 13 April 1809, TCC.

59. Stackhouse to Turner, 10 Oct. 1803, TCC.

60. Stackhouse to Turner, 27 Sept. 1806, TCC; Daniel Matthias Heinrich Mohr and Friedrich Weber proposed a subdivision of the genus *Fucus* in their *Beiträge zur Naturkunde* (Kiel: In der neuen akademischen Buchhandlung, 1805–10).

61. Stackhouse to Smith,17 May 1816, LSL; Stackhouse to Robert Brown, 6 Nov. 1815, British Library, Add. 32440, fol. 93. For Stackhouse's genera, see "Tentamen marino-cryptogamicum," *Mémoires de la Société impériale des naturalistes de Moscou* 2 (1809): 50–97; George G. Papenfuss, "Review of the Genera of Algae described by Stackhouse," *Hydrobiologia* 2 (1949–50): 181–208.

62. Jean Vincent Félix Lamouroux, "Essai sur les genres de la famille des Thalassiophytes non articulées," *Annales du Muséum d'histoire naturelle* 20 (1813): 21–47, 115–39, 267–93.

63. Woodward to Turner, 26 Dec. 1797, and Goodenough to Turner, 21 Oct. 1801, TCC.

64. Turner, *Fuci*, 4: 4.

65. Turner's "Journal of a Tour to France in 1815," 78 (privately owned manuscript).

66. Turner, *Fuci*, 1: 14.

67. Goodenough and Woodward, "Observations," 93.

68. Turner, *Synopsis*, xv.

69. Brewster, *Edinburgh Encyclopaedia*, 10: 2–3. Although the volume bears a publication date of 1830, the article (by Patrick Neill) was written in 1815.

70. L. W. Dillwyn to Turner, 1 Feb. 1814, and Brown to Turner, 24 Nov. 1815, TCC.

71. John Galpine, *Synoptical Compend of British Botany*, 2d ed. (London: Samuel Bagster, 1819), viii.

72. Turner, *Fuci*, 4: 120.

73. Ibid., 4: advertisement.

74. Willard Bissell Pope, ed., *The Diary of Benjamin Robert Haydon* (Cambridge, Mass.: Harvard University Press, 1960–63), 2: 128.

75. Katharine Lyell, ed., *Life, Letters and Journals of Sir Charles Lyell* (London: John Murray, 1881), 1: 42.

76. Pleasance Smith, ed., *Memoir and Correspondence of the late Sir James Edward Smith* (London: Longman, 1832), 2: 115–16.

77. Colley, *Britons*, 189, 151–52.

78. Boyd Hilton, *A Mad, Bad, & Dangerous People? England 1783–1846* (Oxford: Clarendon Press, 2006), 177, 187.

79. James Hervey, *Theron and Aspasio*, Kelley's edition (London: Thomas Kelly, 1813), 2: 190, 192–93.

80. "Journal of the first ten days of a Tour made by Dawson Turner, Esq. in company with

James Sowerby, Esq. Author of 'British Fungi,' to the Land's End, in the months of June & July 1799," Natural History Department, Norwich Castle Museum, NWHCM:1970.483.1, entries for 9th and 11th June. Turner quoted from James Beattie, "The Minstrel," book 1, lines 50–54.

81. Stackhouse to Turner, 10 Oct. 1803, TCC.

82. Goodenough and Woodward, "Observations," 207, 223; Turner, *Fuci*, 1: 18, 114; 2: 147; 3: 67, 78, 100, 128; 4: 4, 92, 129.

83. W. J. T. Mitchell, *Iconology: Image, Text, Ideology* (Chicago: University of Chicago Press, 1987), 126.

84. Turner, *Fuci*, 4: 16.

85. Ibid,. 4: advertisement.

86. Goodenough and Woodward, "Observations," 84.

87. Goodenough to Turner, 1 March 1799, TCC.

88. Gillian Beer, *Darwin's Plots* (London: Routledge, 1983), 81.

89. Stackhouse to Dawson Turner, 4 Feb. 1802, TCC; Stackhouse to J. E. Smith, 11 Nov. 1802, LSL.

90. Turner, *Fuci*, 2: 60; 4: 152.

Contributors

DOMENICO BERTOLONI MELI teaches the history of science at Indiana University, Bloomington. He is the author of *Equivalence and Priority: Leibniz versus Newton* (1993) and *Thinking with Objects: The Transformation of Mechanics in the 17th Century* (2006). He is currently working on anatomy in the seventeenth century.

CHARLOTTE BIGG is a historian at the Centre National de la Recherche Scientifique, Centre Alexandre Koyré in Paris. She has recently coedited "The Laboratory of Nature: Science in the Mountains, Mountains in Science, from the Late Eighteenth to the Early Twentieth Century," a special issue of *Science in Context* 22 (Sept. 2009), as well as *Atombilder: Ikonografie des Atoms in Wissenschaft und Gesellschaft des 20. Jahrhunderts* (2009) and *The Heavens on Earth: Observatories and Astronomy in Nineteenth-Century Science and Culture* (2010).

DANIELA BLEICHMAR is assistant professor in the Departments of Art History and History at the University of Southern California. She is the author of multiple essays on science and visual culture in the Hispanic empire and of *Visible Empire: Colonial Botany and Visual Culture in the Hispanic Enlightenment* (2011), as well as coeditor of *Science in the Spanish and Portuguese Empires (1500–1800)* (2009) and *Collecting across Cultures: Material Trajectories in the Early Modern World* (2011).

JIMENA CANALES is associate professor at Harvard University. She specializes in the history and philosophy of the physical sciences. Areas of interests include epistemology, science and representation, and theories of modernity and postmodernity. She is the author of *A Tenth of a Second: A History* (2010) and of numerous articles on the history of science, film, architecture, and philosophy.

LORRAINE DASTON is director at the Max Planck Institute for the History of Science in Berlin and visiting professor in the Committee on Social Thought at the

University of Chicago. Her most recent book, coauthored with Peter Galison, is *Objectivity* (2007).

OTNIEL E. DROR is the Joel Wilbush Chair in Medical Anthropology and is head of the Department of the History of Medicine at the Hebrew University of Jerusalem. His research focuses on the history of the study of pain, pleasure, and the emotions during the nineteenth and twentieth centuries. His current book project is titled: *The Adrenaline Century, 1900–2000.*

MICHAEL D. GORDIN is professor of history at Princeton University. He is the author of *A Well-Ordered Thing: Dmitrii Mendeleev and the Shadow of the Periodic Table* (2004), *Five Days in August: How World War II Became a Nuclear War* (2007), and *Red Cloud at Dawn: Truman, Stalin, and the End of the Atomic Monopoly* (2009).

ELIZABETH LUNBECK is Nelson Tyrone, Jr., Professor of History and professor of psychiatry at Vanderbilt University. She is the author of *The Psychiatric Persuasion* and *Family Romance, Family Secrets*, and has edited several volumes. She is currently finishing *The Americanization of Narcissism.*

HARRO MAAS is associate professor in history and methodology of economics at the University of Amsterdam. He is the author of *William Stanley Jevons and the Making of Modern Economics* (2005). He is currently leading a research project on the history of observation in economics.

J. ANDREW MENDELSOHN teaches at Imperial College London, where he is head of the Centre for the History of Science, Technology and Medicine. His previous contributions to the history of scientific observation include essays "The Microscopist of Modern Life," *Osiris* 18 (2003): 150–170, and "Lives of the Cell," *Journal of the History of Biology* 36 (2003): 1–37.

MARY S. MORGAN is professor of the history and philosophy of economics at the London School of Economics and at the University of Amsterdam. Her major works include *The History of Econometric Ideas* (1990), *Models as Mediators* (1999 with M. Morrison), and *The World in the Model* (forthcoming). She has recently been leading a major research project that crosses over between the sciences and humanities: *How Well Do Facts Travel?* (with W. P. Howlett, forthcoming) and is now, on a grant from the British Academy and Wolfson Foundation, investigating the role of case studies across the social sciences.

KATHARINE PARK is Samuel Zemurray, Jr., and Doris Zemurray Stone Radcliffe Professor in the Department of the History of Science at Harvard University, where she works on the history of science and medicine in medieval and early modern Europe. Her most recent books are *The Cambridge History of Science*, vol. 3, *Early*

Modern Science (2006, coedited with Lorraine Daston), and *Secrets of Women: Gender, Generation, and the Origins of Human Dissection* (2006).

GIANNA POMATA is professor of the history of medicine at Johns Hopkins University. She is the author of *Contracting a Cure: Patients, Healers, and the Law in Early Modern Bologna* (1998) and the coeditor of *The Faces of Nature in Enlightenment Europe* (2003, with Lorraine Daston), and *Historia: Empiricism and Erudition in Early Modern Europe*, (2005, with Nancy Siraisi).

THEODORE M. PORTER is professor of history of science in the University of California, Los Angeles History Department. His books include *Karl Pearson: The Scientific Life in a Statistical Age* (2005) and *Trust in Numbers: The Pursuit of Objectivity in Science and Public Life* (1996) He also coedited the *Cambridge History of Science* volume on the social sciences (2003, with Dorothy Ross).

ANNE SECORD is an affiliated research scholar in the Department of History and Philosophy of Science, Cambridge University. Her research and writings focus on popular, particularly working-class, natural history in nineteenth-century Britain, and on horticulture, medicine, and consumption in the eighteenth century.

MARY TERRALL teaches history in the Department of History at the University of California, Los Angeles. She is the author of *The Man Who Flattened the Earth: Maupertuis and the Sciences in the Enlightenment* (2002). She is currently writing a book on natural history as practiced by Réaumur and his many correspondents in the mid-eighteenth century.

KELLEY WILDER is senior research fellow at the Centre for Photographic Research, De Montfort University, where she leads the Masters Program in Photographic History and Practice. She was a research scholar from 2005 to 2008 at the Max Planck Institute for the History of Science, Department II, where she wrote *Photography and Science* (2009).

Index